Nonparametric Sequential Selection Procedures

H. Büringer
H. Martin
K.-H. Schriever

Birkhäuser
Boston · Basel · Stuttgart

Dr. H. Büringer
Dr. H. Martin
Dr. K.- H. Schriever
Institut für Statistik und Mathematische Wirtschaftstheorie
Universität Karlsruhe
Postfach 6380
D-7500 Karlsruhe 1
Federal Republic of Germany

Library of Congress Cataloging in Publication Data
Büringer, H 1951-
Nonparametric sequential selection procedures.

Bibliography: p.
1. Ranking and selection (Statistics)
2. Sequential analysis. 3. Nonparametric statistics.
I. Martin, Herbert, 1947- joint author.
II. Schriever, K.-H., 1948- joint author.
III. Title.
QA278.75.B83 519.5'35 80-18568
ISBN 3-7643-3021-X

CIP—Kurztitelaufnahme der Deutschen Bibliothek
Büringer, Helmut:
Nonparametric sequential selection procedures/
H. Büringer; H. Martin; K.-H. Schriever.—
Boston, Basel, Stuttgart: Birkhäuser. 1980.

ISBN 3-7643-3021-X

NE: Martin, Herbert; Schriever, Karl-Heinz:

ISBN 3-7643-3021-X

Printed in USA

Preface

New (statistical) methods spread abroad only if they are available in an extensive self-contained presentation. It was this perception that made us write down the main ideas of a new branch of sequential statistical theory, called "Nonparametric Sequential Selection Procedures".

What do we associate with this notion? Well, a selection procedure or selection model is nothing but a statistical experiment designed for identifying the best of several alternatives, a problem we are faced with almost everyday. The statistical methods included in this book allow to select the best alternative whenever there is any substantial difference among them.

The book contains essentially the results of the past ten years, including our own research only partially published in statistical journals. It is neither a collection of statistical receipes nor a treatise exclusively theoretical. It is designed primarily for scientists with strong mathematical backgrounds interested in new developments of applied statistics.

The general framework is developed in detail so that the reader confronted for the first time with that field of statistics has no difficulties in getting familiar with the specific notions of the theory.

All selection models have been supplemented by detailed and examined numerical results, nevertheless, it was not possible to include all available numerical material in this book. A subject index has been omitted, because the basic notions are explained in the general introduction, and in the introduction of part 1, and part 2.

The authors wish to thank Prof. C. Asano from the Research Institute of Fundamental Information Science, Kyushu University, Fukuoka ,Japan, for stimulating their research work.
Our thanks are also due to Mrs. H. Koch, and Mrs. W. Toussaint for carefully typing our manuscript.

Karlsruhe

May, 1980

H. Büringer

H. Martin

K.-H. Schriever

Contents

NONPARAMETRIC SEQUENTIAL SELECTION PROCEDURES

General Introduction 1

PART 1

SEQUENTIAL PROCEDURES FOR SELECTING THE BEST OF $k \geq 2$ BINOMIAL POPULATIONS

Introduction 5

CHAPTER 1

SELECTION PROCEDURES WITH UNRESTRICTED PATIENT HORIZON

1 The Selection Model [2; PW; $|S_A-S_B| = r$]

1.1 Derivation of the critical r-value 18

1.2 Derivation of the expectations 22

2 The Selection Model [2; VT; $|S_A-S_B| = s$]

2.1 Derivation of the critical s-value 25

2.2 Derivation of the expectations 26

3 The Selection Model [2; PL; $|F_A-F_B| = r$]

3.1 Derivation of the critical r-value 27

3.2 Derivation of the expectations 28

4 Comparison of the Selection Procedures No.1-No.3

4.1 Some general remarks 29

4.2 Comparison of PW- and VT-sampling 30

4.3 Comparison of PW- and PL-sampling 31

4.4 Comparison of PL- and VT-sampling 31

4.5 Recapitulation 32

4.6 Concluding remarks 33

4.7 Numerical results 33

5 Two-Stage Selection Procedures

5.1 The structure of two-stage selection procedures 40

5.2 Derivation of P(CS) 41

5.3 Derivation of $s(m,n,k_c,M)$ and $r(m,n,k_c,M)$ 41

5.4 Derivation of the expectations 44
5.5 Addition of PL-sampling 46
5.6 Concluding remarks 48
5.7 Numerical results 50

6 The Selection Model [2; PW; $\max\{S_A, S_B\} = r$]
6.1 Derivation of the critical r-value 54
6.2 Derivation of the expectations 57
6.3 Expectations for large r 62

7 The Selection Model [2; VT; $\max\{S_A, S_B\} = r$]
7.1 Derivation of the critical r-value 63
7.2 Derivation of the expectations 65
7.3 Expectations for large r 67

8 Comparison of the Selection Models No.6 and No.7 - Some Modifications of these Models
8.1 Comparison of the selection models no.6 and no.7 67
8.2 The selection model [2; PW; $\min\{F_A, F_B\} = r$] 68
8.3 The selection model [2; PL; $\max\{F_A, F_B\} = r$] 70
8.4 Concluding remarks 73
8.5 Numerical results 74

9 The Nature of Termination of a Classical Sequential Selection Procedure
9.1 Basic notions 78
9.2 The stopping-behaviour of selection model no.1 79
9.3 The stopping-behaviour of selection model no.2 80
9.4 The stopping-behaviour of the selection models no.3 and no.5 80
9.5 The stopping-behaviour of selection model no.6 81
9.6 The stopping-behaviour of selection model no.7 81

10 The Selection Model [k; PW; $\max\{S_1, \ldots, S_k\} = r$]
10.1 Introductory remarks 82
10.2 Derivation of the critical r-value 82
10.3 Derivation of the expectations 86
10.4 Expectations for large r 88

11 The Selection Model [k; VT; $\max\{S_1, \ldots, S_k\} = r$]
11.1 Derivation of the critical r-value 90
11.2 Derivation of the expectations 95
11.3 Expectations for large r 96
11.4 Concluding remarks 97
11.5 Numerical results 98

12 Expected Truncation Points 101

13 The Selection Model $[2;PW;|S_A-S_B|=r$ or $|\hat{p}_A-\hat{p}_B| \geq c/(F_A+F_B)]$
13.1 Introduction 139
13.2 Derivation of the critical r- and c-values 139
13.3 Numerical results 141

14 The Selection Model $[2;VT;|S_A-S_B|=s$ or $|\hat{p}_A-\hat{p}_B| \geq d/(F_A+F_B)]$
14.1 Derivation of the critical s- and d-values 144
14.2 Numerical results 145

15 The Selection Models $[k;PW;el.A_i$ if $S_j-S_i=r]$ and $[k;VT;el.A_i$ if $S_j-S_i=s]$
15.1 The PW-elimination procedure 148
15.2 The VT-elimination procedure 149
15.3 Numerical results for the PW-procedure 149
15.4 Numerical results for the VT-procedure 155
15.5 Comparison of selection models 161

CHAPTER 2

SELECTION PROCEDURES WITH RESTRICTED PATIENT HORIZON

1 The Selection Model $[2;\ PW;\ \max\{S_A+F_B,\ S_B+F_A\} = r]$
1.1 Introduction 163
1.2 Derivation of the P(CS)-value 164
1.3 Determination of the LFC 166
1.4 Derivation of the critical r-value 168
1.5 Derivation of the expectations 169
1.6 Numerical results 172

2 The Selection Model $[2;\ PW;\ \max\{S_A,S_B\} = r$ or $F_A=F_B = c]$
2.1 Derivation of the P(CS)-value 173
2.2 Derivation of the critical r- and c-values 176
2.3 Derivation of the expectations 180
2.4 Numerical results 183

3 The Selection Model $[2;\ VT;\ \max\{S_A,S_B\} = r$ or $\min\{F_A,F_B\} = c]$
3.1 Derivation of the P(CS)-value 185
3.2 Derivation of the critical r- and c-values 187
3.3 Derivation of the expectations 188
3.4 Numerical results 190

4 The Selection Model $[2;\ VT;\ \max\{S_A,S_B\} = r$ or $\max\{F_A,F_B\} = c]$
4.1 Derivation of the P(CS)-value 191

4.2 Derivation of the critical r- and c-values 192
4.3 Derivation of the expectations 194
4.4 Numerical results 195

5 The Selection Model [2; PW; $|S_A-S_B|$ = r or F_A+F_B = s]
5.1 Derivation of the P(CS)-value 196
5.2 Derivation of the critical r- and s-values 198
5.3 Derivation of the expectations 199
5.4 Numerical results 201

6 The Selection Model [k;PW;max$\{S_1,\dots,S_k\}$=r or min$\{F_1,\dots,F_k\}$=c]
6.1 Introductory remarks 203
6.2 Derivation of the P(CS)-value 204
6.3 Derivation of the critical r- and c-values 205
6.4 Derivation of the expectations 208
6.5 Numerical results 211

7 The Selection Model [k;VT;max$\{S_1,\dots,S_k\}$=r or min$\{F_1,\dots,F_k\}$=c]
7.1 Derivation of the P(CS)-value 214
7.2 Derivation of the critical r- and c-values 217
7.3 Derivation of the expectations 218
7.4 Numerical results 221

8 The Selection Model [k;VT;max$\{S_1,\dots,S_k\}$=r or el.A_i if F_i=c]
8.1 Derivation of the P(CS)-value 224
8.2 Derivation of the critical r- and c-values 227
8.3 Numerical results 230

9 Further Elimination Procedures
9.1 The selection model [k;PW;el.A_i if S_j-S_i=r or if F_i=c] 233
9.2 The selection model [k;VT;el.A_i if S_j-S_i=r or if F_i=c] 233
9.3 The selection model
[k;PW;el.A_i if S_j-S_i=r or stop if $F_1+\dots+F_k$=s] 233
9.4 The selection model
[k;PW;el.A_i if S_j-S_i=r or el.A_i if $\hat{p}_j-\hat{p}_i \geq c/(F_i+F_j)$] 234
9.5 The selection model
[k;VT;el.A_i if S_j-S_i=r or el.A_i if $\hat{p}_j-\hat{p}_i \geq d/(F_i+F_j)$] 234
9.6 Numerical results 235
9.7 Comparison of selection models 237
9.8 Further selection procedures 239

CHAPTER 3

SELECTION PROCEDURES WITH FIXED PATIENT HORIZON

1 Historical Remarks 240

2 The Zelen Selection Model

2.1 Definition of the model 242

2.2 Comparison with a VT-sampling procedure 244

2.3 Determination of the optimal value of n 246

3 The Selection Models [2;PW;fixed N] and [2;VT;fixed N]

3.1 Introduction 248

3.2 Comparison of the P(CS)-values 249

3.3 Comparison of the expectations 253

3.4 Exact and asymptotic formulae for $E(N_B)$ 254

3.5 Extension of the selection models to odd N 256

3.6 Numerical results 258

3.7 Equivalence to Hoel's selection model 260

4 The Selection Models [2;PW;fixed N] and [2;VT;fixed N] with Curtailment

4.1 Introductory remarks 263

4.2 The PW-sampling procedure with curtailment 263

4.3 The VT-sampling procedure with curtailment 267

4.4 Numerical results 268

5 The Selection Model [2;VT;fixed N] with Truncation Based on $|S_A-S_B|$

5.1 Description of the model 272

5.2 Derivation of the P(CS)-value 272

5.3 Derivation of the probability of declaring the two treatments equal 275

5.4 Derivation of an upper bound for $E(N_B)$ 276

5.5 Derivation of the truncation points and of the patient horizon N 277

5.6 Numerical results 279

6 The Selection Model [2;PW;fixed N] with Truncation Based on $|S_A-S_B|$

6.1 Description of the model 284

6.2 Derivation of the P(CS)-value 284

6.3 Derivation of the probability of declaring the two treatments equal 285

6.4 Derivation of $E(N_B)$ 286

6.5 Numerical results 288

6.6 Comparison of the selection models 289

7 Selection Models Based on the Randomized Play-the-Winner Rule
7.1 Introductory remarks 290
7.2 Expected number of patients on the better treatment within n trials 290
7.3 Derivation of the P(CS)-values 292
7.4 Derivation of the expectations 294
7.5 Numerical results 295
8 Supplementary Investigations —— Topics Requiring Further Research ... 297

PART 2

CONTINUOUS RESPONSE SELECTION MODELS

Introduction 299

CHAPTER 1

SUBSET-SELECTION PROCEDURES BASED ON LINEAR RANK-ORDER STATISTICS

1 Linear Rank-Order Statistics and their Asymptotic Distributions
1.1 The general linear rank-order statistic 310
1.2 Some special linear rank-order statistics 313
1.3 The joint asymptotic distribution of the vector of rank-order statistics $(S_1,\ldots,S_k)$ based on joint ranks 322
1.4 The treatment of ties 323
2 Two Subset-Selection Procedures in One-Factor-Designs Including the General Behrens-Fisher-Problem
2.1 The selection rule R1 328
2.2 The infimum of the probability P(CS|R1) 333
2.3 The asymptotic distributions of two special rank-order statistics in case of consistent estimation of the unknown parameters 334
2.4 The probability P(CS|R1) in the LFC 339
2.5 A numerical example 340
2.6 Some Monte-Carlo studies 343
3 The Selection Rule R1 in the Case of Equal Scale-Parameters
3.1 The probability P(CS|R1) in the LFC 350
3.2 The Haga-statistic 351
3.3 A numerical example 353
3.4 Some Monte-Carlo studies 354

4 A Further Class of Subset-Selection Procedures in One-Factor Designs
4.1 The selection rule R2 361
4.2 The infimum of the probability P(CS|R2) 361
4.3 Exact and asymptotic distribution of max $S_j - S_1$ for identically distributed populations 362
4.4 A numerical example 368
4.5 Some Monte-Carlo studies 368

5 Some Properties of Optimality and a Brief Comparison of the Procedures No.2 — No.4
5.1 Local optimality of selection rule R1 375
5.2 Influence of the scorefunctions on the efficiency of procedures based on rules R1 and R2 379
5.3 Comparison of the procedures given in sections 2,3,and 4 381

6 The Selection Rules R1 and R2 in Case of Randomized-Block-Designs
6.1 Modified definition of ranks and the distribution of the resulting rank-order statistics 382
6.2 The rules R1 and R2 383
6.3 The asymptotic and the exact distribution of max $S_\ell - S_1$ for identical parameters 384
6.4 A numerical example 387
6.5 Some Monte-Carlo studies 389

CHAPTER 2

ASYMPTOTIC DISTRIBUTION-FREE SEQUENTIAL SELECTION PROCEDURES BASED ON AN INDIFFERENCE-ZONE MODEL

1 Introduction 395
2 A Class of Estimators of the Functions $f_i(\tau_1,\ldots,\tau_k)$
2.1 General one-sample rank-order statistics 397
2.2 The one-sample rank-order statistics based on median-scores 400
2.3 The general Hodges-Lehmann-estimator 401
2.4 A class of compatible estimators of the functions $f_i(\tau_1,\ldots,\tau_k)$ 403

3 Several Strongly Consistent Estimators
3.1 An estimator of $(B^2(G))^{-1}$ 406
3.2 An estimator of $(g(0)^2)^{-1}$ 407
3.3 Two estimators of $\lambda_j(G)$ and $G^*(0,0)$ 408

4 A Class of Sequential Selection Procedures
4.1 Definition of the selection procedures 413
4.2 Some important properties of the sequential selection procedures 414

5 A Numerical Example and some Remarks Concerning the Practical Working with the Sequential Procedures
5.1 An example 418
5.2 The implementation of the procedures 421

6 Another Class of Sequential Selection Procedures 427

7 Asymptotic Efficiency and some Monte-Carlo Studies
7.1 The asymptotic efficiency of the procedures of section 4 with respect to the procedures of section 6 429
7.2 Some Monte-Carlo studies of the procedures given in section 4 and 6 430
7.3 Some remarks concerning the application of general scorefunctions 433

CHAPTER 3

METHODS FOR SELECTING AN OPTIMAL SCOREFUNCTION

1 The Basic Idea 440

2 Two Statistics for Characterizing a Distribution
2.1 An estimator for the skewness of some distribution 441
2.2 An estimator for the peakedness of some distribution 442

3 Two Methods for Selecting a Scorefunction
3.1 Selection based on the joint sample 442
3.2 Selection based on the single samples 445

Appendix 1 447
Appendix 2 451
Appendix 3 454
Appendix 4 455
Appendix 5 459
Appendix 6 461

Abbreviations 469

References 471

General Introduction

In numerous situations of the everyday-life we are forced to solve the problem which one of several alternatives is the best, where the attribute "best" must be interpret as the case may be.
A wine-grower for example may be faced with the problem of choosing among several varieties of grapes, and the best choice he can take is to select that variety which yields the highest amount of wine. The same problem must be solved by a farmer who has to select the best of several varieties of wheat in order to maximize his net gain which itself is highly dependent on the wheat-harvest. The owner of an industrial plant must sometimes choose among several types of machines that are all suitable for manufactoring a special product. The best choice he can take is to select that machine which produces the least number of defective items, and last not least a physician must know which one of several therapies or drugs has the most beneficial effect in clinical medicine; that may be for example the drug with the highest probability of curing a disease. The latter question may also run as follows; which is the most beneficial dose of a new drug or is at least one among several new drugs better than a standard one?

These problems and all similar questions that may arise in practice can be shortly expressed in the language of statistics as "which one of $k \geq 2$ populations is the best, in a well-defined sense of best?". Concerning our examples, the k populations $\Pi_1, \Pi_2, \ldots, \Pi_k$ are represented by the k varieties of grapes, the k varieties of wheat, the k types of machines, and the k drugs, respectively, i.e. for example that in the industrial plant example, there are k different populations, that are the k fictitious populations consisting of all items that could be produced by the k machines, respectively.

In order to select the best population we must assume that there exists a complete ordering among the populations. For all following investigations, we additionally assume that the "quality" of a population can be expressed by a single real-valued parameter, i.e. that for example the "quality" of the fictitious population of all patients that could be treated with drug no. i is reflected by the single real number p_i, the success probability of drug (treatment) no. i, and the best population is that which is associated with the largest p_i, i.e. the highest success parameter. The best variety of wheat is that which is associated with the greatest average yield per acre.

The examples just outlined illustrate what will be our task in the following, that is to design a statistical experiment allowing us to select the best population at a preassigned significance level P^*, i.e. the experiment must be so that the best population is selected with a probability of at least P^*. Further details concerning the size of P^* are given in the introductions of part 1 and part 2, respectively.

To be able to take the decision which is the best population we have to procure some information about each population, and that is usually realized in drawing random samples from each population. Recalling our beforementioned examples, to draw an unit from the population associated with machine no. i means nothing but to draw an item produced by machine no. i or, if no item is available, to produce a new item with machine no. i, and to draw an unit from the population associated with drug no. i means consequently nothing but to treat a patient with that drug. The drawing of an item from the respective populations of the wine-grower and farmer example is to be conceived correspondingly.

The judgement which population is the best is naturally based on the realizations of the random samples, we usually speak of the response obtained from each population instead of realizations of random samples. The structure of the statistical experiment to be used to select the best population - we speak of a selection model in the following - is essentially dependent on the type of response. We consider the so-called dichotomous-response-(selection)-models and the continuous-response-(selection)-models. The dichotomous response models are adequate if the response can be described as failure (F) or success (S), i.e. the response can be expressed by a binomial chance variable with index 1 and successparameter p. A typical example for a dichotomous-response-model has been outlined above, that is the testing which of several drugs (treatments) is the best, if we judge the application of a special treatment as success/

failure whenever there exists any beneficial/no beneficial effect.

As for the farmer example, it seems to be obvious to use a continuous-response-model, because the average yield of wheat per acre can be expressed by a continuous random variable. The statistical analysis is essentially based on rank methods.

According to the possible response, the textbook is divided into two parts. Part 1 deals with dichotomous-response-models and part 2 is consequently reserved to continuous-response-models. The main elements of these models are described in the introductory chapters of each part, that is why we have outlined here only the fundamental ideas on which the construction of a selection model must be based.
The organization of the whole textbook is briefly summarized in the diagram below.

Classification of selection models according to the possible response

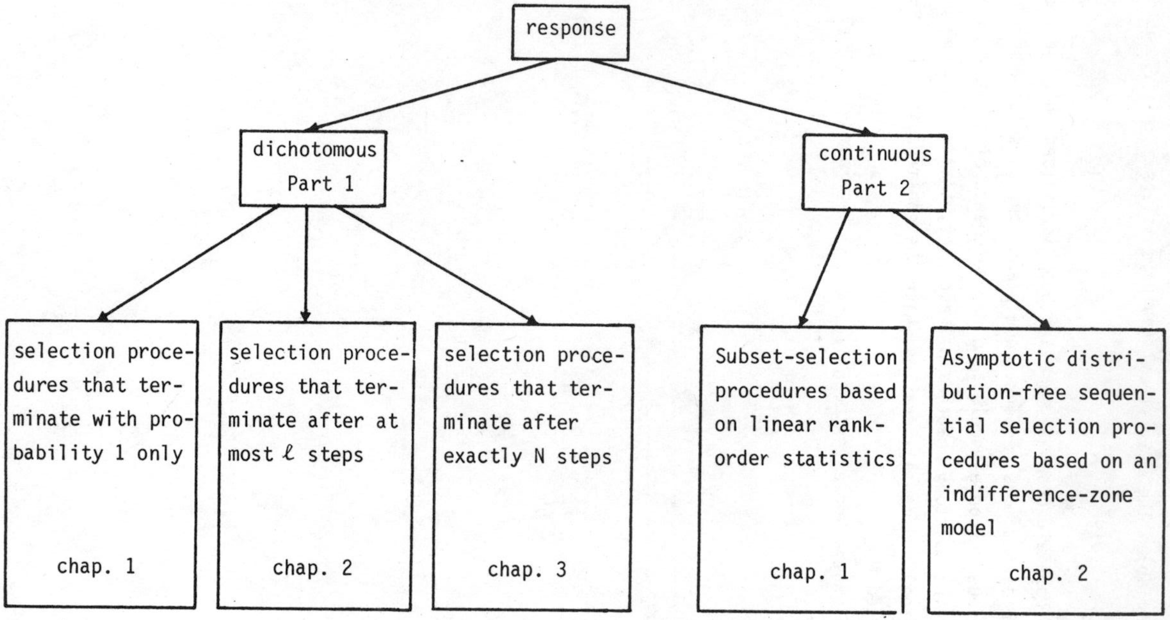

Part 1

Sequential Procedures for Selecting the Best of $k \geqslant 2$ Binomial Populations

Introduction

As already outlined in the general introduction, the first part of this textbook is reserved to the so-called dichotomous-response-models, i.e. our task is to design selection procedures that enable us to select the best of several binomial populations, where the best population is that which is associated with that success parameter which is maximal according to a complete ordering on the parameter space. For almost all practical purposes, we are faced with the problem to determine the population associated with the greatest or smallest success parameter. The fundamental notions needed in constructing a sequential selection model will be first of all explained for the special case that only two populations are available. Later on we generalize these notions to more than two populations.
Having in mind that the development of sequential selection models has been strongly influenced by medical formulations of the questions, - we refer the reader especially to P.Armitage's monograph , Sequential Medical Trials [8] - we make use of a medical description for selection models in the following.

Instead of two populations Π_1 and Π_2, we speak of two medicaments or two treatments A and B; the associated success parameters are p_A and p_B, respectively. Selecting an unit from population A, B is to be interpreted as treating a patient with medicament A, B, respectively. Recalling the industrial plant example, giving treatment A to a patient means nothing but to produce an item with machine A; we sometimes use the expression "a trial is carried out with treatment A".
Having given these short explanations, it seems to be clear how the medical notions are to be understood in non-medical situations. Nevertheless we want to emphasize again that all selection models to be considered in the following are applicable if the corresponding populations are binomial ones, i.e. the notions "medicament", "treatment", "patient" etc. are not always to be interpreted in a strongly medical sense, although these notions are used throughout.

Experimental Conditions

In the following, the best of the two treatments in research is always denoted by A, i.e. there are two treatments no. 1 and no. 2 and the best of them is called treatment A but we do not know whether no. 1 or no. 2 is treatment A. Our goal is to design a statistical experiment so that treatment A can be identified with a probability of at least P^*, where P^* is greater than 1/2, because a selection of A with probability $P^* = 1/2$ can be already realized by randomization. However, before applying statistical methods, we must guarantee that the following experimental conditions are fulfilled.

(1) The response is obtainable without delay and moreover recognizable exactly, i.e. in a reasonable time after the administration of a treatment, the experimenter knows clearly whether the response is a success or a failure.

(2) The success parameters p_A and p_B are constant during the whole experiment.

(3) The single trials are independent of each others, i.e. the result of any trial has no influence on the result of any other trial carried out at the same time or later.

In case that these three conditions are known to be true, the experiment is usually called statistically controlled; this is a fundamental notion from quality control. The statistical experiment is hence given by a sequence of random variables that have all a binomial distribution with index 1 and success parameter p_A or p_B, according to the treatment on which a single trial is just being carried out.
Condition (1) is considered to hold in most practical situations. The experimenter himself has to decide what is meant by "reasonable time" and how to judge the result of a single trial; those decisions are by no means statistical ones. The problem how to proceed in case of delayed response will be discussed in chap. 3.(cf. sec.7 and the concluding remarks in sec. 8).
The validity of condition (2) can be examined in applying appropriate statistical tests. If p_A and p_B denote the success probabilities of machine A and machine B, respectively, i.e. p_A and p_B are the probabilities of producing a correct item, then it is obvious that these probabilities cannot held constant over a very long time period. Wear and tear of the components of the machines usually causes an increase of the failure rates and hence a decrease of the success probabilities. However, in most practical cases the occurrence of wear and tear can be neglected because the duration of the statistical experiment

is only short with regard to the time period in which a marked increase of the failure rates could arise, i.e. with a safe conscience, the success parameters can be considered as constant or at least as approximately constant during the whole testing-phase. In case that p_A and p_B actually denote the success probabilities of two medicaments, we may assume that these parameters are constants. The experimenter must however guarantee that the patients, involved in the experiment, differ only in the treatment received, i.e. all factors such as age, sex, weight etc. that could have an influence on the type of response must be held at (approximately) the same level, in other,words non-treatment effects should be eliminated from the proper experiment.

Condition (3) means that the two sequences of trials carried out with treatment A, B, respectively, can be expressed by sequences of independent identically distributed binomial chance variables with index 1 and success parameter p_A, p_B, respectively.

Selection Principle

Our goal is to select the best treatment, i.e. to select treatment A. In this case, we speak of a "Correct Selection" (CS). However, it is not reasonable to distinguish between two treatments at any price,i.e. the treatments are considered as similar if there are only slight differences between the success parameters. The notion "correct selection" is hence of importance only if the success parameters differ of at least Δ^*, i.e. a correct selection should take place if the true difference of the success parameters, $\Delta := p_A - p_B$, is not less than a preassigned constant Δ^*, where Δ^* must be specified by the experimenter.

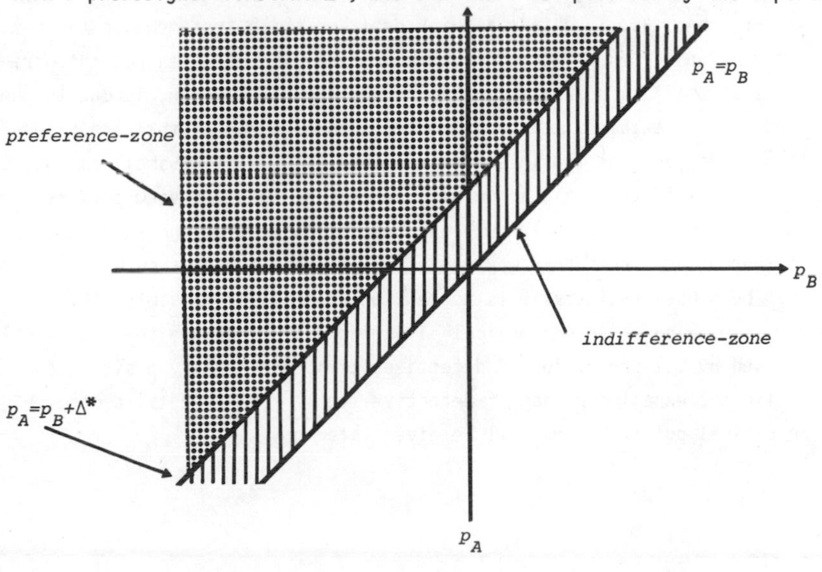

In case that $p_A-p_B=:\Delta\geq\Delta^*$, i.e. if (p_B,p_A) lies in the preference-zone (cf. the above fig.), we require that the probability of a correct selection is at least P^*, i.e.

$$P(CS) \geq P^* \text{ whenever } p_A-p_B =: \Delta \geq \Delta^*,$$

this requirement is usually referred to as $(P^*;\Delta^*)$-condition. For parameter configurations of the indifference-zone, the above requirement is not satisfied in general. Recall that this is not necessary, because treatments associated with parameters, lying in the indifference-zone, are considered as similar.

As indicated above, the distinction between the treatments is expressed by the difference p_A-p_B of their success parameters; it is however possible to define the $(P^*;\Delta^*)$-condition in terms of the ratio of the success probabilities instead of the difference. The corresponding probability requirement is then given by

$$P(CS) \geq P^* \text{ whenever } p_A/p_B =: \delta \geq \delta^*$$

We do not want to follow up this idea; the reader interested in this subject is referred to the corresponding literature (cf. for example Taheri/Young [194]).

Ethical Point of View

The number of patients that have received treatment A, B is denoted by N_A, N_B, respectively. The number of all patients involved in the experiment is hence given by $N=N_A+N_B$. N is usually referred to as patient horizon. Note that this is not the original meaning of patient horizon as for example defined by Colton [49] (cf. chap. 3). To obtain a quick decision which treatment is the best, we have to design our experiment in such a way that the expected patient horizon E(N) is as small as possible. The main requirement, however, is due to the ethical point of view, i.e. the experimenter should guarantee that the expected number of patients on the inferior treatment $(E(N_B))$ is as small as possible. Using a decision theoretic argument, the ethical point of view prescribes that the expected loss $L(p_A,p_B) := (p_A-p_B)E(N_B)$ that denotes the expected additional number of successes that would have been occurred if the superior treatment A had been used exclusively is to be kept as small as possible. This requirement is meaningful in non-medical situations, too, because the economical costs, caused by the production of defective items, are usually positive correlated with the expected number of defective items. Further details concerning the ethical point of view will be given later on.

Components of a Selection Model

A Selection Model consists of two principal components, that are the sampling rule (or the sampling scheme) and the termination rule (or the stopping rule). The sampling rule prescibes how to assign the treatments to the patients, and the termination rule specifies the moment when enough information is available to take a correct decision at a significance level $\geq P^*$ whenever $p_A - p_B =: \Delta \geq \Delta^*$ holds. The most important sampling rules and termination rules are given in detail hereafter.
We want to select correctly with a probability of at least P^* whenever $\Delta \geq \Delta^*$, however, we do not have any knowledge about the magnitude of the success parameters. Having in mind that the probability of a correct selection is a function of p_A and p_B (or of p_A and Δ; $p_B = p_A - \Delta$), the above requirement, i.e. the $(P^*;\Delta^*)$-condition, is satisfied for all parameter configurations of the preference-zone if it is satisfied for that parameter configuration of the preference-zone which yields the lowest probability for a correct selection. That special parameter configuration is called the <u>L</u>east <u>F</u>avorable <u>C</u>onfiguration = LFC. Making use of this notion, the main requirement that each selection model has to satisfy, can be expressed as follows: the $(P^*;\Delta^*)$-condition must be satisfied if the parameters are in the LFC.

Comparison of Selection Models

The beforementioned ideas have already indicated how to compare distinct selection models. An exact comparison of two or more than two selection models is possible only if their P(CS)-values coincide for each parameter configuration of the preference-zone. The selection models no. 6 and no. 7 of chap. 2 for example have identical P(CS)-functions. Since this strong requirement is not always fulfilled we have to weaken it in order to get a possibility to judge the quality of distinct selection models. We therefore speak already of comparable selection models if they have approximately the same P(CS)-value in their LFC's. Using this weak definition of comparability, two procedures can be compared in general; it should, however, emphasized that a comparison of procedures is essentially more predicatory if the P(CS)-functions coincide not only in their LFC's but for each parameter configuration of the preference-zone.
Within this meaning, a selection model $\mathcal{E}_1$ is uniformly better than a selection model $\mathcal{E}_2$ with respect to $F(N_B)(F(N))$ if $E(N_B|\mathcal{E}_1) < E(N_B|\mathcal{E}_2)(E(N|\mathcal{E}_1) \leq E(N|\mathcal{E}_2))$ for all (p_B, p_A)-values of the preference-zone, and if $E(N_B|\mathcal{E}_1) < E(N_B|\mathcal{E}_2)$

$(E(N|\mathcal{E}_1)<E(N|\mathcal{E}_2))$ for at least one configuration (p_B,p_A) of the preference-zone; $E(N_B|*)(E(N|*))$ denotes the expectation of $N_B(N)$ if selection model * is used.
In case that an overall comparison is not possible, we have to compare distinct procedures locally, i.e. we must look for those parts of the preference-zone in which either procedure is better (cf. for example chap. 1, sec. 4).

Sampling Rules

(a) Vector-at-a-time sampling (VT-sampling)

The most obvious way of assigning the treatments to the patients consists in carrying out a single trial with both treatments at each stage of the experiment. Denoting an A,B-success by A,B and an A,B-failure by $\bar{A},\bar{B}$, respectively, the response at any stage of the experiment is given by (A,B); (A,$\bar{B}$); ($\bar{A}$,B) or ($\bar{A}$,$\bar{B}$). Recalling that the trials are carried out at the same time and that the response can be simply expressed by a (two-dimensional) vector, it is easy to understand why this special way of assigning treatments to patients is called vector-at-a-time sampling, we usually make use of the abbreviation VT-sampling. The VT-sampling scheme can be illustrated as follows.

Trial no.	1	2	3	4	. . .
	A	$\bar{A}$	$\bar{A}$	A	. . .
	$\bar{B}$	B	B	B	. . .

A great advantage of the VT-sampling scheme that is often used in medical trials consists in the possibility of matching the experimental units, i.e. the experimenter can form homogeneous pairs in such a way that the patients differ only in the treatment received, non-treatment effects are thus almost eliminated from the proper experiment.

(b) Play-the-Winner sampling (PW-sampling)

The Play-the-Winner rule was first suggested by H.Robbins [155] in 1956, in a discussion of the so-called two-armed bandit problem. The treatments are given one at a time, and the first treatment at the outset is determined randomly. A success with a given treatment generates a further trial on the same treatment whereas a failure generates a switch to the other treatment. The PW-sampling scheme can be illustrated as follows:

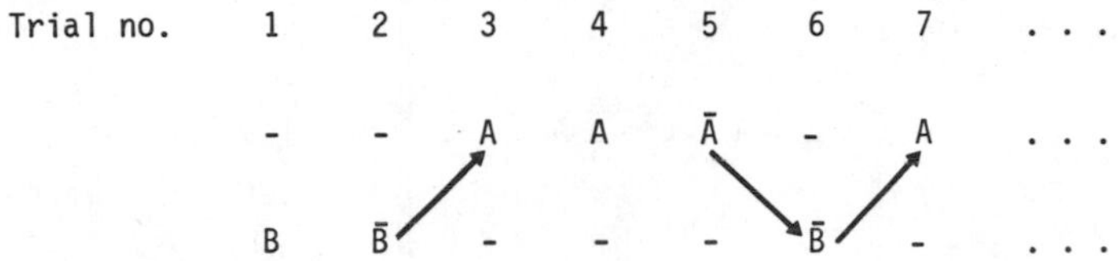

The use of the PW-rule causes a bias in favor of testing the better treatment; however, that does not imply by no means that PW-sampling is in any case better than VT-sampling (cf. chap. 1, sec. 4).

(c) Play-the-Loser sampling (PL-sampling)

Just as PW-sampling, the PL-sampling rule prescribes that the first treatment at the outset is determined randomly and that the trials are carried out one at a time. Contrary to PW-sampling, PL-sampling requires that a failure with a given treatment generates a further trial on the same treatment whereas a success generates a switch to the alternative treatment. The PL-sampling scheme can be illustrated as follows:

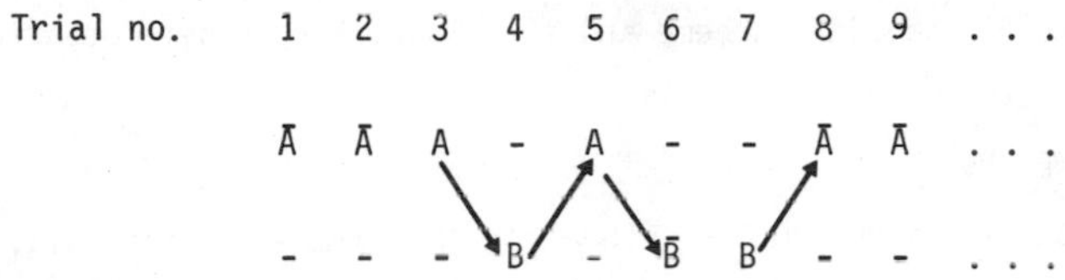

The use of the PL-rule causes a bias in favor of testing the worse treatment. It has turned out that PL-sampling is better than PW-sampling if the success parameters p_A and p_B are small. Some critics raise the objection that the use of PL-sampling contradicts the ethical point of view. We will discuss this objection in detail after having introduced the termination rules, however, it should be already mentioned here that this point of criticism escapes when no human beings are involved in the experiment.

(d) Follow-the-Leader sampling (FL-sampling)

A modification of the PW-sampling scheme is the so-called Follow-the-Leader sampling scheme. The FL-sampling scheme works essentially in the same way as the PW-sampling scheme; in addition, FL-sampling prescribes that in case of $F_A=F_B$ and $S_A=S_B$, the next treatment has to be determined randomly, where S_A, S_B and F_A,F_B denote the current number of successes and failures of treatment A,B, respectively. The FL-sampling scheme can be illustrated as follows:

Trial no.	1	2	3	4	5	6	7	8	9
	A	A	A	$\bar{A}$	-	-	-	-	
	-	-	-	-	B	B	B	$\bar{B}$	

the next treatment is determined randomly because of $F_A=F_B=1$ and $S_A=S_B=3$

(e) Randomized PW-sampling (RPW-sampling)

A sampling rule that allows delayed response without forcing the experimenter to stop the performance of the next trial is the so-called randomized Play-the-Winner sampling rule (RPW-sampling). The first treatment at the outset is determined randomly. If a success of treatment no. 1 or a failure of treatment no. 2 occurs, a white ball is placed into a "statistical hat". In case that a failure of treatment no. 1 or a success of treatment no. 2 occurs, a red ball is placed into the "statistical hat". The treatment assignment for a trial is determined by drawing a ball randomly from the "statistical hat". In case of an empty hat, the next treatment is determined randomly. If no delayed response arises, the RPW-sampling scheme and the PW-sampling scheme coincide.

(f) PW-mF-sampling

A modification of the PW-sampling scheme is the PW-mF-sampling-rule. The first treatment at the outset is determined randomly and the treatment assignment is changed whenever m failures on the applied treatment have occurred. In case that m=1, the PW-mF-rule reduces to the original PW-rule, described in (b).

Another possibility to assign patients to the treatments that is also based on the PW-rule consists in changing the treatments if m failures in succession occur Sobel and Weiss [180] mention that Nebenzahl has shown - a reference of his investigation has not been given - that in using the above described sampling rule, essentially based on runs of failures, together with the so-called inverse stopping rule, i.e. sampling terminates whenever $\max\{S_A,S_B\}=r$, then the best selection model is obtained for m=1, i.e. when the original PW-sampling scheme is used.

Termination Rules

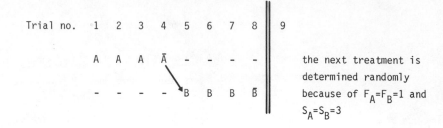

The second principal component of a selection model is the termination rule (or stopping rule), i.e. a rule that prescribes when enough "information" is available to stop the experiment and to select the best treatment, so that the $(P^*;\Delta^*)$-condition is satisfied in the LFC. We distinguish two types of termination rules, that are rules based on either successes or failures, and rules based on successes <u>and</u> failures.

(a) Typical termination rules based on either successes or failures are for example:

$|S_A-S_B| = r$; $|F_A-F_B| = r$; $\max\{S_A,S_B\} = r$; $\max\{F_A,F_B\} = r$; S_A-run or S_B-run of length r. S_A, S_B, F_A, F_B have the same meaning as before. $|S_A-S_B|=r$ means for example that sampling has to be stopped if the absolute difference of the current number of A- and B-successes is equal to r, where the natural number r must be chosen such that the probability requirements of the selection model are satisfied. The treatment associated with the larger number of successes, when sampling has been finished, is declared to be the best; in case of an equal number of successes - which cannot always occur - the best treatment is determined randomly.

(b) Typical sampling rules based on successes and failures are for example:

$\max\{R_A,R_B\} = r$, $R_A := S_A + F_B$ and $R_B := S_B + F_A$

$\max\{S_A,S_B\} = r$ or $\min\{F_A,F_B\} = c$

$\max\{S_A,S_B\} = r$ or $\max\{F_A,F_B\} = c$

$|S_A-S_B| = r$ or $F_A + F_B = s$

$\max\{S_A,S_B\}=r$ or $\min\{F_A,F_B\}=c$ means for example that sampling terminates whenever one treatment yields its r-th success or all treatments yield at least c failures. The best treatment is determined according to the observed numbers of successes.

The main difference between the two types of termination rules is that the rules summarized in (a) usually effect that sampling terminates with probability one only whereas the rules summarized in (b) generally lead to a decision after at most ℓ steps, where ℓ is a preassigned constant known to the experimenter. If only a bounded number of experimental units is available, which will be usually the case in medical situations, it may happen that sampling must be terminated because of a lack of experimental units. In this special case a decision which treatment is the best can be only taken at a significance level below the preassigned P^*. We will make later on some suggestions how to proceed in cases when the used selection model is such that sampling terminates with probability one only.

It seems to be obvious that only meaningful combinations of sampling- and termination rules lead to satisfactory selection models, i.e. for example that the PW-sampling scheme should be used with a termination rule based on successes rather than on failures, whereas the PL-sampling scheme seems to be meaningful only in connection with a termination rule based on failures. We are now able to take up the discussion in context with the ethical point of view. As already mentioned, the PL-sampling scheme causes a bias in favor of testing the worse treatment, and this fact immediately clears which combinations of PL-sampling scheme and termination rules are meaningful, because a bias in favor of testing the worse treatment suggests to use a termination rule based on failures. In particular when the successparameters are small, the use of the PL-sampling scheme in connection with a stopping rule based on failures leads to a substantial reduction of the expected number of patients involved in the experiment (cf. for example table 1 of chap. 1). It should be mentioned, however, that the expected number of patients on the inferior treatment is greater than the expected number of patients on the superior treatment, and this fact could be indeed a point of criticism. Nevertheless, we must emphasize that for small success probabilities, the expected number of patients on the inferior treatment is in any case smaller, and often substantially smaller, when the PL-sampling scheme is used instead of the PW- or VT-sampling scheme. This can be also seen from table 1 of chap. 1. We may conclude from these thoughts that PL-sampling is a useful sampling-rule, at least in case when no human beings are involved in the experiment. We have outlined advantages and disadvantages of this special sampling scheme, but it seems to be impossible for us, to give an overall accepted advice how to proceed, when human beings are actually involved in the experiment. A profound discussion of the ethical point of view cannot be the task of a statistician.

In our textbook only selection models with a termination rule based on failures or successes are considered. The reader who is interested in termination rules that are based on likelihood ratios is referred to the corresponding literature. Bechhofer, Kiefer and Sobel [31] investigate likelihood ratio termination rules in connection with the VT-sampling scheme, further results are given by Hoel and Sobel [95], a new likelihood stopping rule is proposed by Hoel, Simon and Weiss [94], and B.P.Hsi and Louis [105] consider a modified PW-sampling scheme that allows a direct application of the sequential probability ratio test.

More than 2 Treatments

In case of more than two treatments the sampling- and stopping rules have to be generalized correspondingly. The generalization of the VT-sampling scheme to $k > 2$ treatments seems to be obvious, i.e. at each stage of the experiment

a single trial with each treatment is carried out. The PW-sampling scheme is generalized as follows: at the outset the k treatments are put in a random order, where A_1 denotes the best treatment, A_2 the one following A_1 in the initial randomization etc., continuing in cyclic order; the success parameter of treatment A_i is denoted by p_i, and the second largest success parameter is denoted by p_2^*. (Note that p_2^* will be usually different from p_2!). The experiment begins with the first treatment in the random order; a success generates a further trial on the same treatment and a failure generates a switch to the next treatment in the random order etc., continuing in cyclic order. The PL-sampling scheme is generalized correspondingly.
The generalization of the termination-rules to more than two treatments is not always possible. We mainly use the termination rules

"$\max\{S_1,...,S_k\} = r$", and

"$\max\{S_1,...,S_k\} = r$ or $\min\{F_1,...,F_k\} = c$", where

S_i and F_i denote the current numbers of successes and failures of treatment A_i, respectively.
A comparison of selection procedures is possible if they have approximately the same P(CS)-value in their LFC's. The $(P^*;\Delta^*)$-condition is altered as follows:

$P(CS) \geq P^*$ whenever $p_1 - p_2^* \geq \Delta^*$, where

$p_2^* = \max\{p_2,...,p_k\}$, and P^* is greater than $1/k$,

because $P^* = 1/k$ can be realized by randomization only.

The comparison of competing selection models can be based either on E(N), the expected number of patients involved in the experiment, or on the expected loss $\sum_{i=2}^{k} (p_1-p_i)E(N_i)$, i.e. the expected additional number of successes that would have been occurred if the best treatment A_1 had been used throughout.

In order to recognize the structure of a selection model at the first glance, we introduce the following nomenclature:

[*; **; ***],

where * denotes the number of competing treatments, ** the sampling rule, and *** the termination rule. $[k;PW;\max\{S_1,...,S_k\}=r]$ denotes e.g. a selection model, where k treatments are available, the assignment of treatments to the patients is carried out according to the PW-sampling scheme, and sampling terminates whenever one treatment yields r successes.

Contents of Part 1

As can be seen from the diagram at the end of the general introduction, part 1 is divided into three chapters that result in a natural way from the subject considered. Chapter 1 deals with selection models that have a termination-rule preponderantly based on either successes or failures, and sampling consequently terminates with probability one only. We present selection models proposed by Sobel/Weiss [176], [177], [179], Hoel/Sobel/Weiss [96], Nordbrock [131], Hoel/Sobel [95] and by ourselves [164]. Among other things a two-stage selection procedure, where only PW- and VT-sampling is possible in the second stage, is generalized so that PL-sampling can be also applied in the second stage (cf.sec.5). In sec.10 and 11 selection models allowing to select the best treatment out of $k \geq 3$ competing treatments are considered, and detailed tables are given for various values of k. The main disadvantage of all selection models considered in part 1 is that sampling terminates only with probability one. In applying such a selection model it may be possible that sampling must be terminated because of a lack of experimental units without reaching a decision. In such situations, the only thing we can do is to analyse the available information and to take a decision at a significance level that is below the desired P^*. To avoid the occurrence of testing-sequences that do not lead to a terminal decision that meets the probability requirements given by the $(P^*;\Delta^*)$-condition, it is necessary to investigate how many experimental units are needed so that in almost all cases the experiment can be terminated as originally designed. For this purpose we have carried out numerous simulation studies in order to determine upper bounds N^* for each selection model that give the maximal number of trials that have to be conducted. N^* has been chosen so that the probability requirements for the corresponding selection models are met, and so that the number of truncated sequences - that are sequences of length N^* that allow to identify the best treatment only at a significance level less than P^* - is within reasonable bounds. In practice N^* experimental units must be available to be almost sure that the experiment can be carried out as designed. Detailed tables are given in section 12. In section 13 we present a PW-selection model first proposed by Nordbrock [131], the corresponding VT-selection model has been investigated by us and is presented in section 14. Two selection procedures designed for $k \geq 3$ treatments are considered in sec. 15. These so-called elimination procedures allow to eliminate treatments from the experiment that have turned out to be poor. The first chapter is concluded by a comparison of the relative worth of the procedures presented so far.

In using termination-rules based on successes and failures, the problem of a never ending sampling does not arise in most practical situations. - The selection models given in sec. 13 and 14 of the first chapter have also a termination rule based on successes and failures but nevertheless sampling terminates with probability one only - Chapter 2 deals with such selection models that provide a restricted (not fixed) patient horizon, i.e. the number of trials needed to take the terminal decision is again represented by a random variable N but sampling terminates in any case after at most ℓ steps, where the constant ℓ is known to the experimenter, and the decision which treatment is to be declared as best can be taken at the preassigned significance level.
We consider selection models due to Hoel [92], Berry/Sobel [35], Schriever [160] and [161], Fushimi [63], and Schriever [162] and [163]. In particular it is pointed out there that the selection models given in Schriever [162] and [163] are uniformly better than the corresponding ones investigated by Sobel/Weiss [179]. A new selection model constructed for identifying the best of $k \geq 3$ treatments is presented in sec. 8. This VT-sampling selection procedure allows to eliminate treatments that turn out to be poor. For all selection models detailed numerical results are given at the end of each section. We conclude chapter 2 by presenting 5 new elimination procedures not yet published in statistical literature.

After all the third chapter is reserved to selection models with fixed patient horizon. In section 1 we give a brief historical review of what have been done in the field of sequential clinical trials from 1951 to 1969. Because of its importance for the development of sequential selection procedures using adaptive sampling schemes, the selection model proposed by Zelen [202] is discussed in detail.(cf. sec. 2) PW- and VT sampling selection models due to Nebenzahl/Sobel [130] and the corresponding truncated models proposed by Pradhan/Sathe [139] are considered in the sections 3 and 4. In sec. 5 we present a VT-selection model, proposed by Kiefer/Weiss [112], that allows in addition to decide that there are no differences between the treatments. The corresponding PW-model is given in sec. 6.The following sec.7 of chap.3 deals with selection models that use the so-called randomized play-the-winner rule. We consider a selection model due to Durham and Wei [55] that uses the termination rule of the Hoel selection model given in sec. 1 of chap. 2. In addition selection models using the RPW-rule and various termination-rules that have been investigated by simulation are presented there. The first part is concluded by some remarks concerning topics of further research.

CHAPTER 1

Selection Procedures with Unrestricted Patient Horizon

1 The Selection Model $[2; PW; |S_A - S_B| = r]$

1.1 Derivation of the critical r-value

The following procedure uses PW-sampling and the stopping-rule $|S_A - S_B| = r$, where S_A, S_B denote the current number of successes of treatment A,B, respectively. The procedure terminates if the absolute difference of the current number of A- and B successes equals r. The procedure declares treatment A to be the better one when $S_A - S_B = r$. The integer r is chosen to be the smallest value such that

$$P(\text{Correct Selection}) =: P(CS) \geq P^* \text{ whenever } \Delta \geq \Delta^* \tag{1.1}$$

P^* and Δ^* are preassigned constants specified by the experimenter and $\Delta := p_A - p_B$ is the true unknown difference between the successparameter p_A of treatment A and p_B of treatment B. (1.1) is sometimes referred to as $(P^*;\Delta^*)$-condition. CS denotes the event "correct selection" which is the choice of treatment A when $\Delta > 0$.
A correct selection takes place only if $S_A - S_B = r$. In case of $S_A - S_B = -r$, the inferior treatment is wrongly denoted as best.
In order to derive the exact formula for P(CS), we define

$$P_n := P(CS | S_A - S_B = n, NT = A); \; Q_n := P(CS | S_A - S_B = n, NT = B) \tag{1.2}$$

where "NT=A" means that the next trial is carried out with treatment A.
Using (1.2) and letting $q_A := 1 - p_A$, and $q_B := 1 - p_B$, we obtain:

$$P_n = p_A P_{n+1} + q_A Q_n \quad ; \quad Q_n = p_B Q_{n-1} + q_B P_n \tag{1.3}$$

and the boundary conditions are

$$P_r=1;\ Q_{-r}=0 \text{ or } Q_r=p_BQ_{r-1}+q_B \tag{1.4}$$

Solving for P_n in the second equation of (1.3) and substituting it into the first equation of (1.3), we obtain:

$$p_AQ_{n+1}-(p_A+p_B)Q_n+p_BQ_{n-1}=0 \tag{1.5}$$

The roots of the characteristic equation of (1.5) are

$$m_1=1 \text{ and } m_2=\frac{p_B}{p_A}\ =:\lambda<1 \tag{1.6}$$

The solution to (1.5) satisfying the boundary conditions (1.4) is easily shown to be

$$Q_n = \frac{q_B(1-\lambda^{n+r})}{q_B-q_A\lambda^{2r}} \tag{1.7}$$

The corresponding expression for P_n follows immediately by using the first equation of (1.3) and the above formula.

$$P_n = \frac{q_B-q_A\lambda^{n+r}}{q_B-q_A\lambda^{2r}} \tag{1.8}$$

The difference between the current number of A and B successes at the outset is equal to zero and the first treatment is chosen at random, hence, the desired P(CS) is given by

$$P(CS) = \frac{1}{2}(P_0+Q_0) = \frac{1}{2}\,\frac{2q_B-(q_A+q_B)\lambda^r}{q_B-q_A\lambda^{2r}} \tag{1.9}$$

It is easily seen that $\lim_{r\to\infty} P(CS)=1$; that is why we can always find an integer $r\geq 0$ so that $P(CS)\geq P^*$ for $P^*\epsilon(0,1)$.
The typical shape of the P(CS) graph as a function of r , p_A,Δ fixed, can be seen from the figure below.
Setting $P(CS)\geq P^*$, we obtain the following inequality quadratic in λ^r.

$$q_B - \frac{1}{2}(q_A+q_B)\lambda^r \geq P^*(q_B-q_A\lambda^{2r}) \tag{1.10}$$

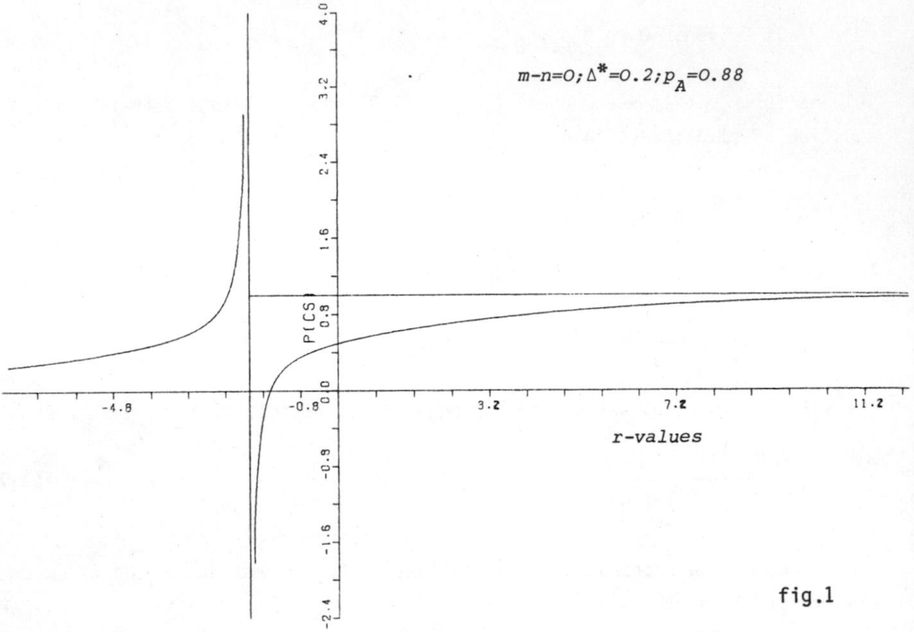

fig.1

The two solutions r_1, r_2 of (1.10) are determined by

$$\lambda^{r_1} \leq \frac{q_A+q_B}{4P^*q_A} - \left(\left(\frac{q_A+q_B}{4P^*q_A}\right)^2 - \frac{q_B(1-P^*)}{q_AP^*}\right)^{\frac{1}{2}} \tag{1.11}$$

$$\lambda^{r_2} \geq \frac{q_A+q_B}{4P^*q_A} + \left(\left(\frac{q_A+q_B}{4P^*q_A}\right)^2 - \frac{q_B(1-P^*)}{q_AP^*}\right)^{\frac{1}{2}}$$

Choosing $P^*>\frac{1}{2}$ - $P(CS) \geq P^* = \frac{1}{2}$ can be realized in selecting the best treatment at random -, we see that only r_1 is a meaningful solution since $r_2<0$. Hence, the desired r value is obtained from the first equation of (1.11)

$$r = \left]\frac{1}{\ln\lambda} \ln\left(\frac{q_A+q_B}{4P^*q_A} - \left(\left(\frac{q_A+q_B}{4P^*q_A}\right)^2 - \frac{q_B(1-P^*)}{q_AP^*}\right)^{\frac{1}{2}}\right)\right[\tag{1.12}$$

where]x[denotes the smallest integral number $\geq$ x.

Equation (1.12) does not give a numerical solution if p_A and p_B are unknown. Therefore it is necessary to look for that configuration of parameters p_A, p_B which minimizes the probability of a correct selection; this special configuration of parameters is referred to as the least favorable configuration (LFC). For this purpose we define $\overline{p}$ to be the arithmetical mean of the successparameters p_A and p_B, i.e.

$$\overline{p} := \frac{1}{2}(p_A + p_B) \tag{1.13}$$

Recalling that $p_A - p_B = \Delta$, we obtain:

$$p_A = \overline{p} + \frac{1}{2}\Delta \; ; \; p_B = \overline{p} - \frac{1}{2}\Delta \tag{1.14}$$

Because of the inequality $\Delta < p_A < 1$, $\overline{p}$ lies in the open interval $(\frac{1}{2}\Delta, 1 - \frac{1}{2}\Delta)$. Replacing "$\leq$" by "=" in the first inequality of (1.11) and then substituting q_A, q_B corresponding to (1.14), we obtain:

$$\lambda^r = \frac{1-\overline{p}-((1-\overline{p})^2-(4(1-\overline{p})^2-\Delta^2)P^*(1-P^*))^{\frac{1}{2}}}{2P^*(1-\overline{p})-P^*\Delta} =: g(\Delta) \tag{1.15}$$

Holding $\overline{p}$ fixed, the right-hand side of (1.15) is a function of Δ alone. Differentiating $g(\Delta)$, we see that $\frac{dg(\Delta)}{d\Delta} > 0$, provided that $P^* > \frac{1}{2}$; therefore, g is strictly increasing with Δ and $\lambda = \lambda(\Delta) = \frac{p_B}{p_A} = \frac{\overline{p} - \frac{1}{2}\Delta}{\overline{p} + \frac{1}{2}\Delta}$ is a strictly decreasing function of Δ.

These two statements together with $(\lambda(\Delta))^{r(\Delta)} = g(\Delta)$ **imply** immediately that r is strictly decreasing with Δ; i.e. $r(\Delta)$ is maximized by the smallest admissible Δ. Thus, in the LFC, we obtain $\Delta = \Delta^*$. Using this first result and the fact that (for practical purposes) P^* is close to 1, we deduce from (1.15):

$$(\lambda(\Delta^*))^r = \left(\frac{2\overline{p}-\Delta^*}{2\overline{p}+\Delta^*}\right)^r \approx \left(1 + \frac{\Delta^*}{2(1-\overline{p})}\right) \cdot (1-P^*), \tag{1.16}$$

with an error in $(\lambda(\Delta^*))^r$ that is $O((1-P^*)^2)$.
The desired $\overline{p}$-value in the LFC can now be derived by maximizing r with respect to $\overline{p}$.
Taking logarithms in (1.16) and using the inequality $\ell n(1+\Delta^*/(2(1-\overline{p}))) < \ell n 2$, we see that $\ell n(1-P^*)$ dominates the right side. Hence, the value of $\overline{p}$ which maximizes r lies close to that $\overline{p}$ which maximizes $\ell n \lambda(\Delta^*) = \ell n((2\overline{p}-\Delta^*)/(2\overline{p}+\Delta^*))$, and this value

is easily shown to be $\overline{p}=1-\frac{1}{2}\Delta^*$, i.e. in the LFC, we have:

$$p_A=1,\ p_B=1-\Delta^* \tag{1.17}$$

Substituting these values into (1.16) and denoting the resulting maximal r by r_{max}, we finally obtain:

$$r_{max} \approx \left] \frac{\ln 2(1-P^*)}{\ln(1-\Delta^*)} \right[\tag{1.18}$$

remark: r_{max} is sometimes referred to as the critical value of r.

1.2 Derivation of the expectations

In a first step, we determine $E(N_B)$ the expected number of patients on the inferior treatment B. For this purpose, we define

$$U_n:=E(N_B|S_A-S_B=n,NT=A);\ V_n:=E(N_B|S_A-S_B=n,NT=B) \tag{1.19}$$

From this, we deduce the following system of difference equations

$$U_n=p_AU_{n+1}+q_AV_n;\ V_n=p_BV_{n-1}+q_BU_n+1 \tag{1.20}$$

with the boundary conditions

$$U_r=V_{-r}=0 \tag{1.21}$$

remark: We must add one to the second equation of (1.20) since a B trial is carried out.

The single difference equation

$$U_{n+1}-\frac{p_A+p_B}{p_A}U_n+\frac{p_B}{p_A}U_{n-1}+\frac{q_A}{p_A}=0, \tag{1.22}$$

obtained from (1.19), possesses the partial solution $U_n=-nq_A/(p_A-p_B)$; and comparing the homogeneous part of (1.22) with (1.5), the general solution of (1.20) satisfying the boundary conditions (1.21), is given by

$$U_n = \frac{q_A(r-n)}{p_A-p_B} - \frac{(2rq_A+p_A)q_A\lambda^r(\lambda^r-\lambda^n)}{(p_A-p_B)(q_A\lambda^{2r}+p_A\lambda-1)} \tag{1.23}$$

$$V_n = \frac{(r-n)q_A+p_A}{p_A-p_B} - \frac{(2rq_A+p_A)\lambda^r(q_A\lambda^r-(1-p_A\lambda)\lambda^n)}{(p_A-p_B)(q_A\lambda^{2r}+p_A\lambda-1)} ,$$

and $E(N_B)$ is found to be

$$E(N_B) = \frac{1}{2}(U_0+V_0) = \frac{(2rq_A+p_A)(1-\lambda^r)(q_A\lambda^r+p_A\lambda-1)}{2(p_A-p_B)(q_A\lambda^{2r}+p_A\lambda-1)} \tag{1.24}$$

If P^* is close to 1, the resulting r will be large enough to neglect λ^r and λ^{2r}. Thus (1.24) reduces to

$$E(N_B) \approx \frac{2rq_A+p_A}{2(p_A-p_B)} \tag{1.25}$$

and the expected loss $L(p_A,p_B)$ is given by

$$L(p_A,p_B)=(p_A-p_B)E(N_B)\approx \frac{1}{2}p_A+q_A r_{max} \approx \Big] q_A \frac{\ell n 2(1-P^*)}{\ell n(1-\Delta^*)} \Big[\tag{1.26}$$

<u>remark:</u> $L(p_A,p_B)$ is the expected additional number of successes that would have been occurred if the superior treatment A had been used exclusively.

The exact formula for $E(N_A)$ is quickly derived in the following manner. Defining R_n and T_n by

$$R_n := E(N_A \mid S_B-S_A=n, NT=A)$$
$$T_n := E(N_A \mid S_B-S_A=n, NT=B) , \tag{1.27}$$

we obtain:

$$R_n=p_AR_{n-1}+q_AT_n+1$$
$$T_n=p_BT_{n+1}+q_BR_n \tag{1.28}$$

$$R_{-r}=T_r=0 \tag{1.29}$$

The above system of difference equations and boundary conditions results from (1.20) and (1.21) by interchanging p_A and p_B as well as q_A and q_B, (λ is replaced by $1/\lambda$), i.e.

$$R_n=E(N_A|S_A-S_B=-n,NT=A)=V_n(p_A\leftrightarrows p_B)=$$
$$=E(N_B|S_A-S_B=n,NT=B,p_A\leftrightarrows p_B) \tag{1.30}$$

$$T_n=E(N_A|S_A-S_B=-n,NT=B)=U_n(p_A\leftrightarrows p_B)=$$
$$=E(N_B|S_A-S_B=n,NT=A,p_A\leftrightarrows p_B) \tag{1.31}$$

"$p_A\leftrightarrows p_B$" means that p_A and p_B as well as q_A and q_B are interchanged. $E(N_A)$ is now obtainable from (1.24),i.e.

$$E(N_A)=\frac{1}{2}(R_0+T_0)=\frac{1}{2}(V_0(p_A\leftrightarrows p_B)+U_0(p_A\leftrightarrows p_B))=$$

$$=E(N_B|p_A\leftrightarrows p_B)=\frac{(rq_B+\frac{1}{2}p_B)(1-\lambda^r)(1-p_A\lambda-q_A\lambda^r)}{p_A(1-\lambda)(1-p_A\lambda-q_A\lambda^{2r})} \tag{1.32}$$

The expected patient horizon E(N) is the sum of (1.24) and (1.32).

$$E(N)=E(N_A)+E(N_B)=\frac{(1-\lambda^r)(1-p_A\lambda-q_A\lambda^r)(r(q_A+q_B)+\frac{1}{2}(p_A+p_B))}{p_A(1-\lambda)(1-p_A\lambda-q_A\lambda^{2r})} \tag{1.33}$$

The comparison of (1.32) and (1.21) gives that $E(N_B)$ is less than $E(N_A)$ if $r>\frac{1}{2}$.

E(N) can be calculated directly by solving the following system of difference equations and boundary conditions.

$$C_n:=E(N|S_A-S_B=n,NT=A)$$
$$D_n:=E(N|S_A-S_B=n,NT=B) \tag{1.34}$$

$$C_n=p_AC_{n+1}+q_AD_n+1,\ D_n=p_BD_{n-1}+q_BC_n+1,\ C_r=D_{-r}=0 \tag{1.35}$$

Numerical results are given at the end of (4.6).

2 The Selection Model $[2;VT;|S_A-S_B|=s]$

2.1 Derivation of the critical s-value

We now use the VT sampling scheme, i.e. at each stage of the experiment, both treatments are given simultaneously. Moreover, this implies an equal number of observations on each treatment. The stopping rule of the preceding procedure is adopted. In order to derive the exact formula for P(CS), we next define

$$P_n:=P(CS|S_A-S_B=n) \tag{2.1}$$

P_n satisfies the following difference equation

$$P_n=p_Aq_BP_{n+1}+q_Ap_BP_{n-1}+(p_Ap_B+q_Aq_B)P_n \tag{2.2}$$

with boundary conditions

$$P_s=1;\ P_{-s}=0 \tag{2.3}$$

The solution of (2.2) satisfying (2.3) is easily shown to be

$$P_n=\frac{1-\delta^{n+s}}{1-\delta^{2s}}\ ,\ \text{where}\ \delta:=\frac{p_Bq_A}{p_Aq_B}<1 \tag{2.4}$$

Hence, the probability of a correct selection is given by

$$P(CS)=P_0=\frac{1}{1+\delta^s} \tag{2.5}$$

Using (1.13), δ can be rewritten as a function of $\overline{p}$ and Δ, i.e.

$$\delta=\frac{p_Bq_A}{p_Aq_B}=\frac{(\overline{p}-\frac{1}{2}\Delta)(1-\overline{p}-\frac{1}{2}\Delta)}{(\overline{p}+\frac{1}{2}\Delta)(1-\overline{p}+\frac{1}{2}\Delta)} \tag{2.6}$$

It is obvious that fixed values of $\overline{p}$ imply that δ is strictly decreasing with Δ. Furthermore, we see from (2.5) that P(CS) is strictly increasing with Δ and strictly increasing with s by holding s and Δ fixed, respectively. Therefore the LFC is obtained for the smallest admissible value of Δ, i.e. for $\Delta=\Delta^*$.

Because of $P(CS)=\frac{1}{1+\delta^s}\geq P^* \iff \delta^s\leq\frac{1-P^*}{P^*}$, we obtain

$$P(CS) \geq P^* \iff (\delta(\Delta^*))^s = \left(\frac{(2\bar{p}-\Delta^*)(2(1-\bar{p})-\Delta^*)}{(2\bar{p}+\Delta^*)(2(1-\bar{p})+\Delta^*)}\right)^s \leq \frac{1-P^*}{P^*} \tag{2.7}$$

From this follows that the maximal value of s, needed to satisfy the $(P^*;\Delta^*)$-condition (1.1), is generated by that value of $\bar{p}$ which maximizes $\delta(\Delta^*)$, and this is rapidly shown to be $\bar{p} = \frac{1}{2}$. The LFC is given by

$$p_A = \frac{1}{2} + \frac{1}{2}\Delta^*;\ p_B = \frac{1}{2} - \frac{1}{2}\Delta^* \tag{2.8}$$

Substituting these values into (2.7), we obtain

$$s \geq \frac{\ln(1-P^*)-\ln P^*}{2\ln\left(\frac{1-\Delta^*}{1+\Delta^*}\right)} \tag{2.9}$$

Using the approximation $\ln P^* \approx 0$ if P^* is close to 1, the desired critical value of s is found to be

$$s_{max} \approx \left]\frac{\ln(1-P^*)}{2\ln\left(\frac{1-\Delta^*}{1+\Delta^*}\right)}\right[\tag{2.10}$$

2.2 Derivation of the expectations

The use of VT-sampling implies an equal number of patients on each treatment, i.e.

$$E(N_A)=E(N_B);\ E(N)=2E(N_B) \tag{2.11}$$

If U_n denotes $E(N_B|S_A-S_B=n)$ then U_n satisfies the following difference equation and boundary conditions

$$U_n=p_Aq_BU_{n+1}+q_Ap_BU_{n-1}+(p_Ap_B+q_Aq_B)U_n+1 \tag{2.12}$$

$$U_s=U_{-s}=0 \tag{2.13}$$

The homogeneous part of (2.12) is solved in the same way as (2.2) and a partial solution of (2.12) is given by $U_n=-n/(p_Aq_B-q_Ap_B)$. The general solution of the above system is then given by

$$U_n = \frac{s(1+\delta^{2s}-2\delta^{n+s})}{(1-\delta^{2s})\Delta} - \frac{n}{\Delta}\ ;\ \Delta=p_A-p_B \tag{2.14}$$

$$E(N_B)=U_0 = \frac{s(1-\delta^s)}{\Delta(1+\delta^s)} \; ; \; L(p_A,p_B) = \frac{s(1-\delta^s)}{1+\delta^s} = s(2P_0-1) \qquad (2.15)$$

Using $P_0 \approx 1$ if P^* is close to one, $L(p_A,p_B)$ reduces to

$$L(p_A,p_B) \approx s_{max} \approx \left] \frac{\ell n(1-P^*)}{2\ell n\left(\frac{1-\Delta^*}{1+\Delta^*}\right)} \right[\qquad (2.16)$$

The total number of patients involved in the experiment is

$$E(N) = \frac{2s(1-\delta^s)}{\Delta(1+\delta^s)} \qquad (2.17)$$

Numerical results are given at the end of (4.6).

3 The Selection Model $[2; PL; |F_A-F_B|=r]$

3.1 Derivation of the critical r-value

The following procedure uses PL-sampling and the stopping rule $|F_A-F_B|=r$, where F_A,F_B denote the current number of failures of treatment A,B respectively. We stop if the absolute difference of the current number of A- and B-failures is equal to r, and declare that treatment to be the better one which possesses the least failures. A correct selection, therefore, takes place only if $F_A-F_B=-r$. Noting that the first treatment is chosen randomly and defining P_n^* and Q_n^* as follows

$$P_n^*:=P(CS|F_B-F_A=n,NT=B); \; Q_n^*:=P(CS|F_B-F_A=n,NT=A), \qquad (3.1)$$

we obtain $P(CS) = \frac{1}{2}(P_0^*+Q_0^*)$ by solving the following system of difference equations and boundary conditions.

$$P_n^*=q_B P_{n+1}^*+p_B Q_n^* \; ; \; Q_n^*=q_A Q_{n-1}^*+p_A P_n^* \qquad (3.2)$$

$$P_r^*=1; \; Q_{-r}^*=0 \qquad (3.3)$$

The above system results from (1.3), (1.4) by interchanging p_A and q_B as well as q_A and p_B, i.e.

$$\begin{aligned} P^*_n &= P(CS|F_B-F_A=n,NT=B)=P(CS|PW,S_A-S_B=n,NT=A,p_A \leftrightarrows q_B) \\ Q^*_n &= P(CS|F_B-F_A=n,NT=A)=P(CS|PW,S_A-S_B=n,NT=B,p_A \leftrightarrows q_B) \end{aligned} \tag{3.4}$$

Together with (1.7) and (1.8), the exact expression for P(CS) is given by

$$Q^*_n = \frac{p_A(1-\lambda^{*n+r})}{p_A-p_B\lambda^{*2r}} \; ; \; P^*_n = \frac{p_A-p_B\lambda^{*n+r}}{p_A-p_B\lambda^{*2r}} \; ; \; \lambda^* := \frac{q_A}{q_B} < 1 \tag{3.5}$$

$$P(CS) = \frac{1}{2}(P^*_0+Q^*_0) = \frac{1}{2} \; \frac{2p_A-(p_A+p_B)\lambda^{*r}}{p_A-p_B\lambda^{*2r}} \tag{3.6}$$

Noting that $P(CS|PL)=P(CS|PW,p_A \leftrightarrows q_B)$, it is obvious that the critical r-value of this procedure and of that one proposed in no. 1 coincide, i.e.

$$r_{max} \approx \left] \frac{\ell n 2(1-P^*)}{\ell n(1-\Delta^*)} \right[\tag{3.7}$$

3.2 Derivation of the expectations

Defining U^*_n and V^*_n by

$$U^*_n := E(N_A|F_B-F_A=n,NT=B); \; V^*_n := E(N_A|F_B-F_A=n,NT=A), \tag{3.8}$$

we obtain the following system of difference equations and boundary conditions.

$$U^*_n = q_B U^*_{n+1} + p_B V^*_n ; \; V^*_n = q_A V^*_{n-1} + p_A U^*_n + 1 \tag{3.9}$$

$$U^*_r = V^*_{-r} = 0 \tag{3.10}$$

The above system results from (1.20), (1.21) by interchanging p_A and q_B as well as q_A and p_B, i.e.

$$\begin{aligned} U^*_n &= E(N_A|F_B-F_A=n,NT=B)=E(N_B|PW,S_A-S_B=n,NT=A,p_A \leftrightarrows q_B) \\ V^*_n &= E(N_A|F_B-F_A=n,NT=A)=E(N_B|PW,S_A-S_B=n,NT=B,p_A \leftrightarrows q_B) \end{aligned} \tag{3.11}$$

$E(N_A)$ is now given by

$$E(N_A)=E(N_A|PL) = \frac{1}{2}(U^*_0+V^*_0)=E(N_B|PW,p_A \leftrightarrows p_B) \tag{3.12}$$

$E(N_B)$ is derived in a similar way. For this purpose we define

$$R_n^*:=E(N_B|F_A-F_B=n,NT=B);\quad T_n^*:=E(N_B|F_A-F_B=n,NT=A) \tag{3.13}$$

The resulting system of difference equations and boundary conditions is as follows

$$R_n^*=q_BR_{n-1}^*+p_BT_n^*+1;\quad T_n^*=q_AT_{n+1}^*+p_AR_n^* \tag{3.14}$$

$$R_{-r}^*=T_r^*=0 \tag{3.15}$$

A comparison with (1.28), (1.29) and the use of the interchanging technique, demonstrated above, gives the desired result.

$$\begin{aligned} E(N_B)&=E(N_B|PL)=E(N_A|PW,p_A\leftrightarrows q_B)\\ E(N)\ &=E(N\ |PL)=E(N\ |PW,p_A\leftrightarrows q_B)\end{aligned} \tag{3.16}$$

$$\begin{aligned} E(N_B|PL)&=\frac{(rp_A+\frac{1}{2}q_A)(1-\lambda^{*r})(1-q_B\lambda^*-p_B\lambda^{*r})}{q_B(1-\lambda^*)(1-q_B\lambda^*-p_B\lambda^{*2r})}\approx\frac{2rp_A+q_A}{2(q_B-q_A)}\\ E(N|PL)&=\frac{(1-\lambda^{*r})(1-q_B\lambda^*-p_B\lambda^{*r})(r(p_A+p_B)+\frac{1}{2}(q_A+q_B))}{q_B(1-\lambda^*)(1-q_B\lambda^*-p_B\lambda^{*2r})}\approx\frac{2r\overline{p}+\overline{q}}{q_B-q_A}\end{aligned} \tag{3.17}$$

where $\overline{q}:=1-\overline{p}$ and $\overline{p}=\frac{1}{2}(p_A+p_B)$. Numerical results are given in the next section.

4 Comparison of the Selection Procedures No.1 - No. 3

4.1 Some general remarks

The quality of a selection procedure is reflected by its expected number of patients on the poorer treatment and by the expected patient horizon, the total number of patients involved in the experiment. An exact comparison of two different selection procedures is possible only if the probabilities of a correct selection are the same for either procedure. Since this strong requirement isn't always fulfilled, we have to weaken it to get some statement about the quality of competitive procedures. Two procedures are, therefore, considered as comparable if they have the same or, at least, approximately the same probability of a correct selection in the least fa-

vorable configuration. Using integers for the critical r- and s-values, we cannot expect the probabilities of a correct selection to coincide exactly, even in the LFC. Our further investigations, however, will show that no disadvantage arises from such an approximate comparison.
In the following, we first compare the procedures with respect to $E(N_B)$ and then with respect to $E(N)$.

4.2 Comparison of PW- and VT-sampling

The selection procedure of no.1 is better than that one of no. 2 if $E(N_B|PW)<E(N_B|VT)$. Using (1.24) and (2.15), we obtain:

$$\frac{E(N_B|PW)}{E(N_B|VT)} \approx \frac{q_A \ell n(2(1-P^*))\cdot 2\ell n\left(\frac{1-\Delta^*}{1+\Delta^*}\right)}{\ell n(1-\Delta^*)\cdot \ell n(1-P^*)} \approx 2q_A \frac{\ell n\left(\frac{1-\Delta^*}{1+\Delta^*}\right)}{\ell n(1-\Delta^*)} < 1 \tag{4.1}$$

$$\iff p_A > 1 - \frac{\ell n(1-\Delta^*)}{2\ell n\left(\frac{1-\Delta^*}{1+\Delta^*}\right)} =: k(\Delta^*)$$

Using the power series expansion of $\ell n(1-x)$, we obtain

$$k(\Delta^*) = 1 - \frac{1}{2}\,\frac{\sum_{\nu=1}^{3} \frac{\Delta^{*\nu}}{\nu} + O(\Delta^{*4})}{\sum_{\nu=1}^{3} \frac{1}{\nu}\left(\frac{2\Delta^*}{1+\Delta^*}\right)^{\nu} + O(\Delta^{*4})} = \frac{3}{4} - \frac{1}{8}\Delta^* + O(\Delta^{*3}), \text{ i.e.} \tag{4.2}$$

$$E(N_B|PW)<E(N_B|VT) \iff p_A > k(\Delta^*) = \frac{3}{4} - \frac{1}{8}\Delta^* + O(\Delta^{*3}) \tag{4.3}$$

We now turn to the comparison of PW- and VT-sampling with respect to $E(N)$. Using the approximations $E(N|PW)\approx\frac{1}{\Delta}(q_A+q_B)r_{max}$ and $E(N|VT)\approx\frac{2}{\Delta}s_{max}$, we obtain

$$\frac{E(N|PW)}{E(N|VT)} \approx (q_A+q_B)\,\frac{\ell n\left(\frac{1-\Delta^*}{1+\Delta^*}\right)}{\ell n(1-\Delta^*)} < 1 \iff \bar{q} = \frac{1}{2}(q_A+q_B) < 1-k(\Delta^*) \tag{4.4}$$

$$E(N|PW)<E(N|VT) \iff \bar{p} > k(\Delta^*) = \frac{3}{4} - \frac{1}{8}\Delta^* + O(\Delta^{*3}) \tag{4.5}$$

Calculating the expectations only for $\Delta=\Delta^*$, i.e. $p_B=p_A-\Delta^*$, (4.5) may be replaced by

$$E(N|PW)<E(N|VT) \iff p_A > \frac{3}{4} + \frac{3}{8}\Delta^* + O(\Delta^{*3}) \tag{4.6}$$

4.3 Comparison of PW- and PL-sampling

The following comparisons are exact within the meaning of section 4.1, i.e. the P(CS)-values of both procedures coincide exactly in the LFC. Using (1.25) and (3.17), we obtain

$$\frac{E(N_B|PL)}{E(N_B|PW)} \approx \frac{2r_{max}p_A+q_A}{2r_{max}q_A+p_A} < 1 \iff 2p_A(2r_{max}-1)<2r_{max}-1 \tag{4.7}$$

Noting that $r_{max}>\frac{1}{2}$, it is evident that

$$E(N_B|PL)<E(N_B|PW) \iff p_A<\frac{1}{2} \tag{4.8}$$

$p_A=\frac{1}{2}$ implies no difference in the quality of PW- and VT-sampling.
The comparison with respect to E(N) gives

$$\frac{E(N|PL)}{E(N|PW)} \approx \frac{2r_{max}\overline{p}+\overline{q}}{2r_{max}\overline{q}+\overline{p}} <1 \iff 2r_{max}(\overline{p}-\overline{q})<\overline{p}-\overline{q} \tag{4.9}$$

The last inequality is correct only if $\overline{p}-\overline{q}<0$, and from this follows, using $p_A-p_B=\Delta^*$,

$$E(N|PL)<E(N|PW) \iff p_A<\frac{1}{2}+\frac{\Delta^*}{2} \tag{4.10}$$

4.4 Comparison of PL- and VT-sampling

The comparisons of PL- and VT-sampling with respect to $E(N_B)$ and E(N) are carried out by using the expectations given in (2.15), (3.17) and (2.17), (3.17) respectively. We next obtain

$$\frac{E(N_B|PL)}{E(N_B|VT)} \approx p_A\frac{r_{max}}{s_{max}} \approx 2p_A\frac{\ln\left(\frac{1-\Delta^*}{1+\Delta^*}\right)}{\ln(1-\Delta^*)} <1 \iff 1-p_A>k(\Delta^*),\text{i.e.} \tag{4.11}$$

$$E(N_B|PL)<E(N_B|VT) \iff p_A<\frac{1}{4}+\frac{1}{8}\Delta^*+0(\Delta^{*3}),$$

and the last comparison gives

$$\frac{E(N|PL)}{E(N|VT)} \approx (p_A+p_B)\,\frac{r_{max}}{s_{max}} \approx (p_A+p_B)\,\frac{\ell n\left(\frac{1-\Delta^*}{1+\Delta^*}\right)}{\ell n(1-\Delta^*)} < 1$$

$$\Longleftrightarrow\ p_A+p_B<2(1-k(\Delta^*)) = \frac{1}{2}+\frac{1}{4}\Delta^*+0(\Delta^{*3}) \qquad (4.12)$$

Using $p_A-p_B=\Delta^*$, the result is as follows

$$E(N|PL)<E(N|VT)\ \Longleftrightarrow\ p_A<\frac{1}{4}+\frac{5}{8}\Delta^*+0(\Delta^{*3}) \qquad (4.13)$$

4.5 Recapitulation

For practical purposes, Δ^* can be considered as small. Terms of order Δ^{*3} may, therefore, be neglected. From the preceding sections, the following rules are deduced.

(a) comparison with respect to $E(N_B)$:

$E(N_B|PL)<E(N_B|PW)$, if $p_A<\frac{1}{2}$ (see (4.8))

$E(N_B|PL)<E(N_B|VT)$, if $p_A\leq\frac{1}{4}$ (see (4.11))

$E(N_B|PW)<E(N_B|VT)$, if $p_A\geq\frac{3}{4}$ (see (4.3))

Rule of thumb

$p_A\varepsilon[0,\frac{1}{4}]$: prefer PL-sampling

$p_A\varepsilon(\frac{1}{4},\frac{3}{4})$: prefer VT-sampling (4.14)

$p_A\varepsilon[\frac{3}{4},1]$: prefer PW-sampling

We see that PL- and PW-sampling are to be preferred to VT-sampling only within the region of extreme values of the success parameter p_A.

(b) comparison with respect to E(N)

$E(N|PL)<E(N|PW)$, if $p_A<\frac{1}{2}$ (see (4.10))

$E(N|PL)<E(N|VT)$, if $p_A\leq\frac{1}{4}$ (see (4.13))

$E(N|PW)<E(N|VT)$, if $p_A\geq\frac{9}{10}$ (see (4.6))

Rule of thumb

$p_A \varepsilon [0,\frac{1}{4}]$: prefer PL-sampling

$p_A \varepsilon (\frac{1}{4},\frac{9}{10})$: prefer VT-sampling (4.15)

$p_A \varepsilon [\frac{9}{10},1]$: prefer PW-sampling

The difference between PW- and VT-sampling is not substantial, if $\Delta^*>0.4$.

4.6 Concluding remarks

Detailed numerical results are given on the following pages. The P(CS)-values, the expectations, and the expected loss of the selection procedures from no.1 - no.3 have been computed for $P^*=0.90$, 0.95, 0.99 and Δ^* from 0.1 to 0.4. Using PW-sampling, the LFC is given by $p_A=1$ and $p_B=1-\Delta^*$ (see (1.17)). This fact is illustrated by the decreasing P(CS)-value, shown in the tables. Using PL-sampling, the LFC is reached for $p_A=\Delta^*$, and the VT-sampling procedure has its LFC for $p_A = 1/2 + \Delta^*/2$.

The rules of thumb, given in 4.5, are easily verified by the numerical results. It can also be seen from the tables that the expected patient horizon is rapidly increasing when p_A lies in the extreme parts of the parameter space (0,1). The selection procedures of no.1 - no.3 can be modified if some prior information about the successparameters p_A and p_B is available. Knowing in advance that the range of p_A and p_B is restricted by the constants a and b, i.e. $a<p_B\leq p_A<b$, we must not look for the LFC in (0,1) but in the interval (a,b). The restriction of the parameterspace from (0,1) to (a,b) may result in a reduction of the critical values r_{max} and s_{max}, and that happens if and only if the LFC determined over (0,1) and the LFC determined over (a,b) do not coincide. (See e.g. Sobel/Weiss [178]).

The slight modification of a given selection procedure, described above, is always possible on principle; the essential point, however, consists in the procurement of the prior information.

4.7 Numerical results

In the following tables L(PW) denotes the expected loss when PW-sampling has been used.

$]L(\cdot)[$ and $]E(\cdot|\cdot)[$ have been tabulated throughout.

Table 1

p_A	0.2	0.3	0.4	0.5	0.6	0.7	0.8	0.9
P*=0.90; Δ*=0.1; r=16; s=6								
P(CS|PW)	1.000	0.999	0.991	0.975	0.954	0.931	0.910	0.896
$E(N_B\|PW)$	129	114	97	79	61	45	30	17
E(N|PW)	274	242	208	172	136	103	72	45
L(PW)	13	12	10	8	7	5	3	2
P(CS|VT)	0.992	0.962	0.934	0.919	0.919	0.934	0.962	0.992
$E(N_B\|VT)$	60	56	53	51	51	53	56	60
E(N|VT)	119	111	105	101	101	105	111	119
L(VT)	6	6	6	6	6	6	6	6
P(CS|PL)	0.896	0.910	0.931	0.954	0.975	0.991	0.999	1.000
$E(N_B\|PL)$	29	43	58	75	94	112	129	145
E(N|PL)	45	72	103	136	172	208	242	274
L(PL)	3	5	6	8	10	12	13	15
P*=0.90; Δ*=0.2; r=8; s=3								
P(CS|PW)	-	1.000	0.997	0.986	0.968	0.948	0.930	0.916
$E(N_B\|PW)$	-	29	25	21	17	13	9	6
E(N|PW)	-	65	58	49	40	32	24	17
L(PW)	-	6	5	5	4	3	2	2
P(CS|VT)	-	0.983	0.950	0.927	0.919	0.927	0.950	0.983
$E(N_B\|VT)$	-	15	14	13	13	13	14	15
E(N|VT)	-	29	27	26	26	26	27	29
L(VT)	-	3	3	3	3	3	3	3
P(CS|PL)	-	0.916	0.930	0.948	0.968	0.986	0.997	1.000
$E(N_B\|PL)$	-	12	16	20	24	28	33	37
E(N|PL)	-	17	24	32	40	49	58	65
L(PL)	-	3	4	4	5	6	7	8

Table 1 (continued)

p_A	0.2	0.3	0.4	0.5	0.6	0.7	0.8	0.9
			P*=0.90; Δ*=0.3; r=5; s=2					
P(CS|PW)	-	-	0.999	0.992	0.976	0.956	0.937	0.922
$E(N_B\|PW)$	-	-	11	10	8	6	5	3
E(N|PW)	-	-	26	23	19	16	13	10
L(PW)	-	-	4	3	3	2	2	1
P(CS|VT)	-	-	0.973	0.941	0.925	0.925	0.941	0.973
$E(N_B\|VT)$	-	-	7	6	6	6	6	7
E(N|VT)	-	-	13	12	12	12	12	13
L(VT)	-	-	2	2	2	2	2	2
P(CS|PL)	-	-	0.922	0.937	0.956	0.976	0.992	0.999
$E(N_B\|PL)$	-	-	7	9	10	12	14	16
E(N|PL)	-	-	10	13	16	19	23	26
L(PL)	-	-	2	3	3	4	5	5
			P*=0.90; Δ*=0.4; r=4; s=2					
P(CS|PW)	-	-	-	0.999	0.991	0.976	0.960	0.945
$E(N_B\|PW)$	-	-	-	6	5	4	3	2
E(N|PW)	-	-	-	15	13	11	9	7
L(PW)	-	-	-	3	2	2	2	1
P(CS|VT)	-	-	-	0.988	0.973	0.967	0.973	0.988
$E(N_B\|VT)$	-	-	-	5	5	5	5	5
E(N|VT)	-	-	-	10	10	10	10	10
L(VT)	-	-	-	2	2	2	2	2
P(CS|PL)	-	-	-	0.945	0.960	0.976	0.991	0.999
$E(N_B\|PL)$	-	-	-	6	6	8	9	10
E(N|PL)	-	-	-	7	9	11	13	15
L(PL)	-	-	-	3	3	3	4	4

Table 1 (continued)

p_A	0.2	0.3	0.4	0.5	0.6	0.7	0.8	0.9
P*=0.95; Δ*=0.1; r=22; s=8								
P(CS| PW)	1.000	1.000	0.998	0.993	0.984	0.971	0.958	0.946
$E(N_B\|PW)$	177	156	134	111	89	66	44	24
E(N| PW)	376	333	289	244	197	152	108	67
L(PW)	18	16	14	12	9	7	5	3
P(CS| VT)	0.999	0.987	0.972	0.962	0.962	0.972	0.987	0.999
$E(N_B\|VT)$	80	78	76	74	74	76	78	80
E(N| VT)	160	156	151	148	148	151	156	160
L(VT)	8	8	8	8	8	8	8	8
P(CS| PL)	0.946	0.958	0.971	0.984	0.993	0.998	1.000	1.000
$E(N_B\|PL)$	43	64	86	109	133	155	177	199
E(N| PL)	67	108	152	197	244	289	333	376
L(PL)	5	7	9	11	14	16	18	20
P*=0.95; Δ*=0.2; r=11; s=4								
P(CS|PW)	-	1.000	1.000	0.997	0.990	0.981	0.969	0.959
$E(N_B\|PW)$	-	40	34	29	24	18	13	8
E(N|PW)	-	89	79	68	57	46	35	24
L(PW)	-	8	7	6	5	4	3	2
P(CS|VT)	-	0.996	0.981	0.967	0.962	0.967	0.981	0.996
$E(N_B\|VT)$	-	20	20	19	19	19	20	20
E(N|VT)	-	40	39	38	37	38	39	40
L(VT)	-	4	4	4	4	4	4	4
P(CS|PL)	-	0.959	0.969	0.981	o.990	0.997	1.000	1.000
$E(N_B\|PL)$	-	17	23	28	34	40	45	50
E(N|PL)	-	24	35	46	57	68	79	89
L(PL)	-	4	5	6	7	8	9	10

Table 1 (continued)

p_A	0.2	0.3	0.4	0.5	0.6	0.7	0.8	0.9
	$P^*=0.95$; $\Delta^*=0.3$; $r=7$; $s=3$							
$P(CS\|PW)$	-	-	1.000	0.999	0.994	0.985	0.974	0.964
$E(N_B\|PW)$	-	-	15	13	11	8	6	4
$E(N\|PW)$	-	-	36	32	27	23	18	14
$L(PW)$	-	-	5	4	4	3	2	2
$P(CS\|VT)$	-	-	0.995	0.985	0.977	0.977	0.985	0.995
$E(N_B\|VT)$	-	-	10	10	10	10	10	10
$E(N\|VT)$	-	-	20	20	20	20	20	20
$L(VT)$	-	-	3	3	3	3	3	3
$P(CS\|PL)$	-	-	0.964	0.974	0.985	0.994	0.999	1.000
$E(N_B\|PL)$	-	-	10	12	15	17	19	22
$E(N\|PL)$	-	-	14	18	23	27	32	36
$L(PL)$	-	-	3	4	5	6	6	7
	$P^*=0.95$; $\Delta^*=0.4$; $r=5$; $s=2$							
$P(CS\|PW)$	-	-	-	1.000	0.997	0.990	0.979	0.969
$E(N_B\|PW)$	-	-	-	7	6	5	4	3
$E(N\|PW)$		-	-	19	16	14	12	9
$L(PW)$	-	-	-	3	3	2	2	1
$P(CS\|VT)$	-	-	-	0.988	0.973	0.967	0.973	0.988
$E(N_B\|VT)$	-	-	-	5	5	5	5	5
$E(N\|VT)$	-	-	-	10	10	10	10	10
$L(VT)$	-	-	-	2	2	2	2	2
$P(CS\|PL)$	-	-	-	0.969	0.979	0.990	0.997	1.000
$E(N_B\|PL)$	-	-	-	7	8	9	11	12
$E(N\|PL)$	-	-	-	9	12	14	16	19
$L(PL)$	-	-	-	3	4	4	5	5

Table 1 (continued)

p_A	0.2	0.3	0.4	0.5	0.6	0.7	0.8	0.9
$P^*=0.99;\ \Delta^*=0.1;\ r=38;\ s=12$								
P(CS|PW)	1.000	1.000	1.000	1.000	0.999	0.998	0.995	0.992
$E(N_B\|PW)$	305	268	230	193	155	117	80	42
E(N|PW)	648	573	498	423	347	272	196	121
L(PW)	31	27	23	20	16	12	8	5
P(CS|VT)	1.000	1.000	0.995	0.992	0.992	0.995	1.000	1.000
$E(N_B\|VT)$	120	120	119	119	119	119	120	120
E(N|VT)	240	240	238	237	237	238	240	240
L(VT)	12	12	12	12	12	12	12	12
P(CS|PL)	0.992	0.995	0.998	0.999	1.000	1.000	1.000	1.000
$E(N_B\|PL)$	79	117	155	193	230	268	305	343
E(N|PL)	121	196	272	347	423	498	573	648
L(PL)	8	12	16	20	23	27	31	35
$P^*=0.99;\ \Delta^*=0.2;\ r=18;\ s=6$								
P(CS|PW)	-	1.000	1.000	1.000	0.999	0.998	0.996	0.993
$E(N_B\|PW)$	-	64	55	47	38	29	20	12
E(N|PW)	-	145	128	110	93	75	58	40
L(PW)	-	13	11	10	8	6	4	3
P(CS|VT)	-	1.000	0.997	0.994	0.992	0.994	0.997	1.000
$E(N_B\|VT)$	-	30	30	30	30	30	30	30
E(N|VT)	-	60	60	60	60	60	60	60
L(VT)	-	6	6	6	6	6	6	6
P(CS|PL)	-	0.993	0.996	0.998	0.999	1.000	1.000	1.000
$E(N_B\|PL)$	-	29	38	47	55	64	73	82
E(N|PL)	-	40	58	75	93	110	128	145
L(PL)	-	6	8	10	11	13	15	17

Table 1 (continued)

p_A	0.2	0.3	0.4	0.5	0.6	0.7	0.8	0.9
P*=0.99; Δ*=0.3; r=11; s=4								
P(CS|PW)	-	-	1.000	1.000	1.000	0.998	0.996	0.993
$E(N_B\|PW)$	-	-	23	20	16	13	9	6
E(N|PW)	-	-	56	49	42	35	28	21
L(PW)	-	-	7	6	5	4	3	2
P(CS|VT)	-	-	0.999	0.996	0.993	0.993	0.996	0.999
$E(N_B\|VT)$	-	-	14	14	14	14	14	14
E(N|VT)	-	-	27	27	27	27	27	27
L(VT)	-	-	4	4	4	4	4	4
P(CS|PL)	-	-	0.993	0.996	0.998	1.000	1.000	1.000
$E(N_B\|PL)$	-	-	16	20	23	27	30	34
E(N|PL)	-	-	21	28	35	42	49	56
L(PL)	-	-	5	6	7	8	9	10
P*=0.99; Δ*=0.4; r=8; s=3								
P(CS|PW)	-	-	-	1.000	1.000	0.999	0.997	0.995
$E(N_B\|PW)$	-	-	-	11	9	7	5	4
E(N|PW)	-	-	-	29	25	22	18	14
L(PW)	-	-	-	5	4	3	2	2
P(CS|VT)	-	-	-	0.999	0.995	0.994	0.995	0.999
$E(N_B\|VT)$	-	-	-	8	8	8	8	8
E(N|VT)	-	-	-	15	15	15	15	15
L(VT)	-	-	-	3	3	3	3	3
P(CS|PL)	-	-	-	0.995	0.997	0.999	1.000	1.000
$E(N_B\|PL)$	-	-	-	11	13	15	17	19
E(N|PL)	-	-	-	14	18	22	25	29
L(VT)	-	-	-	5	5	6	7	8

5 Two-Stage Selection Procedures

5.1 The structure of two-stage selection procedures

In the preceding section, we have seen that neither sampling scheme is uniformly better than the others with respect to $E(N_B)$ or $E(N)$. Recalling the rules of thumb, given in 4.5, it is obvious to design a two-stage selection procedure in which the first stage is used to obtain a rough estimate of the successparameter of the better treatment, and the second to carry out the remainder of the trial with the sampling rule judged to be the best according to (4.14) or (4.15).

In a first step, we design a two-stage selection procedure which uses either PW- or VT-sampling in the second stage; in a second step, we add PL-sampling; i.e. PW-, VT- or PL-sampling can be used in the second stage. It is shown by the numerical results that a fairly small estimation stage allows us to benefit considerably from the choice between the two or the three sampling rules, available in the second stage, respectively.

In the first stage, M trials are carried out with each treatment, where M is a preassigned constant, i.e. the sampling rule of the first stage is VT-sampling. Having observed m successes of one treatment and n successes of the other one, we choose the relative frequencies of successes to be estimates of p_A and p_B, i.e. we estimate p_A by $\frac{1}{M}\max\{m,n\}$ and p_B by $\frac{1}{M}\min\{m,n\}$. The PW-sampling rule is chose in the second stage whenever $\max\{m,n\}\geq k_c$ while in the contrary case, VT-sampling is used. The two-stage procedure terminates after the first stage if the absolute difference of successes is not less than s_{max}, i.e. the two-stage procedure reduces to a single-stage procedure with a fixed patient horizon if $\max\{m,n\}-\min\{m,n\}\geq s_{max}$. It may be possible, however, that the absolute difference of the current number of successes is already $\geq s_{max}$ during the first stage; but nevertheless all M trials are carried out, and then a further decision is taken. A modification allowing termination during the first stage is possible but will not be considered here.

Comparing the procedures with respect to $E(N_B)$, we choose $k_c =](\frac{3}{4}-\frac{1}{8}\Delta^*)M[$ accordi to (4.3). Applying PW-sampling in the second stage, the first trial is carried out with that treatment which yields the most successes up to the end of the first stage; in case of an equal number of successes, the first treatment, given by the experimenter, is selected at random. The procedure terminates whenever

$|S_A-S_B|=s(m,n,k_c,M)$, if VT-sampling has been used in the second stage,

or (5.1)

$|S_A-S_B|=r(m,n,k_c,M)$, if PW-sampling has been used in the second stage.

In either case, the treatment with the largest number of successes is declared to be the better one.

5.2 Derivation of P(CS)

Let R_{mn} denote the probability that m successes of treatment A and n successes of treatment B have occurred in the first stage, i.e.

$$R_{mn}=\binom{M}{m}\binom{M}{n}p_A^m q_A^{M-m} p_B^n q_B^{M-n} \tag{5.2}$$

The probability of a correct selection is thus given by

$$P(CS) = \sum_{\substack{m \\ \max\{m,n\}<k_c}} \sum_{n} R_{mn}P_{VT}(CS|(m,n))+\sum_{\substack{m \\ \max\{m,n\}\geq k_c}} \sum_{n} R_{mn}P_{PW}(CS|(m,n)), \tag{5.3}$$

where $P_{VT}(CS|(m,n))$ denotes the probability of a correct selection if m A-successes and n B-successes have occurred in the first stage and if VT-sampling has been used in the second stage; $P_{PW}(CS|(m,n))$ is defined accordingly. Without loss of generality, the maximum of {m,n} will be denoted by m. Hence, the first double series of (5.3) can be rewritten as

$$\sum_{\substack{m \\ n<m<k_c}} \sum_{n} (R_{mn}P_{VT}(CS|(m,n))+R_{nm}P_{VT}(CS|(n,m)))+\sum_{m<k_c} R_{mm}P_{VT}(CS|(m,m)), \tag{5.4}$$

and the second double series of (5.3) can be rewritten in the same way.

5.3 Derivation of $s(m,n,k_c,M)$ and $r(m,n,k_c,M)$

Using (2.4) and the following identity for $m \neq n$

$(R_{mn}+R_{nm})P_{VT}(CS|(m,n) \text{ or } (n,m))=P_{VT}(CS|(m,n))R_{mn}+P_{VT}(CS|(n,m))R_{nm}=(R_{mn}+R_{nm})\frac{1}{1+\delta^s}$,

(5.4) can be rewritten as

$$\sum_{\substack{m \\ n<m<k_c}} \sum_{n} (R_{mn}+R_{nm}) \frac{1}{1+\delta^s} + \sum_{m<k_c} R_{mm} \frac{1}{1+\delta^s} \qquad (5.5$$

In the same way, the second double series of (5.3) can be rewritten as

$$\sum_{\substack{m \\ m\geq k_c,n-1}} \sum_{n} (R_{mn}+R_{nm}) f(r;m-n,p_A,\Delta) + \sum_{m\geq k_c} R_{mm} f(r;0,p_A,\Delta), \qquad (5.6$$

where $f(r;m-n,p_A,\Delta)$ is defined as follows:

$$f(r;m-n,p_A,\Delta) := \frac{P_{m-n}R_{mn}+Q_{n-m}R_{nm}}{R_{mn}+R_{nm}} = \frac{\delta^{n-m}(q_B-q_A\lambda^{r+m-n})+q_B(1-\lambda^{r+n-m})}{(q_B-q_A\lambda^{2r})(1+\delta^{n-m})} \qquad (5.7$$

and P_{m-n} and Q_{n-m} are given by (1.8) and (1.7) respectively. It is easily seen from (5.5) and (5.6) that P(CS) will be $\geq$ P* if $\frac{1}{1+\delta^s}$ and $f(r;m-n,p_A,\Delta)$ are both $\geq$ P* for all possible $m-n\geq 0$ and $p_A \varepsilon (\Delta,1)$. The comparison with (2.10) gives us as a first result

$$s(m,n,k_c,M) = s_{max} \approx \left] \frac{\ln(1-P^*)}{2\ln\left(\frac{1-\Delta^*}{1+\Delta^*}\right)} \right[\qquad (5.8$$

For m=n, $r(m,m,k_c,M)$ is given by (1.18). In case of m>n, a detailed discussion of $f(r;m-n,p_A,\Delta)$ is needed to determine the critical $r(m,n,k_c,M)$ value.

Setting $f(r;m-n,p_A,\Delta)$ equal to P*, an equation quadratic in λ^r is obtained from (5.7). The roots r_1,r_2 of this equation are given by

$$r_{1,2} = \frac{1}{\ln\lambda} \ln \left(\frac{\delta^{n-m}q_A\lambda^{m-n}+q_B\lambda^{n-m}}{2q_AP^*(1+\delta^{n-m})} \mp \left(\left(\frac{\delta^{n-m}q_A\lambda^{m-n}+q_B\lambda^{n-m}}{2q_AP^*(1+\delta^{n-m})} \right)^2 - \frac{q_B(1-P^*)}{q_AP^*} \right)^{\frac{1}{2}} \right) \qquad (5.$$

Using the following abbreviations

$$\beta := \frac{\delta^{n-m}q_A\lambda^{m-n}+q_B\lambda^{n-m}}{2q_AP^*(1+\delta^{n-m})} \; ; \; \alpha := \frac{q_B(1+\delta^{n-m})}{q_A\lambda^{m-n}\delta^{n-m}+q_B\lambda^{n-m}} \qquad (5.$$

it can be shown that:

(a) (5.9) gives complex roots if $\beta^2 < \frac{q_B(1-P^*)}{q_A P^*}$; i.e. $f(r;m-n,p_A,\Delta) \geq P^*$ whenever $\beta^2 \leq \frac{q_B(1-P^*)}{q_A P^*}$. The desired r-value to stop the selection procedure is thus given by $r(p_A;P^*,\Delta^*,m-n)=m-n$. $(p_B:=p_A-\Delta^*)$

(b) $\beta > \frac{q_B(1-P^*)}{q_A P^*}$ and $\alpha^2 < \frac{q_B}{q_A}$ jointly imply that the typical shape of the function $f(r;m-n,p_A,\Delta)$ is exactly that one of fig. 1. Hence we choose $r(p_A;P^*,\Delta^*,m-n)=r_1$

(c) $\beta > \frac{q_B(1-P^*)}{q_A P^*}$ and $\alpha^2 > \frac{q_B}{q_A}$ jointly imply that $r(p_A;P^*,\Delta^*,m-n)$ must be chosen equal to r_1, even if $f(m-n;m-n,p_A,\Delta) \geq P^*$.
(A detailed discussion is carried out in Schriever [164]).

The typical shape of $f(r;m-n,p_A,\Delta)$ in case that (a) or (c) hold can be seen from the figure below. (p_A, Δ and m-n are fixed).

fig.2

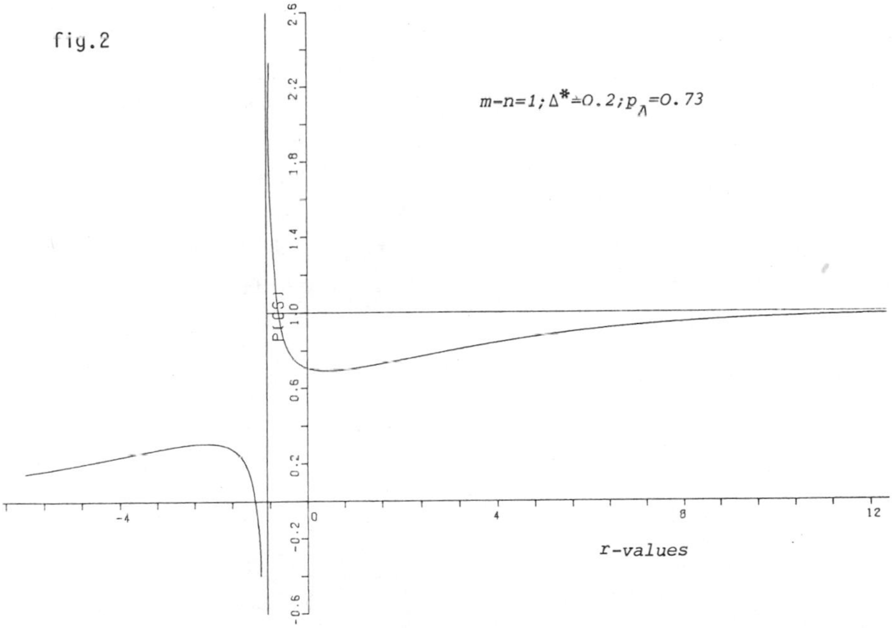

According to (a)-(c), $r(p_A;P^*,\Delta^*,m-n)$ is obtained as follows:

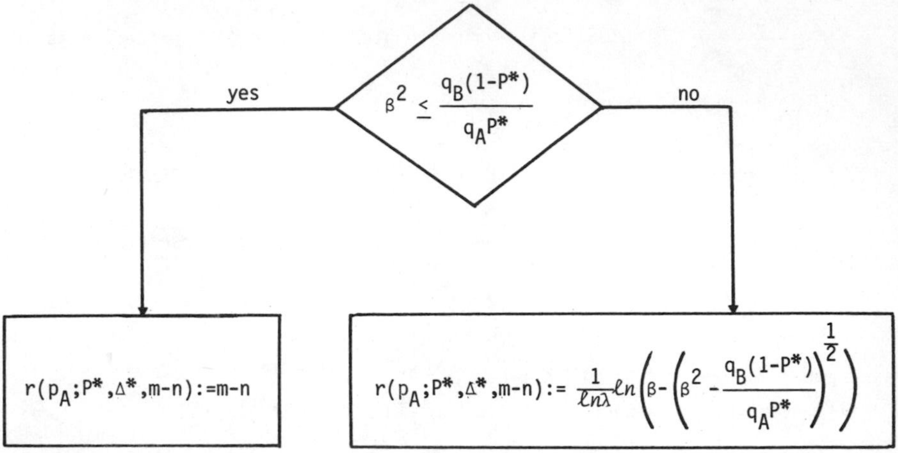

The critical r-value is thus given by:

$$r(m-n):=r(m,n,k_c,M)=\max_{p_A\varepsilon(\Delta^*,1)} r(p_A;P^*,\Delta^*,m-n) \tag{5.11}$$

An explicit calculation of the LFC seems to be impossible, therefore, we determin $r(m-n)$ numerically, i.e. for preassigned Δ^* and P^*, $r(1-\frac{\ell}{k};P^*,\Delta^*,m-n)$ is calculat according to (5.11), where $\ell\varepsilon\{1,2,\dots,[(1-\Delta^*)k]-1\}$ and $k\varepsilon\mathbb{N}$. ([x] denotes the integer part of x). The maximum of all those r-values will be a good approximatio of $r(m-n)$. This numerical procedure must be carried out for all m-n that may aris in the first stage of the selection procedure; i.e. $r(m-n)$ must be calculated for $0<m-n\le s_{max}-1$.

5.4 Derivation of the expectations

Noting that in the first stage of the selection procedure M trials are carried out with treatment B, and that $|m-n|\le s_{max}-1$ must hold, the expression for $E(N_B)$

$$E(N_B)=M+\sum_{n=0}^{k_c-1}\sum_{m=n+1}^{\min\{k_c-1,n+s_{max}-1\}} R_{mn}E_{VT}(N_B|(m,n))+\sum_{m=0}^{k_c-1}\sum_{n=m}^{\min\{k_c-1,m+s_{max}-1\}} R_{mn}E_{VT}(N_B|(m,n))+$$

$$+\sum_{n=k_c}^{M}\sum_{m=n+1}^{\min\{M,n+s_{max}-1\}}(R_{mn}E_{PW}(N_B|(m,n))+R_{nm}E_{PW}(N_B|(n,m)))+$$

$$+\sum_{n=0}^{k_c-1}\sum_{m=k_c}^{\min\{M,n+s_{max}-1\}}(R_{mn}E_{PW}(N_B|(m,n))+R_{nm}E_{PW}(N_B|(n,m)))+ \qquad (5.12)$$

$$+\sum_{m=k_c}^{M}R_{mm}E_{PW}(N_B|(m,m))$$

in which

$$E_{VT}(N_B|(m,n)) = \frac{1}{p_A-p_B}\left(\frac{s(1+\delta^{2s}-2\delta^{m-n+s})}{1-\delta^{2s}} - (m-n)\right);\ (s:=s_{max}) \qquad (5.13)$$

$$E_{PW}(N_B|(m,n)) = \frac{q_A}{p_A-p_B}\left(r-(m-n) - \frac{(2rq_A+p_A)\lambda^r(\lambda^r-\lambda^{m-n})}{q_A\lambda^{2r}-q_B}\right);(r:=r(m-n)) \qquad (5.14)$$

$$E_{PW}(N_B|(n,m)) = \frac{1}{p_A-p_B}\left((r-(n-m))q_A+p_A - \frac{(2rq_A+p_A)\lambda^r(q_A\lambda^r-q_B\lambda^{n-m})}{q_A\lambda^{2r}-q_B}\right) \qquad (5.15)$$

$$E_{PW}(N_B|(m,m)) = \frac{1}{2}(U_o+V_o) = \frac{(2rq_A+p_A)(1-\lambda^r)(q_A\lambda^r-q_B)}{2(p_A-p_B)(q_A\lambda^{2r}-q_B)};\ (r:=r_{max}) \qquad (5.16)$$

The above formulae follow directly from (2.14) and (1.23).
$E(N_A)$ is obtained from (5.13) by replacing N_B by N_A. Because of $E_{VT}(N_B|(m,n))= E_{VT}(N_A|(m,n))$, the first and the second double series of (5.13) remain unchanged. The contribution of PW-sampling to $E(N_A)$ is obtained from (1.30) and (1.31) as follows:

$$E_{PW}(N_A|(m,n)):=E_{PW}(N_A|S_A-S_B=m-n,NT=A)=$$

$$=E_{PW}(N_B|S_A-S_B=-(m-n),NT=B,p_A \leftrightarrows p_B)=E_{PW}(N_B|(n,m),p_A \leftrightarrows p_B) \qquad (5.17)$$

$$E_{PW}(N_A|(n,m)):=E_{PW}(N_A|S_A-S_B=n-m,NT=B)=$$

$$=E_{PW}(N_B|S_A-S_B=m-n,NT=A,p_A \leftrightarrows p_B)=E_{PW}(N_B|(m,n),p_A \leftrightarrows p_B) \qquad (5.18)$$

$$E_{PW}(N_A|(m,m)):=E_{PW}(N_A)=E_{PW}(N_B|p_A \leftrightarrows p_B)=E_{PW}(N_B|(m,m),p_A \leftrightarrows p_B) \qquad (5.19)$$

Noting in addition the following identities

$$R_{mn}=R_{nm}(p_A \leftrightarrows p_B);R_{nm}=R_{mn}(p_A \leftrightarrows p_B);R_{mm}=R_{mm}(p_A \leftrightarrows p_B), \tag{5.20}$$

we finally obtain:

$$R_{mn}E_{PW}(N_A|(m,n))+R_{nm}E_{PW}(N_A|(n,m)) = \tag{5.21}$$

$$=R_{nm}(p_A \leftrightarrows p_B)\ E_{PW}(N_B|(n,m),p_A \leftrightarrows p_B)+R_{mn}(p_A \leftrightarrows p_B)E_{PW}(N_B|(m,n),p_A \leftrightarrows p_B)$$

$$R_{mm}E_{PW}(N_A|(m,m))=R_{mm}(p_A \leftrightarrows p_B)E_{PW}(N_B|(m,m),p_A \leftrightarrows p_B), \tag{5.22}$$

i.e. the contribution of PW-sampling to $E(N_A)$ is obtained from the contribution of PW-sampling to $E(N_B)$ by interchanging p_A with p_B. The interchange must be carried out in the coefficients R_{mn} and R_{nm}, too.
In order to compare the two-stage procedure with a single-stage VT-sampling selection procedure, we have to calculate $E(N_B|VT)$, the expected number of patients on the poorer treatment if VT-sampling has been used within both stages. The expression for $E(N_B|VT)$ is

$$E(N_B|VT)=M+\sum_{\substack{m\ n\\ m>n}}\sum R_{mn}E_{VT}(N_B|(m,n))+\sum_{\substack{m\ n\\ n\geq m}}\sum R_{mn}E_{VT}(N_B|(m,n))=$$

$$=M+\sum_{n=0}^{M}\ \sum_{m=n+1}^{\min\{n+s_{max}-1,M\}} R_{mn}E_{VT}(N_B|(m,n))+\sum_{m=0}^{M}\ \sum_{n=m}^{\min\{m+s_{max}-1,M\}} R_{mn}E_{VT}(N_B|(m,n)), \tag{5.23}$$

where $E_{VT}(N_B|(m,n))$ is given by (5.13)

A comparison of the two-stage procedure with a selection procedure based on PW-sampling exclusively, has been omitted, because the single-stage PW-sampling selection procedure is considerably worse (with respect to $E(N_B)$ and $E(N)$) than the single-stage VT-sampling selection procedure for $p_A \leq 0.6$, as can be seen from the tables following 4.6.

5.5 Addition of PL-sampling

According to (4.11), VT-sampling should be replaced by PL-sampling if p_A is less than $\frac{1}{4}+\frac{1}{8}\Delta^*$. We will profit by this fact extending the two-stage selection procedure just outlined. The first stage is carried out as before and the sampling scheme for the second stage is determined by using the following decision rule:

$$\begin{array}{ll} \max\{m,n\}\geq k_c: & \text{choose PW-sampling} \\ M-k_c<\max\{m,n\}<k_c: & \text{choose VT-sampling} \\ \max\{m,n\}\leq M-k_c: & \text{choose PL-sampling} \end{array} \tag{5.24}$$

Using (3.4) and the identity "$S_A-S_B=m-n \Longleftrightarrow F_B-F_A=m-n$", where m denotes the higher number of successes, we obtain

$$\begin{aligned} P_{PL}(CS|(m,n))&=P_{PL}(CS|F_B-F_A=m-n,NT=B)=P_{m-n}(p_A \overset{\leftarrow}{\to} q_B) \\ P_{PL}(CS|(n,m))&=P_{PL}(CS|F_B-F_A=n-m,NT=A)=Q_{n-m}(p_A \overset{\leftarrow}{\to} q_B) \end{aligned} \tag{5.25}$$

Similar to (5.6), (5.7) we deduce

$$f^*(r;m-n,p_A,\Delta):=P_{PL}(CS|(m,n) \text{ or } (n,m)) - \frac{P_{m-n}(p_A \overset{\leftarrow}{\to} q_B)R_{mn}+Q_{n-m}(p_A \overset{\leftarrow}{\to} q_B)R_{nm}}{R_{mn}+R_{nm}} \tag{5.26}$$

Noting that $\delta=\delta(p_A \overset{\leftarrow}{\to} q_B)$, (5.26) can be rewritten as

$$f^*(r;m-n,p_A,\Delta) = \frac{(\delta(p_A \overset{\leftarrow}{\to} q_B))^{n-m}P_{m-n}(p_A \overset{\leftarrow}{\to} q_B)+Q_{n-m}(p_A \overset{\leftarrow}{\to} q_B)}{1+(\delta(p_A \overset{\leftarrow}{\to} q_B))^{n-m}}, \tag{5.27}$$

$$\text{i.e. } f^*(r;m-n,p_A,\Delta)=f(r;m-n,p_A,\Delta,p_A \overset{\leftarrow}{\to} q_B) \tag{5.28}$$

Because of the equal range of p_A and q_B ($p_A,q_B \in (\Delta^*,1)$), the critical value $r^*(m-n):=r^*(m,n,k_c,M)$ is just the same as $r(m-n)$. The addition of PL-sampling to the second stage of our selection procedure does not result in additional computations to determine the critical $r^*(m-n)$ value.

Taking into consideration that PL-sampling may arise in the second stage, the expression for $E(N_B)$ is derived similar as before.

$$E(N_B)=M+\sum_{n=0}^{M-k_c}\sum_{m=n+1}^{\min\{M-k_c,n+s_{max}-1\}}(R_{mn}E_{PL}(N_B|(m,n))+R_{nm}E_{PL}(N_B|(n,m)))+$$

$$+\sum_{m=0}^{M-k_c}R_{mm}E_{PL}(N_B|(m,m))+\sum_{n=0}^{M-k_c}\sum_{m=M-k_c+1}^{\min\{k_c-1,n+s_{max}-1\}}R_{mn}E_{VT}(N_B|(m,n))+$$

$$+\sum_{n=M-k_c+1}^{k_c-1}\sum_{m=n+1}^{\min\{k_c-1,n+s_{max}-1\}} R_{mn}E_{VT}(N_B|(m,n))+\sum_{m=0}^{M-k_c}\sum_{n=M-k_c+1}^{\min\{k_c-1,m+s_{max}-1\}} R_{mn}E_{VT}(N_B|(m,n))+$$

$$+\sum_{m=M-k_c+1}^{k_c-1}\sum_{n=m}^{\min\{k_c-1,m+s_{max}-1\}} R_{mn}E_{VT}(N_B|(m,n))+\sum_{n=k_c}^{M}\sum_{m=n+1}^{\min\{M,n+s_{max}-1\}}(R_{mn}E_{PW}(N_B|(m,n))+R_{nm}E_{PW}(N_B|(n,m)))+$$

$$+\sum_{n=0}^{k_c-1}\sum_{m=k_c}^{\min\{M,n+s_{max}-1\}}(R_{mn}E_{PW}(N_B|(m,n))+R_{nm}E_{PW}(N_B|(n,m)))+\sum_{m=k_c}^{M} R_{mm}E_{PW}(N_B|(m,m)) \quad (5$$

$E_{VT}(N_B|(m,n))$, $E_{PW}(N_B|(m,n))$ and $E_{PW}(N_B|(n,m))$ are given by (5.13)-(5.15), and using the interchanging technique again, the remaining expectations in (5.29) are easily shown to be

$$E_{PL}(N_B|(m,n))=V_{n-m}(p_A \leftrightarrows q_A);E_{PL}(N_B|(n,m))=U_{m-n}(p_A \leftrightarrows q_A)$$
$$E_{PL}(N_B|(m,m))=E_{PW}(N_A|p_A \leftrightarrows q_B)=E_{PW}(N_B|p_A \leftrightarrows q_A), \quad (5$$

where U_k and V_k are given by (1.23).

The expression for $E(N_A)$ is obtained just as the corresponding one in case where only PW- or VT-sampling may arise in the second stage of experimentation, i.e. th contribution of VT-sampling to $E(N_A)$ is the same as to $E(N_B)$, and $E_{PW}(N_A)$ is obtained from $E_{PW}(N_B)$ by interchanging p_A with p_B as demonstrated in (5.17)-(5.19). The contribution of PL-sampling to $E(N_A)$ is obtained from the contribution of PL-sampling to $E(N_B)$ by interchanging p_A with p_B. (Note that the interchange must be carried out in the coefficients R_{mn} and R_{nm}, too).

5.6 Concluding remarks

In the following tables, r(m-n) and the corresponding p_A-values are given for $P^*=0.90$, 0.95, 0.99 and $\Delta^*=0.1$, 0.2. Furthermore, the expectations of general interest are tabulated for M=5, $k_c=3$; M=5,10,15,20 and $k_c=](\frac{3}{4}-\frac{1}{8}\Delta^*)M[$.

$E(\cdot|2)$ and $E(\cdot|3)$ denote the expectations if two and three sampling rules are available in the second stage of experimentation respectively, and $E(N|VT)$ denote the expected patient horizon if VT-sampling is used in both stages.

We have already mentioned that a fairly small estimation stage, approximately fiv or ten tests with each treatment, allows us to benefit considerably from the choi between the sampling rules in the second stage. Naturally, the two-stage procedur are inferior to the single-stage selection procedures in those parts of the parameter space where they operate in the best way. The main advantage of a two-stage procedure in comparison with a single-stage procedure is due to the avoidance of an extremely great patient horizon and to the fact that the expected number of patients on the poorer treatment increases only slightly if the unknown success parameter p_A is less than 0.6. In the latter case, a single-stage VT-procedure would be preferred, provided that $p_A<0.6$ is known to the experimenter.

<u>Some comments concerning the use of the tables</u>

$s_{max}=(x,y)$ means that $s_{max}=x$ is needed if $\Delta^*=0.1$ and $s_{max}=y$ is needed if $\Delta^*=0.2$. Using $M=15$, $k_c=12$ is needed if $\Delta^*=0.1$, and $k_c=11$ is needed if $\Delta^*=0.2$.

5.7 Numerical results

Table 2

$\begin{pmatrix} r(m-n) \\ p_{LFC} \end{pmatrix}$ -values for $P^*=0.90,\ 0.95,\ 0.99$ and $\Delta^*=0.1,\ 0.2$

P^*	0.90		0.95		0.99	
Δ^*	0.1	0.2	0.1	0.2	0.1	0.2
m-n						
0	17 0.87	8 0.85	23 0.87	11 0.88	38 0.92	18 0.93
1	16 0.85	6 0.64	22 0.85	9 0.73	36 0.88	16 0.85
2	14 0.74	5 0.61	20 0.78	8 0.72	34 0.85	14 0.78
3	13 0.76	-	19 0.79	6 0.62	33 0.88	12 0.71
4	11 0.69	-	17 0.73	-	31 0.83	11 0.75
5	8 0.56	-	15 0.69	-	29 0.80	9 0.71
6	-	-	13 0.66	-	28 0.83	-
7	-	-	10 0.59	-	26 0.79	-
8	-	-	-	-	25 0.83	-
9	-	-	-	-	23 0.80	-
10	-	-	-	-	21 0.80	-
11	-	-	-	-	18 0.82	-

Table 3

$P^*=0.90;\ \Delta^*=(0.1,\ 0.2);\ s_{max}=(6,3)$											
		$E(N_B\|2)$		$E(N_B\|3)$		$E(N_B\|2)$		$E(N_B\|3)$		$E(N\|VT)$	
M k_c	p_A	Δ^*=0.1	0.2	0.1	0.2	0.1	0.2	0.1	0.2	0.1	0.2
5 3	0.2	63	-	37	-	126	-	63	-	119	-
	0.3	66	15	57	13	134	31	106	21	111	30
	0.4	69	15	73	15	142	31	143	27	105	28
	0.5	67	15	76	16	142	31	155	31	101	27
	0.6	59	14	66	15	128	30	140	32	101	26
	0.7	46	12	49	13	104	27	109	28	105	27
	0.8	33	10	34	10	77	23	78	24	111	28
	0.9	20	8	20	8	52	20	53	20	119	30
5 4	0.2	60	-	41	-	119	-	73	-	119	-
	0.3	57	15	53	13	115	30	101	23	111	30
	0.4	56	14	58	14	113	28	114	27	105	28
	0.5	56	14	58	14	115	28	118	28	101	27
	0.6	55	13	55	14	114	27	115	28	101	26
	0.7	48	13	48	13	103	27	103	27	105	27
	0.8	35	11	35	11	81	24	81	24	111	28
	0.9	21	9	21	9	54	20	54	20	119	30
10 8	0.2	60	-	42	-	119	-	76	-	119	-
	0.3	56	17	53	15	111	33	102	28	111	33
	0.4	53	16	53	16	105	32	105	31	105	32
	0.5	52	16	52	16	103	31	104	31	101	31
	0.6	52	16	52	16	105	31	105	31	101	31
	0.7	50	16	49	16	103	31	103	31	105	31
	0.8	39	14	39	14	86	30	86	30	111	32
	0.9	25	12	25	12	59	27	59	27	119	33
15 12/11	0.2	60	-	44	-	119	-	80	-	119	-
	0.3	56	19	54	18	112	38	105	35	112	38
	0.4	53	19	53	19	106	38	106	37	106	38
	0.5	51	19	52	19	102	38	103	38	102	38
	0.6	52	19	52	19	103	38	103	38	102	38
	0.7	51	19	51	19	104	38	104	38	106	38
	0.8	42	18	42	18	91	36	91	37	112	38
	0.9	28	17	28	17	65	34	65	34	119	38
20 15	0.2	60	-	42	-	119	-	75	-	119	-
	0.3	57	23	53	22	113	45	102	43	113	45
	0.4	54	23	54	23	108	46	108	45	108	46
	0.5	53	23	53	23	105	46	105	46	105	46
	0.6	53	23	53	23	106	46	106	46	105	46
	0.7	52	23	52	23	106	46	106	46	108	46
	0.8	42	22	42	22	91	45	91	45	113	46
	0.9	31	21	31	21	70	43	70	43	119	46

Table 3 (continued)

P*=0.95; Δ*=(0.1, 0.2); s_{max}=(8,4)												
			$E(N_B\|2)$		$E(N\|2)$		$E(N_B\|3)$		$E(N\|3)$		$E(N\|VT)$	
M	k_c	p_A	Δ*=0.1	0.2	0.1	0.2	0.1	0.2	0.1	0.2	0.1	0.2
5	3	0.2	86	-	54	-	172	-	90	-	160	-
		0.3	92	22	83	18	188	43	153	31	156	40
		0.4	98	22	105	22	204	45	206	41	151	39
		0.5	97	22	109	24	206	46	223	48	148	38
		0.6	85	20	95	23	187	45	203	48	148	38
		0.7	67	17	72	18	152	40	160	42	151	38
		0.8	47	13	48	14	112	33	114	34	156	39
		0.9	28	10	28	10	74	26	74	26	160	40
5	4	0.2	81	-	57	-	161	-	101	-	160	-
		0.3	81	20	76	18	161	40	144	32	156	40
		0.4	81	20	84	20	164	40	165	38	151	39
		0.5	82	20	85	21	169	40	173	41	148	38
		0.6	80	19	81	20	168	40	170	41	148	38
		0.7	69	18	70	18	151	39	152	39	151	38
		0.8	51	15	51	15	117	34	117	34	156	39
		0.9	29	10	29	10	75	27	75	27	160	40
10	8	0.2	80	-	59	-	160	-	105	-	160	-
		0.3	78	21	75	19	156	41	145	36	156	41
		0.4	76	21	77	21	152	41	153	40	151	41
		0.5	76	20	76	20	152	40	153	40	148	40
		0.6	76	20	76	20	155	40	156	40	148	40
		0.7	72	20	72	20	151	40	151	40	151	40
		0.8	55	18	55	18	123	38	123	38	156	41
		0.9	32	14	32	14	81	32	81	32	160	41
15	12 11	0.2	80	-	60	-	160	-	108	-	160	-
		0.3	78	22	76	21	156	44	148	39	156	44
		0.4	76	23	76	23	152	45	152	44	152	45
		0.5	75	23	75	23	149	45	149	45	149	45
		0.6	75	22	75	23	151	45	151	45	149	44
		0.7	73	22	73	22	151	45	151	45	152	45
		0.8	58	20	58	20	127	42	127	42	156	45
		0.9	35	18	35	18	86	38	86	38	160	44
20	15	0.2	80	-	56	-	160	-	99	-	160	-
		0.3	78	25	74	24	157	50	142	46	157	50
		0.4	76	26	77	26	152	51	152	50	153	51
		0.5	75	26	75	26	150	51	150	51	149	51
		0.6	76	26	76	26	152	51	152	51	149	51
		0.7	73	25	73	25	151	51	151	51	153	51
		0.8	56	24	56	24	125	49	125	49	157	51
		0.9	38	22	38	22	90	46	90	46	160	50

Table 3 (continued)

		$P^*=0.99$; $\Delta^*=(0.1, 0.2)$; $s_{max}=(12,6)$									
		$E(N_B\|2)$		$E(N\|2)$		$E(N_B\|3)$		$E(N\|3)$		$E(N\|VT)$	
M k_c	p_A	Δ^*=0.1	0.2	0.1	0.2	0.1	0.2	0.1	0.2	0.1	0.2
5 3	0.2	131	-	91	-	263	-	150	-	240	-
	0.3	146	34	141	31	299	69	260	51	240	60
	0.4	159	36	176	39	332	75	346	71	238	60
	0.5	159	36	182	41	340	79	374	83	237	60
	0.6	142	33	159	38	313	76	341	82	237	60
	0.7	113	28	120	30	258	76	271	71	238	60
	0.8	79	21	80	21	191	55	194	56	240	60
	0.9	44	14	44	14	121	41	122	41	240	60
5 4	0.2	121	-	93	-	243	-	161	-	240	-
	0.3	124	31	122	30	250	62	231	51	240	60
	0.4	130	31	135	33	262	63	267	62	238	60
	0.5	134	32	138	33	276	66	282	67	237	60
	0.6	131	31	133	32	277	67	280	68	237	60
	0.7	114	28	115	29	252	64	252	64	238	60
	0.8	83	22	83	22	196	55	196	55	240	60
	0.9	45	14	45	14	124	42	124	42	240	60
10 8	0.2	120	-	94	-	240	-	165	-	240	-
	0.3	120	31	119	29	240	61	227	53	240	61
	0.4	120	30	122	31	241	60	242	60	238	60
	0.5	122	30	123	31	245	61	246	61	237	60
	0.6	124	31	124	31	254	62	254	62	237	60
	0.7	116	30	116	30	247	62	247	62	238	60
	0.8	88	25	88	25	201	57	201	57	240	60
	0.9	48	18	48	18	129	46	129	46	240	61
15 12	0.2	120	-	95	-	240	-	169	-	240	-
11	0.3	120	31	119	30	240	61	230	52	240	61
	0.4	119	31	120	32	238	61	239	60	238	61
	0.5	120	31	120	31	239	62	239	62	237	61
	0.6	121	31	121	31	245	63	245	63	237	61
	0.7	117	30	117	30	245	64	245	64	238	61
	0.8	92	26	92	26	206	59	206	59	240	61
	0.9	51	20	51	20	133	49	133	49	240	61
20 15	0.2	120	-	89	-	240	-	152	-	240	-
	0.3	120	32	118	31	240	63	223	57	240	63
	0.4	119	32	121	33	238	64	240	64	238	64
	0.5	120	32	120	33	239	64	239	64	237	64
	0.6	122	33	122	33	247	65	247	65	237	64
	0.7	117	32	117	32	248	65	248	65	238	64
	0.8	88	29	88	29	203	62	203	62	240	64
	0.9	53	24	53	24	137	55	137	55	240	63

6 The Selection Model $[2;PW;\max\{S_A,S_B\}=r]$

6.1 Derivation of the critical r-value

The first treatment at the outset is determined randomly, and then the tests are carried out, one at a time, according to the PW-sampling scheme. Our stopping-rule is no longer based on the absolute difference of successes but on the individual numbers of successes, i.e. the selection procedure terminates if one treatment yields its r-th success and the other does not yield more than r-1 successes, in other words sampling is finished when any one population attains r successes. The highest success parameter p_A is associated with that treatment which has the most successes when sampling terminates. Using PW-sampling, an equal number of successes at the end of the selection procedure cannot arise, this may be possible only during the testing-phase. The stopping rule $\max\{S_A,S_B\}=r$ is sometimes called "inverse sampling".

Denoting by $\tilde{F}_A$ the number of faulty A-trials - we use the abbreviation $\bar{A}$-trial - preceding the r-th success of treatment A, we have

$$P(CS|\tilde{F}_A=j)=\tfrac{1}{2}(P(CS|\tilde{F}_A=j,\text{ begin.with A})+P(CS|\tilde{F}_A=j,\text{ begin.with B}))$$

$$P(CS)=\sum_{j=0}^{\infty}P(CS|\tilde{F}_A=j)P(\tilde{F}_A=j)\ , \tag{6.1}$$

where "beginning with A,B" means that the first treatment, given by the experimenter, is A,B respectively.

Beginning with A and selecting correctly, the structure of a testing-sequence with $\tilde{F}_A=j$ can be illustrated by the following diagram.

(6.2)

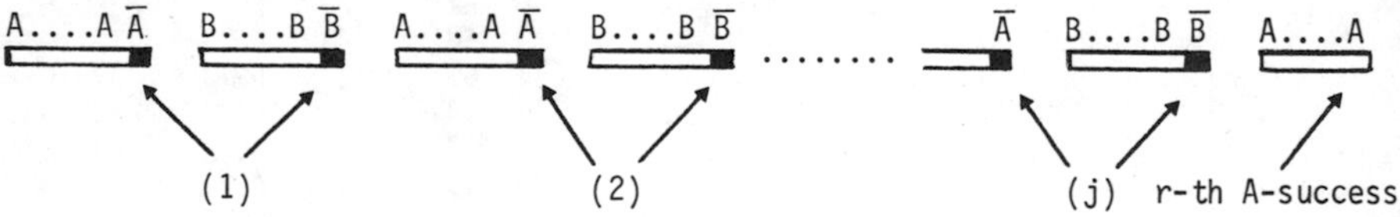

A......A Ā denotes a maximal section of the experiment, where treatment A is used exclusively; such a section is called an A-block. The first block-symbol is an A and the last block-symbol is an $\bar{A}$, where $\bar{A}$ denotes a failure of treatment A. It may be possible that an A-block consists of one $\bar{A}$ only; this happens if the first trial with treatment A, following a switch from treatment B to treatment A, is a failure. This special case is to be included in the above block-symbol, too. A testing-sequence leading to a correct selection terminates with an A-run, i.e. a sequence of successive A-successes. (A B-block is defined in the same way) It is immediately seen from (6.2) that a beginning with A and a final selection of A, i.e. a correct selection, imply that $\tilde{F}_A$ and $\tilde{F}_B$ coincide, where $\tilde{F}_B$ denotes the number of B-failures preceding the r-th A success. - Note that we have to distinguish plainly between $\tilde{F}_A,\tilde{F}_B$ on the one hand and between F_A,F_B on the other hand! -

A beginning with treatment A and $\tilde{F}_B=\tilde{F}_A=j$ imply that j B-failures and $k \leq r-1$ B-successes have occurred, provided that the best treatment has been actually selected. Our first result is consequently:

$$P(\tilde{F}_A=j)P(CS|\tilde{F}_A=j,\ \text{begin. w.A}) = \binom{j+r-1}{j} p_A^r\, q_A^j \sum_{k=0}^{r-1} \binom{j+k-1}{k} p_B^k\, q_B^j \tag{6.3}$$

Beginning with treatment B and selecting correctly, the structure of a testing-sequence with $\tilde{F}_A=j$ can be illustrated as follows:

(6.4)

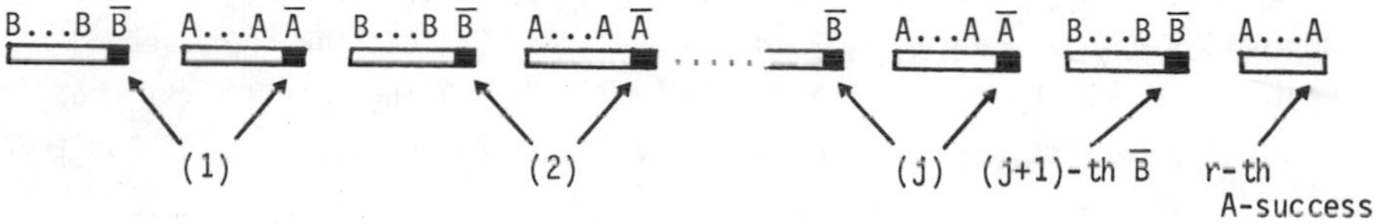

In this case, $\tilde{F}_B=\tilde{F}_A+1=j+1$ holds. Up to the termination of the experiment j+k+1 B-trials are carried out, and the second term of the first formula in (6.1) is thus found to be

$$P(\tilde{F}_A=j)P(CS|\tilde{F}_A=j,\ \text{begin. w.B}) = \binom{j+r-1}{j} p_A^r\, q_A^j \sum_{k=0}^{r-1} \binom{j+k}{k} p_B^k\, q_B^{j+1} \tag{6.5}$$

The expression for P(CS) now follows from (6.3) and (6.5)

$$P(CS) = \frac{1}{2} \sum_{j=0}^{\infty} \binom{j+r-1}{j} p_A^r\, q_A^j \sum_{k=0}^{r-1} \left(\binom{j+k-1}{k} p_B^k\, q_B^j + \binom{j+k}{k} p_B^k\, q_B^{j+1} \right) \tag{6.6}$$

Denoting by $X_{r,p}$ a negative binomial chance variable with index r and success parameter p(cf A1/6), the above formula can be rewritten as

$$P(CS) = \frac{1}{2} E\left(J_{q_B}(X_{r,p_A},r) + J_{q_B}(X_{r,p_A}+1,r)\right), \tag{6.7}$$

where $J_q(.,.)$ denotes an incomplete beta-function (cf A1/4).

In the next section we deal with a selection procedure using inverse sampling and the VT-sampling scheme. It will be proved there that the P(CS) is exactly the same for both procedures. That is why the determination of r_{max} is dropped here.

In using (6.6) to calculate the exact P(CS) values $-r, p_A$ and $p_B = p_A - \Delta^*$ are present - we have to cut off the infinite series in (6.6) at any term, for example $j=\alpha$, where α must be chosen such that the resulting infinite sum, starting with $j=\alpha+1$, will be less than a preassigned $\varepsilon>0$. The resulting P(CS) values, based on the first α terms of (6.6), is exact up to a correction term of at most ε. Defining

$$U(\alpha) := \frac{1}{2} \sum_{j=0}^{\alpha} \binom{j+r-1}{j} p_A^r q_A^j \left(J_{q_B}(j,r) + J_{q_B}(j+1,r)\right) , \qquad (6.8)$$

the following inequalities are **easily** shown to hold

$$U(\alpha) \leq P(CS) \leq U(\alpha) + J_{q_B}(\alpha+1,r) J_{q_A}(\alpha+1,r) =: S(\alpha), \qquad (6.9)$$

i.e. $U(\alpha)$ is a lower bound for the P(CS), and $S(\alpha)$ is an upper bound for the P(CS)-value. It is found without a hitch that $U(\alpha)$, $S(\alpha)$ are strictly increasing, decreasing with α, respectively. Noting in addition that for fixed second argument the incomplete beta-function is strictly decreasing in the first argument, it is obvious that $\{[U(\alpha),S(\alpha)] \mid \alpha \in \mathbb{N}\}$ is a system of nested intervals for the P(CS)-value.

In order to calculate an approximate P(CS)-value, the terms of the infinite series in (6.6) are added up as long as $J_{q_B}(\alpha+1,r) \cdot J_{q_A}(\alpha+1,r)$ exceeds the preassigned ε for the first time. We get a rough idea of the magnitude of α by using the approximation of the incomplete beta-function described below (cf A1/9)

$$J_q(\alpha+1,r) = \sum_{\tau=0}^{r-1} \binom{\alpha+r}{\tau} q^{\alpha+r-\tau} p^{\tau} \leq \left(\sum_{\ell=0}^{r-1} \binom{r}{\ell} q^{r-\ell} p^{\ell}\right)\left(\sum_{k=0}^{r-1} \binom{\alpha}{k} q^{\alpha-k} p^{k}\right) =$$

$$= (1-p^r) \sum_{k=0}^{r-1} \binom{\alpha}{k} q^{\alpha-k} p^k \approx (1-p^r) P(\hat{X}_{\alpha,p} \leq r-1) \approx (1-p^r) \cdot \Phi\left(\frac{r-1-\alpha p}{\sqrt{\alpha p q}}\right) , \qquad (6.10)$$

where $\hat{X}_{\alpha,p}$ denotes a binomial chance variable with index α and success parameter p (Note that $r-1 \leq \alpha$ is tacitly assumed.)

A rough estimate for α is now easily derived by using (6.10) and the inequality $J_{q_A}(\alpha+1,r)J_{q_B}(\alpha+1,r) \leq (J_{q_B}(\alpha+1,r))^2$.
Another possibility to obtain an approximate expression for α arises by using the approximation of the incomplete beta-function suggested in A2/5. A comparison of the estimates of α with the α value actually needed shows that the first approximation is the better one.

The expression for P(CS) can also be derived by solving a system of difference equations. Let $T=(T_A,T_B)$, *where* T_A, T_B *denote the additional numbers of successes needed to declare A,B to be the best treatment, respectively, and let*

$$U(m,n) := P(CS|T=(m,n), NT=A) \; ; \; V(m,n) := P(CS|T=(m,n), NT=B) \; , \qquad (6.11)$$

then, the following system of difference equations and boundary conditions is easily derived:

$$\begin{aligned} &U(m,n) = p_A U(m-1,n) + q_A V(m,n) \; ; \; V(m,n) = p_B V(m,n-1) + q_B U(m,n) \\ &\quad U(0,n) = 1, \; V(m,0) = 0 \quad \text{provided } m,n > 0; \end{aligned} \qquad (6.12)$$

the above system can be solved by using generating functions. We will make frequently use of this method in the following, therefore, a detailed description is omitted here, but will be given, if generating functions are actually used. The expression for P(CS), based on the solution of (6.12), is

$$P(CS) = \tfrac{1}{2}(U(r,r) + V(r,r)) \qquad (6.13)$$

6.2 Derivation of the expectations

We next determine $E(N_B)$. For this purpose, we define

$$\Omega_S := \{\omega=(\alpha_1,\alpha_2,\ldots,\alpha_n) | n \in \mathbb{N} \setminus \{1,\ldots,r-1\}\} \text{ — where } \alpha_i \text{ may be} \qquad (6.14)$$

a success or a failure of treatment A or treatment B, i.e. $\alpha_i \in \{A,\bar{A},B,\bar{B}\}$ — to be the subset of all testing-sequences that may lead to a termination of the selection procedure. An element ω of Ω_S is called a stopping-sequence. The length of a stopping-sequence is at least r, and the last component of each $\omega \in \Omega_S$ is a success of either treatment A or treatment B. Furthermore, we define the random variables $N_{B,CS}$ and $N_{B,FS}$ as follows:

$$N_{B,CS}(\omega) := \begin{cases} \ell_B(\omega) & \text{, if sampling terminates with a correct selection} \\ 0 & \text{, if sampling terminates with a false selection} \end{cases}$$

$$N_{B,FS}(\omega) := \begin{cases} 0 & \text{, if sampling terminates with a correct selection} \\ \ell_B(\omega) & \text{, if sampling terminates with a false selection} \end{cases} \tag{6.15}$$

($N_{A,CS}$ and $N_{A,FS}$ are defined in the same way)

$\ell_B(\omega)$ denotes the number of trials (components) of ω which have been carried out with treatment B.
It follows directly from the above definitions that $P(N_{B,CS}=0) = P(FS)$ and $P(N_{B,FS}=0) = P(CS)$ hold, where "FS" means "False Selection". Moreover, it can be shown, that $P(CS) + P(FS) = 1$. The last identity is by no means trivial but holds only if the probability is one that sampling will eventually terminate, in other words, the set $\complement\Omega_S$ of all sequences that will not lead to a final decision is only a null set with respect to the probability measure P induced by the selection model under investigation. It will be shown afterwards that this important fact holds true for all selection procedures considered so far. The expected number of patients on the poorer treatment can be calculated by using the identity $N_B = N_{B,CS} + N_{B,FS}$. We obtain:

$$E(N_B) = E(N_{B,CS}) + E(N_{B,FS}) \tag{6.16}$$

$$E(N_{B,CS}) = \frac{1}{2} \sum_{j=0}^{\infty} P(\tilde{F}_A=j)(E(N_{B,CS}|\tilde{F}_A=j,\text{b.w. A}) + E(N_{B,CS}|\tilde{F}_A=j,\text{b.w. B})) =$$

$$= \frac{1}{2} \sum_{j=0}^{\infty} \binom{j+r-1}{j} p_A^r q_A^j \left(\sum_{n=0}^{r-1} (n+j) \binom{n+j-1}{j-1} p_B^n q_B^j + \sum_{n=0}^{r-1} (n+j+1) \binom{n+j}{j} p_B^n q_B^{j+1} \right) =$$

$$= \frac{r q_A}{2 q_B p_A} p_A^{r+1} \sum_{j=0}^{\infty} \binom{j+r}{j} q_A^j (J_{q_B}(j+2,r) + J_{q_B}(j+3,r)) + \tag{6.17}$$

$$+ \frac{1}{2 q_B} p_A^r \sum_{j=0}^{\infty} \binom{j+r-1}{j} q_A^j J_{q_B}(j+2,r) =$$

$$= \frac{rq_A}{2q_B p_A} E(J_{q_B}(X_{r+1,p_A}+2,r) + J_{q_B}(X_{r+1,p_A}+3,r)) + \frac{1}{2q_B} E(J_{q_B}(X_{r,p_A}+2,r))$$

(b.w.A means "beginning with A")

The expression for $E(N_{B,FS})$ is easily derived from (6.17) by using the following identity:

$$\begin{aligned} &N_{B,FS} \quad \text{(where } p_A \text{ is associated with treatment A, and } p_B \text{ is associated with treatment B)} \\ \equiv\ &N_{A,CS} \quad \text{(where } p_B \text{ is associated with treatment A, and } p_A \text{ is associated with treatment B)} \end{aligned} \tag{6.18}$$

We use the abbreviation: $N_{B,FS} = N_{A,CS}(p_A \leftrightarrows p_B)$

The validity of (6.18) is immediately evident if we imagine that in either case the trials associated with the success parameter p_B are considered, provided sampling terminates with the selection of treatment B. Associating p_A with treatment A, a selection of B means a false selection, and associating p_A with B, a selection of B means nothing but a correct selection.
Using (6.6),(6.18) and noting that $N_{A,CS}$ takes the value r+j iff $\tilde{F}_A$ takes the value j <u>and</u> sampling terminates with a correct selection, i.e.

$$P(N_{A,CS} = r+j) = P(\tilde{F}_A = j, CS) \,, \tag{6.19}$$

the expression for $E(N_{B,FS})$ can be derived as follows:

$$E(N_{B,FS}) = E(N_{A,CS}(p_A \leftrightarrows p_B)) = \sum_{j=0}^{\infty} (r+j)P(CS|\tilde{F}_A=j, p_A \leftrightarrows p_B)P(\tilde{F}_A=j, p_A \leftrightarrows p_B) =$$

$$= \frac{1}{2} \sum_{j=0}^{\infty} (r+j)\cdot \binom{j+r-1}{j} p_B^r q_B^j \sum_{k=0}^{r-1} \left(\binom{j+k-1}{k} p_A^k q_A^j + \binom{j+k}{k} p_A^k q_A^{j+1} \right) =$$

$$= \frac{r}{2p_B} \sum_{j=0}^{\infty} \binom{j+r}{j} p_B^{r+1} q_B^j (J_{q_A}(j,r) + J_{q_A}(j+1,r)) \,, \text{ i.e.}$$

$$B(p_A,p_B) := E(N_{B,FS}) = \frac{r}{2p_B} E(J_{q_A}(X_{r+1,p_B},r) + J_{q_A}(X_{r+1,p_B}+1,r)) \tag{6.20}$$

Denoting $E(N_{B,CS})$ by $A(p_A,p_B)$, we finally obtain:

$$E(N_B) = A(p_A,p_B) + B(p_A,p_B) \tag{6.21}$$

Because of $N_{A,CS}=N_{B,FS}(p_A \leftrightarrows p_B)$ and $N_{A,FS}=N_{B,CS}(p_A \leftrightarrows p_B)$, $E(N_A)$ and $E(N)$ are easily shown to be

$$\begin{aligned} E(N_A) &= E(N_{A,CS})+E(N_{A,FS})=E(N_{B,FS}(p_A \leftrightarrows p_B))+E(N_{B,CS}(p_A \leftrightarrows p_B))= \\ &= B(p_B,p_A)+A(p_B,p_A)=E(N_B|\, p_A \leftrightarrows p_B) \qquad (6.22) \\ E(N) &= A(p_A,p_B)+A(p_B,p_A)+B(p_A,p_B)+B(p_B,p_A) \end{aligned}$$

In order to calculate $E(N_B)$, we cut off the infinite series, given by (6.21), at $j=\alpha$. The sum of the first α terms is a lower bound for $E(N_B)$; we denote it by $U(\alpha;N_B)$. An upper bound for $E(N_B)$ is rapidly seen to be

$$S(\alpha;N_B):=U(\alpha;N_B)+\frac{rq_A}{q_Bp_A} J_{q_B}(\alpha+3,r)J_{q_A}(\alpha+1,r+1)+\frac{1}{2q_B} J_{q_B}(\alpha+3,r)J_{q_A}(\alpha+1,r)+$$
$$+\frac{r}{p_B} J_{q_A}(\alpha+1,r)J_{q_B}(\alpha+1,r+1) \qquad (6.23)$$

Moreover, it is easily found that $\{[U(\alpha;N_B),S(\alpha;N_B)]\,|\,\alpha\in\mathbb{N}\}$ is a system of nested intervals for $E(N_B)$. A rough estimate of α can be derived in the same way as before; for details, we refer to Schriever [164].
Lower and upper bounds for $E(N_A)$ are obtained by interchanging the success parameters, i.e.

$$U(\alpha;N_A)=U(\alpha;N_B,p_A \leftrightarrows p_B);\; S(\alpha;N_A)=S(\alpha;N_B,p_A \leftrightarrows p_B) \qquad (6.24)$$

$E(N_B)$ and $E(N_A)$ can also be derived as solutions of systems of difference equations. For this purpose, we define

$$\begin{aligned} &R_A(m,n):=E(N_A|T=(m,n),NT=A);\; S_A(m,n):=E(N_A|T=(m,n),NT=B) \\ &R_B(m,n):=E(N_B|T=(m,n),NT=A);\; S_B(m,n):=E(N_B|T=(m,n),NT=B), \end{aligned} \qquad (6.25)$$

where T has the same meaning as in (6.11).
Noting that the trials are carried out according to the PW-sampling scheme, the following systems of difference equations and boundary conditions are quickly derived.

$$\begin{aligned} &R_A(m,n)=p_AR_A(m-1,n)+q_AS_A(m,n)+1 \\ &S_A(m,n)=p_BS_A(m,n-1)+q_BR_A(m,n) \qquad (6.26) \\ &R_A(0,n)=S_A(m,0)=0 \text{ for } m,n>0 \end{aligned}$$

$$R_B(m,n)=p_A R_B(m-1,n)+q_A S_A(m,n)$$
$$S_B(m,n)=p_B S_B(m,n-1)+q_B R_B(m,n)+1 \qquad (6.27)$$
$$R_B(0,n)=S_B(m,0)=0 \text{ for } m,n>0$$

The expectations are, hence, given by

$$E(N_A) = \tfrac{1}{2}(R_A(r,r)+S_A(r,r));E(N_B) = \tfrac{1}{2}(R_B(r,r)+S_B(r,r)) \qquad (6.28)$$

Using A2/11, the solution of (6.27) can be written as

$$E(N_B) = \frac{rq_A}{p_A q_B} + \frac{r}{q_B p_B} E(J_{p_B}(r+1,X_{r,p_A}))- \qquad (6.29)$$

$$- \frac{r}{p_A q_B} E(J_{p_B}(r,X_{r+1,p_A})) + \frac{1}{2q_B} E(J_{q_B}(X_{r,p_A}+1,r))$$

It isn't reasonable to solve (6.26) since $E(N_A)$ can be derived by using the well-known interchanging technique. To see this, we have to establish another system of difference equations and boundary conditions, that is

$$\tilde{R}_A(m,n):=E(N_A|T=(n,m),NT=B);\ \tilde{S}_A(m,n):=E(N_A|T=(n,m),NT=A)$$
$$\tilde{R}_A(m,n)=p_B\tilde{R}_A(m-1,n)+q_B\tilde{S}_A(m,n)$$
$$\tilde{S}_A(m,n)=p_A\tilde{S}_A(m,n-1)+q_A\tilde{R}_A(m,n)+1 \qquad (6.30)$$
$$\tilde{R}_A(0,n)=\tilde{S}_A(m,0)=0 \quad \text{for } m,n>0$$

$E(N_A)$ is given by:

$$E(N_A) = \tfrac{1}{2}(\tilde{R}_A(r,r)+\tilde{S}_A(r,r)) \qquad (6.31)$$

Comparing (6.30) with the system of difference equations and boundary conditions given in (6.27), we see that (6.30) can be deduced from (6.27) by interchanging p_A with p_B as well as q_A with q_B, i.e.

$$E(N_A)=E(N_B|p_A \leftrightarrows p_B) \qquad (6.32)$$

This identity has been derived already in (6.22). Interchanging the parameters in (6.29) and using A1/8 and A1/12, we finally obtain:

$$E(N_A) = \frac{r}{p_A} + \frac{r}{p_B q_A} E(J_{p_B}(r+1, X_{r,p_A})) - \\ - \frac{r}{q_A p_A} E(J_{p_B}(r, X_{r+1,p_A})) + \frac{1}{2q_A} E(J_{p_B}(r, X_{r,p_A})) \tag{6.33}$$

Comparing this result with (6.29), we have the following approximative relation:

$$E(N) \approx (1 + \frac{q_B}{q_A}) E(N_B) \tag{6.34}$$

6.3 Expectations for large r

For large values of r, simpler approximative expressions for the expectations can be derived. Using A1/13, $E(N_B)$ and $E(N)$ are easily shown to be

$$E(N_B) \approx \frac{rq_A}{q_B p_A} + \frac{1}{2q_B} \; ; \; E(N) \approx (\frac{1}{q_A} + \frac{1}{q_B}) \frac{rq_A}{p_A} + \frac{1}{2q_B} \tag{6.35}$$

Further approximative expressions for the expectations are available by using the approximation of P(CS), carried out in the following section (cf. (7.4)), that is

$$E(N_B) = E(N_B|CS)P(CS) + E(N_B|FS)P(FS) \approx E(\tilde{F}_B + S_B|CS)\Phi(D) + \\ + E(r + X_{r,p_B})(1-\Phi(D)) = (E(\tilde{F}_B|CS) + E(S_B|CS))\Phi(D) + (r + \frac{rq_B}{p_B})(1-\Phi(D)) \tag{6.36}$$

Noting that $\tilde{F}_B = \tilde{F}_A + 1$ when the PW-sampling scheme has been used and a correct selection took place and sampling has begun with treatment B (cf. (6.4)), we have

$$E(\tilde{F}_B|CS) = \frac{1}{2}(E(\tilde{F}_A|CS) + E(\tilde{F}_A + 1|CS)) = E(\tilde{F}_A|CS) + \frac{1}{2} = \frac{rq_A}{p_A} + \frac{1}{2} \tag{6.37}$$

Denoting by X the number of B-successes preceding a B-failure, i.e. $P(X=k) = p_B^k q_B$, $k \in \mathbb{N}_0$; we obtain

$$E(S_B|CS) = E(X_1 + X_2 + \ldots + X_{\tilde{F}_B}|CS) = \sum_{\tau=0}^{\infty} E(X_1 + \ldots + X_\tau) P(\tilde{F}_B = \tau|CS) = \\ = E(X) \cdot E(\tilde{F}_B|CS) = \frac{p_B}{q_B}(\frac{rq_A}{p_A} + \frac{1}{2}), \text{ where} \tag{6.38}$$

the random variables X_i are distributed like X. Hence, the approximative expression for $E(N_B)$ is

$$E(N_B)\approx\frac{r}{q_B}\left(\frac{q_A}{p_A}\Phi(D)+\frac{q_B}{p_B}(1-\Phi(D))\right)+\frac{1}{2q_B}\Phi(D) \tag{6.39}$$

If P* is close to one, i.e. $\Phi(D)=P(CS) \geq P^*\approx 1$, (6.40) reduces to the first expression given in (6.35). An approximative representation for $E(N_A)$ is obtained in the following way.

$$\begin{aligned}E(N_A)&=E(N_A|CS)P(CS)+E(N_A|FS)P(FS)\approx\left(r+\frac{rq_A}{p_A}\right)\Phi(D)+\\&+E(N_{A,CS}+N_{A,FS}|FS)(1-\Phi(D))\end{aligned} \tag{6.40}$$

Noting that $E(N_{A,CS}|FS)=0$, we have to do nothing but to calculate $E(N_{A,FS}|FS)$. Using the interchanging technique for parameters again, we obtain:

$$E(N_{A,FS}|FS)=E(N_{B,CS}(p_A\leftrightarrows p_B)|CS)=E(N_B(p_A\leftrightarrows p_B)|CS)=\left(\frac{rq_B}{p_B}+\frac{1}{2}\right)\frac{1}{q_A} \tag{6.41}$$

$$\begin{aligned}E(N_A)&\approx\frac{r}{p_A}\Phi(D)+\left(\frac{rq_B}{p_B}+\frac{1}{2}\right)\frac{1}{q_A}(1-\Phi(D))\\E(N)&\approx\left(\frac{r(q_A+q_B)}{p_Aq_B}+\frac{1}{2q_B}\right)\Phi(D)+\left(\frac{r(q_A+q_B)}{p_Bq_A}+\frac{1}{2q_A}\right)(1-\Phi(D))\end{aligned} \tag{6.42}$$

If P* is close to one, E(N) reduces to the second formula of (6.35). Neglecting in addition the term $1/2q_B$, we obtain the approximative relation between $E(N_B)$ and E(N) already mentioned in (6.34).

7 The Selection Model $[2;VT;\max\{S_A,S_B\}=r]$

7.1 Derivation of the critical r-value

We use VT-sampling and the same stopping-rule as before, i.e. sampling terminates whenever at least one treatment yields its r-th success. The treatment with the larger number of successes at the end of the experiment is called the best. If both treatments have the same number of successes, the best is selected randomly. P(CS) is the sum of two terms P_1 and P_2, where P_1 denotes the probability that treatment A yields r successes at first, and P_2 denotes the probability that both treatments yield their r-th success in the same trial and that sampling terminates with a correct selection. Using A2/9, P_1 and P_2 are easily seen to be

$$P_1=\sum_{j=0}^{\infty}\left(\binom{j+r-1}{j}p_A^r q_A^j \sum_{i=0}^{r-1}\binom{j+r}{i}p_B^i q_B^{j+r-i}\right)=E(J_{q_B}(X_{r,p_A}+1,r)) \tag{7.1}$$

$$P_2=\frac{1}{2}\sum_{j=0}^{\infty}\binom{j+r-1}{j}^2 (p_A p_B)^r (q_A q_B)^j=\frac{1}{2}(E(J_{q_B}(X_{r,p_A},r))-E(J_{q_B}(X_{r,p_A}+1,r)))$$

Hence, P(CS) is given by

$$P(CS)=\frac{1}{2}E(J_{q_B}(X_{r,p_A},r)+J_{q_B}(X_{r,p_A}+1,r)) \tag{7.2}$$

The comparison with (6.7) gives, that the two selection models coincide in their P(CS)-value. Therefore, the same critical r-value is needed for either procedure. P(CS) can be written as

$$P(CS)=P(CS|VT)=P(X_{r,p_A}<X_{r,p_B})+\frac{1}{2}P(X_{r,p_A}=X_{r,p_B}), \tag{7.3}$$

where X_{r,p_A}, X_{r,p_B} are mutually independent negative binomial chance variables. Using the central limit theorem (cf A2/1), the second term of (7.3) tends to 0 and may therefore be neglected. The standardized difference $X_{r,p_A}-X_{r,p_B}$ tends to a standard normal distribution as r becomes large, i.e. we have the approximative result

$$P(CS)\approx P(X_{r,p_A}<X_{r,p_B})=\Phi\left(\frac{\Delta\sqrt{r}}{\sqrt{p_A^2 q_B+p_B^2 q_A}}\right)=:\Phi(D), \tag{7.4}$$

where $\Delta := p_A-p_B$ is the true unknown difference between the success parameters.

Rewriting D as a function of Δ and $\overline{p}$ (cf (1.13)), holding $\overline{p}$ fixed and differentiating

$$D=D(\Delta)=\frac{\Delta\sqrt{r}}{\sqrt{(\overline{p}-\frac{1}{2}\Delta)^2(1-\overline{p}-\frac{1}{2}\Delta)+(\overline{p}+\frac{1}{2}\Delta)^2(1-\overline{p}+\frac{1}{2}\Delta)}} \tag{7.5}$$

with respect to Δ, we obtain that $D(\Delta)$ is strictly increasing with Δ. Noting that $P(CS)\approx\Phi(D(\Delta))$, we have as a first result $\Delta = \Delta^*$ in the LFC. Considering now D as a function of $\overline{p},\Delta$ replaced by Δ^*, it is evident immediately that the LFC is reached when $D=D(\overline{p},\Delta^*)$ becomes minimal, and that happens if

$$\overline{p} = \frac{1}{3} \pm \frac{1}{3}(1+\frac{3}{4}\Delta^{*2})^{1/2} = \frac{1}{3} \pm \frac{1}{3}(1+O(\Delta^{*2})) \tag{7.6}$$

Noting that $D(\overline{p},\Delta^*)$ becomes minimal if the denominator of (7.5) is maximized, the desired $\overline{p}$-value in the LFC is quickly shown to be

$$\overline{p} = \frac{2}{3} + O(\Delta^{*2}); \quad D(\frac{2}{3} + O(\Delta^{*2}),\Delta^*) \approx \Delta^* \sqrt{\frac{27r}{8}} \tag{7.7}$$

The critical r-value is now available from

$$\min_{\Delta,\overline{p}} P(CS) \approx \Phi(\Delta^* \sqrt{\frac{27r}{8}}) = P^*; \quad \Phi^{-1}(P^*) =: \lambda(P^*), \text{ i.e.}$$

$$r_{max} = \left]\frac{8}{27}\left(\frac{\lambda(P^*)}{\Delta^*}\right)^2\right[\tag{7.8}$$

7.2 Derivation of the Expectations

To derive an exact formula for $E(N_B)$, we use again the relation $E(N_B)=E(N_{B,CS})+$ $+E(N_{B,FS})$ (cf (6.15)), and using the VT-sampling scheme, the following identities are rapidly shown to hold.

$$\begin{aligned} N_{B,CS} &= N_{A,FS}(p_A \leftrightarrows p_B) = N_{B,FS}(p_A \leftrightarrows p_B) \\ N_{B,FS} &= N_{A,CS}(p_A \leftrightarrows p_B) = N_{B,CS}(p_A \leftrightarrows p_B) \end{aligned} \tag{7.9}$$

Note that in either case the treatment associated with p_A is selected. $E(N_{B,FS})$ is thus given by

$$E(N_{B,FS}) = E(N_{A,CS}(p_A \leftrightarrows p_B)) = \sum_{j=0}^{\infty} E(N_{A,CS}(p_A \leftrightarrows p_B) \mid \tilde{F}_A=j,\ 0\le S_B \le r-1,\ CS)\cdot$$

$$\cdot \underbrace{P(CS \mid \tilde{F}_A=j,}_{1}\ 0\le S_B \le r-1)\cdot P(\tilde{F}_A=j)P(0\le S_B\le r-1) \; +$$

$$+ \sum_{j=0}^{\infty} E(N_{A,CS}(p_A \leftrightarrows p_B) \mid \tilde{F}_A=j,S_B=r,CS)\underbrace{P(CS \mid \tilde{F}_A=j,S_B=r)}_{=\,1/2}P(S_B=r \mid \tilde{F}_A=j)P(\tilde{F}_A=j) =$$

$$= \sum_{j=0}^{\infty} ((r+j)\underbrace{P(N_{A,CS}(p_A \mathrel{\substack{\leftarrow\\\rightarrow}} p_B)=r+j\,|\widetilde{F}_A=j,\ 0\leq S_B\leq r-1,CS)}_{=1}\binom{j+r-1}{j}p_B^r q_B^j \cdot \tag{7.10}$$

$$\cdot \sum_{k=0}^{r-1}\binom{j+r}{k}p_A^k q_A^{j+r-k}) + \sum_{j=0}^{\infty}((r+j)\underbrace{P(N_{A,CS}(p_A \mathrel{\substack{\leftarrow\\\rightarrow}} p_B)=r+j\,|\widetilde{F}_A=j,S_B=r,CS)}_{=1}\cdot$$

$$\cdot \frac{1}{2}\binom{j+r-1}{j}p_B^r q_B^j\binom{j+r-1}{j}p_A^r q_A^j) = \frac{r}{2p_B}E(J_{q_A}(X_{r+1,p_B}+1,r)+J_{q_A}(X_{r+1,p_B},r))$$

The last term of (7.10) is equal to $B(p_A,p_B)$; cf (6.20). Using the interchanging technique, we obtain:

$$\begin{aligned} &E(N_{B,CS})=E(N_{B,FS}(p_A \mathrel{\substack{\leftarrow\\\rightarrow}} p_B))=B(p_B,p_A)\\ &E(N_B)=B(p_A,p_B)+B(p_B,p_A);\ E(N)=2E(N_B)\end{aligned} \tag{7.11}$$

Noting A1/12, $E(N_B)$ can be rewritten as

$$\begin{aligned} E(N_B) &= \frac{r}{2p_B}(E(J_{p_B}(r+1,X_{r,p_A}+1))+E(J_{p_B}(r+1,X_{r,p_A}))) +\\ &+ \frac{r}{p_A} - \frac{r}{2p_A}(E(J_{p_B}(r,X_{r+1,p_A}))+E(J_{p_B}(r,X_{r+1,p_A}+1))),\end{aligned} \tag{7.12}$$

and using A1/13, we obtain the approximative results:

$$E(N_B)\approx\frac{r}{p_A} \quad ;\ E(N)\approx\frac{2r}{p_A} \tag{7.13}$$

It is seen, just as before, that the sum of the first α terms of the infinite series in (7.11) is a lower bound for $E(N_B)$, denoted by $U(\alpha;N_B)$; moreover $S(\alpha;N_B):=U(\alpha,N_B)+r(\frac{1}{p_A}+\frac{1}{p_B})J_{q_A}(\alpha+1,r+1)J_{q_B}(\alpha+1,r+1)$ is an upper bound for $E(N_B)$, and $\{[U(\alpha;N_B),S(\alpha;N_B)]\,|\alpha\in\mathbb{N}\}$ is a system of nested intervals for $E(N_B)$. A rough estimate for α, $\varepsilon > 0$ preassigned, can be derived in the same way, demonstrated before; details are omitted.

P(CS) and $E(N_B)$ can be obtained as solutions of systems of difference equations. Denoting by T the same random variable as defined in (6.11), we have

$$U(m,n):=P(CS|T=(m,n))$$

$$U(m,n)=p_Ap_BU(m-1,n-1)+p_Aq_BU(m-1,n)+q_Ap_BU(m,n-1)+q_Aq_BU(m,n) \tag{7.14}$$

$$U(0,n)=1;\ U(m,0)=0,\ \text{for } m,n>0;\ U(0,0)=1/2$$

$E(N_B)$ can be derived as follows

$$V(m,n):=E(N_B|T=(m,n))$$

$$V(m,n)=p_Ap_BV(m-1,n-1)+p_Aq_BV(m-1,n)+q_Ap_BV(m,n-1)+q_Aq_BV(m,n)+1 \tag{7.15}$$

$$V(0,m)=V(n,0)=V(0,0)=0;\ m,n>0$$

7.3 Expectations for large r

Similar to section 6, $E(N_B)$ can be expressed by using (7.4), i.e.

$$E(N_B)=E(N_B|CS)P(CS)+E(N_B|FS)P(FS)\approx E(N_A|CS)\Phi(D)+$$
$$+E(N_B|FS)(1-\Phi(D))=E(r+X_{r,p_A})\Phi(D)+E(r+X_{r,p_B})(1-\Phi(D))= \tag{7.16}$$
$$=(r+\frac{rq_A}{p_A})\Phi(D)+(r+\frac{rq_B}{p_B})(1-\Phi(D))=\frac{r}{p_A}\Phi(D)+\frac{r}{p_B}(1-\Phi(D))$$

If P^* is close to one, $\Phi(D)=P(CS)$ is close to one, too, and thus $E(N_B)\approx r/p_A$. This is the same result as stated in (7.13).

8 Comparison of the Selection Models No.6 and No.7 — Some Modifications of These Models

8.1 Comparison of the Selection Models No. 6 and No. 7

As usual we begin with the comparison of both procedures with respect to $E(N_B)$. For this purpose we recapitulate:

$$E(N_B|PW)=A(p_A,p_B)+B(p_A,p_B);\ E(N_B|VT)=B(p_B,p_A)+B(p_A,p_B) \tag{8.1}$$

(cf.(6.21) and (7.11)). Using the second line of (6.17) and the following inequalities: $n+j<r+j$ and $n+j+1\leq r+j$ if $n\epsilon\{0,1,\dots,r-1\}$, we find

$$A(p_A,p_B)<\frac{1}{2}\sum_{j=0}^{\infty}(r+j)\binom{j+r-1}{j}p_A^r q_A^j\left(\sum_{n=0}^{r-1}\binom{n+j-1}{j-1}p_B^n q_B^j+\binom{n+j}{j}p_B^n q_B^{j+1}\right)=B(p_B,p_A) \quad (8.2)$$

(8.1) and (8.2) jointly imply that $E(N_B|PW)<E(N_B|VT)$, i.e.
The selection model using inverse sampling and the PW-sampling scheme is uniformly better with respect to $E(N_B)$ than the selection model using the same termination rule and the VT-sampling scheme.

Interchanging p_A with p_B in the inequality (8.2), it is easily found that $A(p_B,p_A)<B(p_A,p_B)$. Noting that

$$E(N|PW)=A(p_A,p_B)+A(p_B,p_A)+B(p_A,p_B)+B(p_B,p_A)$$

$E(N|VT)=2(B(p_A,p_B)+B(p_B,p_A))$, it is seen immediately that $E(N|PW)<E(N|VT)$, i.e. the comparison of both procedures with respect to E(N) gives the same result as above. A comparison can be based also on the approximations (6.39) and (7.16); we obtain:

$$E(N_B|PW)=\frac{r}{p_A}\left(\frac{q_A}{q_B}+\frac{p_A}{2rq_B}\right)\Phi(D)+\frac{r}{p_B}(1-\Phi(D))<E(N_B|VT)$$

if and only if $\frac{q_A}{q_B}+\frac{p_A}{2rq_B}<1$, and this is the case if $r>p_A/2\Delta$. The last inequality usually holds as can be seen from the numerical results presented at the end of this section.

8.2 The Selection Model [2;PW;min{F_A,F_B}=r]

The expectations $E(N_B|PW)$ and $E(N|PW)$ become very large if the success probability of the best treatment is substantially smaller than 1/2. In this special case, the probability for a single A-failure is higher than the probability for a single A-success. Therefore, it seems to be obvious that a termination rule based on the failures of both treatments is more convenient than the inverse sampling rule, used so far. Sampling now terminates if both treatments yield at least r failures. The one with the larger number of successes is declared as best. In case of an equal number of successes, the best is determined randomly. Denoting by $S_A(r)$, $S_B(r)$ the number of A-successes, B-successes preceding the r-th A-failure, B-failure, respectively, P(CS) is given by:

$$P(CS)=P(S_B(r)<S_A(r))+\tfrac{1}{2}P(S_B(r)=S_A(r)) \tag{8.3}$$

Using the fact that $S_A(r)$, $S_B(r)$ are negative binomial chance variables with index r and success parameter q_A, q_B, respectively, and applying A1/8, the exact formula for P(CS) is found to be

$$\begin{aligned} P(CS) &= \sum_{j=0}^{\infty}\binom{j+r-1}{j}p_A^j q_A^r \sum_{\tau=0}^{j-1}\binom{r+\tau-1}{\tau}p_B^\tau q_B^r + \frac{1}{2}\sum_{j=0}^{\infty}\binom{j+r-1}{j}p_A^j q_A^r \binom{j+r-1}{j}p_B^j q_B^r = \\ &= \sum_{j=0}^{\infty}\binom{j+r-1}{j}p_A^j q_A^r (J_{q_B}(r,j)+\tfrac{1}{2}(J_{q_B}(r,j+1)-J_{q_B}(r,j))) = \\ &= \tfrac{1}{2}E(J_{q_B}(r,X_{r,q_A})+J_{q_B}(r,X_{r,q_A}+1)) \end{aligned} \tag{8.4}$$

To derive the critical r-value, we make use of the approximation carried out in section 7, i.e. $P(CS)\approx P(S_B(r)<S_A(r))$. Standardizing the random variable $S_B(r)-S_A(r)$ and applying the central limit theorem, just as before, we find

$$P(CS)\approx \Phi\left(\frac{\Delta\sqrt{r}}{\sqrt{q_A^2 p_B+q_B^2 p_A}}\right) \tag{8.5}$$

The LFC is determined in the same way as the one in section 7. We obtain that in the LFC $\Delta=\Delta^*$ and $\bar{q}=\frac{2}{3}+O(\Delta^{*2})$ hold. The critical r-value is hence given by

$$r_{max} = \left]\frac{8}{27}\left(\frac{\lambda(P^*)}{\Delta^*}\right)^2\right[\tag{8.6}$$

Since both treatments yield r failures when sampling terminates, a comparison of this modified procedure with others should be based on E(N) only. Using the identity $N=S_A(r)+S_B(r)+2r$ and the above comments concerning $S_A(r)$ and $S_B(r)$, E(N) is found to be

$$E(N)=E(S_A(r)+S_B(r)+2r)=r\left(\frac{p_A}{q_A}+\frac{p_B}{q_B}+2\right)=r\left(\frac{1}{q_A}+\frac{1}{q_B}\right) \tag{8.7}$$

Comparing this procedure with the PW-selection procedure of section 6, the following approximative statement is obtained if the term $1/2q_B$ in (6.35) is neglected:

$$E(N|PW)>E(N) \text{ if } q_A>p_A, \text{ i.e. if } p_A<\frac{1}{2} \tag{8.8}$$

This result is not surprising, quite on the contrary, it seems to be obvious that in counting the absolute number of successes a quicker decision may be expected if the "event" having the highest probability of realization ("occurrence of a failure") is involved in the termination rule.

8.3 The Selection Model $[2;PL;\max\{F_A,F_B\}=r]$

Another modification of the original procedure consists in the application of the PL-sampling-scheme and the stopping rule based on the absolute numbers of failures i.e. sampling terminates whenever either treatment yields its r-th failure. This treatment is declared to be the worst. Defining

$$U^*(m,n):=P(CS|T^*=(m,n),NT=B);\; V^*(m,n):=P(CS|T^*=(m,n),NT=A), \tag{8.9}$$

where $T^*:=(T_B^*,T_A^*)$ denotes the additional numbers of failures needed to select B,A as the worst treatment, respectively, the following system of difference equations and boundary conditions is easily derived.

$$U^*(m,n)=q_BU^*(m-1,n)+p_BV^*(m,n);\; V^*(m,n)=q_AV^*(m,n-1)+p_AU^*(m,n) \tag{8.10}$$
$$U^*(0,n)=1;\; V^*(m,0)=0 \text{ for } m,n>0$$

P(CS) is given by: $P(CS)=\frac{1}{2}(U^*(r,r)+V^*(r,r))$. (8.10) is obtained from (6.12) by interchanging p_A and q_B as well as p_B and q_A. Hence the probability of a correct selection is

$$P(CS|PL)=P(CS|PW,p_A\leftrightarrows q_B) \tag{8.11}$$

The LFC coincides with the one of the preceding selection procedure, i.e. we find again

$$r_{max}=\left]\frac{8}{27}\left(\frac{\lambda(P^*)}{\Delta^*}\right)^2\right[\tag{8.12}$$

In order to derive expressions for $E(N_A)$ and $E(N_B)$, we define

$$\begin{aligned}R_B^*(m,n)&:=E(N_B|T^*=(m,n),NT=B);\; S_B^*(m,n):=E(N_B|T^*=(m,n),NT=A)\\ R_A^*(m,n)&:=E(N_A|T^*=(m,n),NT=B);\; S_A^*(m,n):=E(N_A|T^*=(m,n),NT=A)\end{aligned} \tag{8.13}$$

Using PL-sampling, the following systems of difference equations and boundary conditions are quickly shown to hold.

$$R_B^*(m,n)=q_B R_B^*(m-1,n)+p_B S_B^*(m,n)+1;\quad S_B^*(m,n)=q_A S_B^*(m,n-1)+p_A R_B^*(m,n)$$
$$R_B^*(0,n)=S_B^*(m,0)=0 \quad \text{for } m,n>0 \tag{8.14}$$

$$R_A^*(m,n)=q_B R_A^*(m-1,n)+p_B S_A^*(m,n);\quad S_A^*(m,n)=q_A S_A^*(m,n-1)+p_A R_A^*(m,n)+1$$
$$R_A^*(0,n)=S_A^*(m,0)=0 \text{ for } m,n>0 \tag{8.15}$$

Since (8.14), (8.15) is obtained from (6.26), (6.27), respectively, by interchanging p_A and q_B as well as p_B and q_A, a complete evaluation of the expectations can be avoided. Noting that $E(N_A|PL)=E(N_B|PW,p_A \leftrightarrows q_B)$; $E(N_B|PL)=E(N_A|PW,p_A \leftrightarrows q_B)$, we finally obtain (cf (6.21)):

$$E(N_A|PL)=A(q_B,q_A)+B(q_B,q_A)$$
$$E(N_B|PL)=A(q_A,q_B)+B(q_A,q_B)=E(N_A|PL,q_A \leftrightarrows q_B) \tag{8.16}$$

Because of the identical range of p_A and q_B; i.e. $p_A,q_B \varepsilon(\Delta^*,1)$; and because of the relation $p_A-p_B=q_B-q_A$, we must not calculate any expectation.

$$E(N_A|PL,p_A=p)=E(N_B|PW,p_A=1-p+\Delta^*)$$
$$E(N_B|PL,p_A=p)=E(N_A|PW,p_A=1-p+\Delta^*) \tag{8.17}$$

Note that $p_A=p$ implies $q_B=1-p+\Delta^*$; $p\varepsilon(\Delta^*,1)$!

If r is large enough, the following approximations are shown to hold by using (6.35) and (8.16):

$$E(N_B|PL)\approx\frac{r}{q_B};\quad E(N|PL)\approx\left(\frac{1}{p_B}+\frac{1}{p_A}\right)\frac{rp_B}{q_B}+\frac{1}{2p_A} \tag{8.18}$$

Neglecting the term $1/2p_A$, the comparison of the above procedure with the PW-selection procedure of No. 6 gives that the PL-sampling procedure is approximately better than the PW-sampling procedure with respect to $E(N_B)(E(N))$ if $p_A<\frac{1}{2}$ $(p_A<\frac{1}{2}+\Delta^*)$.

P(CS)-function for various values of P^* and Δ^* (cf.(6.7))

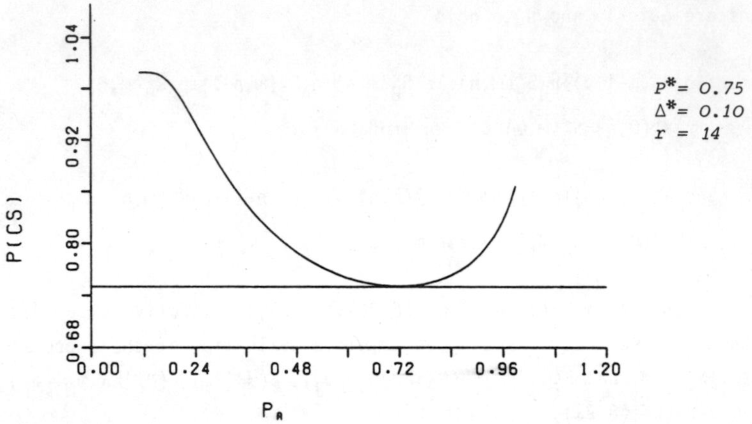

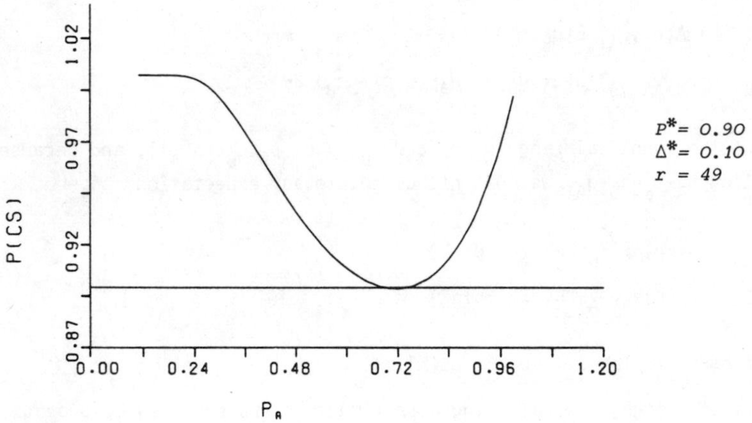

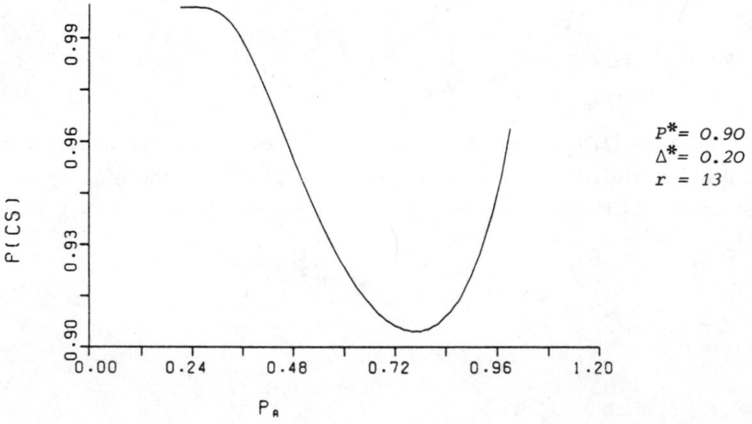

8.4 Concluding remarks

The construction of a two-stage selection procedure similar to that one of section 5 is not meaningful because the VT-sampling procedure of section 7, which had to be applied in the first stage of experimentation to get a rough estimate of the highest success parameter, is uniformly worse than the corresponding PW-sampling procedure of section 6. Applying such a two-stage procedure means to use in any case the worse sampling rule. The result would be an expected number of patients on the poorer treatment unnecessarily enlarged, a proceeding that would differ considerably from the ethical point of view.

Some comments concerning the use of the tables

The α-values in table 4 can be calculated by using (6.10). n is that natural number for which $J_{q_B}(n+1,r)\cdot J_{q_A}(n+1,r)<\varepsilon$ for the first time.

$\varepsilon=1/1000$ has been used for all calculations.
$E(N_B|PW)$ denotes the exact expression of $E(N_B)$ when the PW-sampling scheme has been used. (cf section 6)

$E(N|\cdot)$, $E(N_B|\cdot)$ denote the approximative expectations given in (6.35),(6.39) and (6.42).

The exact expectation of N_B when the VT-sampling scheme has been used is denoted by $E(N_B|VT)$.
The corresponding approximations are denoted by $E(N|\cdot)$ and $E(N_B|\cdot)$ and can be obtained from (7.13) and (7.16).

It is immediately seen from the tables that the estimates of α and α_B are only rough ones whereas the approximations of the expectations are beyond all doubt good ones.

8.5 Numerical results

Table 4

p_A	0.3	0.4	0.5	0.6	0.7	0.8	0.9
$P^*=0.90,\ \Delta^*=0.2,\ r=13$							
α	72.54	37.31	25.63	19.86	16.46	14.27	12.80
n	70	46	32	23	16	11	7
P(CS)	0.998	0.979	0.949	0.924	0.908	0.905	0.921
$P^*=0.90,\ \Delta^*=0.3,\ r=6$							
α	-	23.17	12.09	8.43	6.64	5.62	4.99
n	-	30	21	15	11	8	5
P(CS)	-	0.994	0.968	0.938	0.916	0.906	0.911
$P^*=0.95,\ \Delta^*=0.2,\ r=21$							
α	135.21	69.12	47.18	36.29	29.85	25.65	22.77
n	98	65	45	32	22	15	9
P(CS)	0.999	0.995	0.981	0.966	0.955	0.953	0.965
$P^*=0.95,\ \Delta^*=0.3,\ r=9$							
α	-	43.32	22.41	15.50	12.09	10.10	8.84
n	-	38	27	19	14	10	6
P(CS)	-	0.998	0.988	0.971	0.956	0.949	0.954
$P^*=0.99,\ \Delta^*=0.2,\ r=41$							
α	303.00	154.91	104.35	79.69	65.02	55.38	48.66
n	160	106	74	52	36	24	14
P(CS)	0.999	0.999	0.998	0.994	0.991	0.990	0.994
$P^*=0.99,\ \Delta^*=0.3,\ r=18$							
α	-	111.23	56.96	38.96	30.03	27.76	21.33
n	-	59	42	30	21	15	9
P(CS)	-	0.999	0.999	0.996	0.992	0.990	0.992

Table 5

P*=0.90; Δ*=0.2; r=13								
p_A	α_B	n_B	$E(N_B\|PW)$	$E(N_B\|6.35)$	$E(N_B\|6.40)$	$E(N\|PW)$	$E(N\|6.35)$	$E(N\|6.43)$
0.3	63.04	93	34.25	34.26	34.98	77.57	77.59	79.25
0.4	33.56	60	24.86	25.00	26.29	57.20	57.50	60.56
0.5	23.64	41	18.96	19.29	20.78	44.61	45.29	48.99
0.6	18.70	28	14.82	15.28	16.77	35.98	36.94	40.88
0.7	15.80	20	11.61	12.14	13.55	29.60	30.71	34.79
0.8	13.95	14	8.84	9.38	10.65	24.48	25.63	29.98
0.9	12.75	9	6.03	6.48	7.58	19.92	20.93	26.24
p_A	α_B	n_B	$E(N_B\|VT)$	$E(N_B\|7.13)$	$E(N_B\|7.16)$	$E(N\|VT)$	$E(N\|7.13)$	$E(N\|7.16)$
0.3	62.90	96	43.32	43.33	43.99	86.65	86.67	87.97
0.4	33.46	62	32.40	32.50	33.55	64.80	65.00	67.09
0.5	23.55	42	25.80	26.00	27.08	51.60	52.00	54.15
0.6	13.61	30	21.41	21.67	22.60	42.87	43.33	45.21
0.7	15.71	21	18.36	18.57	19.32	36.72	37.14	38.65
0.8	13.86	15	16.09	16.25	16.81	32.19	32.50	33.63
0.9	12.66	9	14.36	14.44	14.82	28.73	28.79	29.64

Table 5 (continued)

P*=0.95; Δ*=0.2; r=21

p_A	α_B	n_B	$E(N_B\|PW)$	$E(N_B\|6.35)$	$E(N_B\|6.40)$	$E(N\|PW)$	$E(N\|6.35)$	$E(N\|6.43)$
0.3	116.50	127	55.00	55.00	55.16	125.00	125.00	125.36
0.4	61.30	82	39.97	40.00	40.61	92.43	92.50	93.94
0.5	42.73	56	30.59	30.71	31.71	72.44	72.71	75.16
0.6	33.47	39	23.94	24.17	25.35	58.69	59.17	62.22
0.7	27.99	28	18.71	19.00	20.21	48.38	49.00	52.40
0.8	24.45	19	14.08	14.38	15.50	39.99	40.63	44.29
0.9	22.08	12	9.23	9.44	10.37	32.27	32.78	36.92
p_A	α_B	n_B	$E(N_B\|VT)$	$E(N_B\|7.13)$	$E(N_B\|7.16)$	$E(N\|VT)$	$E(N\|7.13)$	$E(N\|7.16)$
0.3	116.30	130	70.00	70.00	70.14	140.00	140.00	140.28
0.4	61.14	84	52.48	52.50	52.99	104.95	105.00	105.98
0.5	42.59	58	41.92	42.00	42.71	83.84	84.00	85.42
0.6	33.35	40	34.88	35.00	35.73	69.77	70.00	71.46
0.7	27.87	29	29.88	30.00	30.63	59.76	60.00	61.26
0.8	24.33	20	26.16	26.25	26.73	52.32	52.50	53.46
0.9	21.95	12	23.29	23.33	23.63	46.58	46.77	47.27

Table 5 (continued)

P*=0.99; Δ*=0.2; r=41

p_A	α_B	n_B	$E(N_B \mid PW)$	$E(N_B \mid 6.35)$	$E(N_B \mid 6.40)$	$E(N \mid PW)$	$E(N \mid 6.35)$	$E(N \mid 6.43)$
0.3	264.17	200	106.85	106.85	106.85	243.52	243.52	243.52
0.4	137.14	132	77.50	77.50	77.57	180.00	180.00	180.15
0.5	94.50	91	59.27	59.29	59.53	141.26	141.29	141.88
0.6	73.21	63	46.35	46.39	46.82	114.64	114.72	115.32
0.7	60.54	44	36.08	36.14	36.68	94.58	94.51	96.20
0.8	52.26	30	26.81	26.88	27.40	77.98	78.13	79.76
0.9	46.58	18	16.82	16.85	17.22	62.32	62.41	63.98
p_A	α_B	n_B	$E(N_B \mid VT)$	$E(N_B \mid 7.13)$	$E(N_B \mid 7.16)$	$E(N \mid VT)$	$E(N \mid 7.13)$	$E(N \mid 7.16)$
0.3	263.84	203	136.67	136.67	136.67	273.33	273.33	273.34
0.4	136.89	133	102.50	102.50	102.55	205.00	205.00	205.10
0.5	94.29	92	81.99	82.00	82.17	163.98	164.00	164.35
0.6	73.01	64	68.31	68.33	68.60	136.63	136.67	137.20
0.7	60.36	45	58.55	58.57	58.85	117.09	117.14	117.70
0.8	52.07	30	51.23	51.25	51.47	102.46	102.50	102.93
0.9	46.37	18	45.55	45.56	45.67	91.10	91.11	91.34

9 The Nature of Termination of a Classical Sequential Selection Procedure

9.1 Basic notions

An essential feature of a classical sequential selection procedure, as distinguished from a current selection procedure, is that the number of observations required by a classical sequential selection procedure depends on the outcome of the observations only and is, therefore, not predetermined, but a random variable N with an unrestricted range; i.e. N may assume any natural number with a positive probability. This is the general notion of a sequential test outlined in A. Wald's fundamental monograph "Sequential Analysis". The selection models investigated so far may be considered as classical sequential procedures in a wider sense since the patient horizon N does not depend on the observations only but also on the sampling scheme and the termination rule adopted by the experimenter. It can be shown, and we will do so in Chapter 2, that a skilful choice of the termination rule and of the sampling scheme results in a restricted patient horizon N, i.e the testing-procedure is carried out sequentially but terminates after at most ℓ steps, where ℓ is a preassigned constant, known by the experimenter.

The selection procedures considered up to now possess an unrestricted patient horizon N, and our main task, therefore, consists in the verification that the probability is one that all these procedures will eventually terminate. For this purpose we define Ω_S to be the set of all stopping-sequences, i.e. Ω_S is the set of all testing-sequences that may lead to a termination of the selection procedure under consideration. The set of all testing-sequences that do not lead to a termination of sampling will be denoted by Ω_∞. We conceive Ω_∞ as a set consisting of exactly one point ω_∞ ; i.e. $\Omega_\infty = \{\omega_\infty\}$. The event "sampling never terminates" will be referred to as "Ω_∞ will be realized". The probability space associated with a special selection procedure is hence $(\Omega,\sigma(\Omega),P)$, where $\Omega=\Omega_S\cup\{\omega_\infty\}$, $\sigma(\Omega)$ denotes the σ-field generated by the class of all simple events, and $P(\{\omega\})$ is the probability that the simple event $\{\omega\}$ occurs, i.e. for all $\omega\varepsilon\Omega_S$, $P(\{\omega\})$ denotes the probability that the stopping-sequence ω occurs, and the probability that sampling never terminates is hence given by $P(\{\omega_\infty\}):=1-\Sigma P(\{\omega\})$, where the summation is extended over all $\omega\varepsilon\Omega_S$. The random variable N is, therefore, represented by the following probability function:

$P(N=n)=\sum_{\omega\varepsilon\Omega_S^n} P(\{\omega\})$, where Ω_S^n denotes the subset of all stopping-sequences of length n, and $P(N=\infty)=1-\sum_{n=1}^{\infty} P(N=n)=1-P(\Omega_S)$ (9.1)

Using the above notation, we can say that a selection procedure terminates with probability one after an unknown but finite number of steps if and only if

$P(N<\infty)=\sum_{n=1}^{\infty} P(N=n)=1$, in other words, the probability is one that sampling will eventually terminate iff the probability that ω_∞ occurs equals zero. Since at least r trials must be carried out before a decision can be taken, it is obvious that N may assume the integers from 0 to r-1 only with probability zero. The probability function of N is consequently given by:

$P(N=r+j)=P(N=r+j,CS)+P(N=r+j,FS)+P(N=r+j,NS)$, where CS,FS,NS means Correct Selection, False Selection, None Selection, respectively. Since the probability $P(N=r+j,NS)$ is equal to zero, we finally obtain

$$P(N<\infty)=\sum_{j=0}^{\infty} P(N=r+j)=\sum_{j=0}^{\infty} P(CS|N=r+j)P(N=r+j)+\sum_{j=0}^{\infty} P(FS|N=r+j)P(N=r+j)$$

$$\text{i.e. } P(N<\infty)=P(CS)+P(FS) \tag{9.2}$$

The above statement is intuitively clear; stopping with probability one means to take a decision with probability one and this can be either a correct selection or a false one. In order to verify that all selection procedures considered so far will terminate with probability one, we have to do nothing but to evaluate the corresponding expressions for P(FS). Because of the same considerations needed to derive the desired formulae, only the main steps and the final results for P(FS) are presented in the following. Details are left as an exercise to the reader.

9.2 The stopping-behaviour of selection model no. 1

Defining $\overline{P}_n$ and $\overline{Q}_n$ by

$$\overline{P}_n:=P(FS|S_A-S_B=n,\ NT=A);\ \overline{Q}_n:=P(FS|S_A-S_B=n,\ NT=B), \tag{9.3}$$

the following system of difference equations and boundary conditions is easily shown to hold:

$$\overline{P}_n=p_A\overline{P}_{n+1}+q_A\overline{Q}_n;\ \overline{Q}_n=p_B\overline{Q}_{n-1}+q_B\overline{P}_n;\ \overline{P}_r=0,\overline{Q}_{-r}=1 \tag{9.4}$$

In the same way as in section 1, a solution of (9.4) is found to be

$$\bar{Q}_n=\lambda^{n+r}-\frac{q_A\lambda^{2r}(1-\lambda^{n+r})}{q_B-q_A\lambda^{2r}};\ \bar{P}_n=\frac{1}{q_B}\cdot\left(q_A\lambda^{n+r}-\frac{q_A\lambda^{2r}(q_B-q_A\lambda^{n+r})}{q_B-q_A\lambda^{2r}}\right) \qquad (9.5)$$

and P(FS) is hence given by

$$P(FS)=\tfrac{1}{2}(\bar{Q}_0+\bar{P}_0)=\frac{1}{2}\ \frac{(q_A+q_B)\lambda^r-2q_A\lambda^{2r}}{q_B-q_A\lambda^{2r}} \qquad (9.6)$$

(1.9) and (9.6) add up to one, i.e. the selection procedure of section 1 terminates with probability one.

9.3 The stopping-behaviour of selection model no. 2

In order to derive P(FS), we define

$$\bar{P}_n:=P(FS|\,S_A-S_B=n), \qquad (9.7)$$

and obtain the following difference equation and boundary conditions:

$$\bar{P}_n=p_Aq_B\bar{P}_{n+1}+q_Ap_B\bar{P}_{n-1}+(p_Ap_B+q_Aq_B)\bar{P}_n;\ \bar{P}_s=0,\bar{P}_{-s}=1 \qquad (9.8)$$

The solution of (9.8) is

$$\bar{P}_n=\frac{\delta^s(\delta^n-\delta^s)}{1-\delta^{2s}}\ ;\ \delta:=\frac{p_Bq_A}{p_Aq_B}\ ;\ P(FS)=P_0=\frac{\delta^s}{1+\delta^s} \qquad (9.9)$$

Using (2.5) and the above result, $P(N<\infty)=1$ is immediately seen to hold.

9.4 The stopping-behaviour of the selection models no. 3 and no. 5

The PL-sampling selection procedure of section 3 terminates with probability one, too. This is directly clear if we make use of the fact that P(CS|PL) and P(FS|PL) may be obtained from the corresponding probabilities of the PW-selection procedure of section 1 by interchanging p_A and q_B as well as p_B and q_A.
The two-stage procedures also terminate with probability one because of a fixed number of trials in the first stage and the use of the PW-, VT- or PL-selection

model, just proved to terminate with probability one, in the second stage.

9.5 The stopping-behaviour of selection model no. 6

To determine P(FS), we next define F_B^* to be the number of B failures preceding the r-th B success and F_A^* to be the number of A failures preceding the r-th B success. Interchanging A and B in (6.2) and (6.4), $F_B^*=F_A^*$ and $F_B^*=F_A^*+1$ are immediately seen to hold. Just in the same way as in section 6, we find:

$$P(FS) = \frac{1}{2}\sum_{j=0}^{\infty}\binom{j+r-1}{j}p_B^r q_B^j \sum_{k=0}^{r-1}\left(\binom{j+k-1}{k}p_A^k q_A^j + \binom{j+k}{k}p_A^k q_A^{j+1}\right) = \frac{1}{2}E(J_{q_A}(X_{r,p_B},r)+J_{q_A}(X_{r,p_B}+1,r)) \qquad (9.10)$$

Making use of A1/7 and A1/12, we finally obtain:

$$P(FS)=1-\frac{1}{2}E(J_{q_B}(X_{r,p_A},r)+J_{q_B}(X_{r,p_A}+1,r)) \qquad (9.11)$$

(6.7) and the above result jointly imply that the selection procedure considered in section 6 will terminate with probability one. The correctness of this fact for the selection procedures given in 8.2 and 8.3 is proved quite analogously.

9.6 The stopping-behaviour of selection model no. 7

P(FS) consists of two terms P_1^* and P_2^*, where P_1^* denotes the probability that treatment B yields its r-th success at first, and P_2^* denotes the probability that both treatments yield their r-th success in the same trial <u>and</u> that treatment B is selected. We obtain

$$P_1^*=\sum_{j=0}^{\infty}\binom{j+r-1}{j}p_B^r q_B^j \sum_{i=0}^{r-1}\binom{j+r}{i}p_A^i q_A^{j+r-i}=E(J_{q_A}(X_{r,p_B}+1,r)), \qquad (9.12)$$

and noting that $P_2^*=P_2$ (cf (7.1)), P(FS) is found to be

$$P(FS)=1-\frac{1}{2}E(J_{q_B}(X_{r,p_A},r)+J_{q_B}(X_{r,p_A}+1,r)) \qquad (9.13)$$

This result and the expression for P(CS), given in (7.2), jointly imply that sampling terminates with probability one.

10 The Selection Model $[k;PW;\max\{S_1,\ldots,S_k\}=r]$

10.1 Introductory remarks

In the following section we will extend the PW-selection procedure of section 6 to $k\geq 3$ treatments. At the outset we put the treatments in a random order. Let A_1 denote the best treatment, i.e. the treatment with the largest success parameter, A_2 the one following A_1 in the initial randomization, etc. (continuing in cyclic order) and let p_i denote the success parameter of treatment A_i, $i\varepsilon\{1,2,\ldots,k\}$. We use the PW-sampling scheme with these ordered treatments, i.e.

$$\begin{array}{ccccccccc} & & & & & \text{best treatment} & & & \\ & & & & & \blacktriangledown & & & \\ 1 & 2 & 3 & \ldots\ldots\ldots & i-1 & i & i+1 & \ldots\ldots\ldots & k \\ A_{k-i+2} & A_{k-i+3} & A_{k-i+4} & \ldots\ldots\ldots & A_k & A_1 & A_2 & \ldots\ldots\ldots & A_{k-i+1} \end{array}$$

we start with that treatment which is the first in the random order, a success generates a further trial on the same treatment and a failure generates a switch to the next treatment in the random order, etc. If the last treatment in the random order has been given to a patient and the response is known to be a failure, then the cycle gets repeated, i.e. we start again with the leading treatment of the random order. Sampling terminates when one of the k treatments yields its r-th success. That treatment is selected as best. Similar to (1.1), we have to specify constants P^* and Δ^* such that

$$P(CS)\geq P^* \text{ whenever } \Delta\geq\Delta^*, \tag{10.1}$$

where Δ ist the true unknown difference between the largest and second largest success parameter. We choose $\Delta^*\varepsilon(0,1)$ and $\frac{1}{k}<P^*<1$, where the last inequality is obvious by noting that $P(CS)=1/k$ may be realized by randomization only. (10.1) is sometimes referred to as $(P^*;\Delta^*)$-condition. In the following we have to determine r such that the $(P^*;\Delta^*)$-condition holds for all possible p-values.

10.2 Derivation of the critical r-value

Let $T:=(T_1,\ldots,T_k)$ denote the additional numbers of successes needed to select treatment no. i as best, $i\varepsilon\{1,2,\ldots,k\}$, and let

$$U_i(n_1,\ldots,n_k):=P(CS|T=(n_1,\ldots,n_k),\ NT=A_i), \tag{10.2}$$

then the following system of difference equations and boundary conditions is quickly derived.

$$\begin{aligned}&U_i(n_1,\ldots,n_k)=p_iU_i(n_1,\ldots,n_{i-1},n_i-1,n_{i+1},\ldots,n_k)+q_iU_{i+1}(n_1,\ldots,n_k),\\&\text{where } U_{k+1}(\ldots)\equiv U_1(\ldots)\\&U_1(0,n_2,\ldots,n_k)=1,\ \text{if } n_j>0 \text{ for all } j\varepsilon\{2,\ldots,k\}\\&U_i(n_1,\ldots,n_{i-1},0,n_{i+1},\ldots,n_k)=0 \text{ if } n_j>0 \text{ for all } j\neq i;\ i>1\end{aligned} \tag{10.3}$$

The exact expression for P(CS) is hence given by

$$P(CS) = \frac{1}{K}\sum_{i=1}^{k} U_i(r,\ldots,r) \tag{10.4}$$

The difference equations in (10.3) are to be solved by using generating functions. For this purpose, we define

$$U_i := \sum_{n_1=1}^{\infty} \cdots \sum_{n_k=1}^{\infty} U_i(n_1,\ldots,n_k)x_1^{n_1}\ldots x_k^{n_k} \tag{10.5}$$

Using this definition in (10.3), we obtain:

(a) i=1

$$\begin{aligned}&U_1=p_1x_1\sum_{n_1=1}^{\infty}\cdots\sum_{n_k=1}^{\infty}U_1(n_1-1,n_2,\ldots,n_k)x_1^{n_1-1}x_2^{n_2}\ldots x_k^{n_k}+q_1U_2\\&U_1(1-p_1x_1)-q_1U_2=p_1x_1\prod_{j=2}^{k}\left(\frac{x_j}{1-x_j}\right)\end{aligned} \tag{10.6}$$

(b) i>1

$U_i(1-p_ix_i)-q_iU_{i+1}=0$, and by iteration follows

$$U_i = \frac{q_i}{1-p_ix_i}U_{i+1};\ U_2=U_1\prod_{j=2}^{k}\left(\frac{q_j}{1-p_jx_j}\right) \tag{10.7}$$

Making use of (10.6), we have

$$U_1(1-p_1x_1)-q_1U_1\prod_{j=2}^{k}\left(\frac{q_j}{1-p_jx_j}\right)=p_1x_1\prod_{j=2}^{k}\left(\frac{x_j}{1-x_j}\right)$$

$$U_1 = \frac{p_1x_1}{G}\prod_{j=2}^{k}\left(\frac{x_j(1-p_jx_j)}{1-x_j}\right),\ \text{where G is defined by} \tag{10.8}$$

$$G:=(1-p_1x_1)\dots(1-p_kx_k)-q_1q_2\dots q_k \tag{10.9}$$

Applying the iteration technique to (10.7), we finally get

$$U_i=\frac{p_1x_1}{G}\prod_{j=2}^{k}\left(\frac{q_jx_j}{1-x_j}\right)\prod_{j=2}^{i-1}\left(\frac{1-p_jx_j}{q_j}\right) \tag{10.10}$$

(Note that $\prod_{j=2}^{\ell}(\cdot):=0$, if $\ell<2$)

To get an explicit expression for the P(CS)-value, we have to evaluate the power series expansion of each U_i and to determine the coefficient of $x_1^r x_2^r \dots x_k^r$. The expansion of 1/G is given by

$$\frac{1}{G}=\frac{1}{\prod_{j=1}^{k}(1-p_jx_j)\left(1-\frac{q_1\dots q_k}{\prod_{j=1}^{k}(1-p_jx_j)}\right)}=\sum_{\ell=0}^{\infty}\frac{(q_1\dots q_k)^{\ell}}{\left(\prod_{j=1}^{k}(1-p_jx_j)\right)^{\ell+1}}=$$

$$=\sum_{\ell=0}^{\infty}\left(\prod_{j=1}^{k}\left(\sum_{n_j=0}^{\infty}(p_jx_j)^{n_j}\right)^{\ell+1}\right)(q_1\dots q_k)^{\ell}= \tag{10.11}$$

$$=\sum_{n_1=0}^{\infty}\dots\sum_{n_k=0}^{\infty}\left(\sum_{\ell=0}^{\infty}\binom{n_1+\ell}{n_1}\dots\binom{n_k+\ell}{n_k}p_1^{n_1}\dots p_k^{n_k}(q_1\dots q_k)^{\ell}\right)x_1^{n_1}\dots x_k^{n_k}$$

U_1, given by (10.8), is a product of four factors, that is

$$U_1=\underbrace{\frac{1}{G}\,p_1x_1\dots x_k}_{=:C}\underbrace{\left(\prod_{j=2}^{k}\left(\frac{1}{1-x_j}\right)\right)}_{=:A}\underbrace{\left(\prod_{j=2}^{k}(1-p_jx_j)\right)}_{=:B}, \tag{10.12}$$

and the expansions of A and B are as follows:

$$A=\sum_{\alpha_2=0}^{\infty}\dots\sum_{\alpha_k=0}^{\infty}x_2^{\alpha_2}\dots x_k^{\alpha_k}$$

$$B=\sum_{\beta_2=0}^{1}\dots\sum_{\beta_k=0}^{1}(-1)^{\beta_2+\dots+\beta_k}p_2^{\beta_2}\dots p_k^{\beta_k}x_2^{\beta_2}\dots x_k^{\beta_k} \tag{10.13}$$

Multiplying C with the expansions of A,B and 1/G and using A1/8, the coefficient of $x_1^r \ldots x_k^r$ in the resulting power series is found to be

$$U_1(r,\ldots,r)=\sum_{\ell=0}^{\infty}\left(\binom{r+\ell-1}{\ell}p_1^r q_1^{\ell}\prod_{j=2}^{k}J_{q_j}(\ell,r)\right) \tag{10.14}$$

The desired formula for $U_i(r,\ldots,r)$ in case of $i\geq 2$ is derived in the same way. Details are omitted here but are left as an exercise to the reader. We get

$$U_i(r,\ldots,r)=\sum_{\ell=0}^{\infty}\left(\binom{r+\ell-1}{\ell}p_1^r q_1^{\ell}\prod_{j=2}^{i-1}J_{q_j}(\ell,r)\prod_{j=i}^{k}J_{q_j}(\ell+1,r)\right) \tag{10.15}$$

P(CS) is hence given by

$$P(CS)=\frac{1}{k}E\left(\prod_{j=2}^{k}J_{q_j}(X_{r,p_1},r)+\sum_{i=2}^{k}\left(\prod_{j=2}^{i-1}J_{q_j}(X_{r,p_1},r)\right)\left(\prod_{j=i}^{k}J_{q_j}(X_{r,p_1}+1,r)\right)\right) \tag{10.16}$$

where X_{r,p_1} denotes a negative binomial chance variable with index r and success parameter p_1 (see e.g. A1/6).

To calculate the P(CS)-values exact up to a correction term of at most ε, we must determine upper and lower bounds for the expression given in (10.16) (cf section 6, (6.8) and (6.9)). Noting A1/4, it is immediately seen that we get a minimal P(CS)-value if each q_j, $j\geq 2$, is replaced by the second smallest "failure probability", that is q_2^*. - Note that q_2^* will be usually different from q_2, where q_2 denotes the "failure probability" of the treatment following A_1 in the initial randomization. - The P(CS)-value, therefore, is calculated for the parameter values p_1 and $p_j=p_2^*$; $j\geq 2$; only. A lower bound for P(CS) is hence given by

$$U(\alpha):=\frac{1}{k}\sum_{\ell=0}^{\alpha}\binom{r+\ell-1}{\ell}p_1^r q_1^{\ell}((J_{q_2^*}(\ell,r))^{k-1}+\sum_{i=2}^{k}(J_{q_2^*}(\ell,r))^{i-2}(J_{q_2^*}(\ell+1,r))^{k-i+1}) \tag{10.17}$$

Recalling that $J_q(\ell,r)$ is strictly decreasing in ℓ, provided r is fixed,

$$S(\alpha):=U(\alpha)+(J_{q_2^*}(\alpha+1,r))^{k-1}J_{q_1}(\alpha+1,r) \tag{10.18}$$

is found to be an upper bound for P(CS). Just as in section 6,$\{[U(\alpha),S(\alpha)]\,|\,\alpha\in\mathbb{N}\}$ is shown to be a system of nested intervals for the P(CS)-value, specified by $p_j=p_2^*$ if $j\geq 2$. The terms of the infinite series in (10.16) are added up as long as

$(J_{q_2^*}(\alpha+1,r))^{k-1}J_{q_1}(\alpha+1,r)$ exceeds the preassigned ε. A rough estimate of α may be derived in the same way as demonstrated in section 6 (cf (6.10)).
In the next section we deal with a selection procedure using the same stopping rule as the procedure just regarded and the VT-sampling scheme. It will be pointed out there that both selection procedures need the same critical r-value. That is why the determination of r_{max} is dropped here.

10.3 Derivation of the expectations

Let N_i denote the number of patients on treatment no. i, $i\varepsilon\{1,2,\ldots,k\}$. In order to get explicit expressions for the expectations $E(N_i)$, we next define

$$R_{ij}(n_1,\ldots,n_k):=E(N_i | T=(n_1,\ldots,n_k),\ NT=A_j);\ 1\leq i,\ j\leq k;\ A_{k+1}\equiv A_1 \tag{10.19}$$

$R_{ij}(n_1,\ldots,n_k)$ denotes the conditional expectation of N_i if $n_1,\ldots,n_k$ additional successes are needed to select $A_1,\ldots,A_k$ as best, respectively, and the next trial will be carried out with treatment A_j. From (10.19) the following system of difference equations and boundary conditions is quickly derived.

$$R_{ij}(n_1,\ldots,n_k)=p_jR_{ij}(n_1,\ldots,n_{j-1},n_j-1,n_{j+1},\ldots,n_k)+$$
$$+q_jR_{i,j+1}(n_1,\ldots,n_k)+\delta_{ij},\ \text{where } \delta_{ij}=1 \text{ for } i=j \tag{10.20}$$
$$\text{and zero otherwise, } R_{i,k+1}\equiv R_{i1}$$
$$R_{ij}(n_1,\ldots,n_{j-1},0,n_{j+1},\ldots,n_k)=0,\ \text{if } n_\lambda>0 \text{ for } \lambda\neq j$$

The term δ_{ij} is intuitively obvious, that is, for i=j a further trial is carried out on treatment A_i and, therefore, we have to add 1 to $E(N_i)$; for $i\neq j$ a trial is carried out on treatment A_j that has no influence over $E(N_i)$. The desired formula for the expectations are

$$E(N_i)=\frac{1}{k}\sum_{j=1}^{k}R_{ij}(r,\ldots,r);\ E(N)=\sum_{i=1}^{k}E(N_i) \tag{10.21}$$

Using the generating functions

$$R_{ij}:=\sum_{n_1=1}^{\infty}\ldots\sum_{n_k=1}^{\infty}R_{ij}(n_1,\ldots,n_k)x_1^{n_1}\ldots x_k^{n_k}, \tag{10.22}$$

we derive from (10.20):

$$R_{ii}(1-p_i x_i)=q_i R_{i,i+1}+\prod_{\ell=1}^{k}\left(\frac{x_\ell}{1-x_\ell}\right)\ ;\ R_{ij}=\frac{q_j}{1-p_j x_j}R_{i,j+1},\ \text{if } i\neq j \tag{10.23}$$

By iteration, we obtain:

$$R_{ij}=R_{ii}\prod_{\ell=j}^{k}\left(\frac{q_\ell}{1-p_\ell x_\ell}\right)\prod_{\ell=1}^{i-1}\left(\frac{q_\ell}{1-p_\ell x_\ell}\right),\ \text{if } i<j \tag{10.24}$$

$$R_{ij}=R_{ii}\prod_{\ell=j}^{i-1}\left(\frac{q_\ell}{1-p_\ell x_\ell}\right),\ \text{if } i>j \tag{10.25}$$

$$R_{ii}=\frac{1}{G}\prod_{\substack{\ell=1\\ \ell\neq i}}^{k}(1-p_\ell x_\ell)\prod_{\ell=1}^{k}\left(\frac{x_\ell}{1-x_\ell}\right),\ \text{where G denotes} \tag{10.26}$$

the expression given by (10.9). The next step consists in the evaluation of the powerseries expansions of the R_{ij}. Because of the similar analysis as before, details are omitted here. The desired formulae are found to be

$$R_{ij}(r,\ldots,r)=\frac{1}{q_i}\sum_{\ell=0}^{\infty}J_{q_i}(\ell+1,r)\left(\prod_{\lambda=j}^{i-1}J_{q_\lambda}(\ell+1,r)\right)\left(\prod_{\substack{\lambda<j\\ \lambda>i}}J_{q_\lambda}(\ell,r)\right), \tag{10.27}$$

provided that $j\leq i$

$$R_{ij}(r,\ldots,r)=\frac{1}{q_i}\sum_{\ell=0}^{\infty}J_{q_i}(\ell+1,r)\left(\prod_{\substack{\lambda<i\\ \lambda\geq j}}J_{q_\lambda}(\ell+1,r)\right)\left(\prod_{\lambda=i+1}^{j-1}J_{q_\lambda}(\ell,r)\right), \tag{10.28}$$

provided that $j>i$

The expectations are now available by using (10.21) and the above formulae.

Setting in (1.41) and (1.42) all incomplete beta functions after the first one equal to unity and using A1/4, we obtain:

$$R_{ij}(r,\ldots,r)\leq\frac{1}{q_i}\sum_{\ell=0}^{\infty}J_{q_i}(\ell+1,r)=\frac{1}{q_i}\sum_{\ell=0}^{\infty}\frac{1}{B(\ell+1,r)}\int_0^{q_i}x^\ell(1-x)^{r-1}dx=$$

$$=\frac{r}{q_i}\int_0^{q_i}(1-x)^{r-1}\sum_{\ell=0}^{\infty}\binom{-(r+1)}{\ell}(-1)^\ell x^\ell dx=\frac{r}{q_i}\int_0^{q_i}\frac{dx}{(1-x)^2}=\frac{r}{p_i} \tag{10.29}$$

Upper bounds for the expectations are hence given by

$$E(N_i) \leq \frac{1}{k} \sum_{j=1}^{k} \frac{r}{p_i} = \frac{r}{p_i} \; ; \; E(N) \leq r \sum_{i=1}^{k} \frac{1}{p_i} \,, \tag{10.30}$$

i.e. the expectations are finite if the p_i are all bounded away from zero.

Following (10.16), we emphazised that in the LFC all p_j, $j \geq 2$, are equal to p_2^*. The expectations are calculated, therefore, only for that special configuration of parameters. Lower and upper bounds for the expectations are found to be

<u>i=1</u>

$$U(\alpha;N_1) := \frac{1}{kq_1} \sum_{j=2}^{k} \sum_{\ell=0}^{\alpha} J_{q_1}(\ell+1,r)(J_{q_2^*}(\ell+1,r))^{k-j+1}(J_{q_2^*}(\ell,r))^{j-2} +$$

$$+ \frac{1}{kq_1} \sum_{\ell=0}^{\alpha} J_{q_1}(\ell+1,r)(J_{q_2^*}(\ell,r))^{k-1} \tag{10.31}$$

$$S(\alpha;N_1) := U(\alpha;N_1) + \frac{r}{p_1}(J_{q_2^*}(\alpha+1,r))^{k-1}$$

<u>i>1</u>

$$U(\alpha;N_i) := \frac{1}{kq_2^*} \sum_{\ell=0}^{\alpha} J_{q_2^*}(\ell+1,r) J_{q_1}(\ell+1,r)(J_{q_2^*}(\ell+1,r))^{i-2}(J_{q_2^*}(\ell,r))^{k-i} +$$

$$+ \frac{1}{kq_2^*} \sum_{j=2}^{i} \sum_{\ell=0}^{\alpha} J_{q_2^*}(\ell+1,r)(J_{q_2^*}(\ell+1,r))^{i-j}(J_{q_2^*}(\ell,r))^{k+j-i-2} J_{q_1}(\ell,r) + \tag{10.32}$$

$$+ \frac{1}{kq_2^*} \sum_{j=i+1}^{k} \sum_{\ell=0}^{\alpha} J_{q_2^*}(\ell+1,r) J_{q_1}(\ell+1,r)(J_{q_2^*}(\ell+1,r))^{k+i-j-1}(J_{q_2^*}(\ell,r))^{j-i-1}$$

$$S(\alpha;N_i) := U(\alpha;N_i) + \frac{q_1 r}{q_2^* p_1} (J_{q_2^*}(\alpha+1,r))^{k-1}$$

$\{[U(\alpha;N_i),S(\alpha;N_i)] \,|\, \alpha \in \mathbb{N}\}$ is a system of nested intervals for $E(N_i)$; $i \in \{1,2,\ldots,k\}$. We will not check the truth of this because the same analysis as before is required (see e.g. (6.8),(6.9)). Rough estimates for α can be derived in the usual manner.

10.4 Expectations for large r

Noting A1/10 if r is large enough, $E(N_i)$ is approximately given by

$$E(N_i) = \frac{1}{q_i} \sum_{\ell=0}^{\infty} \prod_{\lambda=1}^{k} J_{q_\lambda}(\ell+1,r) =: \frac{1}{q_i} R; \; i \varepsilon \{1,2,\dots,k\} \tag{10.33}$$

Using the second equality in A1/8, we find: (we emphasize that R is independent of i)

$$\begin{aligned} R &= \sum_{\ell=0}^{k} \sum_{\tau=\ell+1}^{\infty} \binom{r+\tau-1}{\tau} p_1^r q_1^\tau \prod_{\lambda=2}^{k} J_{q_\lambda}(\ell+1,r) = \\ &= \sum_{\tau=0}^{\infty} \sum_{\ell=0}^{\tau-1} \binom{r+\tau-1}{\tau} p_1^r q_1^\tau \prod_{\lambda=2}^{k} (1-J_{p_\lambda}(r,\ell+1)) = \\ &= \frac{rq_1}{p_1} + \sum_{\tau=0}^{\infty} \binom{r+\tau-1}{\tau} p_1^r q_1^\tau \sum_{\ell=0}^{\tau-1} \sum_{n=1}^{k-1} \sum_{\beta_2+\dots+\beta_k=n} (-1)^n \prod_{\lambda=2}^{k} (J_{p_\lambda}(r,\ell+1))^{\beta_\lambda}, \end{aligned} \tag{10.34}$$

where $\beta_2,\dots,\beta_k \varepsilon \{0,1\}$.

Making use of $J_p(\cdot,\cdot) \leq 1$ and applying A1/11, we obtain from (10.34):

$$\begin{aligned} \left|R - \frac{rq_1}{p_1}\right| &\leq \sum_{\tau=0}^{\infty} \binom{r+\tau-1}{\tau} p_1^r q_1^\tau \left(\sum_{\lambda=2}^{k} \sum_{\ell=0}^{\tau-1} J_{p_\lambda}(r,\ell+1) + \sum_{n=2}^{k-1} \binom{k-2}{n-1} \cdot \right. \\ &\left. \cdot \sum_{\lambda=2}^{k} \sum_{\ell=0}^{\tau-1} J_{p_\lambda}(r,\ell+1) \right) = \sum_{\tau=0}^{\infty} \binom{r+\tau-1}{\tau} p_1^r q_1^\tau 2^{k-2} \sum_{\lambda=2}^{k} \sum_{\ell=0}^{\tau-1} J_{p_\lambda}(r,\ell+1) = \\ &= 2^{k-2} \sum_{\lambda=2}^{k} \sum_{\tau=0}^{\infty} \binom{r+\tau-1}{\tau} p_1^r q_1^\tau \left((r+\tau) J_{p_\lambda}(r,\tau) - \frac{r}{p_\lambda} J_{p_\lambda}(r+1,\tau) \right) = \\ &= 2^{k-2} \sum_{\lambda=2}^{k} \left(\frac{r}{p_1} E(J_{p_\lambda}(r, X_{r+1,p_1})) - \frac{r}{p_\lambda} E(J_{p_\lambda}(r+1, X_{r,p_1})) \right) \end{aligned} \tag{10.35}$$

Because of A1/13, the right hand side of the above inequality tends to 0 as r becomes large. The expectations are hence approximately

$$E(N_i) \approx \frac{rq_1}{p_1} \cdot \frac{1}{q_i} \; ; \; E(N) \approx \frac{rq_1}{p_1} \sum_{i=1}^{k} \frac{1}{q_i} \tag{10.36}$$

For k=2, (10.36) reduces to the formulae given in (6.35) with the exception of the term $1/2q_B$.

11 The Selection Model $[k;VT;\max\{S_1,\dots,S_k\}=r]$

11.1 Derivation of the critical r-value

The following procedure uses VT-sampling and the termination rule $\max\{S_1,\dots,S_k\}=r$; i.e. all treatments are given at each stage of experimentation and sampling terminates if at least one treatment yields its r-th success. The stopping rule used here is sometimes referred to as inverse sampling. The treatment yielding the r-th success at first will be selected as best. Ties are decided by randomization, i.e. each of the treatments that reach the r-th success at the final stage has the same probability to be selected as best. The exact expression for P(CS) is found to be

$$
\begin{aligned}
P(CS)=&\sum_{j=0}^{\infty}\left(\binom{j+r-1}{j}p_1^r q_1^j\prod_{\lambda=2}^{k}\left(\sum_{\tau=0}^{r-1}\binom{j+r}{\tau}p_\lambda^\tau q_\lambda^{j+r-\tau}\right)\right)+\\
&+\sum_{\ell=1}^{k-1}\sum_{\omega\in S_\ell}\sum_{j=0}^{\infty}\left(\frac{1}{\ell+1}\binom{j+r-1}{j}p_1^r q_1^j\prod_{\lambda=1}^{\ell}\left(\binom{j+r-1}{j}p_{i_\lambda}^r q_{i_\lambda}^j\right)\cdot\right.\\
&\left.\cdot\prod_{\lambda=1}^{k-\ell-1}\left(\sum_{\tau=0}^{r-1}\binom{j+r}{\tau}p_{j_\lambda}^\tau q_{j_\lambda}^{j+r-\tau}\right)\right)=E\left(\prod_{\lambda=2}^{k}J_{q_\lambda}(X_{r,p_1}+1,r)\right)+\\
&+\sum_{\ell=1}^{k-1}\sum_{\omega\in S_\ell}\frac{1}{\ell+1}E\left(\prod_{\lambda=1}^{\ell}\left(J_{q_{i_\lambda}}(X_{r,p_1},r)-J_{q_{i_\lambda}}(X_{r,p_1}+1,r)\right)\prod_{\lambda=1}^{k-\ell-1}J_{q_{j_\lambda}}(X_{r,p_1}+1,r)\right),
\end{aligned}
\tag{11.1}
$$

where S_ℓ is defined as follows:

$$
\begin{aligned}
S_\ell:=\{&\omega=(\omega_1,\omega_2)\,|\,\omega_1=(i_1,\dots,i_\ell),\omega_2=(j_1,\dots,j_{k-\ell-1});\ i_1<\dots<i_\ell;\\
&j_1<\dots<j_{k-\ell-1};\ i_1,\dots,i_\ell,j_1,\dots,j_{k-\ell-1}\in\{2,\dots,k\}\text{ and}\\
&\{i_1,\dots,i_\ell\}\cap\{j_1,\dots,j_{k-\ell-1}\}=\emptyset\}
\end{aligned}
\tag{11.2}
$$

We now multiply out the product of differences in (11.1) and combine the two series to obtain for P(CS)

$$
\begin{aligned}
P(CS)=&\sum_{j=0}^{\infty}\binom{j+r-1}{j}p_1^r q_1^j\sum_{\ell=0}^{k-1}\sum_{\omega\in S_\ell}\sum_{\nu=0}^{\ell}\sum_{\beta_1+\dots+\beta_\ell=\nu}\left(\frac{1}{\ell+1}(-1)^\nu\cdot\right.\\
&\cdot\prod_{\tau=1}^{\ell}(J_{q_{i_\tau}}(j+1,r))^{\beta_\tau}(J_{q_{i_\tau}}(j,r))^{1-\beta_\tau}\prod_{\lambda=1}^{k-\ell-1}J_{q_{j_\lambda}}(j+1,r)\Big),
\end{aligned}
\tag{11.3}
$$

where β_τ equals either 0 or 1.
We next combine all terms that are associated with the factor

$$\prod_{\lambda=1}^{k-1-h} J_{q_{i_\lambda}}(j,r) \prod_{\lambda=1}^{h} J_{q_{j_\lambda}}(j+1,r), \quad h:=k-\ell-1 \tag{11.4}$$

Holding ν of the h components of the second product in (11.4) fixed, $\binom{h}{h-\nu}=\binom{h}{\nu}$ elements of $S_{k-1-\nu}$ may be shown to generate the same factor. The contribution of those elements is $(-1)^{h-\nu}\binom{h}{\nu}\frac{1}{k-\nu}$. Noting this reordering of terms and using A3/2, P(CS) is found to be

$$P(CS)=\sum_{j=0}^{\infty}\left(\binom{j+r-1}{j}p_1^r q_1^j \sum_{h=0}^{k-1}\left(\sum_{\nu=0}^{h}(-1)^{h-\nu}\binom{h}{\nu}\frac{1}{k-\nu}\right)\cdot\right.$$

$$\left.\cdot\sum_{\omega\varepsilon S_{k-1-h}}\sum_{\lambda=1}^{k-1-h} J_{q_{i_\lambda}}(j,r)\prod_{\lambda=1}^{h} J_{q_{j_\lambda}}(j+1,r)\right) = \tag{11.5}$$

$$=E\left(\sum_{h=0}^{k-1}\frac{1}{k\binom{k-1}{h}}\sum_{\omega\varepsilon S_{k-1-h}}\prod_{\lambda=1}^{k-1-h} J_{q_{i_\lambda}}(X_{r,p_1},r)\prod_{\lambda=1}^{h} J_{q_{j_\lambda}}(X_{r,p_1}+1,r)\right)$$

It is immediately obvious that we get a minimal P(CS) value if all q_j for $j\geq 2$ are replaced by the second smallest failure probability q_2^*. Knowing in addition that S_{k-1-h} consists of $\binom{k-1}{h}$ elements, we have

$$P(CS|q_j=q_2^*;j\geq 2)=\frac{1}{k}E\left(\sum_{h=0}^{k-1}(J_{q_2^*}(X_{r,p_1},r))^{k-1-h}(J_{q_2^*}(X_{r,p_1}+1,r))^h\right)=$$

$$=\frac{1}{k}E\left(\frac{(J_{q_2^*}(X_{r,p_1},r))^k-(J_{q_2^*}(X_{r,p_1}+1,r))^k}{J_{q_2^*}(X_{r,p_1},r)-J_{q_2^*}(X_{r,p_1}+1,r)}\right) \tag{11.6}$$

An elementary analysis shows that the P(CS)-value of the preceding selection procedure, given by (10.16), and that one of the selection procedure to be considered in this section coincide if $q_j=q_2^*$ for all $j\geq 2$. Moreover, the P(CS)-values of both procedures can be proved to coincide for every configuration of parameters. This result, due to D.A.Berry and D.H. Young [34] , is by no means surprising but has been expected on the contrary because of a similar result in case of two treatments (cf. (7.2)). To determine the critical r-value, we only need the fact that both procedures require the same r_{max} in their LFC, and this is implied by that which has been pointed out above.

Let $F_i(r)$ denote the number of A_i-failures preceding the r-th A_i-success. Then P(CS) can be expressed as follows

$$P(CS)=P(F_1(r)<F_2(r),\dots,F_1(r)<F_k(r))+ \tag{11.7}$$

$$+\sum_{\ell=1}^{k-1}\sum_{\omega\varepsilon S_\ell}\frac{1}{\ell+1}P(F_1(r)=F_{i_1}(r)=\dots=F_{i_\ell}(r),F_1(r)<F_{j_1}(r),\dots,F_1(r)<F_{j_{k-\ell-1}}(r))$$

Applying the central limit theorem to the negative binomial distributed random variables $F_i(r)$, we find the approximation

$$P(CS)\approx P(F_1(r)<F_2(r),\dots,F_1(r)<F_k(r)) \tag{11.8}$$

Standardizing the random variables $F_i(r)$ for all $i\geq 2$, we obtain

$$P(CS)\approx P\left(\underbrace{\frac{F_\lambda(r)-rq_\lambda/p_\lambda}{\sqrt{rq_\lambda}\;/p_\lambda}}_{=:Y_{\lambda r}}>\underbrace{\frac{F_1(r)-rq_1/p_1}{\sqrt{rq_1}\;/p_1}}_{=:Y_{1r}}\cdot\frac{p_\lambda}{p_1}\sqrt{\frac{q_1}{q_\lambda}}-\frac{\Delta_{1\lambda}}{p_1}\sqrt{\frac{r}{q_\lambda}}\;;\lambda\varepsilon\{2,.,k\}\right), \tag{11.9}$$

where $\Delta_{1\lambda}:=p_1-p_\lambda$.

The first possibility to get the desired r_{max}-value arises if we take no notice of the dependence between $Z_{\alpha r}$ and $Z_{\beta r}$, where $Z_{\lambda r}:=Y_{1r}\frac{p_\lambda}{p_1}\sqrt{\frac{q_1}{q_\lambda}}-Y_{\lambda r}$; $\alpha,\beta\varepsilon\{2,\dots,k\}$; $\alpha\neq\beta$. (11.9) then reduces to

$$P(CS)\approx\prod_{\lambda=2}^{k}P\left(Z_{\lambda r}<\frac{\Delta_{1\lambda}}{p_1}\sqrt{\frac{r}{q_\lambda}}\right) \tag{11.10}$$

Applying the central limit theorem to each factor of the product in (11.10) and noting that $E(Z_{\lambda r})=0$ and $Var(Z_{\lambda r})=(p_1^2q_\lambda+q_1p_\lambda^2)/p_1^2q_\lambda$, P(CS) is found to be

$$P(CS)\approx\prod_{\lambda=2}^{k}\Phi\left(\frac{\Delta_{1\lambda}\sqrt{r}}{\sqrt{p_1^2q_\lambda+q_1p_\lambda^2}}\right) \tag{11.11}$$

Making now use of the beforementioned fact that in the LFC $q_j=q_2^*$ for all $j\geq 2$, and carrying out the same minimization as in section 7, we obtain

$$P(CS) \approx \left(\Phi\left(\frac{\Delta^*\sqrt{r}}{\sqrt{p_1^2 q_2^* + q_1 p_2^{*2}}}\right)\right)^{k-1} ; \min_{p_1\varepsilon(\Delta^*,1)} P(CS) \approx \left(\Phi\left(\Delta^*\sqrt{\frac{27r}{8}}\right)\right)^{k-1} \qquad (11.12)$$

the critical r-value is hence given by

$$r_{max} = \left]\frac{8}{27}\left(\frac{\lambda_k(P^*)}{\Delta^*}\right)^2\right[\quad \text{, where} \qquad (11.13)$$

$\lambda_k(P^*) := \Phi^{-1}\left(\sqrt[k-1]{P^*}\right) = \Delta^*\sqrt{\frac{27r}{8}}$. In (7.8), $\lambda_2(P^*)$ has been denoted by $\lambda(P^*)$.

Another possibility to determine r_{max} arises if we take into consideration that the random variables $Z_{\alpha r}$ and $Z_{\beta r}$ are not independent in general; $(\alpha\neq\beta)$. (11.9) can be rewritten as follows:

$$P(CS) \approx \int_{-\infty}^{+\infty} P\left(Y_{\lambda r} > Y_{1r}\frac{p_\lambda}{p_1}\sqrt{\frac{q_1}{q_\lambda}} - \frac{\Delta_{1\lambda}}{p_1}\sqrt{\frac{r}{q_\lambda}} ; \lambda\varepsilon\{2,\dots,k\} \,\middle|\, Y_{1r}=y_1\right) dV_{1r}(y_1) =$$

$$= \int_{-\infty}^{+\infty} \prod_{\lambda=2}^{k}\left(1 - V_{\lambda r}\left(y_1\frac{p_\lambda}{p_1}\sqrt{\frac{q_1}{q_\lambda}} - \frac{\Delta_{1\lambda}}{p_1}\sqrt{\frac{r}{q_\lambda}}\right)\right) dV_{1r}(y_1), \text{ where} \qquad (11.14)$$

$V_{\lambda r}$ denotes the distribution function of $Y_{\lambda r}$; $\lambda\varepsilon\{1,\dots,k\}$. Noting that each $Y_{\lambda r}$ has a limiting distribution which is normal with mean zero and variance one and applying A2/4 to (11.14), we obtain

$$P(CS) \approx \int_{-\infty}^{+\infty} \prod_{\lambda=2}^{k} \Phi\left(\frac{y_1 p_\lambda\sqrt{q_1} + \Delta_{1\lambda}\sqrt{r}}{p_1\sqrt{q_\lambda}}\right) d\Phi(y_1), \qquad (11.15)$$

and from this formula follows immediately that $\Delta_{1\lambda}=\Delta^*$ holds for each $\lambda\varepsilon\{2,\dots,k\}$ if the LFC is reached. Noting in addition that all success parameters p_j, $j\geq 2$, are equal to p_2^* in the LFC, we finally get

$$P(CS) \approx \int_{-\infty}^{+\infty} \left(\Phi\left(\frac{y_1 p_2^*\sqrt{q_1} + \Delta^*\sqrt{r}}{p_1\sqrt{q_2^*}}\right)\right)^{k-1} d\Phi(y_1), \qquad (11.16)$$

Using the abbreviations

$$D^* := \frac{\Delta^*\sqrt{r}}{\sqrt{q_1 p_2^{*2} + q_2^* p_1^2}} \quad \text{and} \quad \rho := \frac{q_1 p_2^{*2}}{\sqrt{q_1 p_2^{*2} + q_2^* p_1^2}} \quad \text{, where} \qquad (11.17)$$

ρ is the correlation coefficient between $F_1(r)-F_\lambda(r)$ and $F_1(r)-F_\tau(r)$; $\lambda,\tau\varepsilon\{2,\ldots,k\}$; if $F_\lambda(r)$ and $F_\tau(r)$ have the success parameter p_2^*(cf. A1/14), (11.17) can be rewritten as

$$P(CS) \approx \int_{-\infty}^{+\infty} \left(\Phi\left(\frac{y\sqrt{\rho}+D^*}{\sqrt{1-\rho}}\right)\right)^{k-1} d\Phi(y) =: A_{k-1}(\rho,D^*) \qquad (11.18)$$

Disregarding the fact that ρ is varying if p_1 is varying, an approximate minimum of P(CS) is obtained by minimizing D^* in (11.18). Differentiating $D^*(p_1)$ with respect to p_1, the minimum is given for $p_1 = \frac{2}{3} + \frac{\Delta^*}{2} + O(\Delta^{*2})$, and the corresponding D^*-and ρ-values are

$$D^*\left(\frac{2}{3} + \frac{\Delta^*}{2} + O(\Delta^{*2})\right) \approx \Delta^* \sqrt{\frac{27r}{8}} \; ; \; \rho = \frac{1}{2} - \frac{3}{2}\Delta^* + O(\Delta^{*2}) \qquad (11.19)$$

(11.19) indicates that for $\rho = \frac{1}{2}$ we get a first approximation of the minimal value of P(CS). Using the tables of S.S.Gupta [77] , the value of D^*, satisfying the equation

$$(P(CS)\approx)A_{k-1}\left(\tfrac{1}{2},D^*\right) = P^*, \qquad (11.20)$$

can be obtained; that special value of D^* will be denoted by λ^*. Setting now λ^* equal to the minimal D^* given in (11.19), the critical r_{max}-value is found to be (approximately)

$$r_{max} = \left]\frac{8}{27}\left(\frac{\lambda^*}{\Delta^*}\right)^2\right[\qquad (11.21)$$

To get a better approximation, we expand the right side of (11.18) about $\rho = \frac{1}{2}$. Applying A3/3, we obtain

$$\begin{aligned} P(CS) &\approx A_{k-1}\left(\tfrac{1}{2},D^*\right) + \left(\rho - \tfrac{1}{2}\right)\cdot\left[\frac{d}{d\rho} A_{k-1}(\rho,D^*)\right]_{\rho=\frac{1}{2}} \approx \\ &\approx A_{k-1}\left(\tfrac{1}{2},D^*\right) - \frac{3}{2}\Delta^* \frac{(k-1)(k-2)}{\sqrt{6}} \varphi(D^*)\varphi\left(\frac{D^*}{\sqrt{3}}\right)A_{k-3}\left(\frac{1}{4},\frac{D^*}{\sqrt{6}}\right), \end{aligned} \qquad (11.22)$$

where φ denotes the standard normal density function. The critical r_{max}-value is now determined as follows

(a) The equation $A_{k-1}(\frac{1}{2},D^*)=P^*$ is solved for D^* by using the before mentioned tables of S.S.Gupta.

(b) P(CS) given in (11.22) will be less than P^* for that special D^* because the correction term is subtracted from $A_{k-1}(\frac{1}{2},D^*)$. Noting that $A_{k-1}(\rho,D^*)$ is increasing in D^*, we have to enlarge D^* as long as P(CS), given in (11.22), exceeds P^* for the first time. The D^* found by this method will be denoted by λ^*. Setting λ^* equal to $\Delta^*\sqrt{\frac{27r}{8}}$, the critical r-value is seen to be

$$r_{max} = \left]\frac{8}{27}\left(\frac{\lambda^*}{\Delta^*}\right)^2\right[\tag{11.23}$$

Note that the λ^* in the above formula and that one in (11.21) have different meanings.

11.2 Derivation of the expectations

Similar as in (6.15), we introduce random variables $N_{\lambda,i}$; $\lambda, i\varepsilon\{1,\ldots,k\}$ by

$$N_{\lambda,i}(\omega_S) := \begin{cases} \ell(\omega_S), & \text{if treatment } A_i \text{ is selected} \\ 0, & \text{otherwise,} \end{cases} \tag{11.24}$$

where $\ell(\omega_S)$ denotes the length of the stopping-sequence $\omega_S \varepsilon \Omega_S$; (cf. section 9), and the first index λ indicates that treatment no.λ is just considered. The whole number of patients on treatment A_i is denoted by N_i. Hence N_1 is given by

$$N_1 = N_{1,1} + N_{1,2} + \ldots + N_{1,k} \tag{11.25}$$

An exact expression for $E(N_1)$ can now be derived by using (11.5); we find

$$E(N_{1,1}) = \frac{r}{p_1}\, E\left(\sum_{h=0}^{k-1} \frac{1}{k\binom{k-1}{h}} \sum_{\omega\varepsilon S_{k-1-h}} \prod_{\lambda=1}^{k-1-h} J_{q_{i_\lambda}}(X_{r+1,p_1}, r) \prod_{\lambda=1}^{h} J_{q_{j_\lambda}}(X_{r+1,p_1}+1, r)\right) \tag{11.26}$$

The other expectations are obtained by applying the well-known interchanging technique to $E(N_{1,1})$, i.e.

$$N_{1,i} = N_{i,1}(p_1 \leftrightarrows p_i);\ i\varepsilon\{1,2,\ldots,k\}, \tag{11.27}$$

(cf. (6.18)), and recalling that the VT-sampling scheme is used, we have $N_{i,1} = N_{1,1}$ for each $i\varepsilon\{1,2,\ldots,k\}$, and this implies

$$N_{1,i} = N_{1,1}(p_1 \leftrightarrows p_i);\ i\varepsilon\{1,2,\ldots,k\}. \tag{11.28}$$

Denoting $E(N_{1,1})$ by $E(p_1,p_2,\ldots,p_k)$, then $E(N_1)$ can be calculated from

$$E(N_1) = E(p_1,p_2,\ldots,p_k) + E(p_2,p_1,p_3,\ldots,p_k) + \ldots + E(p_k,p_2,p_3,\ldots,p_{k-1},p_1), \tag{11.29}$$

and $E(N)$ is just the k-fold of $E(N_1)$.

We have already mentioned that all p_j are equal to p_2^*, $j\varepsilon\{2,\dots,k\}$ in the LFC. (11.26) thus reduces to

$$E(N_{1,1}) = \frac{r}{p_1} E\Big(\frac{1}{k}\sum_{h=0}^{k-1} (J_{q_2^*}(X_{r+1,p_1},r))^{k-1-h}(J_{q_2^*}(X_{r+1,p_1}+1,r))^h\Big) \tag{11.30}$$

A straightforward calculation shows that

$$\begin{aligned} U(\alpha;N_{1,1}) &:= \frac{r}{p_1}\sum_{j=0}^{\alpha}\binom{j+r-1}{j} p_1^r q_1^j (J_{q_2^*}(j+1,r))^{k-1} \text{ and} \\ S(\alpha;N_{1,1}) &:= U(\alpha;N_{1,1}) + \frac{r}{p_1}(J_{q_2^*}(\alpha+1,r))^{k-1}J_{q_1}(\alpha+1,r) \end{aligned} \tag{11.31}$$

are a lower and upper bound for $E(N_{1,1})$, respectively. The corresponding bounds for $E(N_{1,i})$ are readily shown to be

$$\begin{aligned} U(\alpha;N_{1,i}) &:= \frac{r}{p_2^*}\sum_{j=0}^{\alpha}\binom{j+r-1}{j} p_2^{*r} q_2^{*j}\, J_{q_1}(j+1,r)(J_{q_2^*}(j+1,r))^{k-2} \\ S(\alpha;N_{1,i}) &:= U(\alpha;N_{1,i}) + \frac{r}{p_2^*}(J_{q_2^*}(\alpha+1,r))^{k-1}J_{q_1}(\alpha+1,r). \end{aligned} \tag{11.32}$$

Each expectation $E(N_{1,i})$ can now be calculated exactly up to a correction term of at most ε. Rough estimates of α may be obtained in the same way as before.

11.3 Expectations for large r

The following inequalities for $E(N_{1,1})$ are easily derived from (11.30):

$$\frac{r}{p_1} E((J_{q_2^*}(X_{r+1,p_1}+1,r))^{k-1}) \le E(N_{1,1}) \le \frac{r}{p_1} E((J_{q_2^*}(X_{r+1,p_1},r))^{k-1}) \tag{11.33}$$

Let γ be equal to zero or one, then A1/13 can be applied to the expectations of the left and right side of (11.33), and we obtain

$$\begin{aligned} &\frac{r}{p_1} E((J_{q_2^*}(X_{r+1,p_1}+\gamma,r))^{k-1}) = \frac{r}{p_1} E((1-J_{p_2^*}(r,X_{r+1,p_1}+\gamma))^{k-1}) = \\ &= \frac{r}{p_1} + \frac{1}{p_1}\sum_{\ell=1}^{k-1}\binom{k-1}{\ell}(-1)^\ell r E((J_{p_2^*}(r,X_{r+1,p_1}+\gamma))^\ell) \xrightarrow[r\to\infty]{} \frac{r}{p_1}, \end{aligned} \tag{11.34}$$

i.e. $E(N_{1,1}) \approx r/p_1$, if r is large enough.

Using A1/12 and A1/13, we obtain for $E(N_{1,i})$:

$$0 \le \frac{r}{p_2^*} E((J_{q_2^*}(X_{r+1,p_2^*}+1,r))^{k-2} J_{q_1}(X_{r+1,p_2^*}+1,r)) \le E(N_{1,i}) \le$$

$$\leq \frac{r}{p_2^*} E((J_{q_2^*}(X_{r+1,p_2^*},r))^{k-2} J_{q_1}(X_{r+1,p_2^*},r)) \leq \frac{r}{p_2^*} E(J_{q_1}(X_{r+1,p_2^*},r)) = \tag{11.35}$$

$$= \frac{r}{p_2^*} E(J_{p_2^*}(r+1, X_{r,p_1}+1)) \xrightarrow[r\to\infty]{} 0$$

(11.34) and (11.35) jointly imply that

$$E(N_1) \approx \frac{r}{p_1} \quad \text{and } E(N) \approx \frac{kr}{p_1}, \tag{11.36}$$

provided that r is large enough.

11.4 Concluding remarks

The selection models no. 10 and no. 11 are the generalizations of the procedures considered in section 6 and 7, respectively. Modifications of these models may be obtained in a similar way as the ones in section 8 in case of two treatments only. A detailed description of those modifications is omitted here but may be carried out without difficulties by the reader himself. The main reason to drop the discussion of modifications becomes clear only in the next chapter; we shall present there procedures that are better than those given in sections 10 and 11 of this chapter. The verification that the selection procedures considered in section 10 and 11 will terminate with probability one is carried out in the same way as in case of two treatments (cf. section 9). Details are omitted. The comparison of the two procedures yields that the PW-sampling procedure of section 10 is uniformly better with respect to E(N) than the VT-sampling procedure just outlined.

Some comments concerning the use of the tables

The P(CS)-values have been calculated for $\varepsilon = 1/1000$.
The expectations have been calculated for $\varepsilon = 1/1000$, too.
Note that $\varepsilon = 1/100$ corresponds to the calculation of a single expectation. If E(N) is the sum of ℓ expectations, then E(N) is exact up to a correction term of at most $\ell\varepsilon$. We have tabulated $]E(\cdot|\cdot)[$ only.
The calculations of $P(CS|\cdot)$ and $E(N|\cdot)$ are based on the r_{max}-values given in (11.13). The corresponding results, obtained by using the r_{max}-values of (11.21), yield only slight differences to the tabulated values.
All calculations are based on the parameter configuration $p_1, p_j = p_1 - \Delta^*$ for all $j \geq 2$.
The P(CS)-values for the VT-sampling procedure (cf. table 7) have not been ta-

bulated; recall that the P(CS)-values of the PW- and those of the VT-sampling procedure coincide. According to our calculations, tables for the PW-sampling procedure could be established up to k=4 and P*=0.95 only by using the computer of the computer center of the university of Karlsruhe whereas the tables for the VT-sampling procedure could be calculated for more than 4 treatments.

11.5 Numerical results

Table 6

k = 3; P* = 0.90; PW									
p_1	0.2	0.3	0.4	0.5	0.6	0.7	0.8	0.9	r
P(CS|Δ*=0.1)	0.999	0.996	0.975	0.945	0.920	0.908	0.915	0.949	79
E(N |Δ*=0.1)	1099	726	537	421	342	281	231	179	
P(CS|Δ*=0.2)	-	0.999	0.989	0.961	0.931	0.912	0.908	0.930	20
E(N |Δ*=0.2)	-	172	127	98	79	64	51	40	
P(CS|Δ*=0.3)	-	-	0.998	0.978	0.946	0.919	0.906	0.915	9
E(N |Δ*=0.3)	-	-	54	42	33	27	22	27	
P(CS|Δ*=0.4)	-	-	-	0.993	0.962	0.928	0.905	0.901	5
E(N |Δ*=0.4)	-	-	-	23	18	15	12	9	
k = 3; P* =0.95; PW									
p_1	0.2	0.3	0.4	0.5	0.6	0.7	0.8	0.9	r
P(CS|Δ*=0.1)	0.999	0.999	0.992	0.977	0.962	0.954	0.959	0.980	114
E(N |Δ*=0.1)	1585	1047	775	669	495	408	335	258	
P(CS|Δ*=0.2)	-	0.999	0.997	0.986	0.970	0.957	0.955	0.970	29
E(N |Δ*=0.2)	-	249	183	142	114	93	75	57	
P(CS|Δ*=0.3)	-	-	0.999	0.994	0.979	0.963	0.955	0.961	13
E(N |Δ*=0.3)	-	-	77	60	48	39	31	24	
P(CS|Δ*=0.4)	-	-	-	0.999	0.991	0.977	0.965	0.964	8
E(N |Δ*=0.4)	-	-	-	35	28	23	18	15	

Table 6 (continued)

k = 3; P* = 0.99; PW									
p_1	0.2	0.3	0.4	0.5	0.6	0.7	0.8	0.9	r
P(CS|Δ*=0.1)	0.999	0.999	0.999	0.996	0.992	0.990	0.991	0.997	197
E(N |Δ*=0.1)	2738	1808	1339	1053	856	706	578	443	
P(CS|Δ*=0.2)	-	0.999	0.999	0.998	0.995	0.991	0.991	0.995	50
E(N |Δ*=0.2)	-	428	314	245	196	159	128	96	
P(CS|Δ*=0.3)	-	-	0.999	0.999	0.997	0.993	0.990	0.993	22
E(N |Δ*=0.3)	-	-	130	101	80	65	52	40	
P(CS|Δ*=0.4)	-	-	-	0.999	0.999	0.996	0.993	0.993	13
E(N |Δ*=0.4)	-	-	-	56	45	36	29	23	

k = 4; P* = 0.90; PW									
p_1	0.2	0.3	0.4	0.5	0.6	0.7	0.8	0.9	r
P(CS|Δ*=0.1)	0.999	0.998	0.981	0.952	0.926	0.913	0.921	0.956	98
E(N |Δ*=0.1)	1799	1186	877	686	555	454	369	278	
P(CS|Δ*=0.2)	-	0.999	0.993	0.969	0.939	0.918	0.915	0.938	25
E(N |Δ*=0.2)	-	280	205	159	126	101	80	60	
P(CS|Δ*=0.3)	-	-	0.998	0.984	0.952	0.923	0.909	0.919	11
E(N |Δ*=0.3)	-	-	85	65	52	41	32	24	
P(CS|Δ*=0.4)	-	-	-	0.998	0.980	0.952	0.931	0.929	7
E(N |Δ*=0.4)	-	-	-	39	31	25	20	15	

k = 4; P* = 0.95; PW									
p_1	0.2	0.3	0.4	0.5	0.6	0.7	0.8	0.9	r
P(CS|Δ*=0.1)	0.999	0.999	0.994	0.980	0.964	0.956	0.961	0.982	134
E(N |Δ*=0.1)	2459	1622	1199	940	761	624	506	379	
P(CS|Δ*=0.2)	-	0.999	0.998	0.989	0.972	0.959	0.957	0.972	34
E(N |Δ*=0.2)	-	380	279	216	172	138	109	80	
P(CS|Δ*=0.3)	-	-	0.999	0.995	0.981	0.963	0.954	0.962	15
E(N |Δ*=0.3)	-	-	115	89	70	56	44	33	
P(CS|Δ*=0.4)	-	-	-	0.999	0.992	0.977	0.964	0.963	9
E(N |Δ*=0.4)	-	-	-	50	40	32	25	19	

Table 7

p_1	0.2	0.3	0.4	0.5	0.6	0.7	0.8	0.9	r
k = 3; P* = 0.90; VT									
$E(N\|\Delta^*=0.1)$	1185	787	579	452	370	315	278	253	79
$E(N\|\Delta^*=0.2)$	-	200	149	116	94	80	70	64	20
$E(N\|\Delta^*=0.3)$	-	-	68	53	43	36	32	29	9
$E(N\|\Delta^*=0.4)$	-	-	-	30	24	20	18	16	5
k = 3; P* = 0.95; VT									
$E(N\|\Delta^*=0.1)$	1710	1140	849	670	552	470	413	374	114
$E(N\|\Delta^*=0.2)$	-	290	217	172	141	120	105	95	29
$E(N\|\Delta^*=0.3)$	-	-	98	78	64	54	47	42	13
$E(N\|\Delta^*=0.4)$	-	-	-	48	40	34	30	26	8
k = 3; P* = 0.99; VT									
$E(N\|\Delta^*=0.1)$	2955	1970	1477	1179	979	837	734	656	197
$E(N\|\Delta^*=0.2)$	-	500	375	300	249	213	186	166	50
$E(N\|\Delta^*=0.3)$	-	-	165	132	110	94	82	73	22
$E(N\|\Delta^*=0.4)$	-	-	-	78	65	56	49	44	13
k = 4; P* = 0.90; VT									
$E(N\|\Delta^*=0.1)$	1960	1305	961	746	606	514	454	418	98
$E(N\|\Delta^*=0.2)$	-	334	248	193	156	131	115	105	25
$E(N\|\Delta^*=0.3)$	-	-	110	87	70	58	50	46	11
$E(N\|\Delta^*=0.4)$	-	-	-	56	46	38	33	24	7
k = 4; P* = 0.95; VT									
$E(N\|\Delta^*=0.1)$	2680	1787	1333	1050	861	732	645	586	134
$E(N\|\Delta^*=0.2)$	-	454	340	269	220	186	163	147	34
$E(N\|\Delta^*=0.3)$	-	-	150	120	98	83	72	65	16
$E(N\|\Delta^*=0.4)$	-	-	-	72	60	50	44	39	9

12 Expected Truncation Points

A common characteristic of all selection models presented so far - that may turn out to be disadvantageous to practical applications - is their termination with probability one only. We have already mentioned that in almost all selection situations only a restricted number of experimental units is available. To guarantee that the identification of the best treatment at the preassigned significance level P^* is practically always possible in the long run, we have to determine the maximal number N^* of experimental units needed to select the best treatment by virtue of untruncated testing sequences, i.e. the experimenter has to provide N^* experimental units to be practically sure that a decision can be taken at the preassigned significance level P^*. The constant N^* depends essentially on the used selection model and on the probability requirements. We have determined these "truncation points" N^* by simulation, where the results, presented in the following, are based on 1000 simulation runs. N^* has been chosen so that at most one among 1000 simulation runs did not lead to a decision. The following tables contain the exact P(CS)-values (P(Cs;cal.)), the simulated P(CS)-values (P(Cs;sim.)); the exact $E(N_B)$-values (E(NB;cal.)), the simulated $E(N_B)$-values (E(NB;sim.)), the upper bound N^*, and the respective numbers of truncated sequences among 1000 simulation runs.

Our results show that in case of a very small $\Delta^*(\Delta^*=0.1)$ the number of experimental units is often extremely large, i.e. the practical application of the selection models is essentially restricted to such situations where Δ^* is known to be at least 0.2.

In case that more than two treatments are available, $E(N_B)$ denotes the average expected number of trials carried out on the inferior treatments, i.e. $E(N_B):=(E(N)-E(N_1))/(k-1)$, where N_1 denotes the number of trials carried out with the best treatment.

remark

If the upper bound N^* is not large enough, then many truncated sequences occur, and that effects that the expected patient horizon E(N) obtained by simulation is smaller than the actual one.

The Selection Model $[2;PW;|S_A-S_B|=r]$, cf. sec. 1

	0.2	0.3	0.4	0.5	0.6	0.7	0.8	0.9
P*=0.90,Δ*=0.1 r=16								
P(Cs; cal.)	1.000	0.999	0.991	0.975	0.954	0.931	0.910	0.896
P(Cs; sim.)	1.000	1.000	0.997	0.976	0.954	0.932	0.916	0.903
E(NB; cal.)	129	114	97	79	61	45	30	17
E(NB; sim.)	127	116	98	77	61	45	30	18
E(N; cal.)	274	242	208	172	136	103	72	45
E(N; sim.)	268	247	211	168	136	103	73	47
upper bound	1640	1450	1240	1030	820	620	430	270
trunc seq.	0	0	0	1	0	0	0	0
P*=0.90,Δ*=0.2 r=8								
P(Cs; cal.)	-	1.000	0.997	0.986	0.968	0.948	0.930	0.916
P(Cs; sim.)	-	1.000	0.994	0.989	0.967	0.950	0.933	0.927
E(NB; cal.)	-	29	25	21	17	13	9	6
E(NB; sim.)	-	30	26	21	17	13	9	6
E(N; cal.)	-	65	58	49	40	32	24	17
E(N; sim.)	-	66	59	49	40	32	24	17
upper bound	-	350	310	270	220	180	130	100
trunc. seq.	-	0	0	0	0	0	0	0
P*=0.90,Δ*=0.3 r=5								
P(Cs; cal.)	-	-	0.999	0.992	0.976	0.956	0.937	0.922
P(Cs; sim.)	-	-	0.999	0.994	0.974	0.959	0.931	0.934
E(NB; cal.)	-	-	11	10	8	6	5	3
E(NB; sim.)	-	-	12	10	8	6	5	4
E(N; cal.)	-	-	26	23	19	16	13	10
E(N; sim.)	-	-	27	23	20	16	14	10
upper bound	-	-	160	140	120	100	80	60
trunc. seq.	-	-	0	0	0	0	0	0
P*=0.90,Δ*=0.4 r=4								
P(Cs; cal.)	-	-	-	0.999	0.991	0.976	0.960	0.945
P(Cs; sim.)	-	-	-	1.000	0.991	0.979	0.963	0.944
E(NB; cal.)	-	-	-	6	5	4	3	2
E(NB; sim.)	-	-	-	7	6	5	4	3
E(N; cal.)	-	-	-	15	13	11	9	7
E(N; sim.)	-	-	-	16	14	12	10	8
upper bound	-	-	-	80	70	60	50	40
trunc. seq.	-	-	-	0	0	0	0	0

The Selection Model [2;PW;$|S_A-S_B|=r$], cf. sec. 1

	0.2	0.3	0.4	0.5	0.6	0.7	0.8	0.9
$P^*=0.95, \Delta^*=0.1$ r=22								
P(Cs; cal.)	1.000	1.000	0.998	0.993	0.984	0.971	0.958	0.946
P(Cs; sim.)	1.000	1.000	1.000	0.995	0.978	0.968	0.957	0.959
E(NB; cal.)	177	156	134	111	89	66	44	24
E(NB; sim.)	181	158	129	113	90	66	45	25
E(N; cal.)	376	333	289	244	197	152	108	67
E(N; sim.)	383	337	279	246	200	151	108	68
upper bound	2070	1830	1590	1340	1085	840	595	370
trunc. seq.	0	0	0	0	1	1	1	2
$P^*=0.95, \Delta^*=0.2$ r=11								
P(Cs; cal.)	-	1.000	1.000	0.997	0.990	0.981	0.969	0.959
P(Cs; sim.)	-	1.000	1.000	0.998	0.994	0.986	0.963	0.963
E(NB; cal.)	-	40	34	29	24	18	13	8
E(NB; sim.)	-	40	35	29	25	19	13	8
E(N; cal.)	-	89	79	68	57	46	35	24
E(N; sim.)	-	89	80	68	58	46	35	25
upper bound	-	480	430	370	320	250	200	140
trunc. seq.	-	0	0	0	0	0	0	0
$P^*=0.95, \Delta^*=0.3$ r=7								
P(Cs; cal.)	-	-	1.000	0.994	0.994	0.985	0.974	0.964
P(Cs; sim.)	-	-	1.000	1.000	0.992	0.985	0.972	0.966
E(NB; cal.)	-	-	15	13	11	8	6	4
E(NB; sim.)	-	-	16	14	11	9	7	5
E(N; cal.)	-	-	36	32	27	23	18	14
E(N; sim.)	-	-	37	34	27	23	18	14
upper bound	-	-	200	180	150	130	100	80
trunc. seq.	-	-	0	0	0	0	1	0
$P^*=0.95, \Delta^*=0.4$ r=5								
P(Cs; cal.)	-	-	-	1.000	0.997	0.990	0.979	0.969
P(Cs; sim.)	-	-	-	1.000	0.997	0.990	0.980	0.969
E(NB; cal.)	-	-	-	7	6	5	4	3
E(NB; sim.)	-	-	-	8	6	6	4	3
E(N; cal.)	-	-	-	19	16	14	12	9
E(N; sim.)	-	-	-	19	16	15	12	10
upper bound	-	-	-	100	90	80	70	50
trunc. seq.	-	-	-	0	0	1	0	0

The Selection Model [2;PW;$|S_A-S_B|=r$], cf. sec. 1

	0.2	0.3	0.4	0.5	0.6	0.7	0.8	0.9
P*=0.99,Δ*=0.1 r=38								
P(Cs; cal.)	1.000	1.000	1.000	0.999	0.998	0.995	0.992	1.000
P(Cs; sim.)	1.000	1.000	1.000	1.000	0.998	0.999	0.996	0.994
E(NB; cal.)	305	268	230	193	155	117	80	42
E(NB; sim.)	309	265	231	191	160	117	75	42
E(N; cal.)	648	573	498	423	347	272	196	121
E(N; sim.)	654	566	498	419	357	269	187	119
upper bound	3560	3150	2730	2320	1900	1490	1080	700
trunc. seq.	0	0	0	0	0	0	0	0
P*=0.99,Δ*=0.2 r=18								
P(Cs; cal.)	-	1.000	1.000	1.000	0.999	0.998	0.996	0.993
P(Cs; sim.)	-	1.000	1.000	1.000	0.999	0.995	0.996	0.995
E(NB; cal.)	-	64	55	47	38	29	20	12
E(NB; sim.)	-	64	56	48	39	30	21	12
E(N; cal.)	-	145	128	110	93	75	58	40
E(N; sim.)	-	145	128	112	94	76	58	41
upper bound	-	790	700	600	510	410	320	220
trunc. seq.	-	0	0	0	0	0	0	0
P*=0.99,Δ*=0.3 r=11								
P(Cs; cal.)	-	-	1.000	1.000	1.000	0.998	0.996	0.993
P(Cs; sim.)	-	-	1.000	1.000	0.998	0.998	0.996	0.988
E(NB; cal.)	-	-	23	20	16	13	9	6
E(NB; sim.)	-	-	24	21	16	13	9	6
E(N; cal.)	-	-	56	49	42	35	28	21
E(N; sim.)	-	-	58	51	42	35	28	21
upper bound	-	-	250	220	190	160	130	95
trunc. seq.	-	-	0	0	0	0	0	0
P*=0.99,Δ*=0.4 r=11								
P(Cs; cal.)	-	-	-	1.000	1.000	0.999	0.997	0.995
P(Cs; sim.)	-	-	-	1.000	1.000	0.997	0.998	0.997
E(NB; cal.)	-	-	-	11	9	7	5	4
E(NB; sim.)	-	-	-	12	10	8	6	4
E(N; cal.)	-	-	-	29	25	22	18	14
E(N; sim.)	-	-	-	30	26	22	18	15
upper bound	-	-	-	130	115	100	80	60
trunc. seq.	-	-	-	0	0	0	0	0

The Selection Model [2;VT;$|S_A-S_B|=s$], cf. sec. 2

	0.2	0.3	0.4	0.5	0.6	0.7	0.8	0.9
$P^*=0.90, \Delta^*=0.1$ s=6								
P(Cs; cal.) P(Cs; sim.)	0.992 0.994	0.962 0.957	0.934 0.934	0.919 0.921	0.919 0.926	0.934 0.949	0.962 0.962	0.992 0.996
E(NB; cal.) E(NB; sim.)	60 59	56 55	53 54	51 52	51 52	53 51	56 57	60 59
E(N; cal.) E(N; sim.)	119 117	111 109	105 108	101 103	101 103	105 101	111 112	119 116
upper bound	655	610	580	555	555	580	610	655
trunc. seq.	0	0	2	1	0	2	0	0
$P^*=0.90, \Delta^*=0.2$ s=3								
P(Cs; cal.) P(Cs; sim.)	– –	0.983 0.982	0.950 0.945	0.927 0.921	0.919 0.926	0.927 0.928	0.950 0.957	0.983 0.980
E(NB; cal.) E(NB; sim.)	– –	15 15	14 14	13 14	13 13	13 14	14 15	15 15
E(N; cal.) E(N; sim.)	– –	29 30	27 28	26 26	26 25	26 27	27 28	29 30
upper bound	–	160	150	145	145	145	150	160
trunc. seq.	–	1	0	0	0	1	0	0
$P^*=0.90, \Delta^*=0.3$ s=2								
P(Cs; cal.) P(Cs; sim.)	– –	– –	0.973 0.970	0.941 0.942	0.925 0.916	0.925 0.931	0.941 0.942	0.973 0.969
E(NB; cal.) E(NB; sim.)	– –	– –	7 7	6 7	6 7	6 7	6 7	7 7
E(N; cal.) E(N; sim.)	– –	– –	13 14	12 13	12 12	12 13	12 13	13 14
upper bound	–	–	75	70	70	70	70	75
trunc. seq.	–	–	0	0	1	1	1	0
$P^*=0.90, \Delta^*=0.4$ s=2								
P(Cs; cal.) P(Cs; sim.)	– –	– –	– –	0.988 0.989	0.973 0.973	0.967 0.978	0.973 0.968	0.988 0.985
E(NB; cal.) E(NB; sim.)	– –	– –	– –	5 6	5 6	5 6	5 6	5 6
E(N; cal.) E(N; sim.)	– –	– –	– –	10 10	10 11	10 10	10 10	10 11
upper bound	–	–	–	55	55	55	55	55
trunc. seq.	–	–	–	0	0	0	1	0

The Selection Model [2;VT;$|S_A-S_B|$=s], cf. sec. 2

	0.2	0.3	0.4	0.5	0.6	0.7	0.8	0.9
P*=0.95,Δ*=0.1 s=8								
P(Cs; cal.)	0.999	0.987	0.972	0.962	0.962	0.972	0.987	0.999
P(Cs; sim.)	0.998	0.986	0.967	0.954	0.959	0.962	0.986	1.000
E(NB; cal.)	80	78	76	74	74	76	78	80
E(NB; sim.)	83	77	77	76	74	75	79	80
E(N; cal.)	160	156	151	148	148	151	156	160
E(N; sim.)	165	153	154	150	146	148	157	159
upper bound	880	860	830	815	815	830	860	880
trunc. seq.	0	0	0	0	0	0	0	0
P*=0.95,Δ*=0.2 s=4								
P(Cs; cal.)	-	0.996	0.981	0.967	0.962	0.967	0.981	0.996
P(Cs; sim.)	-	0.997	0.986	0.965	0.968	0.961	0.981	0.998
E(NB; cal.)	-	20	20	19	19	19	20	20
E(NB; sim.)	-	21	20	20	20	19	20	21
E(N; cal.)	-	40	39	38	37	38	39	40
E(N; sim.)	-	41	40	39	38	37	40	41
upper bound	-	220	215	210	205	210	215	220
trunc. seq.	-	0	0	1	2	0	1	0
P*=0.95,Δ*=0.3 s=3								
P(Cs; cal.)	-	-	0.995	0.985	0.977	0.977	0.985	0.995
P(Cs; sim.)	-	-	0.997	0.983	0.970	0.971	0.983	0.995
E(NB; cal.)	-	-	10	10	10	10	10	10
E(NB; sim.)	-	-	11	11	10	11	11	11
E(N; cal.)	-	-	20	20	20	20	20	20
E(N; sim.)	-	-	21	20	20	20	20	22
upper bound	-	-	110	110	110	110	110	110
trunc. seq.	-	-	0	0	0	0	1	0
P*=0.95,Δ*=0.4 s=2								
P(Cs; cal.)	-	-	-	0.988	0.973	0.967	0.973	0.988
P(Cs; sim.)	-	-	-	0.991	0.968	0.975	0.986	0.983
E(NB; cal.)	-	-	-	5	5	5	5	5
E(NB; sim.)	-	-	-	6	6	6	6	6
E(N; cal.)	-	-	-	10	10	10	10	10
E(N; sim.)	-	-	-	11	11	10	10	11
upper bound	-	-	-	55	55	55	55	55
trunc. seq.	-	-	-	0	0	0	0	0

The Selection Model $[2;VT;|S_A-S_B|=s]$, cf. sec. 2

	0.2	0.3	0.4	0.5	0.6	0.7	0.8	0.9
$P^*=0.99,\Delta^*=0.1$ s=12								
P(Cs; cal.)	1.000	1.000	0.995	0.992	0.992	0.995	1.000	1.000
P(Cs; sim.)	1.000	0.999	0.993	0.990	0.992	0.999	0.997	1.000
E(NB; cal.)	120	120	119	119	119	119	120	120
E(NB; sim.)	122	121	119	122	120	123	120	123
E(N; cal.)	240	240	238	237	237	238	240	240
E(N; sim.)	243	241	237	242	239	245	239	245
upper bound	1320	1320	1310	1305	1305	1310	1320	1320
trunc. seq.	0	0	0	0	0	0	0	0
$P^*=0.99,\Delta^*=0.2$ s=6								
P(Cs; cal.)	-	1.000	0.997	0.994	0.992	0.994	0.997	1.000
P(Cs; sim.)	-	1.000	1.000	0.995	0.992	0.993	0.998	0.999
E(NB; cal.)	-	30	30	30	30	30	30	30
E(NB; sim.)	-	30	31	32	31	31	31	30
E(N; cal.)	-	60	60	60	60	60	60	60
E(N; sim.)	-	59	60	62	60	61	61	60
upper bound	-	300	300	300	300	300	300	300
trunc. seq.	-	0	0	1	0	0	0	1
$P^*=0.99,\Delta^*=0.3$ s=4								
P(Cs; cal.)	-	-	0.999	0.996	0.993	0.993	0.996	0.999
P(Cs; sim.)	-	-	1.000	0.995	0.994	0.993	0.996	1.000
E(NB; cal.)	-	-	14	14	14	14	14	14
E(NB; sim.)	-	-	15	14	14	14	14	14
E(N; cal.)	-	-	27	27	27	27	27	27
E(N; sim.)	-	-	28	28	27	27	27	27
upper bound	-	-	135	135	135	135	135	135
trunc. seq.	-	-	0	0	0	0	0	0
$P^*=0.99,\Delta^*=0.4$ s=3								
P(Cs; cal.)	-	-	-	0.999	0.995	0.994	0.995	0.999
P(Cs; sim.)	-	-	-	0.999	0.992	0.996	0.998	1.000
E(NB; cal.)	-	-	-	8	8	8	8	8
E(NB; sim.)	-	-	-	9	8	8	8	8
E(N; cal.)	-	-	-	15	15	15	15	15
E(N; sim.)	-	-	-	16	16	16	16	16
upper bound	-	-	-	75	75	75	75	75
trunc. seq.	-	-	-	0	0	0	0	0

The Selection Model $[2;PL;|F_A-F_B|=r]$, cf. sec. 3

	0.2	0.3	0.4	0.5	0.6	0.7	0.8	0.9
P*=0.90,Δ*=0.1 r = 16								
P(Cs; cal.)	0.896	0.910	0.931	0.954	0.975	0.991	0.999	1.000
P(Cs; sim.)	0.896	0.908	0.931	0.956	0.970	0.970	0.990	1.000
E(NB; cal.)	29	43	58	75	94	112	129	145
E(NB; sim.)	29	43	59	73	93	113	132	148
E(N; cal.)	45	72	103	136	172	208	242	274
E(N; sim.)	45	73	104	132	170	210	247	279
upper bound	225	360	515	680	860	1040	1210	1370
trunc. seq.	0	1	1	1	0	0	0	0
P*=0.90,Δ*=0.2 r = 8								
P(Cs; cal.)	-	0.916	0.930	0.948	0.968	0.986	0.997	1.000
P(Cs; sim.)	-	0.915	0.923	0.961	0.966	0.987	0.996	1.000
E(NB; cal.)	-	12	16	20	24	28	33	37
E(NB; sim.)	-	12	16	20	24	29	33	37
E(N; cal.)	-	17	24	32	40	49	58	65
E(N; sim.)	-	17	25	32	40	50	58	66
upper bound	-	85	120	160	200	245	290	325
trunc. seq.	-	1	0	2	1	1	0	0
P*=0.90,Δ*=0.3 r = 5								
P(Cs; cal.)	-	-	0.922	0.937	0.956	0.976	0.992	0.999
P(Cs; sim.)	-	-	0.930	0.932	0.959	0.980	0.988	1.000
E(NB; cal.)	-	-	7	9	10	12	14	16
E(NB; sim.)	-	-	7	9	11	13	15	16
E(N; cal.)	-	-	10	13	16	19	23	26
E(N; sim.)	-	-	10	13	16	20	24	27
upper bound	-	-	50	65	80	95	115	130
trunc. seq.	-	-	0	0	0	1	0	0
P*=0.90,Δ*=0.4 r = 4								
P(Cs; cal.)	-	-	-	0.945	0.960	0.976	0.991	0.999
P(Cs; sim.)	-	-	-	0.952	0.967	0.970	0.992	0.998
E(NB; cal.)	-	-	-	6	6	8	9	10
E(NB; sim.)	-	-	-	6	7	8	9	10
E(N; cal.)	-	-	-	7	9	11	13	15
E(N; sim.)	-	-	-	8	10	12	14	16
upper bound	-	-	-	35	45	55	65	75
trunc. seq.	-	-	-	0	1	0	1	0

The Selection Model [2;PL;$|F_A-F_B|$=r], cf. sec. 3

	0.2	0.3	0.4	0.5	0.6	0.7	0.8	0.9
P*=0.95,Δ*=0.1 r = 22								
P(Cs; cal.)	0.946	0.958	0.971	0.984	0.993	0.998	1.000	1.000
P(Cs; sim.)	0.932	0.947	0.966	0.987	0.992	0.999	1.000	1.000
E(NB; cal.)	43	64	86	109	133	155	177	199
E(NB; sim.)	43	64	84	111	134	156	177	201
E(N; cal.)	67	108	152	197	244	289	233	376
E(N; sim.)	66	107	147	200	246	289	332	379
upper bound	335	540	760	985	1220	1445	1665	1880
trunc. seq.	0	1	1	0	1	0	0	0
P*=0.95,Δ*=0.2 r = 11								
P(Cs; cal.)	-	0.959	0.969	0.981	0.990	0.997	1.000	1.000
P(Cs; sim.)	-	0.961	0.980	0.981	0.995	0.998	0.999	1.000
E(NB; cal.)	-	17	23	28	34	40	45	50
E(NB; sim.)	-	18	24	29	35	40	46	51
E(N; cal.)	-	24	35	46	57	68	79	89
E(N; sim.)	-	25	36	46	58	69	80	90
upper bound	-	120	175	230	285	340	395	445
trunc. seq.	-	0	1	0	0	0	0	0
P*=0.95,Δ*=0.3 r = 7								
P(Cs; cal.)	-	-	0.964	0.974	0.985	0.994	0.999	1.000
P(Cs; sim.)	-	-	0.969	0.973	0.984	0.995	0.999	1.000
E(NB; cal.)	-	-	10	12	15	17	19	22
E(NB; sim.)	-	-	11	13	15	18	20	22
E(N; cal.)	-	-	14	18	23	27	32	36
E(N; sim.)	-	-	15	19	23	28	32	38
upper bound	-	-	70	90	115	135	160	180
trunc. seq.	-	-	0	1	0	0	0	0
P*=0.95,Δ*=0.4 r = 5								
P(Cs; cal.)	-	-	-	0.969	0.979	0.990	1.000	0.992
P(Cs; sim.)	-	-	-	0.971	0.978	0.990	0.998	1.000
E(NB; cal.)	-	-	-	7	8	9	11	12
E(NB; sim.)	-	-	-	7	9	10	11	12
E(N; cal.)	-	-	-	9	12	14	16	19
E(N; sim.)	-	-	-	10	12	14	17	19
upper bound	-	-	-	45	60	70	80	95
trunc. seq.	-	-	-	0	0	1	0	0

The Selection Model [2;PL;$|F_A-F_B|$=r], cf. sec. 3

	0.2	0.3	0.4	0.5	0.6	0.7	0.8	0.9
P*=0.99,Δ*=0.1 r = 38								
P(Cs; cal.) P(Cs; sim.)	0.992 0.990	0.995 0.995	0.998 0.997	0.999 0.999	1.000 1.000	1.000 1.000	1.000 1.000	1.000 1.000
E(NB; cal.) E(NB; sim.)	79 79	117 116	155 151	193 191	230 240	268 267	305 306	343 341
E(N; cal.) E(N; sim.)	121 120	196 194	272 264	347 343	423 442	498 495	575 573	648 644
upper bound	545	885	1225	1565	1905	2245	2580	2920
trunc. seq.	0	0	1	0	0	0	0	0
P*=0.99,Δ*=0.2 r = 18								
P(Cs; cal.) P(Cs; sim.)	- -	0.993 0.989	0.996 0.999	0.998 1.000	0.999 0.999	1.000 1.000	1.000 1.000	1.000 1.000
E(NB; cal.) E(NB; sim.)	- -	29 29	38 38	47 48	55 56	64 65	73 75	82 83
E(N; cal.) E(N; sim.)	- -	40 40	58 57	75 78	93 94	110 111	128 131	145 147
upper bound	-	180	265	340	420	495	580	655
trunc. seq.	-	0	1	1	0	0	0	0
P*=0.99,Δ*=0.3 r = 11								
P(Cs; cal.) P(Cs; sim.)	- -	- -	0.993 0.997	0.996 0.995	0.998 1.000	1.000 0.998	1.000 1.000	1.000 1.000
E(NB; cal.) E(NB; sim.)	- -	- -	16 17	20 21	23 24	27 27	30 30	34 34
E(N; cal.) E(N; sim.)	- -	- -	21 22	28 30	35 36	42 42	49 49	56 57
upper bound	-	-	95	130	160	190	220	255
trunc. seq.	-	-	0	1	0	0	0	0
P*=0.99,Δ*=0.4 r = 8								
P(Cs; cal.) P(Cs; sim.)	- -	- -	- -	0.995 0.995	0.997 0.998	0.999 0.999	1.000 1.000	1.000 1.000
E(NB; cal.) E(NB; sim.)	- -	- -	- -	11 11	13 13	15 15	17 17	19 19
E(N; cal.) E(N; sim.)	- -	- -	- -	14 15	18 18	22 22	25 26	29 30
upper bound	-	-	-	65	85	100	115	130
trunc. seq.	-	-	-	0	0	0	0	0

The Two-Stage Selection Procedure with PW- and VT-sampling in the Second Stage (cf. sec.5)

P*=.90,Δ*=0.1	0.2	0.3	0.4	0.5	0.6	0.7	0.8	0.9
s=6 M=5, k_c=3	stage 1 denotes the number of sequences truncated after the first stage of experimentation							
P(Cs; cal.)	0.993	0.969	0.956	0.952	0.944	0.929	0.914	0.911
P(Cs; sim.)	0.990	0.980	0.947	0.949	0.944	0.931	0.903	0.910
E(NB; cal.)	63	66	69	67	59	46	33	20
E(NB; sim.)	69	69	70	71	61	47	34	21
E(N; cal.)	126	134	142	142	128	104	77	52
E(N; sim.)	137	139	145	149	131	105	79	54
stage 1	0	0	0	0	0	0	0	0
trunc. seq.	0	1	1	0	0	2	0	0
s=6 M=10, k_c=8	the upper bounds of the single stage procedures have been adopted, respectively							
P(Cs; cal.)	0.992	0.962	0.934	0.921	0.922	0.928	0.922	0.916
P(Cs; sim.)	0.993	0.965	0.936	0.919	0.917	0.922	0.915	0.898
E(NB; cal.)	60	56	53	52	52	49	39	25
E(NB; sim.)	71	65	64	61	58	52	43	25
E(N; cal.)	119	111	104	103	105	103	86	59
E(N; sim.)	140	129	127	122	118	108	94	60
stage 1	6	10	18	22	24	18	10	5
trunc. seq.	0	0	1	0	0	2	0	0
s=6 M=15, k_c=12								
P(Cs; cal.)	0.992	0.962	0.935	0.921	0.921	0.930	0.927	0.921
P(Cs; sim.)	0.991	0.957	0.932	0.931	0.931	0.935	0.915	0.918
E(NB; cal.)	60	56	53	51	52	51	42	28
E(NB; sim.)	73	68	65	62	63	57	46	29
E(N; cal.)	119	112	106	102	103	104	91	65
E(N; sim.)	146	134	129	122	125	117	98	68
stage 1	20	45	75	83	71	57	36	18
trunc. seq.	0	0	0	1	0	0	1	0
s=6 M=20, k_c=15								
P(Cs; cal.)	0.992	0.963	0.936	0.922	0.923	0.930	0.924	0.925
P(Cs; sim.)	0.993	0.967	0.909	0.919	0.923	0.925	0.908	0.896
E(NB; cal.)	60	57	54	53	53	52	42	31
E(NB; sim.)	78	71	64	63	63	57	44	32
E(N; cal.)	119	113	108	105	106	106	91	70
E(N; sim.)	155	141	127	126	125	117	96	73
stage 1	53	110	133	144	166	141	126	66
trunc. seq.	0	0	2	1	0	0	0	0

P*=.90,Δ*=0.2	0.2	0.3	0.4	0.5	0.6	0.7	0.8	0.9
s=3 M=5, k_c=3								
P(Cs; cal.) P(Cs; sim.)	- -	0.984 0.985	0.958 0.965	0.943 0.943	0.934 0.929	0.925 0.920	0.920 0.912	0.928 0.934
E(NB; cal.) E(NB; sim.)	- -	15 19	15 19	15 17	14 15	12 13	10 11	8 9
E(N; cal.) E(N; sim.)	- -	31 38	31 37	31 35	30 33	27 29	23 25	20 21
stage 1	-	99	130	178	187	177	166	100
trunc. seq.	-	0	1	0	1	0	0	0
s=3 M=10, k_c=8								
P(Cs; cal.) P(Cs; sim.)	- -	0.986 0.985	0.958 0.962	0.940 0.938	0.934 0.931	0.938 0.937	0.945 0.942	0.952 0.948
E(NB; cal.) E(NB; sim.)	- -	17 19	16 19	16 18	16 18	16 17	14 16	12 13
E(N; cal.) E(N; sim.)	- -	33 38	32 37	31 35	31 35	31 34	30 32	27 28
stage 1	-	396	404	440	431	431	419	388
trunc. seq.	-	0	0	0	0	0	0	0
s=3 M=15, k_c=11								
P(Cs; cal.) P(Cs; sim.)	- -	0.989 0.989	0.967 0.969	0.952 0.947	0.947 0.960	0.951 0.952	0.956 0.950	0.966 0.966
E(NB; cal.) E(NB; sim.)	- -	19 22	19 22	19 21	19 21	19 21	18 19	17 17
E(N; cal.) E(N; sim.)	- -	38 42	38 42	38 40	38 41	38 41	36 39	34 35
stage 1	-	622	588	613	612	594	592	599
trunc. seq.	-	0	0	0	0	0	0	0
s=3 M=20, k_c=15								
P(Cs; cal.) P(Cs; sim.)	- -	0.992 0.984	0.975 0.976	0.962 0.954	0.958 0.963	0.962 0.963	0.968 0.954	0.977 0.966
E(NB; cal.) E(NB; sim.)	- -	23 25	23 25	23 25	23 24	23 24	22 23	21 22
E(N; cal.) E(N; sim.)	- -	45 50	46 49	46 48	46 48	46 47	45 46	43 44
stage 1	-	718	700	717	733	729	728	736
trunc. seq.	-	1	0	0	2	0	0	0

P*=.95,Δ*=0.1	0.2	0.3	0.4	0.5	0.6	0.7	0.8	0.9
s=8 M=5, k_c=3								
P(Cs; cal.)	0.999	0.990	0.983	0.982	0.979	0.970	0.959	0.953
P(Cs; sim.)	0.997	0.991	0.980	0.981	0.981	0.970	0.958	0.953
E(NB; cal.)	86	92	98	97	85	67	47	28
E(NB; sim.)	92	94	102	103	87	69	49	28
E(N; cal.)	172	188	204	206	187	152	112	74
E(N; sim.)	184	191	212	219	189	155	117	75
stage 1	0	0	0	0	0	0	0	0
trunc. seq.	0	0	0	0	0	1	1	0
s=8 M=10, k_c=8								
P(Cs; cal.)	0.998	0.987	0.972	0.964	0.965	0.969	0.963	0.955
P(Cs; sim.)	0.999	0.989	0.971	0.961	0.963	0.967	0.957	0.948
E(NB; cal.)	80	78	76	76	76	72	55	32
E(NB; sim.)	92	89	88	84	83	76	58	33
E(N; cal.)	160	156	152	152	155	151	123	81
E(N; sim.)	183	177	175	168	169	160	129	83
stage 1	0	0	1	0	1	0	0	0
trunc. seq.	0	0	0	0	0	0	0	0
s=8 M=15, k_c=11								
P(Cs; cal.)	0.998	0.987	0.972	0.963	0.963	0.968	0.964	0.956
P(Cs; sim.)	0.997	0.987	0.976	0.976	0.967	0.963	0.958	0.953
E(NB; cal.)	80	78	76	75	75	73	58	35
E(NB; sim.)	96	92	94	91	87	79	60	36
E(N; cal.)	160	156	152	149	151	151	127	86
E(N; sim.)	190	183	187	180	175	162	133	90
stage 1	0	1	10	11	12	6	4	0
trunc. seq.	0	0	0	0	2	0	2	0
s=8 M=20, k_c=15								
P(Cs; cal.)	0.998	0.987	0.972	0.963	0.963	0.967	0.960	0.957
P(Cs; sim.)	1.000	0.983	0.962	0.961	0.955	0.952	0.948	0.945
E(NB; cal.)	80	78	76	75	76	73	56	38
E(NB; sim.)	102	99	92	95	91	82	61	40
E(N; cal.)	160	157	152	150	152	151	125	90
E(N; sim.)	204	196	183	188	183	170	134	97
stage 1	6	23	37	30	47	40	22	10
trunc. seq.	0	0	0	0	0	0	1	0

P*=.95,Δ*=0.2	0.2	0.3	0.4	0.5	0.6	0.7	0.8	0.9
s=4 M=5, k_c=3								
P(Cs; cal.)	-	0.996	0.986	0.979	0.974	0.967	0.960	0.961
P(Cs; sim.)	-	0.999	0.986	0.973	0.974	0.968	0.946	0.962
E(NB; cal.)	-	22	22	22	20	17	13	10
E(NB; sim.)	-	26	26	25	23	19	15	11
E(N; cal.)	-	43	45	46	45	40	33	26
E(N; sim.)	-	52	53	52	50	44	36	29
stage 1	-	18	35	46	59	43	31	24
trunc. seq.	-	0	0	0	0	0	0	0
s=4 M=10, k_c=8								
P(Cs; cal.)	-	0.996	0.982	0.970	0.966	0.969	0.971	0.971
P(Cs; sim.)	-	0.995	0.986	0.972	0.966	0.977	0.957	0.964
E(NB; cal.)	-	21	21	20	20	20	18	14
E(NB; sim.)	-	27	26	25	24	24	20	15
E(N; cal.)	-	41	41	40	40	40	38	32
E(N; sim.)	-	52	52	48	47	48	43	35
stage 1	-	219	225	252	273	258	236	186
trunc. seq.	-	0	3	2	0	0	0	0
s=4 M=15, k_c=11								
P(Cs; cal.)	-	0.996	0.985	0.974	0.971	0.973	0.975	0.978
P(Cs; sim.)	-	0.997	0.984	0.984	0.978	0.967	0.962	0.969
E(NB; cal.)	-	22	23	23	22	22	20	18
E(NB; sim.)	-	28	27	26	26	24	22	19
E(N; cal.)	-	44	45	45	45	45	42	38
E(N; sim.)	-	56	52	51	51	49	46	40
stage 1	-	394	436	461	464	447	419	401
trunc. seq.	-	0	0	0	0	0	0	0
s=4 M=20, k_c=15								
P(Cs; cal.)	-	0.997	0.988	0.979	0.976	0.979	0.981	0.985
P(Cs; sim.)	-	1.000	0.991	0.983	0.977	0.974	0.965	0.973
E(NB; cal.)	-	25	26	26	26	25	24	21
E(NB; sim.)	-	30	29	29	29	29	26	23
E(N; cal.)	-	50	51	51	51	51	49	46
E(N; sim.)	-	59	57	57	56	58	52	47
stage 1	-	548	587	562	576	517	565	564
trunc. seq.	-	0	1	0	0	0	0	0

P*=.99,Δ*=0.1	0.2	0.3	0.4	0.5	0.6	0.7	0.8	0.9
s=12 M=5, k_c=3								
P(Cs; cal.)	1.000	0.999	0.997	0.997	0.998	0.996	0.994	0.991
P(Cs; sim.)	1.000	1.000	0.994	0.999	0.998	0.993	0.993	0.990
E(NB; cal.)	131	146	159	159	142	113	79	44
E(NB; sim.)	138	145	164	163	142	116	79	46
E(N; cal.)	263	299	332	340	313	258	191	121
E(N; sim.)	279	295	342	348	312	264	192	126
stage 1	0	0	0	0	0	0	0	0
trunc. seq.	0	0	0	0	0	1	0	0
s=12 M=10, k_c=8								
P(Cs; cal.)	1.000	0.998	0.995	0.993	0.994	0.995	0.994	0.991
P(Cs; sim.)	1.000	1.000	0.995	0.992	0.995	0.996	0.992	0.993
E(NB; cal.)	120	120	120	122	124	116	88	48
E(NB; sim.)	130	132	131	128	132	123	91	49
E(N; cal.)	240	240	241	245	254	247	201	129
E(N; sim.)	260	263	261	258	269	261	208	131
stage 1	0	0	0	0	0	0	0	0
trunc. seq.	0	0	0	0	0	0	0	0
s=12 M=15, k_c=11								
P(Cs; cal.)	1.000	0.998	0.995	0.992	0.993	0.995	0.994	0.991
P(Cs; sim.)	0.999	0.999	0.996	0.997	0.993	0.997	0.994	0.991
E(NB; cal.)	120	120	119	120	121	117	92	51
E(NB; sim.)	135	133	139	135	137	126	96	53
E(N; cal.)	240	240	238	239	244	245	206	133
E(N; sim.)	270	266	277	270	275	263	216	139
stage 1	0	0	0	0	0	0	0	0
trunc. seq.	0	0	0	0	0	0	0	0
s=12 M=20, k_c=15								
P(Cs; cal.)	1.000	0.998	0.995	0.992	0.993	0.995	0.993	0.991
P(Cs; sim.)	1.000	0.999	0.999	0.989	0.991	0.991	0.985	0.992
E(NB; cal.)	120	120	119	120	122	117	88	53
E(NB; sim.)	141	140	141	142	140	129	95	56
E(N; cal.)	240	240	238	239	247	247	203	137
E(N; sim.)	280	278	280	284	284	273	216	145
stage 1	0	0	0	0	1	2	0	0
trunc. seq.	0	0	0	0	0	0	0	0

P*=.99,Δ*=0.2	0.2	0.3	0.4	0.5	0.6	0.7	0.8	0.9
s=6 M=5, k_c=3								
P(Cs; cal.) P(Cs; sim.)	- -	1.000 1.000	0.998 0.998	0.997 0.996	0.997 0.996	0.996 0.999	0.993 0.990	0.992 0.986
E(NB; cal.) E(NB; sim.)	- -	34 38	36 40	36 40	33 36	28 30	21 22	14 14
E(N; cal.) E(N; sim.)	- -	69 77	75 83	79 87	76 83	67 72	55 58	41 43
stage 1	-	0	0	0	0	0	0	0
trunc. seq.	-	0	0	0	0	0	0	0
s=6 M=10, k_c=8								
P(Cs; cal.) P(Cs; sim.)	- -	1.000 1.000	0.997 0.995	0.994 0.992	0.993 0.989	0.994 0.994	0.994 0.981	0.992 0.986
E(NB; cal.) E(NB; sim.)	- -	31 40	30 40	30 40	31 39	30 36	25 29	18 19
E(N; cal.) E(N; sim.)	- -	61 79	60 78	61 79	62 78	62 75	57 66	46 50
stage 1	-	22	30	52	43	50	36	20
trunc. seq.	-	0	0	1	0	0	0	0
s=6 M=15, k_c=11								
P(Cs; cal.) P(Cs; sim.)	- -	1.000 1.000	0.997 0.999	0.994 0.995	0.993 0.993	0.994 0.987	0.994 0.981	0.993 0.987
E(NB; cal.) E(NB; sim.)	- -	31 43	31 41	31 40	31 39	30 35	26 29	20 22
E(N; cal.) E(N; sim.)	- -	61 86	61 80	62 79	63 79	64 74	59 66	49 53
stage 1	-	102	158	178	193	185	158	123
trunc. seq.	-	0	0	0	0	0	0	0
s=6 M=20, k_c=15								
P(Cs; cal.) P(Cs; sim.)	- -	1.000 1.000	0.998 0.995	0.995 0.998	0.994 0.991	0.995 0.989	0.995 0.988	0.994 0.992
E(NB; cal.) E(NB; sim.)	- -	32 43	32 42	32 41	33 40	32 40	29 32	24 26
E(N; cal.) E(N; sim.)	- -	63 85	64 83	64 81	65 79	65 81	62 70	55 59
stage 1	-	271	291	319	345	296	306	255
trunc. seq.	-	0	0	0	1	0	0	0

The Two-Stage Selection Procedure with PW-,VT-,and PL-sampling in the Second Stage (cf. sec.5)

P*=.90,Δ*=0.1	0.2	0.3	0.4	0.5	0.6	0.7	0.8	0.9
s=6 M=5, k_c=3								
P(Cs; cal.)	0.916	0.932	0.956	0.964	0.953	0.932	0.915	0.911
P(Cs; sim.)	0.897	0.935	0.953	0.960	0.955	0.922	0.908	0.904
E(NB; cal.)	37	57	73	76	66	49	33	20
E(NB; sim.)	40	58	73	79	68	52	34	21
E(N; cal.)	63	106	143	155	140	109	78	53
E(N; sim.)	69	107	143	161	143	114	80	54
stage 1	0	0	0	0	0	0	0	0
trunc. seq.	1	1	1	0	0	1	0	0
s=6 M=10, k_c=8								
P(Cs; cal.)	0.941	0.948	0.926	0.908	0.911	0.923	0.922	0.916
P(Cs; sim.)	0.945	0.949	0.926	0.923	0.913	0.922	0.911	0.899
E(NB; cal.)	42	53	53	52	52	49	39	24
E(NB; sim.)	51	62	65	61	59	52	43	25
E(N; cal.)	76	102	105	104	105	103	86	59
E(N; sim.)	94	120	128	122	118	109	93	60
stage 1	3	12	16	19	26	17	13	3
trunc. seq.	0	0	1	0	0	1	0	0
s=6 M=15, k_c=12								
P(Cs; cal.)	0.943	0.951	0.926	0.908	0.909	0.924	0.926	0.921
P(Cs; sim.)	0.926	0.939	0.937	0.939	0.927	0.932	0.918	0.921
E(NB; cal.)	44	54	53	52	52	51	42	28
E(NB; sim.)	51	64	65	62	62	57	46	29
E(N; cal.)	80	105	106	103	103	104	91	65
E(N; sim.)	94	124	128	123	124	116	99	68
stage 1	19	47	61	79	72	59	33	15
trunc. seq.	2	0	0	2	0	0	1	0
s=6 M=20, k_c=15								
P(Cs; cal.)	0.933	0.945	0.929	0.910	0.912	0.925	0.923	0.925
P(Cs; sim.)	0.926	0.937	0.914	0.920	0.923	0.923	0.905	0.894
E(NB; cal.)	42	53	54	53	53	52	42	31
E(NB; sim.)	49	64	65	63	63	58	45	32
E(N; cal.)	75	102	108	105	106	106	91	70
E(N; sim.)	88	124	127	125	125	118	96	73
stage 1	54	124	135	139	161	143	122	64
trunc. seq.	0	0	2	1	0	0	0	0

P* =.90, Δ* = 0.2	0.2	0.3	0.4	0.5	0.6	0.7	0.8	0.9
s=3 M=5, k_c=3								
P(Cs; cal.)	-	0.932	0.934	0.943	0.940	0.929	0.920	0.928
P(Cs; sim.)	-	0.927	0.921	0.939	0.950	0.921	0.909	0.927
E(NB; cal.)	-	13	15	16	15	13	10	8
E(NB; sim.)	-	14	16	17	17	15	11	9
E(N; cal.)	-	21	27	31	32	28	24	20
E(N; sim.)	-	23	29	34	34	31	25	21
stage 1	-	102	160	173	189	166	157	102
trunc. seq.	-	0	0	0	0	0	0	0
s=3 M=10, k_c=8								
P(Cs; cal.)	-	0.963	0.952	0.934	0.926	0.934	0.944	0.952
P(Cs; sim.)	-	0.958	0.956	0.936	0.935	0.934	0.936	0.949
E(NB; cal.)	-	15	16	16	16	16	14	12
E(NB; sim.)	-	18	19	18	18	17	16	13
E(N; cal.)	-	28	31	31	31	31	30	27
E(N; sim.)	-	32	36	35	35	34	32	29
stage 1	-	395	404	435	431	425	403	382
trunc. seq.	-	0	0	0	0	0	0	0
s=3 M=15, k_c=11								
P(Cs; cal.)	-	0.968	0.960	0.948	0.941	0.947	0.955	0.966
P(Cs; sim.)	-	0.966	0.959	0.945	0.960	0.943	0.951	0.969
E(NB; cal.)	-	18	19	19	19	19	18	17
E(NB; sim.)	-	20	21	21	21	21	19	17
E(N; cal.)	-	35	37	38	38	38	36	34
E(N; sim.)	-	37	40	41	41	41	38	35
stage 1	-	575	605	591	598	599	601	612
trunc. seq.	-	0	0	0	0	0	0	0
s=3 M=20, k_c=15								
P(Cs; cal.)	-	0.979	0.971	0.959	0.953	0.959	0.967	0.977
P(Cs; sim.)	-	0.980	0.968	0.948	0.959	0.966	0.955	0.972
E(NB; cal.)	-	22	23	23	23	23	22	21
E(NB; sim.)	-	23	25	24	25	24	23	22
E(N; cal.)	-	43	45	46	46	46	45	43
E(N; sim.)	-	45	48	48	48	48	46	44
stage 1	-	729	714	725	712	717	719	735
trunc. seq.	-	0	0	0	1	0	0	0

P*=0.95,Δ*=0.1	0.2	0.3	0.4	0.5	0.6	0.7	0.8	0.9
s=8 M=5, k_c=3								
P(Cs; cal.)	0.956	0.968	0.983	0.989	0.984	0.972	0.959	0.953
P(Cs; sim.)	0.946	0.971	0.982	0.988	0.990	0.965	0.959	0.952
E(NB; cal.)	54	83	105	109	95	72	48	28
E(NB; sim.)	58	84	106	113	96	74	50	28
E(N; cal.)	90	153	206	223	203	160	114	74
E(N; sim.)	99	156	207	231	204	165	118	75
stage 1	0	0	0	0	0	0	0	0
trunc. seq.	0	2	1	0	0	1	1	0
s=8 M=10, k_c=8								
P(Cs; cal.)	0.970	0.980	0.970	0.961	0.963	0.968	0.963	0.955
P(Cs; sim.)	0.973	0.977	0.969	0.965	0.961	0.966	0.960	0.947
E(NB; cal.)	59	75	77	76	76	72	55	32
E(NB; sim.)	69	86	90	84	83	76	58	34
E(N; cal.)	105	145	153	153	156	151	123	81
E(N; sim.)	125	167	177	169	169	160	128	85
stage 1	0	0	1	0	3	0	1	0
trunc. seq.	1	0	0	0	0	0	1	0
s=8 M=15, k_c=12								
P(Cs; cal.)	0.971	0.981	0.970	0.960	0.961	0.967	0.964	0.956
P(Cs; sim.)	0.964	0.976	0.984	0.972	0.969	0.961	0.956	0.951
E(NB; cal.)	60	76	76	75	75	73	58	35
E(NB; sim.)	69	90	94	91	88	80	61	36
E(N; cal.)	108	148	152	149	151	151	127	86
E(N; sim.)	126	175	185	180	177	164	134	90
stage 1	0	3	10	9	10	7	3	0
trunc. seq.	1	0	1	0	2	0	2	0
s=8 M=20, k_c=15								
P(Cs; cal.)	0.964	0.977	0.970	0.960	0.961	0.966	0.960	0.957
P(Cs; sim.)	0.954	0.977	0.958	0.960	0.955	0.951	0.951	0.940
E(NB; cal.)	56	74	77	74	76	73	56	38
E(NB; sim.)	67	90	92	94	91	82	61	40
E(N; cal.)	99	142	152	150	152	151	125	90
E(N; sim.)	119	172	182	188	182	169	136	96
stage 1	9	22	35	29	46	42	22	11
trunc. seq.	0	1	1	0	0	0	1	0

P*=0.95, Δ*=0.2	0.2	0.3	0.4	0.5	0.6	0.7	0.8	0.9
s=4 M=5, k_c=3								
P(Cs; cal.)	-	0.966	0.972	0.980	0.979	0.969	0.961	0.961
P(Cs; sim.)	-	0.966	0.977	0.983	0.978	0.964	0.947	0.959
E(NB; cal.)	-	18	22	24	23	18	14	10
E(NB; sim.)	-	21	25	26	25	20	15	11
E(N; cal.)	-	31	41	48	48	42	34	26
E(N; sim.)	-	36	46	52	53	45	36	29
stage 1	-	17	30	51	56	41	35	20
trunc. seq.	-	0	0	1	0	0	0	0
s=4 M=10, k_c=8								
P(Cs; cal.)	-	0.983	0.979	0.969	0.964	0.968	0.970	0.971
P(Cs; sim.)	-	0.977	0.986	0.976	0.965	0.978	0.959	0.969
E(NB; cal.)	-	19	21	20	20	20	18	14
E(NB; sim.)	-	24	26	25	24	24	20	15
E(N; cal.)	-	36	40	40	40	40	38	32
E(N; sim.)	-	45	50	48	48	49	43	35
stage 1	-	196	230	249	250	260	239	192
trunc. seq.	-	0	3	2	0	0	0	0
s=4 M=15, k_c=11								
P(Cs; cal.)	-	0.982	0.981	0.973	0.970	0.973	0.975	0.978
P(Cs; sim.)	-	0.981	0.982	0.981	0.977	0.965	0.960	0.971
E(NB; cal.)	-	21	23	23	23	22	20	18
E(NB; sim.)	-	24	26	26	26	24	22	19
E(N; cal.)	-	39	44	45	45	45	42	38
E(N; sim.)	-	43	49	51	51	49	46	40
stage 1	-	387	415	451	471	447	415	417
trunc. seq.	-	0	0	0	0	0	0	0
s=4 M=20, k_c=15								
P(Cs; cal.)	-	0.987	0.985	0.978	0.975	0.978	0.981	0.985
P(Cs; sim.)	-	0.993	0.985	0.984	0.979	0.974	0.966	0.974
E(NB; cal.)	-	24	26	26	26	25	24	22
E(NB; sim.)	-	27	29	29	29	30	25	23
E(N; cal.)	-	46	50	51	51	51	49	46
E(N; sim.)	-	50	56	57	56	59	52	47
stage 1	-	572	593	571	568	518	569	565
trunc. seq.	-	0	0	0	0	0	0	0

P*=0.99, Δ*=0.1	0.2	0.3	0.4	0.5	0.6	0.7	0.8	0.9
s=12 M=5, k_c=3								
P(Cs; cal.) P(Cs; sim.)	0.992 0.992	0.995 0.996	0.998 0.998	0.999 1.000	0.999 0.999	0.997 0.992	0.994 0.992	0.991 0.991
E(NB; cal.) E(NB; sim.)	91 98	141 141	176 178	182 185	159 158	120 125	80 80	44 46
E(N; cal.) E(N; sim.)	150 165	260 259	346 349	374 378	341 338	271 280	194 193	122 127
stage 1	0	0	0	0	0	0	0	0
trunc. seq.	0	3	0	1	0	1	0	0
s=12 M=10, k_c=8								
P(Cs; cal.) P(Cs; sim.)	0.994 0.995	0.997 0.997	0.995 0.995	0.993 0.992	0.993 0.996	0.995 0.995	0.994 0.993	0.991 0.993
E(NB; cal.) E(NB; sim.)	94 103	119 130	122 135	123 129	124 132	116 122	88 91	48 50
E(N; cal.) E(N; sim.)	165 183	227 251	242 266	246 258	254 269	247 260	201 208	129 133
stage 1	0	0	0	0	0	0	0	0
trunc. seq.	0	0	0	0	0	0	0	0
s=12 M=15, k_c=12								
P(Cs; cal.) P(Cs; sim.)	0.994 0.989	0.998 0.994	0.995 0.998	0.992 0.997	0.993 0.995	0.995 0.997	0.994 0.995	0.991 0.989
E(NB; cal.) E(NB; sim.)	95 104	119 130	120 137	120 135	121 138	117 127	92 97	51 54
E(N; cal.) E(N; sim.)	169 184	230 252	239 273	239 270	245 277	245 265	206 218	133 139
stage 1	0	0	0	0	0	0	0	0
trunc. seq.	1	1	0	0	0	0	0	0
s=12 M=20, k_c=15								
P(Cs; cal.) P(Cs; sim.)	0.993 0.994	0.997 0.996	0.995 0.999	0.992 0.992	0.993 0.991	0.995 0.992	0.993 0.987	0.991 0.993
E(NB; cal.) E(NB; sim.)	89 101	118 135	121 140	120 142	121 140	117 130	88 96	53 56
E(N; cal.) E(N: sim.)	152 175	223 259	240 278	239 283	247 283	247 274	203 218	137 143
stage 1	0	0	0	1	0	2	0	0
trunc. seq.	0	1	0	0	0	0	0	0

P*=0.99,Δ*=0.2	0.2	0.3	0.4	0.5	0.6	0.7	0.8	0.9
s=6 M=3, k_c=3								
P(Cs; cal.) P(Cs; sim.)	- -	0.993 0.995	0.996 0.994	0.998 0.998	0.998 0.997	0.996 0.997	0.993 0.989	0.992 0.988
E(NB; cal.) E(NB; sim.)	- -	31 34	39 42	41 44	38 41	30 33	21 23	14 14
E(N; cal. E(N; sim.)	- -	51 56	71 78	83 88	82 89	71 76	56 59	41 43
stage 1	-	0	0	0	0	0	0	0
trunc. seq.	-	0	0	0	0	0	0	0
s=6 M=10, k_c=8								
P(Cs; cal.) P(Cs; sim.)	- -	0.997 0.997	0.997 0.995	0.994 0.992	0.993 0.991	0.994 0.995	0.994 0.980	0.992 0.988
E(NB; cal.) E(NB; sim.)	- -	29 38	31 41	31 40	31 39	30 36	25 30	18 19
E(N; cal.) E(N; sim.)	- -	53 69	60 79	61 79	62 78	62 75	57 67	46 50
stage 1	-	21	35	60	45	48	42	19
trunc. seq.	-	1	0	0	0	0	0	0
s=6 M=15, k_c=11								
P(Cs; cal.) P(Cs; sim.)	- -	0.996 0.997	0.997 0.998	0.994 0.997	0.993 0.994	0.994 0.989	0.994 0.979	0.993 0.987
E(NB; cal.) E(NB; sim.)	- -	30 39	32 42	31 40	31 40	30 36	26 30	20 22
E(N; cal.) E(N; sim.)	- -	52 69	60 79	62 79	63 79	64 74	59 67	49 53
stage 1	-	105	143	177	194	179	147	125
trunc. seq.	-	0	0	0	0	0	0	0
s=6 M=20, k_c=15								
P(Cs; cal.) P(Cs; sim.)	- -	0.997 0.994	0.997 0.995	0.995 1.000	0.994 0.992	0.995 0.991	0.995 0.989	0.994 0.992
E(NB; cal.) E(NB; sim.)	- -	31 39	33 42	33 40	33 39	32 40	29 32	24 26
E(N; cal.) E(N; sim.)	- -	57 71	64 81	64 80	65 78	65 81	63 69	65 59
stage 1	-	274	309	321	357	296	305	248
trunc. seq.	-	0	0	0	1	0	0	0

The Selection Model [2;PW;max$\{S_A,S_B\}$= r], cf. sec. 6

	0.2	0.3	0.4	0.5	0.6	0.7	0.8	0.9
P*=0.90,Δ*=0.1 r = 49								
P(Cs; cal.)	0.999	0.988	0.960	0.930	0.909	0.900	0.905	0.933
P(Cs; sim.)	1.000	0.987	0.962	0.923	0.918	0.898	0.890	0.919
E(NB; cal.)	219	144	105	82	66	53	42	29
E(NB; sim.)	218	144	106	83	67	54	42	31
E(N; cal.)	464	307	228	179	146	122	102	83
E(N; sim.)	462	307	227	179	147	123	102	84
upper bound	700	460	360	270	220	180	160	130
trunc. seq.	0	0	0	0	0	0	0	0
P*=0.90,Δ*=0.2 r = 13								
P(Cs; cal.)	-	0.998	0.979	0.949	0.924	0.908	0.905	0.921
P(Cs; sim.)	-	0.999	0.977	0.952	0.925	0.916	0.908	0.931
E(NB; cal.)	-	35	25	19	15	12	9	6
E(NB; sim.)	-	35	26	20	16	12	10	7
E(N; cal.)	-	78	58	46	37	31	26	21
E(N; sim.)	-	78	58	46	37	30	26	21
upper bound	-	170	130	90	70	60	50	40
trunc. seq.	-	0	0	0	0	0	0	0
P*=0.90,Δ*=0.3 r = 6								
P(Cs; cal.)	-	-	0.994	0.968	0.938	0.916	0.906	0.911
P(Cs; sim.)	-	-	0.995	0.963	0.949	0.904	0.906	0.909
E(NB; cal.)	-	-	11	8	7	5	4	3
E(NB; sim.)	-	-	11	9	7	6	5	4
E(N; cal.)	-	-	26	20	13	11	9	7
E(N; sim.)	-	-	26	21	17	14	12	10
upper bound	-	-	90	70	45	40	30	25
trunc. seq.	-	-	0	0	0	0	0	0
P*=0.90,Δ*=0.4 r = 4								
P(Cs; cal.)	-	-	-	0.991	0.966	0.941	0.926	0.922
P(Cs; sim.)	-	-	-	0.993	0.970	0.960	0.928	0.920
E(NB; cal.)	-	-	-	5	4	3	3	2
E(NB; sim.)	-	-	-	6	5	4	3	3
E(N; cal.)	-	-	-	13	11	9	7	6
E(N; sim.)	-	-	-	14	11	9	8	7
upper bound	-	-	-	45	40	30	25	20
trunc. seq.	-	-	-	0	0	0	0	0

The Selection Model [2;PW;max{S_A,S_B}=r], cf. sec. 6

	0.2	0.3	0.4	0.5	0.6	0.7	0.8	0.9
P*=0.95,Δ*=0.1 r = 81								
P(Cs; cal.)	0.999	0.998	0.987	0.971	0.957	0.950	0.954	0.974
P(Cs; sim.)	1.000	0.997	0.991	0.969	0.963	0.958	0.955	0.972
E(NB; cal.)	361	237	175	136	109	88	69	48
E(NB; sim.)	360	238	174	136	110	87	69	48
E(N; cal.)	766	507	377	298	244	204	171	138
E(N; sim.)	764	509	377	296	244	202	169	138
upper bound	1150	760	570	450	370	310	260	210
trunc. seq.	0	0	0	0	0	0	0	0
P*=0.95,Δ*=0.2 r = 21								
P(Cs; cal.)	-	0.999	0.995	0.981	0.966	0.955	0.953	0.965
P(Cs; sim.)	-	1.000	0.996	0.973	0.946	0.962	0.952	0.974
E(NB; cal.)	-	55	40	31	24	19	14	10
E(NB; sim.)	-	56	41	32	25	20	15	10
E(N; cal.)	-	125	93	73	59	49	40	33
E(N; sim.)	-	126	93	73	59	49	41	33
upper bound	-	250	190	150	120	100	80	70
trunc. seq.	-	0	0	0	0	0	0	0
P*=0.95,Δ*=0.3 r = 9								
P(Cs; cal.)	-	-	0.998	0.988	0.971	0.956	0.949	0.954
P(Cs; sim.)	-	-	1.000	0.988	0.966	0.959	0.959	0.955
E(NB; cal.)	-	-	16	12	10	7	6	4
E(NB; sim.)	-	-	16	13	10	8	6	5
E(N; cal.)	-	-	38	30	24	20	17	14
E(N; sim.)	-	-	38	31	25	21	17	14
upper bound	-	-	100	75	60	50	45	35
trunc. seq.	-	-	1	0	0	0	0	0
P*=0.95,Δ*=0.4 r = 6								
P(Cs; cal.)	-	-	-	0.998	0.983	0.974	0.964	0.963
P(Cs; sim.)	-	-	-	0.997	0.991	0.978	0.968	0.969
E(NB; cal.)	-	-	-	8	6	5	4	3
P(NB; sim.)	-	-	-	8	7	5	4	3
E(N; cal.)	-	-	-	20	16	13	11	9
E(N; sim.)	-	-	-	20	16	14	12	10
upper bound	-	-	-	60	50	40	35	30
trunc. seq.	-	-	-	0	0	0	0	0

The Selection Model [2;PW;max{S_A,S_B}=r], cf. sec. 6

	0.2	0.3	0.4	0.5	0.6	0.7	0.8	0.9
P*=0.99,Δ*=0.1 r = 161								
P(Cs; cal.)	0.999	0.999	0.998	0.996	0.992	0.989	0.991	0.996
P(Cs; sim.)	1.000	1.000	1.000	0.999	0.995	0.992	0.988	0.997
E(NB; cal.)	717	471	346	270	216	174	136	92
E(NB; sim.)	715	472	347	269	217	175	136	91
E(N; cal.)	1522	1007	749	591	484	404	337	271
E(N; sim.)	1518	1009	750	592	485	404	336	270
upper bound	2280	1510	1120	890	730	610	510	410
trunc. seq.	0	0	0	0	0	0	0	0
P*=0.99,Δ*=0.2 r = 41								
P(Cs; cal.)	-	0.999	0.999	0.998	0.994	0.991	0.990	0.994
P(Cs; sim.)	-	1.000	1.000	1.000	0.997	0.987	0.992	0.993
E(NB; cal.)	-	107	78	60	47	36	27	17
E(NB; sim.)	-	108	79	60	47	37	27	18
E(N; cal.)	-	244	180	142	115	95	78	63
E(N; sim.)	-	244	183	142	115	96	78	64
upper bound	-	490	360	290	230	190	160	130
trunc. seq.	-	0	0	0	0	0	0	0
P*=0.99,Δ*=0.3 r = 18								
P(Cs; cal.)	-	-	0.999	0.999	0.996	0.992	0.990	0.992
P(Cs; sim.)	-	-	1.000	1.000	0.998	0.991	0.996	0.992
E(NB; cal.)	-	-	31	24	18	14	10	7
E(NB; sim.)	-	-	31	24	19	14	11	7
E(N; cal.)	-	-	76	60	48	40	33	27
E(N; sim.)	-	-	76	60	49	40	34	27
upper bound	-	-	160	120	100	80	70	60
trunc. seq.	-	-	1	0	0	0	0	0
P*=0.99,Δ*=0.4 r = 10								
P(Cs; cal.)	-	-	-	0.999	0.998	0.996	0.993	0.993
P(Cs; sim.)	-	-	-	1.000	0.999	0.995	0.987	0.992
E(NB; cal.)	-	-	-	13	10	8	6	4
E(NB; sim.)	-	-	-	13	10	8	6	4
E(N; cal.)	-	-	-	35	29	24	20	16
E(N; sim.)	-	-	-	33	26	22	18	15
upper bound	-	-	-	75	65	50	40	35
trunc. seq.	-	-	-	0	0	0	0	0

The Selection Model [2;VT;max{S_A,S_B}=r], cf. sec. 7

	0.2	0.3	0.4	0.5	0.6	0.7	0.8	0.9
P*=0.90,Δ*=0.1 r=49								
P(Cs; cal.)	0.999	0.988	0.960	0.930	0.909	0.900	0.905	0.933
P(Cs; sim.)	1.000	0.991	0.974	0.933	0.917	0.911	0.912	0.940
E(NB; cal.)	245	164	123	98	82	70	61	55
E(NB; sim.)	245	164	123	98	82	71	62	55
E(N; cal.)	490	327	245	195	163	140	122	109
E(N; sim.)	489	327	245	196	163	140	123	110
upper bound	735	495	370	295	245	210	185	165
trunc. seq.	1	0	0	0	0	0	0	0
P*=0.90,Δ*=0.2 r=13								
P(Cs; cal.)	-	0.998	0.979	0.949	0.924	0.908	0.905	0.921
P(Cs; sim.)	-	0.997	0.987	0.950	0.924	0.897	0.911	0.925
E(NB; cal.)	-	44	33	26	22	19	17	15
E(NB; sim.)	-	44	33	27	22	19	17	15
E(N; cal.)	-	87	65	52	43	37	33	29
E(N; sim.)	-	87	65	52	44	38	33	30
upper bound	-	175	130	105	90	75	70	60
trunc. seq.	-	1	0	0	0	0	0	0
P*=0.90,Δ*=0.3 r=6								
P(Cs; cal.)	-	-	0.994	0.968	0.938	0.916	0.906	0.911
P(Cs; sim.)	-	-	0.993	0.974	0.957	0.920	0.904	0.901
E(NB; cal.)	-	-	15	12	10	9	8	7
E(NB; sim.)	-	-	16	13	11	9	8	8
E(N; cal.)	-	-	30	24	20	17	15	14
E(N; sim.)	-	-	31	25	21	18	16	14
upper bound	-	-	75	60	50	45	40	35
trunc. seq.	-	-	2	0	0	0	0	0
P*=0.90,Δ*=0.4 r=4								
P(Cs; cal.)	-	-	-	0.991	0.966	0.941	0.926	0.922
P(Cs; sim.)	-	-	-	0.991	0.968	0.942	0.931	0.923
E(NB; cal.)	-	-	-	8	7	6	5	5
E(NB; sim.)	-	-	-	9	8	7	6	5
E(N; cal.)	-	-	-	16	14	12	10	9
E(N; sim.)	-	-	-	17	14	12	11	10
upper bound	-	-	-	40	35	30	25	25
trunc. seq.	-	-	-	2	0	0	0	0

The Selection Model [2;VT;max{S_A,S_B}=r], cf. sec. 7

	0.2	0.3	0.4	0.5	0.6	0.7	0.8	0.9
P*=0.95,Δ*=0.1 r=81								
P(Cs; cal.)	0.999	0.998	0.987	0.971	0.957	0.950	0.954	0.974
P(Cs; sim.)	1.000	1.000	0.987	0.973	0.958	0.954	0.958	0.977
E(NB; cal.)	405	270	203	162	135	116	102	90
E(NB; sim.)	404	270	204	163	136	116	102	91
E(N; cal.)	810	540	405	324	270	231	203	180
E(N; sim.)	806	539	407	324	271	232	204	181
upper bound	1215	810	610	490	405	350	305	270
trunc. seq.	0	0	0	0	0	0	0	0
P*=0.95,Δ*=0.2 r=21								
P(Cs; cal.)	-	0.999	0.995	0.981	0.966	0.955	0.953	0.965
P(Cs; sim.)	-	1.000	0.996	0.984	0.967	0.965	0.965	0.948
E(NB; cal.)	-	70	53	42	35	30	27	24
E(NB; sim.)	-	71	53	43	36	31	27	24
E(N; cal.)	-	140	105	84	70	60	53	47
E(N; sim.)	-	141	105	85	70	61	53	48
upper bound	-	280	210	170	140	120	110	95
trunc. seq.	-	0	0	0	0	0	0	0
P*=0.95,Δ*=0.3 r=9								
P(Cs; cal.)	-	-	0.998	0.988	0.971	0.956	0.949	0.954
P(Cs; sim.)	-	-	0.999	0.989	0.968	0.954	0.960	0.955
E(NB; cal.)	-	-	23	18	15	13	12	10
E(NB; sim.)	-	-	23	19	16	14	12	11
E(N; cal.)	-	-	45	36	30	26	23	20
E(N; sim.)	-	-	46	37	31	27	23	21
upper bound	-	-	90	75	60	55	50	40
trunc. seq.	-	-	1	1	0	0	0	0
P*=0.95,Δ*=0.4 r=6								
P(Cs; cal.)	-	-	-	0.998	0.988	0.974	0.964	0.963
P(Cs; sim.)	-	-	-	1.000	0.990	0.976	0.964	0.953
E(NB; cal.)	-	-	-	12	10	9	8	7
E(NB; sim.)	-	-	-	13	11	9	8	8
E(N; cal.)	-	-	-	24	20	18	15	14
E(N; sim.)	-	-	-	26	21	18	16	14
upper bound	-	-	-	60	50	45	40	35
trunc. seq.	-	-	-	0	0	0	0	0

The Selection Model [2;VT;max{S_A,S_B}=r], cf. sec. 7

	0.2	0.3	0.4	0.5	0.6	0.7	0.8	0.9
P*=0.99,Δ*=0.1 r=161								
P(Cs; cal.)	0.999	0.999	0.998	0.996	0.992	0.989	0.991	0.996
P(Cs; sim.)	1.000	1.000	1.000	0.997	0.992	0.988	0.991	0.996
E(NB; cal.)	805	537	403	322	269	230	202	179
E(NB; sim.)	807	539	402	323	269	231	202	180
E(N; cal.)	1610	1074	805	644	537	460	403	358
E(N; sim.)	1612	1077	803	644	536	460	404	359
upper bound	2415	1615	1210	970	810	690	605	540
trunc. seq.	0	0	0	0	0	0	0	0
P*=0.99,Δ*=0.2 r=41								
P(Cs; cal.)	-	0.999	0.999	0.998	0.994	0.991	0.990	0.994
P(Cs; sim.)	-	1.000	0.998	0.999	0.996	0.989	0.988	0.994
E(NB; cal.)	-	137	103	82	69	59	52	46
E(NB; sim.)	-	139	105	83	69	60	52	47
E(N; cal.)	-	274	205	164	137	118	103	92
E(N; sim.)	-	276	208	165	138	118	103	92
upper bound	-	415	310	250	210	180	155	140
trunc. seq.	-	1	0	0	0	0	0	0
P*=0.99,Δ*=0.3 r=18								
P(Cs; cal.)	-	-	0.999	0.999	0.996	0.992	0.990	0.992
P(Cs; sim.)	-	-	1.000	0.998	1.000	0.994	0.995	0.994
E(NB; cal.)	-	-	45	36	30	26	23	20
E(NB; sim.)	-	-	46	37	31	27	24	21
E(N; cal.)	-	-	90	72	60	52	45	40
E(N; sim.)	-	-	91	73	61	52	46	41
upper bound	-	-	180	145	120	105	90	80
trunc. seq.	-	-	0	0	0	0	0	0
P*=0.99,Δ*=0.4 r=10								
P(Cs; cal.)	-	-	-	0.999	0.998	0.996	0.993	0.993
P(Cs; sim.)	-	-	-	1.000	1.000	0.994	0.994	0.996
E(NB; cal.)	-	-	-	22	19	16	14	13
E(NB; sim.)	-	-	-	21	18	15	14	12
E(N; cal.)	-	-	-	44	37	32	28	25
E(N; sim.)	-	-	-	41	35	30	26	23
upper bound	-	-	-	90	75	65	60	50
trunc. seq.	-	-	-	0	0	0	0	0

The Selection Model [k;PW;max{$S_1,\ldots,S_k$}=r], cf. sec. 10

	0.2	0.3	0.4	0.5	0.6	0.7	0.8	0.9
k=3,P*=0.90 Δ*=0.1,r=79								
P(Cs; cal.)	0.999	0.996	0.975	0.945	0.920	0.908	0.915	0.949
P(Cs; sim.)	1.000	0.997	0.968	0.935	0.918	0.908	0.921	0.947
E(NB; cal.)	352	231	170	132	105	85	66	46
E(NB; sim.)	352	231	171	133	107	86	68	46
E(N; cal.)	1099	726	537	421	342	281	231	179
E(N; sim.)	1097	725	537	423	344	283	232	179
upper bound	2200	1455	1075	845	685	565	465	360
trunc. seq.	0	0	0	0	0	0	0	0
k=3,P*=0.90 Δ*=0.2,r=20								
P(Cs; cal.)	-	0.999	0.989	0.961	0.931	0.912	0.908	0.930
P(Cs; sim.)	-	1.000	0.986	0.964	0.930	0.911	0.917	0.936
E(NB; cal.)	-	53	38	29	23	18	13	9
E(NB; sim.)	-	53	39	30	23	19	14	10
E(N; cal.)	-	172	127	98	79	64	51	40
E(N; sim.)	-	171	127	100	79	64	52	40
upper bound	-	345	255	200	160	130	105	80
trunc. seq.	-	0	0	0	0	1	0	0
k=3,P*=0.90 Δ*=0.3,r=9								
P(Cs; cal.)	-	-	0.998	0.978	0.946	0.919	0.906	0.915
P(Cs; sim.)	-	-	0.998	0.976	0.954	0.929	0.890	0.916
E(NB; cal.)	-	-	16	12	9	7	5	4
E(NB; sim.)	-	-	17	13	10	8	6	4
E(N; cal.)	-	-	54	42	33	27	22	17
E(N; sim.)	-	-	55	43	33	27	22	17
upper bound	-	-	135	105	85	70	55	45
trunc. seq.	-	-	0	0	0	0	0	0
k=3,P*=0.90 Δ*=0.4,r=5								
P(Cs; cal.)	-	-	-	0.993	0.962	0.928	0.905	0.901
P(Cs; sim.)	-	-	-	0.991	0.961	0.926	0.915	0.906
E(NB; cal.)	-	-	-	6	5	4	3	2
E(NB; sim.)	-	-	-	7	6	5	4	3
E(N; cal.)	-	-	-	23	18	15	12	9
E(N; sim.)	-	-	-	23	19	15	12	10
upper bound	-	-	-	70	55	45	40	30
trunc. seq.	-	-	-	0	0	0	0	0

The Selection Model $[k;PW;\max\{S_1,\ldots,S_k\}=r]$, cf. sec. 10

	0.2	0.3	0.4	0.5	0.6	0.7	0.8	0.9
k=3,P*=0.95 Δ*=0.1,r=114								
P(Cs; cal.)	0.999	0.999	0.992	0.977	0.962	0.954	0.959	0.980
P(Cs; sim.)	1.000	0.999	0.990	0.980	0.962	0.942	0.965	0.981
E(NB; cal.)	508	333	245	191	153	123	96	65
E(NB; sim.)	511	333	247	191	156	124	98	66
E(N; cal.)	1585	1047	775	609	495	408	335	258
E(N; sim.)	1593	1044	778	609	500	409	337	258
upper bound	3170	2095	1550	1220	990	820	670	520
trunc. seq.	0	0	0	0	0	0	0	0
k=3,P*=0.95 Δ*=0.2,r=29								
P(Cs; cal.)	-	0.999	0.997	0.986	0.970	0.957	0.955	0.970
P(Cs; sim.)	-	1.000	1.000	0.991	0.970	0.954	0.951	0.962
E(NB; cal.)	-	76	55	42	33	26	19	12
E(NB; sim.)	-	76	56	43	34	27	20	14
E(N; cal.)	-	249	183	142	114	93	75	57
E(N; sim.)	-	248	183	143	114	94	74	58
upper bound	-	500	370	285	230	190	150	115
trunc. seq.	-	0	0	0	0	0	0	1
k=3,P*=0.95 Δ*=0.3,r=13								
P(Cs; cal.)	-	-	0.999	0.994	0.979	0.963	0.955	0.961
P(Cs; sim.)	-	-	1.000	0.994	0.976	0.967	0.939	0.964
E(NB; cal.)	-	-	23	17	13	10	8	5
E(NB; sim.)	-	-	23	18	14	11	9	6
E(N; cal.)	-	-	77	60	48	39	31	24
E(N; sim.)	-	-	77	61	48	40	32	25
upper bound	-	-	195	150	120	100	80	60
trunc. seq.	-	-	0	0	0	0	0	0
k=3,P*=0.95 Δ*=0.4,r=8								
P(Cs; cal.)	-	-	-	0.999	0.991	0.977	0.965	0.964
P(Cs; sim.)	-	-	-	1.000	0.994	0.983	0.971	0.970
E(NB; cal.)	-	-	-	10	8	6	4	3
E(NB; sim.)	-	-	-	10	8	7	5	4
E(N; cal.)	-	-	-	35	28	23	18	15
E(N; sim.)	-	-	-	36	29	23	19	15
upper bound	-	-	-	105	85	70	55	45
trunc. seq.	-	-	-	0	0	0	0	0

The Selection Model [k;PW;max{$S_1,\dots,S_k$}=r], cf. sec. 10

	0.2	0.3	0.4	0.5	0.6	0.7	0.8	0.9
k=3,P*=0.99 Δ*=0.1,r=197								
P(Cs; cal.)	0.999	0.999	0.999	0.996	0.992	0.990	0.991	0.997
P(Cs; sim.)	1.000	1.000	1.000	0.996	0.996	0.993	0.993	0.999
E(NB; cal.)	876	575	423	329	264	212	166	112
E(NB; sim.)	878	575	424	330	267	214	167	114
E(N; cal.)	2738	1808	1339	1053	856	706	578	443
E(N; sim.)	2742	1804	1339	1054	862	708	579	445
upper bound	5480	3610	2680	2110	1715	1415	1160	890
trunc. seq.	0	0	0	0	0	0	0	0
k=3,P*=0.99 Δ*=0.2,r=50								
P(Cs; cal.)	-	0.999	0.999	0.998	0.995	0.991	0.991	0.995
P(Cs; sim.)	-	1.000	1.000	1.000	0.997	0.988	0.993	0.996
E(NB; cal.)	-	130	95	72	57	44	33	20
E(NB; sim.)	-	130	96	73	57	45	34	21
E(N; cal.)	-	428	314	245	196	159	128	96
E(N; sim.)	-	426	317	245	196	160	129	97
upper bound	-	860	630	490	395	320	260	195
trunc. seq.	-	0	0	0	0	0	0	0
k=3,P*=0.99 Δ*=0.3,r=22								
P(Cs; cal.)	-	-	0.999	0.999	0.997	0.993	0.990	0.993
P(Cs; sim.)	-	-	1.000	1.000	0.996	0.991	0.985	0.997
E(NB; cal.)	-	-	38	28	22	17	12	7
E(NB; sim.)	-	-	38	28	23	18	13	8
E(N; cal.)	-	-	130	101	80	65	52	40
E(N; sim.)	-	-	131	100	81	66	52	40
upper bound	-	-	325	255	200	165	130	100
trunc. seq.	-	-	0	0	0	0	0	0
k=3,P*=0.99 Δ*=0.4,r=13								
P(Cs; cal.)	-	-	-	0.999	0.999	0.996	0.993	0.993
P(Cs; sim.)	-	-	-	1.000	0.999	0.996	0.992	0.995
E(NB; cal.)	-	-	-	15	12	9	6	4
E(NB; sim.)	-	-	-	16	13	10	7	5
E(N; cal.)	-	-	-	56	45	36	29	23
E(N; sim.)	-	-	-	56	46	37	30	23
upper bound	-	-	-	140	115	90	75	60
trunc. seq.	-	-	-	0	1	0	0	0

The Selection Model [k;PW;max$\{S_1,\ldots,S_k\}$=r], cf. sec. 10

	0.2	0.3	0.4	0.5	0.6	0.7	0.8	0.9
k=4,P*=0.90 Δ*=0.1,r=98								
P(Cs; cal.)	0.999	0.998	0.981	0.952	0.926	0.913	0.921	0.956
P(Cs; sim.)	1.000	0.998	0.982	0.942	0.928	0.910	0.922	0.962
E(NB; cal.)	436	287	211	164	131	105	82	56
E(NB; sim.)	435	288	211	164	132	106	83	57
E(N; cal.)	1799	1186	877	686	555	454	369	278
E(N; sim.)	1792	1188	876	686	557	455	369	278
upper bound	3600	2375	1755	1375	1110	910	740	560
trunc. seq.	0	0	0	0	0	0	0	0
k=4,P*=0.90 Δ*=0.2,r=25								
P(Cs; cal.)	-	0.999	0.993	0.969	0.939	0.918	0.915	0.938
P(Cs; sim.)	-	1.000	0.993	0.970	0.939	0.914	0.907	0.933
E(NB; cal.)	-	66	48	36	28	22	16	10
E(NB; sim.)	-	67	49	37	29	23	17	11
E(N; cal.)	-	280	205	159	126	101	80	60
E(N; sim.)	-	282	207	158	125	102	81	59
upper bound	-	840	615	480	380	305	240	180
trunc. seq.	-	0	0	0	0	0	0	0
k=4,P*=0.90 Δ*=0.3,r=11								
P(Cs; cal.)	-	-	0.998	0.984	0.952	0.923	0.909	0.919
P(Cs; sim.)	-	-	1.000	0.988	0.956	0.932	0.916	0.919
E(NB; cal.)	-	-	19	14	11	9	6	4
E(NB; sim.)	-	-	20	15	12	10	7	5
E(N; cal.)	-	-	85	65	52	41	32	24
E(N; sim.)	-	-	86	65	51	42	32	25
upper bound	-	-	300	230	185	145	115	85
trunc. seq.	-	-	0	0	0	0	0	0
k=4,P*=0.90 Δ*=0.4,r=7								
P(Cs; cal.)	-	-	-	0.998	0.980	0.952	0.931	0.929
P(Cs; sim.)	-	-	-	0.999	0.979	0.957	0.933	0.919
E(NB; cal.)	-	-	-	9	7	5	4	2
E(NB; sim.)	-	-	-	9	7	6	5	3
E(N; cal.)	-	-	-	39	31	25	20	15
E(N; sim.)	-	-	-	39	32	26	20	16
upper bound	-	-	-	140	110	90	70	55
trunc. seq.	-	-	-	0	0	0	0	0

The Selection Model [k;PW;max$\{S_1,\ldots,S_k\}$=r], cf. sec. 10

	0.2	0.3	0.4	0.5	0.6	0.7	0.8	0.9
k=4,P*=0.95 Δ*=0.1,r=134								
P(Cs; cal.)	0.999	0.999	0.994	0.980	0.964	0.956	0.961	0.982
P(Cs; sim.)	1.000	0.999	0.989	0.982	0.966	0.970	0.966	0.981
E(NB; cal.)	596	392	288	224	179	144	113	76
E(NB; sim.)	596	393	289	224	181	144	114	76
E(N; cal.)	2459	1622	1199	940	761	624	506	379
E(N; sim.)	2455	1624	1201	937	764	621	506	375
upper bound	4920	3245	2400	1880	1525	1250	1015	760
trunc. seq.	0	0	0	0	0	0	0	0
k=4,P*=0.95 Δ*=0.2,r=34								
P(Cs; cal.)	-	0.999	0.998	0.989	0.972	0.959	0.957	0.972
P(Cs; sim.)	-	1.000	0.999	0.991	0.975	0.961	0.959	0.976
E(NB; cal.)	-	89	65	49	39	30	22	14
E(NB; sim.)	-	90	65	51	39	31	23	15
E(N; cal.)	-	380	279	216	172	138	109	80
E(N; sim.)	-	382	279	218	173	138	110	82
upper bound	-	950	700	540	430	345	275	200
trunc. seq.	-	0	0	0	0	0	0	0
k=4,P*=0.95 Δ*=0.3,r=15								
P(Cs; cal.)	-	-	0.999	0.995	0.981	0.963	0.954	0.962
P(Cs; sim.)	-	-	1.000	0.997	0.987	0.969	0.943	0.955
E(NB; cal.)	-	-	26	20	15	12	8	5
E(NB; sim.)	-	-	26	20	16	12	9	6
E(N; cal.)	-	-	115	89	70	56	44	33
E(N; sim.)	-	-	114	89	71	56	45	33
upper bound	-	-	345	270	210	170	135	100
trunc. seq.	-	-	0	0	0	0	0	0
k=4,P*=0.95 Δ*=0.4,r=9								
P(Cs; cal.)	-	-	-	0.999	0.992	0.977	0.964	0.963
P(Cs; sim.)	-	-	-	1.000	0.992	0.974	0.953	0.964
E(NB; cal.)	-	-	-	11	8	6	5	3
E(NB; sim.)	-	-	-	12	9	7	6	4
E(N; cal.)	-	-	-	50	40	32	25	19
E(N; sim.)	-	-	-	51	40	32	26	19
upper bound	-	-	-	175	140	115	90	70
trunc. seq.	-	-	-	0	0	0	0	0

The Selection Model [k;VT;max{$S_1,\ldots,S_k$}=r], cf. sec. 11

	0.3	0.4	0.5	0.6	0.7	0.8	0.9
k=3, P*=0.90 Δ*=0.2, r=20							
P(Cs; cal.)	0.999	0.989	0.961	0.931	0.912	0.908	0.930
P(Cs; sim.)	1.000	0.991	0.961	0.932	0.912	0.892	0.927
E(NB; cal.)	67	50	39	32	27	24	22
E(NB; sim.)	68	52	41	34	29	26	23
E(N; cal.)	200	149	116	94	80	70	64
E(N; sim.)	202	153	120	100	86	75	68
upper bound	400	300	235	190	160	140	130
trunc. seq.	0	0	0	0	0	0	0
k=3, P*=0.90 Δ*=0.3, r=9							
P(Cs; cal.)	-	0.998	0.978	0.946	0.919	0.906	0.915
P(Cs; sim.)	-	1.000	0.985	0.945	0.915	0.913	0.914
E(NB; cal.)	-	23	18	15	12	11	10
E(NB; sim.)	-	24	19	16	14	12	11
E(N; cal.)	-	68	53	43	36	32	29
E(N; sim.)	-	69	55	46	39	34	31
upper bound	-	175	135	110	90	80	75
trunc. seq.	-	0	0	0	0	0	0
k=3, P*=0.90 Δ*=0.4, r=5							
P(Cs; cal.)	-	-	0.993	0.962	0.928	0.905	0.901
P(Cs; sim.)	-	-	0.995	0.967	0.924	0.892	0.911
E(NB; cal.)	-	-	10	8	7	6	6
E(NB; sim.)	-	-	11	9	8	7	7
E(N; cal.)	-	-	30	24	20	18	16
E(N; sim.)	-	-	31	26	22	19	18
upper bound	-	-	90	75	60	55	50
trunc. seq.	-	-	0	0	0	0	0

The Selection Model $[k;VT;\max\{S_1,\dots,S_k\}=r]$, cf. sec. 11

	0.3	0.4	0.5	0.6	0.7	0.8	0.9
k=3, P*=0.95 Δ*=0.2, r=29							
P(Cs; cal.)	0.999	0.997	0.986	0.970	0.957	0.955	0.970
P(Cs; sim.)	1.000	0.996	0.988	0.969	0.954	0.950	0.971
E(NB; cal.)	97	73	58	47	40	35	32
E(NB; sim.)	97	74	59	49	42	37	33
E(N; cal.)	290	217	172	141	120	105	95
E(N; sim.)	290	219	175	146	125	110	97
upper bound	580	435	345	285	240	210	190
trunc. seq.	0	0	0	0	0	0	0
k=3, P*=0.95 Δ*=0.3, r=13							
P(Cs; cal.)	-	0.999	0.994	0.979	0.963	0.955	0.961
P(Cs; sim.)	-	1.000	0.994	0.980	0.958	0.951	0.963
E(NB; cal.)	-	33	26	22	18	16	14
E(NB; sim.)	-	33	27	23	19	17	15
E(N; cal.)	-	98	78	64	54	47	42
E(N; sim.)	-	98	78	66	56	50	44
upper bound	-	200	160	130	110	95	85
trunc. seq.	-	0	0	0	0	0	0
k=3, P*=0.95 Δ*0.4, r=8							
P(Cs; cal.)	-	-	0.999	0.991	0.977	0.965	0.964
P(Cs; sim.)	-	-	1.000	0.994	0.975	0.960	0.965
E(NB; cal.)	-	-	16	14	12	10	9
E(NB; sim.)	-	-	17	14	12	11	10
E(N; cal.)	-	-	48	40	34	30	26
E(N; sim.)	-	-	50	41	35	31	28
upper bound	-	-	120	100	85	75	65
trunc. seq.	-	-	0	0	0	0	0

The Selection Model $[k;VT;\max\{S_1,\ldots,S_k\}=r]$, cf. sec. 11

	0.3	0.4	0.5	0.6	0.7	0.8	0.9
k=3, P*=0.99 Δ*=0.2, r=50							
P(Cs; cal.)	0.999	0.999	0.998	0.995	0.991	0.991	0.995
P(Cs; sim.)	1.000	1.000	0.997	0.998	0.991	0.992	0.996
E(NB; cal.)	167	125	100	83	71	62	56
E(NB; sim.)	169	126	101	84	72	63	56
E(N; cal.)	500	375	300	249	213	186	166
E(N; sim.)	504	377	300	249	214	188	167
upper bound	1 000	750	600	500	430	375	335
trunc. seq.	0	0	0	0	0	0	0
k=3, P*=0.99 Δ*=0.3, r=22							
P(Cs; cal.)	-	0.999	0.999	0.997	0.993	0.990	0.993
P(Cs; sim.)	-	1.000	0.999	1.000	0.989	0.991	0.996
E(NB; cal.)	-	55	44	37	32	28	25
E(NB; sim.)	-	56	45	38	32	29	25
E(N; cal.)	-	165	132	110	94	82	73
E(N; sim.)	-	165	134	111	95	84	74
upper bound	-	330	265	220	190	165	150
trunc. seq.	-	0	0	0	0	0	0
k=3, P*=0.99 Δ*=0.4, r=13							
P(Cs; cal.)	-	-	0.999	0.999	0.996	0.993	0.993
P(Cs; sim.)	-	-	1.000	0.999	0.990	0.996	0.989
E(NB; cal.)	-	-	26	22	19	17	15
E(NB; sim.)	-	-	27	22	20	17	15
E(N; cal.)	-	-	78	65	56	49	44
E(N; sim.)	-	-	79	65	57	50	44
upper bound	-	-	160	130	115	100	90
trunc. seq.	-	-	0	0	0	0	0

The Selection Model $[k;VT;\max\{S_1,\dots,S_k\}=r]$, cf. sec. 11

	0.3	0.4	0.5	0.6	0.7	0.8	0.9
k=4, P*=0.90 Δ*=0.2, r=25							
P(Cs; cal.)	0.999	0.993	0.969	0.939	0.918	0.915	0.938
P(Cs; sim.)	1.000	0.986	0.970	0.924	0.923	0.922	0.948
E(NB; cal.)	84	62	49	39	33	29	27
E(NB; sim.)	84	63	51	42	37	32	29
E(N; cal.)	334	246	193	156	131	115	105
E(N; sim.)	332	250	200	167	144	125	112
upper bound	670	500	390	315	265	230	210
trunc. seq.	0	0	0	0	0	0	0
k=4, P*=0.90 Δ*=0.3, r=11							
P(Cs; cal.)	-	0.998	0.984	0.952	0.923	0.909	0.919
P(Cs; sim.)	-	0.998	0.972	0.953	0.918	0.922	0.914
E(NB; cal.)	-	28	22	18	15	13	12
E(NB; sim.)	-	29	23	19	17	15	13
E(N; cal.)	-	110	87	70	58	50	46
E(N; sim.)	-	112	89	74	63	55	50
upper bound	-	275	220	175	145	125	115
trunc. seq.	-	0	0	0	0	0	0
k=4, P*=0.90 Δ*=0.4, r=7							
P(Cs; cal.)	-	-	0.998	0.980	0.952	0.931	0.929
P(Cs; sim.)	-	-	0.999	0.981	0.950	0.930	0.920
E(NB; cal.)	-	-	14	12	10	9	8
E(NB; sim.)	-	-	15	13	11	10	9
E(N; cal.)	-	-	56	46	38	33	29
E(N; sim.)	-	-	56	47	40	36	32
upper bound	-	-	140	115	95	85	75
trunc. seq.	-	-	0	0	0	0	0

The Selection Model [k;VT;max{$S_1,\ldots,S_k$}=r], cf. sec. 11

	0.3	0.4	0.5	0.6	0.7	0.8	0.9
k=4, P*=0.95 Δ*=0.2, r=34							
P(Cs; cal.) P(Cs; sim.)	0.999 1.000	0.998 0.998	0.989 0.990	0.972 0.978	0.959 0.956	0.957 0.945	0.972 0.976
E(NB; cal.) E(NB; sim.)	114 114	85 86	68 69	55 58	47 50	41 44	37 39
E(N; cal.) E(N; sim.)	454 453	340 341	269 274	220 228	186 195	163 171	147 152
upper bound	910	680	540	440	375	330	295
trunc. seq.	0	0	0	0	0	0	0
k=4, P*=0.95 Δ*=0.3, r=15							
P(Cs; cal.) P(Cs; sim.)	- -	0.999 1.000	0.995 0.997	0.981 0.985	0.963 0.966	0.954 0.957	0.962 0.957
E(NB; cal.) E(NB; sim.)	- -	38 39	30 31	25 26	21 22	18 20	17 18
E(N; cal.) E(N; sim.)	- -	150 151	120 121	98 101	83 86	72 75	65 68
upper bound	-	300	240	200	170	145	130
trunc. seq.	-	0	0	0	0	0	0
k=4, P*=0.95 Δ*=0.4, r=9							
P(Cs; cal.) P(Cs; sim.)	- -	- -	0.999 1.000	0.992 0.993	0.977 0.976	0.964 0.960	0.963 0.954
E(NB; cal.) E(NB; sim.)	- -	- -	18 19	15 16	13 14	11 12	10 11
E(N; cal.) E(N; sim.)	- -	- -	72 73	60 61	50 52	44 46	39 41
upper bound	-	-	180	150	125	110	100
trunc. seq.	-	-	0	0	0	0	0

13 The Selection Model

$[2; PW; |S_A - S_B| = r \text{ or } |\hat{p}_A - \hat{p}_B| \geq c/(F_A + F_B)]$

13.1 Introduction

All selection models considered so far have a termination-rule that is based on either successes or failures. The following procedure, due to Nordbrock [131], combines two criteria in a stopping-rule that is based not only on successes but also on failures.

Sampling is performed according to the PW-rule and the experiment will be terminated if either $|S_A - S_B| = r$ or $|\hat{p}_A - \hat{p}_B| \geq c/(F_A + F_B)$, where $\hat{p}_i := S_i/(S_i + F_i)$; $i \in \{A,B\}$, are the (current) proportions of successes of treatment A and B.

The key-note to constructing this termination-rule is as follows. For large parameter values p_A the selection model just being considered provides essentially the same as the selection model of sec. 1, because the terminal decision will be taken on the whole by reason of successes. On the other hand, the selection model of sec. 1 performs poorer and poorer as p_A becomes small. Therefore, we need a second stopping criterion that enables us to terminate sampling in case if only few successes occur. The criterion "stop sampling whenever $|\hat{p}_A - \hat{p}_B| \geq c/(F_A + F_B)$" performs as requested, because in case of small success probabilities there occur many failures and, consequently, $c/(F_A + F_B)$ decreases rapidly so that the absolute difference in the proportions of successes reaches the critical value after only few trials. Nevertheless, it may happen that sampling never terminates. This is for example the case if only failures occur. The absolute difference of current successes is hence equal to zero, and the proportions of successes are zero, too, so that $(r,c>0)$ neither stopping criterion may be applied.

As usual, the treatment associated with the larger number of successes when sampling terminates is declared as best.

13.2 Derivation of the critical r- and c-values

For computing the critical r- and c-values, it is easy to construct a system of difference equations. The essential point, however, is that the corresponding boundary conditions are nonlinear, and analytic solutions are, therefore, scarcely to find. To get still a good conception about the magnitude of the critical values, we investigate the case if one of them is substantially greater than the other.

It is immediately obvious that the selection model reduces to that of sec. 1 as c becomes large. The critical r-value is consequently the same as that of (1.18), i.e.

$$r_{max} \approx \left]\frac{\ell n 2(1-P^*)}{\ell n(1-\Delta^*)}\right[\tag{13.1}$$

When r is infinite sampling terminates whenever $|\hat{p}_A-\hat{p}_B| \geq c/(F_A+F_B)$. Recalling that the current numbers of failures differ by at most one when PW-sampling is used, the above stopping-rule is directly seen to be equivalent (when $F_A=F_B$) or asymptotically equivalent (when $F_A \neq F_B$) to the following stopping-rule

$$|S_A-S_B| \geq c/2(1-\hat{p}_A)(1-\hat{p}_B) \tag{13.2}$$

A high significance level P* implies also a large value of c and this again means that F_A and F_B must take on large values before $|\hat{p}_A-\hat{p}_B|$ may exceed $c/(F_A+F_B)$. The relative frequencies of successes may be considered as good estimates of the success probabilities in this special situation. Therefore, we replace the stopping-rule given in (13.2) by the following one

$$|S_A-S_B| \geq c/2(1-p_A)(1-p_B) =: r^* \tag{13.3}$$

The critical c-value is obtained in essentially the same way as the critical r-value of the selection model presented in sec. 1. Replacing r by r* in (1.16) and taking logarithms, it is easily seen that $\ell n(1-P^*)$ dominates the right side. The value of $\bar{p}$ that maximizes c is, therefore, close to that $\bar{p}$ which maximizes

$$\ell n((2\bar{p}-\Delta^*)/(2\bar{p}+\Delta^*))/2(1-\bar{p}-\Delta^*/2)(1-\bar{p}+\Delta^*/2) \tag{13.4}$$

A straightforward computation shows that c is maximized when $\bar{p}=1/3 + O(\Delta^{*2})$. Using (1.16) again with r replaced by r* (p_A, p_B expressed in terms of $\bar{p}$ and Δ^*) and putting $\bar{p}=1/3$ in this relation, the critical c-value is approximately given by

$$c_{max} \approx \left]2\left(\frac{2}{3}-\frac{\Delta^*}{2}\right)\cdot\left(\frac{2}{3}+\frac{\Delta^*}{2}\right)\cdot\frac{\ell n((1-P^*)(1+3\Delta^*/4))}{\ell n((2-3\Delta^*)/(2+3\Delta^*))}\right[\tag{13.5}$$

We have already mentioned that for extreme parameter values the termination of sampling is based on either successes or failures. In either case the critical stopping values r_{max} or c_{max}, given by (13.1) or (13.5), must be used to guaran-

tee that the $(P^*;\Delta^*)$-condition is satisfied. Since the expectations $E(N_B)$ and $E(N)$ of the reduced procedures are increasing functions of r (cf. sec. 1), the best choice among all pairs (r,c) seems to be (r_{max},c_{max}), given above, provided that the probability requirements are met for all parameter values. In case that the resulting P(CS)-values are smaller than the preassigned P*, the critical values c_{max} or r_{max} must be slightly enlarged.
Exact formulae for $E(N_B)$ and E(N) can be obtained by solving a system of difference equations and nonlinear boundary conditions. However, it seems to be impossible to get these exact expressions. Therefore, we determine the desired values by simulation. The results contained in the following table are, as usual, based on 1000 simulation runs.

13.3 Numerical results

Table 8

	0.2	0.3	0.4	0.5	0.6	0.7	0.8	0.9
$P^*=0.90, \Delta^*=0.1$ r=16, c=7								
P(CS;sim.)	0.978	0.928	0.927	0.939	0.944	0.932	0.915	0.903
E(NB;sim.)	41	41	47	51	57	45	30	18
E(N; sim.)	86	87	100	110	126	103	72	47
upper bound	550	550	550	600	750	600	350	250
trunc. seq.	0	0	0	0	1	0	0	0
$P^*=0.90, \Delta^*=0.2$ r=8, c=4								
P(CS;sim.)	-	0.980	0.942	0.951	0.942	0.941	0.932	0.927
E(NB;sim.)	-	13	14	15	14	13	9	6
E(N; sim.)	-	28	30	34	34	31	24	17
upper bound	-	150	160	200	200	160	120	100
trunc. seq.	-	0	0	0	0	0	0	0

	0.2	0.3	0.4	0.5	0.6	0.7	0.8	0.9
$P^*=0.90, \Delta^*=0.3$ r=5, c=2								
P(CS;sim.)	-	-	0.952	0.920	0.903	0.917	0.917	0.926
E(NB;sim.)	-	-	6	5	5	5	5	3
E(N; sim.)	-	-	12	12	12	12	12	10
upper bound	-	-	70	70	70	70	70	50
trunc. seq.	-	-	0	0	0	0	0	0
$P^*=0.90, \Delta^*=0.4$ r=4, c=2								
P(CS;sim.)	-	-	-	0.979	0.971	0.956	0.947	0.938
E(NB;sim.)	-	-	-	4	4	4	4	3
E(N; sim.)	-	-	-	10	10	10	9	7
upper bound	-	-	-	50	60	50	50	40
trunc. seq.	-	-	-	0	0	0	0	0
$P^*=0.95, \Delta^*=0.1$ r=22, c=9								
P(CS;sim.)	0.991	0.966	0.957	0.962	0.977	0.968	0.957	0.959
E(NB;sim.)	56	58	61	74	85	66	45	25
E(N; sim.)	117	122	131	160	189	150	108	69
upper bound	600	700	800	800	1000	900	700	450
trunc. seq.	0	1	1	0	0	1	1	0
$P^*=0.95, \Delta^*=0.2$ r=11, c=4								
P(CS;sim.)	-	0.981	0.961	0.951	0.955	0.967	0.958	0.963
E(NB;sim.)	-	13	14	15	17	16	13	8
E(N; sim.)	-	28	31	34	40	40	34	25
upper bound	-	180	180	200	250	250	200	150
trunc. seq.	-	0	0	0	0	0	0	0
$P^*=0.95, \Delta^*=0.3$ r=7, c=3								
P(CS;sim.)	-	-	0.982	0.971	0.966	0.968	0.965	0.965
E(NB;sim.)	-	-	8	8	8	8	6	5
E(N; sim.)	-	-	17	19	19	20	18	14
upper bound	-	-	90	90	120	120	120	80
trunc. seq.	-	-	0	0	0	0	0	0

	0.2	0.3	0.4	0.5	0.6	0.7	0.8	0.9
P^*=0.95,Δ^*=0.4 r=5, c=2								
P(CS;sim.)	-	-	-	0.983	0.958	0.951	0.954	0.960
E(NB;sim.)	-	-	-	5	4	4	4	3
E(N; sim.)	-	-	-	10	10	11	10	9
upper bound	-	-	-	50	50	50	50	50
trunc. seq.	-	-	-	0	0	0	0	0
P^*=0.99,Δ^*=0.1 r=38, c=14								
P(CS;sim.)	0.999	0.993	0.992	0.993	0.998	0.999	0.996	0.994
E(NB;sim.)	83	90	99	117	144	116	75	41
E(N; sim.)	174	190	212	256	319	269	187	119
upper bound	900	1000	1100	1250	1650	1250	900	650
trunc. seq.	0	0	0	0	0	0	0	0
P^*=0.99,Δ^*=0.2 r=18, c=7								
P(CS;sim.)	-	0.998	0.998	0.988	0.995	0.993	0.995	0.995
E(NB;sim.)	-	22	24	27	30	29	21	12
E(N; sim.)	-	49	54	63	72	73	58	41
upper bound	-	250	280	300	350	350	350	200
trunc. seq.	-	0	0	0	0	0	0	0
P^*=0.99,Δ^*=0.3 r=11, c=4								
P(CS;sim.)	-	-	0.997	0.981	0.982	0.992	0.992	0.986
E(NB;sim.)	-	-	10	11	11	11	9	6
E(N; sim.)	-	-	22	25	27	29	27	21
upper bound	-	-	120	120	140	150	130	100
trunc. seq.	-	-	0	0	0	0	0	0
P^*-0.99,Δ^*=0.4 r=8, c=3								
P(CS;sim.)	-	-	-	0.993	0.988	0.984	0.995	0.997
E(NB;sim.)	-	-	-	6	6	6	5	4
E(N; sim.)	-	-	-	14	16	17	17	14
upper bound	-	-	-	70	80	90	90	70
trunc. seq.	-	-	-	0	0	0	0	0

14 The Selection Model

$[2; VT; |S_A - S_B| = s \text{ or } |\hat{p}_A - \hat{p}_B| \geq d/(F_A + F_B)]$

14.1 Derivation of the critical s- and d-values

The following selection model uses the VT-sampling scheme and the stopping-rule of the preceding procedure. The critical s- and d-values are obtained by using similar arguments as before. When d is infinite, sampling terminates only if $|S_A - S_B| = s$. The critical s-value is, therefore, the same as that given by (2.10), i.e.

$$s_{max} \approx \left]\frac{\ell n(1-P^*)}{2\ell n\left(\frac{1-\Delta^*}{1+\Delta^*}\right)}\right[\tag{14.1}$$

When s is infinite the termination-rule reduces to "$|\hat{p}_A - \hat{p}_B| \geq d/(F_A+F_B)$". Recalling that for VT-sampling $S_A+F_A = S_B+F_B$, this stopping inequality is equivalent to

$$|S_A - S_B| \geq d/(2-\hat{p}_A-\hat{p}_B) \tag{14.2}$$

Using similar arguments as before, (14.2) can be "approximated" by the following stopping rule

$$|S_A - S_B| \geq d/(2-p_A-p_B) =: d^* \tag{14.3}$$

The critical d-value is obtained in essentially the same way as the critical s-value of the selection procedure considered in sec. 2. Replacing s by d^* in (2.7) and taking logarithms, the value of $\bar{p}$ that maximizes d is obtained by determining that $\bar{p}$-value which maximizes

$$f(\bar{p}) := \frac{1}{(1-\bar{p})} \cdot \ell n \frac{(2\bar{p}-\Delta^*)(2(1-\bar{p})-\Delta^*)}{(2\bar{p}+\Delta^*)(2(1-\bar{p})+\Delta^*)} \tag{14.4}$$

Expanding the logarithm in a power series, we find

$$f(\bar{p}) = \frac{1}{1-\bar{p}} \frac{4\Delta^*}{(2\bar{p}+\Delta^*)(2(1-\bar{p})+\Delta^*)} + O(\Delta^{*2}), \tag{14.5}$$

and this is maximized by

$$\bar{p}_{1,2} = \frac{2}{3} \pm \sqrt{\frac{1}{9} + \frac{2\Delta^*+\Delta^{*2}}{12}} \tag{14.6}$$

The unique solution to our problem is hence given by

$$\bar{p} = \frac{2}{3} - \sqrt{\frac{1}{9} + \frac{2\Delta^*+\Delta^{*2}}{12}} = \frac{1}{3} + \frac{1}{4}\Delta^* + O(\Delta^{*2}) \tag{14.7}$$

A numerical investigation of $f(\bar{p})$ shows that $\bar{p} = \frac{1}{3}$ is a better approximation of that $\bar{p}$ which maximizes $f(\bar{p})$. The critical d-value is consequently given by

$$d_{max} \approx \left] \frac{\frac{4}{3}\ \ell n(1-P^*)}{\ell n\left(\frac{(\frac{2}{3}-\Delta^*)(\frac{4}{3}-\Delta^*)}{(\frac{2}{3}+\Delta^*)(\frac{4}{3}+\Delta^*)}\right)} \right[\tag{14.8}$$

14.2 Numerical results

Table 9

	0.2	0.3	0.4	0.5	0.6	0.7	0.8	0.9
$P^*=0.90, \Delta^*=0.1$ s=6, d=7								
P(CS;sim.)	0.980	0.941	0.925	0.921	0.926	0.949	0.962	0.996
E(NB;sim.)	45	44	52	52	52	51	56	58
E(N; sim.)	89	87	102	103	103	102	112	116
upper bound	680	680	680	680	680	680	680	680
trunc. seq.	0	0	1	1	0	0	0	0
$P^*=0.90, \Delta^*=0.2$ s=3, d=4								
P(CS;sim.)	-	0.982	0.945	0.921	0.924	0.929	0.957	0.980
E(NB;sim.)	-	16	14	14	13	14	15	15
E(N; sim.)	-	30	28	26	25	27	28	30
upper bound	-	200	200	200	200	200	200	200
trunc. seq.	-	0	0	0	0	0	0	0

	0.2	0.3	0.4	0.5	0.6	0.7	0.8	0.9
P*=0.90,Δ*=0.3 s=2, d=3								
P(CS;sim.)	-	-	0.971	0.941	0.916	0.930	0.942	0.969
E(NB;sim.)	-	-	7	7	7	7	7	7
E(N; sim.)	-	-	14	13	12	13	13	14
upper bound	-	-	90	90	90	90	90	90
trunc. seq.	-	-	0	0	0	0	0	0
P*=0.90,Δ*=0.4 s=2, d=2								
P(CS;sim.)	-	-	-	0.989	0.974	0.978	0.968	0.986
E(NB;sim.)	-	-	-	6	6	6	6	6
E(N; sim.)	-	-	-	10	11	10	10	11
upper bound	-	-	-	65	65	65	65	65
trunc. seq.	-	-	-	0	0	0	0	0
P*=0.95,Δ*=0.1 s=8, d=9								
P(CS;sim.)	0.988	0.972	0.955	0.966	0.963	0.969	0.984	1.000
E(NB;sim.)	60	61	68	73	74	78	78	79
E(N; sim.)	119	120	134	145	148	156	154	157
upper bound	850	850	850	850	850	850	850	850
trunc. seq.	0	0	0	0	1	0	0	0
P*=0.95,Δ*=0.2 s=4, d=5								
P(CS;sim.)	-	0.993	0.984	0.972	0.970	0.966	0.987	0.996
E(NB;sim.)	-	19	19	20	19	20	20	22
E(N; sim.)	-	37	38	38	38	39	40	43
upper bound	-	260	260	260	260	260	260	260
trunc. seq.	-	0	0	0	0	0	0	0
P*=0.95,Δ*=0.3 s=3, d=3								
P(CS;sim.)	-	-	0.985	0.966	0.965	0.967	0.983	0.995
E(NB;sim.)	-	-	9	9	10	10	11	11
E(N; sim.)	-	-	16	18	19	19	20	22
upper bound	-	-	120	120	120	120	120	120
trunc. seq.	-	-	0	0	0	0	0	0

	0.2	0.3	0.4	0.5	0.6	0.7	0.8	0.9
P*=0.95,Δ*=0.4 s=2, d=3								
P(CS;sim.)	-	-	-	0.991	0.966	0.975	0.986	0.983
E(NB;sim.)	-	-	-	6	6	6	6	6
E(N; sim.)	-	-	-	11	11	10	10	11
upper bound	-	-	-	65	65	65	65	65
trunc. seq.	-	-	-	0	0	0	0	0
P*=0.99,Δ*=0.1 s=12, d=14								
P(CS;sim.)	1.000	0.994	0.992	0.995	0.997	0.999	0.997	1.000
E(NB;sim.)	89	98	107	116	120	123	120	123
E(N; sim.)	177	195	214	231	238	245	239	245
upper bound	1000	1000	1100	1100	1100	1100	1100	1000
trunc. seq.	0	0	1	2	1	1	0	0
P*=0.99,Δ*=0.2 s=6, d=7								
P(CS;sim.)	-	1.000	0.999	0.995	0.993	0.995	0.998	1.000
E(NB;sim.)	-	25	28	29	32	30	31	31
E(N; sim.)	-	48	54	58	62	60	60	60
upper bound	-	350	350	350	350	350	350	350
trunc. seq.	-	0	0	0	0	0	0	0
P*=0.99,Δ*=0.3 s=4, d=5								
P(CS;sim.)	-	-	0.997	0.995	0.996	0.995	0.996	1.000
E(NB;sim.)	-	-	14	14	14	14	15	14
E(N; sim.)	-	-	26	27	26	27	29	28
upper bound	-	-	160	160	160	160	160	160
trunc. seq.	-	-	0	0	0	0	0	0
P*=0.99,Δ*=0.4 s=3, d=4								
P(CS;sim.)	-	-	-	0.998	0.997	0.997	0.997	0.998
E(NB;sim.)	-	-	-	9	8	8	8	8
E(N; sim.)	-	-	-	16	16	16	16	16
upper bound	-	-	-	75	75	75	75	75
trunc. seq.	-	-	-	0	0	0	0	0

15 The Selection Models

$[k;PW;\text{el.}A_i \text{ if } S_j-S_i=r]$ and

$[k;VT;\text{el.}A_i \text{ if } S_j-S_i=s]$

15.1 The PW-elimination procedure

The selection models considered in section 1 and 2 have a termination-rule that is based on the absolute difference of the current numbers of successes.In case of only two treatments, the difference of successes is well-defined, but in which way can we proceed when there are more than two rival treatments and the termination-rule is to be based essentially on differences of successes? The desired extension of the procedure from sec. 1 will be defined as follows. At the outset the k treatments are put in a random order and sampling is carried out according to the PW-sampling scheme. After each trial - for the first time after the r-th trial - the current numbers of successes are compared. We eliminate (el. means eliminate) treatment A_i if there is a treatment A_j such that $S_j-S_i=r$. Sampling is continued as long as exactly one treatment is left, and that treatment is declared as best. Recalling that the best treatment is denoted by A_1 throughout part 1, the event "correct selection" contains the union of the events "A_1 eliminates A_i" (note that A_i may be also eliminated by A_j, $j\neq 1$). Therefore, we have

$$\begin{aligned} 1-P(CS) &\leq \sum_{i=2}^{k} P(A_i \text{ eliminates } A_1) \leq (k-1)(1-P^*) \\ P(CS) &\geq 1-(k-1)(1-P^*) =: \hat{P}^* \end{aligned} \tag{15.1}$$

For k=2 the critical r-value is approximately given by (1.18). Solving now the second equation of (15.1) for P^* and substituting the result into (1.18), the desired critical r-value for the PW-elimination procedure is seen to be

$$r_{max} \approx \left]\frac{\ell n\left(2\frac{1-\hat{P}^*}{k-1}\right)}{\ell n(1-\Delta^*)}\right[\quad , \tag{15.2}$$

and the probability for a correct selection is at least $\hat{P}^*$ whenever $\Delta \geq \Delta^*$, and the termination-rule is based on r_{max} given above.

15.2 The VT-elimination procedure

An elimination procedure similar to that just described is obtained in using the VT-sampling scheme and the termination-rule of the PW-elimination procedure. The critical s-value needed to satisfy the $(\hat{P}^*;\Delta^*)$-condition in any case is derived from (2.10) by substituting $1-P^*$ by $(1-\hat{P}^*)/(k-1)$. We get

$$s_{max} \approx \left]\frac{\ell n((1-\hat{P}^*)/(k-1))}{2\ell n\left(\frac{1-\Delta^*}{1+\Delta^*}\right)}\right[\tag{15.3}$$

The numerical results presented in the following are based on 1000 simulation runs. $\hat{P}^*$ has been replaced by P^*.

15.3 Numerical results for the PW-procedure

Table 10

p_A	0.3	0.4	0.5	0.6	0.7	0.8	0.9
k=3,P*=0.90 Δ*=0.2,r=11							
P(CS;sim.)	1.000	1.000	0.993	0.984	0.962	0.939	0.915
E(NB;sim.)	39	33	28	24	18	12	7
E(N; sim.)	133	116	101	87	67	50	31
upper bound	420	380	360	400	360	240	190
trunc. seq.	0	0	0	2	2	0	0
k=3,P*=0.90 Δ*=0.3,r=7							
P(CS;sim.)	-	1.000	1.000	0.986	0.976	0.957	0.929
E(NB;sim.)	-	14	12	10	8	6	4
E(N; sim.)	-	50	46	38	32	24	17
upper bound	-	190	180	170	160	160	140
trunc. seq.	-	0	0	0	0	1	0

p_A	0.3	0.4	0.5	0.6	0.7	0.8	0.9
k=3, P^*=0.90 Δ^*=0.4, r=5							
P(CS;sim.)	-	-	1.000	0.999	0.974	0.961	0.952
E(NB;sim.)	-	-	7	6	5	4	2
E(N; sim.)	-	-	26	22	19	15	11
upper bound	-	-	120	110	105	90	60
trunc. seq.	-	-	0	0	0	0	0
k=3, P^*=0.95 Δ^*=0.2, r=14							
P(CS;sim.)	1.000	0.999	0.998	0.994	0.988	0.977	0.968
E(NB;sim.)	49	43	36	29	22	15	9
E(N; sim.)	166	150	128	107	87	62	42
upper bound	530	500	500	440	380	270	240
trunc. seq.	0	0	1	0	0	0	0
k=3, P^*=0.95 Δ^*=0.3, r=9							
P(CS;sim.)	-	1.000	1.000	0.997	0.985	0.984	0.965
E(NB;sim.)	-	19	16	13	10	7	5
E(N; sim.)	-	67	59	52	40	32	23
upper bound	-	230	230	230	220	160	140
trunc. seq.	-	0	0	0	1	0	0
k=3, P^*=0.95 Δ^*=0.4, r=6							
P(CS;sim.)	-	-	1.000	1.000	0.988	0.985	0.961
E(NB;sim.)	-	-	8	7	5	4	3
E(N; sim.)	-	-	31	26	22	18	13
upper bound	-	-	130	130	130	130	100
trunc. seq.	-	-	0	0	3	1	0
k=3, P^*=0.99 Δ^*=0.2, r=21							
P(CS;sim.)	1.000	1.000	1.000	0.999	0.996	0.999	0.988
E(NB;sim.)	73	64	54	44	33	22	13
E(N; sim.)	246	221	190	161	127	93	62
upper bound	850	750	660	580	450	400	350
trunc. seq.	0	0	0	0	0	1	0

p_A	0.3	0.4	0.5	0.6	0.7	0.8	0.9
k=3,P*=0.99 Δ*=0.3,r=13							
P(CS;sim.)	-	1.000	1.000	1.000	0.998	0.999	0.992
E(NB;sim.)	-	26	23	19	14	10	6
E(N; sim.)	-	94	85	72	57	45	31
upper bound	-	330	310	300	210	170	130
trunc. seq.	-	0	0	0	0	0	0
k=3,P*=0.99 Δ*=0.4,r=10							
P(CS;sim.)	-	-	1.000	1.000	1.000	0.998	1.000
E(NB;sim.)	-	-	13	11	8	6	4
E(N; sim.)	-	-	50	43	36	29	22
upper bound	-	-	180	165	135	125	100
trunc. seq.	-	-	0	0	0	0	0
k=4,P*=0.90 Δ*=0.2,r=13							
P(CS;sim.)	1.000	0.999	0.997	0.984	0.967	0.952	0.918
E(NB;sim.)	46	40	34	27	20	14	9
E(N; sim.)	206	182	159	129	101	75	49
upper bound	550	570	520	450	450	390	280
trunc. seq.	0	2	0	0	4	2	0
k=4,P*=0.90 Δ*=0.3,r=8							
P(CS;sim.)	-	1.000	0.998	0.987	0.972	0.963	0.947
E(NB;sim.)	-	16	14	11	9	6	4
E(N; sim.)	-	76	69	57	47	35	25
upper bound	-	250	250	250	250	180	150
trunc. seq.	-	0	0	1	2	0	0
k=4,P*=0.90 Δ*=0.4,r=6							
P(CS;sim.)	-	-	1.000	0.997	0.983	0.967	0.941
E(NB;sim.)	-	-	8	7	5	4	3
E(N; sim.)	-	-	40	36	28	23	16
upper bound	-	-	220	180	180	150	120
trunc. seq.	-	-	1	0	2	0	0

p_A	0.3	0.4	0.5	0.6	0.7	0.8	0.9
k=4,P*=0.95 Δ*=0.2,r=16							
P(CS;sim.)	1.000	1.000	1.000	0.995	0.986	0.973	0.969
E(NB;sim.)	58	50	40	33	26	17	10
E(N; sim.)	256	226	190	158	128	92	60
upper bound	710	670	670	600	560	390	320
trunc. seq.	0	0	2	0	0	1	0
k=4,P*=0.95 Δ*=0.3,r=10							
P(CS;sim.)	-	1.000	1.000	0.996	0.990	0.971	0.966
E(NB;sim.)	-	21	17	14	11	8	5
E(N; sim.)	-	97	84	71	59	44	31
upper bound	-	330	310	290	290	270	180
trunc. seq.	-	0	0	0	0	1	0
k=4,P*=0.95 Δ*=0.4,r=7							
P(CS;sim.)	-	-	1.000	1.000	0.989	0.987	0.970
E(NB;sim.)	-	-	9	8	6	4	3
E(N; sim.)	-	-	46	40	33	26	18
upper bound	-	-	170	170	170	170	120
trunc. seq.	-	-	0	0	0	0	0
k=4,P*=0.99 Δ*=0.2,r=23							
P(CS;sim.)	1.000	1.000	1.000	1.000	0.996	0.996	0.994
E(NB;sim.)	81	69	58	48	36	25	14
E(N; sim.)	357	315	270	229	178	132	84
upper bound	1160	1020	880	740	600	490	400
trunc. seq.	0	0	0	0	0	0	0
k=4,P*=0.99 Δ*=0.3,r=15							
P(CS;sim.)	-	1.000	1.000	0.999	1.000	1.000	0.997
E(NB;sim.)	-	31	26	21	16	11	7
E(N; sim.)	-	142	125	105	83	65	44
upper bound	-	450	390	340	330	280	250
trunc. seq.	-	0	0	0	0	0	0

p_A	0.3	0.4	0.5	0.6	0.7	0.8	0.9
k=4,P*=0.99 Δ*=0.4,r=10							
P(CS;sim.)	-	-	1.000	1.000	0.999	0.999	0.993
E(NB;sim.)	-	-	13	11	8	6	4
E(N; sim.)	-	-	64	55	45	37	26
upper bound	-	-	230	220	220	220	210
trunc. seq.	-	-	0	0	0	1	1
k=5,P*=0.90 Δ*=0.2,r=14							
P(CS;sim.)	1.000	1.000	0.999	0.990	0.987	0.961	0.925
E(NB;sim.)	50	42	37	29	22	15	9
E(N; sim.)	275	240	211	172	136	98	62
upper bound	690	640	630	610	570	490	330
trunc. seq.	0	0	0	1	1	1	1
k=5,P*=0.90 Δ*=0.3,r=9							
P(CS;sim.)	-	1.000	1.000	0.996	0.987	0.969	0.951
E(NB;sim.)	-	19	16	13	10	7	4
E(N; sim.)	-	107	94	77	62	47	33
upper bound	-	290	280	280	280	240	220
trunc. seq.	-	0	0	1	1	0	1
k=5,P*=0.90 Δ*=0.4,r=6							
P(CS;sim.)	-	-	1.000	0.997	0.982	0.967	0.940
E(NB;sim.)	-	-	8	7	5	4	3
E(N; sim.)	-	-	48	43	34	27	20
upper bound	-	-	210	210	200	200	190
trunc. seq.	-	-	0	1	0	1	2
k=5,P*=0.95 Δ*=0.2,r=17							
P(CS;sim.)	1.000	1.000	0.999	0.997	0.992	0.979	0.965
E(NB;sim.)	60	52	44	35	26	18	11
E(N; sim.)	330	292	252	208	162	118	76
upper bound	890	830	740	660	580	560	450
trunc. seq.	0	0	0	0	0	3	2

p_A	0.3	0.4	0.5	0.6	0.7	0.8	0.9
k=5,P*=0.95 Δ*=0.3,r=11							
P(CS;sim.)	-	1.000	1.000	0.998	0.997	0.989	0.971
E(NB;sim.)	-	23	19	16	12	8	5
E(N; sim.)	-	130	112	95	75	56	40
upper bound	-	420	380	320	280	260	250
trunc. seq.	-	0	0	0	0	0	1
k=5,P*=0.95 Δ*=0.4,r=8							
P(CS;sim.)	-	-	1.000	1.000	0.997	0.994	0.985
E(NB;sim.)	-	-	11	9	7	5	3
E(N; sim.)	-	-	63	53	45	35	25
upper bound	-	-	230	230	200	200	200
trunc. seq.	-	-	0	0	0	1	1
k=5,P*=0.99 Δ*=0.2,r=24							
P(CS;sim.)	1.000	1.000	0.999	1.000	1.000	0.997	0.994
E(NB;sim.)	85	73	61	50	37	26	14
E(N; sim.)	461	407	346	294	223	168	102
upper bound	1450	1280	1110	930	750	580	460
trunc. seq.	0	0	0	0	0	0	0
k=5,P*=0.99 Δ*=0.3,r=15							
P(CS;sim.)	-	1.000	1.000	1.000	0.999	0.998	0.996
E(NB;sim.)	-	30	26	21	16	11	6
E(N; sim.)	-	171	149	126	103	77	51
upper bound	-	560	490	420	400	380	280
trunc. seq.	-	0	0	0	1	1	0
k=5,P*=0.99 Δ*=0.4,r=11							
P(CS;sim.)	-	-	1.000	1.000	1.000	0.998	0.996
E(NB;sim.)	-	-	14	12	9	6	4
E(N; sim.)	-	-	86	72	59	46	32
upper bound	-	-	560	490	440	430	410
trunc. seq.	-	-	0	0	0	0	0

15.4 Numerical results for the VT-procedure

Table 11

p_A	0.3	0.4	0.5	0.6	0.7	0.8	0.9
$k=3, P^*=0.90$ $\Delta^*=0.2, s=4$							
P(CS;sim.)	0.988	0.960	0.936	0.945	0.928	0.955	0.995
E(NB;sim.)	20	20	20	19	19	19	21
E(N; sim.)	62	62	61	60	60	61	67
upper bound	280	280	280	280	280	280	280
trunc. seq.	1	1	0	0	2	0	0
$k=3, P^*=0.90$ $\Delta^*=0.3, s=3$							
P(CS;sim.)	-	0.993	0.973	0.960	0.950	0.976	0.991
E(NB;sim.)	-	11	11	10	11	11	11
E(N; sim.)	-	32	32	31	32	33	33
upper bound	-	150	150	150	150	150	150
trunc. seq.	-	0	0	0	1	0	0
$k=3, P^*=0.90$ $\Delta^*=0.4, s=2$							
P(CS;sim.)	-	-	0.982	0.961	0.940	0.947	0.973
E(NB;sim.)	-	-	6	6	6	6	6
E(N; sim.)	-	-	16	16	16	16	17
upper bound	-	-	80	80	80	80	80
trunc. seq.	-	-	1	0	0	0	0
$k=3, P^*=0.95$ $\Delta^*=0.2, s=5$							
P(CS;sim.)	1.000	0.987	0.974	0.976	0.968	0.991	0.999
E(NB;sim.)	26	26	24	25	25	26	26
E(N; sim.)	79	81	74	79	80	82	82
upper bound	350	350	350	350	350	350	350
trunc. seq.	0	0	0	1	0	1	0

p_A	0.3	0.4	0.5	0.6	0.7	0.8	0.9
k=3,P*=0.95 Δ*=0.3,s=3							
P(CS;sim.)	-	0.989	0.957	0.967	0.962	0.965	0.987
E(NB;sim.)	-	11	10	10	10	10	11
E(N; sim.)	-	31	31	31	31	32	33
upper bound	-	150	150	150	150	150	150
trunc. seq.	-	0	0	0	0	0	0
k=3,P*=0.95 Δ*=0.4,s=3							
P(CS;sim.)	-	-	0.998	0.996	0.993	0.993	0.999
E(NB;sim.)	-	-	9	8	8	8	8
E(N; sim.)	-	-	25	24	25	25	25
upper bound	-	-	100	100	100	100	100
trunc. seq.	-	-	0	0	0	0	0
k=3,P*=0.99 Δ*=0.2,s=7							
P(CS;sim.)	0.999	0.999	0.996	0.999	0.996	0.999	1.000
E(NB;sim.)	36	36	36	35	36	36	36
E(N; sim.)	111	113	112	111	114	114	114
upper bound	440	440	440	440	440	440	440
trunc. seq.	0	0	0	0	0	0	0
k=3,P*=0.99 Δ*=0.3,s=5							
P(CS;sim.)	-	0.999	0.998	0.998	0.995	0.999	1.000
E(NB;sim.)	-	17	18	17	18	18	17
E(N; sim.)	-	52	54	54	54	55	53
upper bound	-	200	200	200	200	200	200
trunc. seq.	-	0	0	0	0	0	0
k=3,P*=0.99 Δ*=0.4,s=4							
P(CS;sim.)	-	-	1.000	1.000	0.999	0.999	1.000
E(NB;sim.)	-	-	11	11	11	11	11
E(N; sim.)	-	-	32	33	32	33	34
upper bound	-	-	140	140	140	140	140
trunc. seq.	-	-	0	0	0	0	0

p_A	0.3	0.4	0.5	0.6	0.7	0.8	0.9
k=4,P*=0.90 Δ*=0.2,s=6							
P(CS;sim.)	0.996	0.982	0.966	0.958	0.951	0.974	0.994
E(NB;sim.)	27	25	24	25	25	25	26
E(N; sim.)	110	104	104	104	105	107	109
upper bound	400	400	400	400	400	400	400
trunc. seq.	1	0	1	2	1	0	0
k=4,P*=0.90 Δ*=0.3,s=4							
P(CS;sim.)	-	0.993	0.957	0.945	0.945	0.959	0.986
E(NB;sim.)	-	10	10	10	10	11	11
E(N; sim.)	-	41	42	41	41	43	44
upper bound	-	200	200	200	200	200	200
trunc. seq.	-	0	1	0	0	0	0
k=4,P*=0.90 Δ*=0.4,s=3							
P(CS;sim.)	-	-	0.963	0.932	0.912	0.926	0.974
E(NB;sim.)	-	-	6	6	6	6	6
E(N; sim.)	-	-	21	21	21	21	22
upper bound	-	-	120	120	120	120	120
trunc. seq.	-	-	1	0	0	0	0
k=4,P*=0.95 Δ*=0.2,s=6							
P(CS;sim.)	1.000	0.991	0.981	0.989	0.984	0.995	0.999
E(NB;sim.)	31	30	30	29	30	30	31
E(N; sim.)	126	126	127	126	129	130	131
upper bound	450	450	450	450	450	450	450
trunc. seq.	0	2	1	2	1	0	1
k=4,P*=0.95 Δ*=0.3,s=4							
P(CS;sim.)	-	0.998	0.989	0.990	0.979	0.989	0.997
E(NB;sim.)	-	14	14	14	14	14	14
E(N; sim.)	-	57	57	57	58	57	59
upper bound	-	230	230	230	230	230	230
trunc. seq.	-	1	0	0	0	0	0

p_A	0.3	0.4	0.5	0.6	0.7	0.8	0.9
k=4,P*=0.95 Δ*=0.4,s=3							
P(CS;sim.)	-	-	0.996	0.988	0.985	0.990	0.996
E(NB;sim.)	-	-	8	9	8	8	8
E(N; sim.)	-	-	32	34	33	33	33
upper bound	-	-	140	140	140	140	140
trunc. seq.	-	-	0	1	0	0	0
k=4,P*=0.99 Δ*=0.2,s=8							
P(CS;sim.)	1.000	0.999	0.997	0.996	0.998	1.000	1.000
E(NB;sim.)	41	41	40	40	40	41	41
E(N; sim.)	169	171	170	173	171	174	174
upper bound	580	580	580	580	580	580	580
trunc. seq.	0	0	1	0	1	0	0
k=4,P*=0.99 Δ*=0.3,s=5							
P(CS;sim.)	-	0.999	0.998	0.996	0.993	0.997	1.000
E(NB;sim.)	-	18	18	17	17	17	18
E(N; sim.)	-	71	72	71	73	71	73
upper bound	-	280	280	280	280	280	280
trunc. seq.	-	0	0	0	0	0	0
k=4,P*=0.99 Δ*=0.4,s=4							
P(CS;sim.)	-	-	1.000	0.996	0.998	0.999	0.999
E(NB;sim.)	-	-	11	11	11	11	11
E(N; sim.)	-	-	44	43	44	44	44
upper bound	-	-	160	160	160	160	160
trunc. seq.	-	-	0	0	0	0	0
k=5,P*=0.90 Δ*=0.2,s=5							
P(CS;sim.)	0.998	0.966	0.951	0.952	0.938	0.964	0.998
E(NB;sim.)	26	25	25	24	24	25	25
E(N; sim.)	135	132	130	125	129	136	134
upper bound	500	500	500	500	500	500	500
trunc. seq.	2	0	1	1	0	0	0

p_A	0.3	0.4	0.5	0.6	0.7	0.8	0.9
k=5,P*=0.90 Δ*=0.3,s=3							
P(CS;sim.)	-	0.980	0.951	0.922	0.928	0.953	0.979
E(NB;sim.)	-	11	10	10	10	10	11
E(N; sim.)	-	52	52	51	51	52	55
upper bound	-	230	230	230	230	230	230
trunc. seq.	-	0	0	0	0	0	0
k=5,P*=0.90 Δ*=0.4,s=3							
P(CS;sim.)	-	-	0.997	0.982	0.981	0.986	0.996
E(NB;sim.)	-	-	9	8	8	8	8
E(N; sim.)	-	-	41	40	40	41	42
upper bound	-	-	180	180	180	180	180
trunc. seq.	-	-	0	1	1	0	0
k=5,P*=0.95 Δ*=0.2,s=6							
P(CS;sim.)	0.999	0.995	0.985	0.975	0.982	0.991	0.999
E(NB;sim.)	31	30	29	29	29	30	31
E(N; sim.)	160	157	154	155	156	163	163
upper bound	550	550	550	550	550	550	550
trunc. seq.	0	2	0	1	0	0	0
k=5,P*=0.95 Δ*=0.3,s=4							
P(CS;sim.)	-	0.996	0.986	0.980	0.979	0.983	0.996
E(NB;sim.)	-	14	14	14	14	14	14
E(N; sim.)	-	72	71	70	72	71	73
upper bound	-	300	300	300	300	300	300
trunc. seq.	-	0	0	0	1	0	0
k=5,P*=0.95 Δ*=0.4,s=3							
P(CS;sim.)	-	-	0.998	0.982	0.977	0.985	0.996
E(NB;sim.)	-	-	8	8	8	8	8
E(N; sim.)	-	-	41	40	41	40	41
upper bound	-	-	160	160	160	160	160
trunc. seq.	-	-	0	0	0	0	0

p_A	0.3	0.4	0.5	0.6	0.7	0.8	0.9
k=5,P*=0.99 Δ*=0.2,s=8							
P(CS;sim.)	1.000	0.998	0.998	0.993	1.000	0.999	1.000
E(NB;sim.)	41	41	40	40	40	40	41
E(N; sim.)	210	213	213	212	212	213	216
upper bound	750	750	750	750	750	750	750
trunc. seq.	0	1	1	0	1	0	0
k=5,P*=0.99 Δ*=0.3,s=5							
P(CS;sim.)	-	1.000	0.997	0.992	0.995	1.000	1.000
E(NB;sim.)	-	18	17	17	17	17	18
E(N; sim.)	-	89	88	89	90	89	93
upper bound	-	380	380	380	380	380	380
trunc. seq.	-	0	0	0	1	0	0
k=5,P*=0.99 Δ*=0.4,s=4							
P(CS;sim.)	-	-	1.000	0.997	0.991	0.995	0.999
E(NB;sim.)	-	-	11	11	11	11	11
E(N; sim.)	-	-	53	54	53	55	55
upper bound	-	-	250	250	250	250	250
trunc. seq.	-	-	2	0	0	0	0

15.5 Comparison of selection models

The selection models presented in sec. 1 - sec. 3 have been compared in sec. 4 and a two-stage selection procedure using either PW- and VT-sampling or PW-,VT-, and PL-sampling in the second stage have been consequently constructed in sec.5. In sec. 6 and sec. 7 PW- and VT-sampling selection models using inverse stopping rules have been considered. The comparison of these procedures is given in sec.8 (cf. 8.4). It has been pointed out there that the PW-procedure is uniformly better than the corresponding VT-procedure with respect to both the expected number of patients on the inferior treatment and the whole number of patients involved in the experiment.

The comparison of the PW-procedure of sec. 6 with the VT-procedure of sec. 2 is as follows. The PW-procedure is to be preferred to the VT-procedure only if the success parameter p_A is not less than 0.7 (0.8) when $E(N_B)(E(N))$ is used as criterion for comparison. These limits increase if P^* increases, and that means for instance that the PW-procedure is for the special case $P^*=0.99$ better than the VT procedure with respect to $E(N_B)$ only if $p_A \geq 0.8$. Note in addition that the VT-procedure has higher P(CS)-values in the LFC than the PW-procedure.
The PW-procedure of sec. 1 is always to be preferred to the PW-procedure of sec.6.

Another criterion for choosing a selection model is implied by the expected truncation points given in sec. 12. The experimenter himself must decide whether enough experimental units are available so as to conduct the experiment according to the chosen selection model. In case that the expected truncation points are greater than the available number of experimental units, another selection model, having smaller truncation points, should be applied so that a terminal decision can be taken at the desired significance level P^*. Considering for instance the PW-selection model of sec. 1 that is uniformly better than the PW-selection model of sec.6, we recognize that on the other hand the better selection model requires the larger truncation points.

An exact comparison of the selection models considered in sec. 13 and sec. 14 is not possible, because there are too large differences in the minimal P(CS)-values. In comparing only such expectations that are associated with approximately the same P(CS)-value in the LFC we may conclude that the PW selection model of sec. 13 is better than the VT selection model of sec. 14 if the parameters are either small or large, i.e. the VT-procedure should be preferred if the success parameter p_A lies between 0.4 and 0.6. The comparison of these two procedures suggests to construct a two-stage selection procedure similar to that of sec. 5. We did not carry out this idea.

The PW selection model of sec. 13 is better than the PW selection model of sec.1 with respect to $E(N_B)(E(N))$, if $p_A \leq 0.7$ (≤ 0.8), and the same selection model is better than the VT selection model of sec. 2 with respect to $E(N_B)(E(N))$, if $p_A \leq 0.5$ (≤ 0.4).
The VT selection model of sec. 14 is better than the PW selection model of sec.1 with respect to $E(N_B)(E(N))$, if $p_A \leq 0.6$ (≤ 0.7), and the same selection model is better than the VT selection model of sec. 2 with respect to $E(N_B)(E(N))$, if $p_A \leq 0.5$ (≤ 0.5).
Since none of these selection models is uniformly better than the others, a two-stage selection procedure may be constructed.

Two selection models constructed for selecting the best of $k \geq 3$ treatments have been presented in sec. 10 and sec. 11. The PW selection model of sec. 10 proved to be uniformly better than the corresponding VT selection model of sec. 11, where these models could be compared exactly because of their identical P(CS)-values.

In sec. 15 we have considered two selection models that are the extensions to more than two treatments of the corresponding selection models given in sec.1 and sec.2. The PW selection model proved to be better than the VT selection model with respect to $E(N_B)(E(N))$, if $p_A > 0.6$ (>0.7). (Recall that $E(N_B)=(E(N)-E(N_A))/(k-1)$ if k treatments are considered.)

The PW elimination procedure of sec. 15 seems to be uniformly better than the PW procedure of section 10. An exact comparison of these procedures is of no interest because the selection procedures of sec. 10 and sec. 11 are uniformly worse than the corresponding selection procedures to be presented in sec. 6 and sec. 7 of the following chapter.

CHAPTER 2

Selection Procedures with Restricted Patient Horizon

1 The Selection Model $[2;PW;\max\{S_A+F_B,S_B+F_A\}=r]$

1.1 Introduction

We have found out in chapter 1 that selection procedures, the termination rule of which is based on either successes or failures, lead to an expected number of patients on the inferior treatment which is extremely large whenever the success-parameter of the best treatment is such that the single events "success" or "failure", involved in the termination rule, have only a small probability to occur. It is for example intuitively obvious that a small success probability of the best treatment generates an expected patient horizon (and hence an expected number of patients on the poorer treatment) unnecessarily large if the termination rule is based on the absolute number of successes only. An alternative for this special case, i.e. when some prior information about the success parameters is available, has been found in section 8 of chapter 1.
As already mentioned in 4.6 of the preceding chapter, the essential point consists in the procurement of the prior information. In order to avoid a testing-phase, where each treatment had to be given to a patient, we alter the termination rule in such a way that not only "successes" but also "failures" are considered to take the decision which treatment is to select as best. A very agreable feature, caused by the just mentioned construction of the stopping rule, is the restricted patient horizon, i.e. there exists an upper bound, that gives the maximal number of trials that had to be carried out until a decision is taken. The patient horizon N is moreover a random variable but a restricted one, i.e. N may assume only a finite number of integers with a positive probability whereas

the patient horizon of all procedures, presented in chapter 1, may assume all integers not less than r with a positive probability. It is not necessary to say that selection procedures with a restricted patient horizon are more convenient for practical purposes than selection procedures which terminate only with probability one.

1.2 Derivation of the P(CS)-value

The following procedure uses the PW-sampling scheme and the termination rule $\max\{R_A,R_B\}=r$, where $R_A:=S_A+F_B$, $R_B:=S_B+F_A$ and S_A,S_B and F_A,F_B denote the current number of successes and failures of treatment A and B, respectively. r must be chosen such that the $(P^*;\Delta^*)$-condition holds. Since the experiment is carried out according to the PW-sampling scheme, either R_A or R_B is increased by one at each stage. This implies that the selection procedure is truncated after at most 2r-1 trials, that means, R_A and R_B take on the value r-1 and after the (2r-1)th trial either R_A or R_B takes on the value r, and a decision can be taken. Treatment A is declared as best if $R_A=r$; in case of $R_B=r$, treatment B is wrongly selected. The philosophy on which this proceeding is based seems to be intuitively obvious, that is, many successes of treatment A or many failures of treatment B suggest the selection of treatment A whereas many successes of treatment B or many failures of treatment A indicate that treatment B is the best. In the following, a single trial is said to be a "R_A-trial" if the outcome of the trial is an "A-success" or a "B-failure" and a trial the outcome of which is a "B-success" or an "A-failure" is referred to as a "R_B-trial".
To derive an exact expression for P(CS), we next define the random variable $T=(T_A,T_B)$, where T_A,T_B denote the additional number of R_A-, R_B-trials needed to select A,B as best, respectively. Defining U(m,n) and V(m,n) by

$$U(m,n):=P(CS|T=(m,n),NT=A);V(m,n):=P(CS|T=(m,n),NT=B), \qquad (1.1)$$

and noting that the PW-sampling scheme is used, the following system of difference equations and boundary conditions is readily derived.

$$\begin{gathered}U(m,n)=p_AU(m-1,n)+q_AV(m,n-1);V(m,n)=p_BV(m,n-1)+q_BU(m-1,n)\\ U(0,n)=1;V(m,0)=0 \text{ for } m,n\in\mathbb{N}\end{gathered} \qquad (1.2)$$

Recalling that the first treatment at the outset is determined by randomization, the expression for P(CS) is seen to be

$$P(CS) = \frac{1}{2}\,(U(r,r)+V(r,r)) \qquad (1.3)$$

(1.2) can be solved by using the generating functions

$$U:=\sum_{m=1}^{\infty}\sum_{n=1}^{\infty}U(m,n)x^my^n;\quad V:=\sum_{m=1}^{\infty}\sum_{n=1}^{\infty}V(m,n)x^my^n \tag{1.4}$$

We finally obtain

$$U(1-p_Ax)=\frac{p_Axy}{1-y}+q_AyV;\quad V(1-p_By)=q_BxU+\frac{q_Bxy}{1-y} \tag{1.5}$$

Using the abbreviation $G:=(1-p_Ax)(1-p_By)-q_Aq_Bxy$, (1.6)
we get from (1.5)

$$U=\frac{q_Aq_Bxy^2+p_Axy(1-p_By)}{(1-y)G}\;;\quad V=\frac{q_Bxy}{(1-y)G} \tag{1.7}$$

The power series expansion of $1/G$ is given by

$$\begin{aligned}\frac{1}{G}&=\frac{1}{xy}\sum_{i=0}^{\infty}\left(\frac{x}{1-p_Ax}\right)^{i+1}\left(\frac{y}{1-p_By}\right)^{i+1}(q_Aq_B)^i=\\&=\sum_{j=0}^{\infty}\sum_{\nu=0}^{\infty}\left(\sum_{i=0}^{\infty}\binom{\nu+i}{i}\binom{j+i}{i}(q_Aq_Bxy)^i\right)p_A^{\nu}p_B^{j}\,x^{\nu}y^{j}=:\sum_{\lambda=0}^{\infty}\sum_{\tau=0}^{\infty}\xi_{\lambda\tau}x^{\lambda}y^{\tau}\end{aligned} \tag{1.8}$$

(the exact expression for $\xi_{\lambda\tau}$ will be given in (1.16)).
The expansion of the factor $q_Bxy/(1-y)$ of the second identity in (1.7) is immediately seen to be $\sum_{\ell=0}^{\infty}q_Bxy^{\ell+1}$, and hence the expansion of V is obtained by multiplying the last power series with that one of $1/G$. The only coefficient of interest is $V(r,r)$. Noting that $\nu=r-i-1$ and $j=r-i-1-\ell$, where i runs from 0 to r-1 and ℓ runs from 0 to r-i-1, we obtain

$$\begin{aligned}V(r,r)&=\sum_{i=0}^{r-1}\sum_{\ell=0}^{r-1-i}\binom{\nu+i}{i}\binom{j+i}{i}(q_Aq_B)^ip_A^{\nu}p_B^{j}q_B=\\&=\sum_{i=0}^{r-1}\binom{r-1}{i}q_A^ip_A^{r-1-i}J_{q_B}(i+1,r-i)=E(J_{q_B}(\hat{X}_{r-1,q_A}+1,r-\hat{X}_{r-1,q_A})),\end{aligned} \tag{1.9}$$

where $\hat{X}_{r-1,q_A}$ denotes a binomial chance variable with success parameter q_A and index r-1.
Recalling that we assumed $p_A,p_B\in(0,1)$, the second equation of (1.2) can be used to derive the exact expression for $U(r,r)$. We get

$$U(r,r)=\frac{1}{q_B}V(r+1,r)-\frac{p_B}{q_B}V(r+1,r-1) \tag{1.10}$$

In the same way as above, we obtain

$$V(r+1,r) = \sum_{i=0}^{r} \sum_{\ell=0}^{r-1-i} \binom{r}{i}\binom{r-1-\ell}{i} q_B^{i+1} q_A^i p_A^{r-i} p_B^{r-i-1-\ell} = \tag{1.11}$$
$$= \sum_{i=0}^{r} \binom{r}{i} q_A^i p_A^{r-i} J_{q_B}(i+1,r-i) = E(J_{q_B}(\hat{X}_{r,q_A}+1,r-\hat{X}_{r,q_A})),$$

and the last coefficient needed to evaluate U(r,r) is shown to be

$$V(r+1,r-1) = \sum_{i=0}^{r} \sum_{\ell=0}^{r-2-i} \binom{r}{i}\binom{r-2-\ell}{i} q_A^i q_B^{i+1} p_A^{r-i} p_B^{r-i-2-\ell} = \tag{1.12}$$
$$= \sum_{i=0}^{r} \binom{r}{i} \cdot q_A^i p_A^{r-i} J_{q_B}(i+1,r-1-i) = E(J_{q_B}(\hat{X}_{r,q_A}+1,r-\hat{X}_{r,q_A}-1))$$

The expression for P(CS) is now obtainable from (1.9)-(1.12), we have

$$P(CS) = \frac{1}{2q_B}(q_B E(J_{q_B}(\hat{X}_{r-1,q_A}+1,r-\hat{X}_{r-1,q_A})) + \tag{1.13}$$
$$+ E(J_{q_B}(\hat{X}_{r,q_A}+1,r-\hat{X}_{r,q_A})) - p_B E(J_{q_B}(\hat{X}_{r,q_A}+1,r-\hat{X}_{r,q_A}-1))).$$

1.3 Determination of the LFC

Our goal is the determination of those parameter values which minimize the expression for P(CS) given above. We next verify that V(m,n) and U(m,n) are decreasing functions of p_B and increasing functions of p_A provided p_A,p_B are fixed, respectively. For this purpose we evaluate the derivative of V(m,n) with respect to p_A. From (1.4) and (1.7) follows

$$\frac{\partial V}{\partial p_A} = \sum_{m=1}^{\infty}\sum_{n=1}^{\infty} \frac{\partial V(m,n)}{\partial p_A} x^m y^n; \quad \frac{\partial V}{\partial p_A} = \frac{q_B x^2 y}{G^2} \tag{1.14}$$

It is immediately seen from (1.8) that the coefficients of the expansion of 1/G are all together nonnegative, and the same logically holds for the coefficients of the expansions of $1/G^2$ and $q_B x^2 y/G^2$. Applying now the identity theorem for power series (cf. A3/1) to (1.14), we find that $\frac{\partial V(m,n)}{\partial p_A} \geq 0$ for all $m,n \in \mathbb{N}$, and this again implies that V(m,n) is an increasing function of p_A.

Differentiating V(m,n) with respect to p_B, we obtain from (1.4) and (1.7)

$$\frac{\partial V}{\partial p_B} = \sum_{m=1}^{\infty}\sum_{n=1}^{\infty} \frac{\partial V(m,n)}{\partial p_B} x^m y^n; \quad \frac{\partial V}{\partial p_B} = \frac{xy(p_A x-1)}{G^2} \tag{1.15}$$

Setting $\nu=\lambda-i$ and $j=\tau-i$ the coefficients $\xi_{\lambda\tau}$ of the expansion of 1/G (cf. (1.8)) are found to be

$$\xi_{\lambda\tau} = \sum_{i=0}^{\min\{\tau,\lambda\}} \binom{\lambda}{i}\binom{\tau}{i} (q_A q_B)^i p_A^{\lambda-i} p_B^{\tau-i}, \tag{1.16}$$

and hence

$$\frac{p_A x-1}{G} = \sum_{\lambda=1}^{\infty}\sum_{\tau=0}^{\infty} (p_A\xi_{\lambda-1,\tau}-\xi_{\lambda\tau})x^\lambda y^\tau - \sum_{\tau=0}^{\infty} \xi_{0\tau}y^\tau, \tag{1.17}$$

where $p_A\xi_{\lambda-1,\tau}-\xi_{\lambda\tau}$ is easily shown to be nonpositive by using (1.16) and the inequality $\binom{\lambda-1}{i} \leq \binom{\lambda}{i}$. Noting that the second term in (1.17) is subtracted ($\xi_{0\tau}=1$), we obtain that the coefficients of the expansion of $(p_A x-1)/G$ are all together nonpositive. Recalling in addition that the coefficients of the expansion of $1/G$ are all together nonnegative and applying A3/1 to (1.15), we finally obtain that $\frac{\partial V(m,n)}{\partial p_B} \leq 0$ for all $m,n \in \mathbb{N}$, and this means that $V(m,n)$ is a decreasing function of p_B.

It would be tedious to carry out similar calculations for $U(m,n)$, they are left as an exercise to the reader; we only specify the derivations of $U(m,n)$ with respect to p_A and p_B; we find

$$\frac{\partial U}{\partial p_A} = \frac{xy(1-p_B y)}{G^2} \quad ; \quad \frac{\partial U}{\partial p_B} = -\frac{q_A x y^2}{G^2} \tag{1.18}$$

As usual we define $\bar{p} := \frac{1}{2}\cdot(p_A+p_B)$. Hence, $p_A=\bar{p}+\frac{\Delta}{2}$, $p_B=\bar{p}-\frac{\Delta}{2}$; and p_A, p_B are increasing, decreasing functions of Δ, respectively, provided $\bar{p}$ is fixed. This implies that for fixed $\bar{p}$ $U(m,n)$ and $V(m,n)$ are increasing functions of Δ, in particular $P(CS) = \frac{1}{2}(U(r,r)+V(r,r))$ is an increasing function of Δ, i.e. $P(CS)$ gets its lowest value, - and may be enlarged by an enlargement of r only - if Δ takes on the smallest possible value, that means,

$$\text{the LFC requires that } \Delta = \Delta^* \tag{1.19}$$

We next have to determine that value of $\bar{p}$ which minimizes $P(CS)$. Noting that

$$\frac{\partial U}{\partial \bar{p}} = \frac{\partial U}{\partial p_A} + \frac{\partial U}{\partial p_B} \quad ; \quad \frac{\partial V}{\partial \bar{p}} = \frac{\partial V}{\partial p_A} + \frac{\partial V}{\partial p_B}, \tag{1.20}$$

and using (1.14), (1.15), and (1.18), we obtain

$$\frac{\partial (U+V)}{\partial \bar{p}} = \frac{xy(x(q_B+p_A)-y(p_B+q_A))}{G^2} \tag{1.21}$$

Recalling that $2\frac{\partial}{\partial \bar{p}}P(CS) = \frac{\partial}{\partial \bar{p}}(U(r,r)+V(r,r))$, we have to do nothing but to evaluate the coefficient of $x^r y^r$ in the expansion of (1.21). We get

$$\frac{1}{G^2} := \sum_{m=0}^{\infty}\sum_{n=0}^{\infty} \omega_{mn} x^m y^n, \text{ where } \omega_{mn} = \sum_{\tau=0}^{n}\sum_{\lambda=0}^{m} \xi_{\lambda\tau}\xi_{m-\lambda,n-\tau} \tag{1.22}$$

and ξ_{ij} is given by (1.16). It follows

$$\omega_{mn} = p_A^m p_B^n \sum_{\tau=0}^{n}\sum_{\lambda=0}^{m}\sum_{i=0}^{\min\{\tau,\lambda\}}\sum_{j=0}^{\min\{n-\tau,m-\lambda\}} \binom{\lambda}{i}\binom{\tau}{i}\binom{m-\lambda}{j}\binom{n-\tau}{j}\left(\frac{q_A q_B}{p_A p_B}\right)^{i+j}, \tag{1.23}$$

and using now (1.21), we finally obtain

$$\frac{\partial}{\partial\bar{p}}(U(r,r)+V(r,r)) = (q_B+p_A)\omega_{r-2,r-1} - (p_B+q_A)\omega_{r-1,r-2} =$$

$$= (q_B+p_A)p_A^{r-2}p_B^{r-1}\sum_{\tau=0}^{r-1}\sum_{\lambda=0}^{r-2}\sum_{i=0}^{\min\{\tau,\lambda\}}\sum_{j=0}^{\min\{r-1-\tau,r-2-\lambda\}}\binom{\lambda}{i}\binom{\tau}{i}\binom{r-2-\lambda}{j}\binom{r-1-\tau}{j}\cdot$$

$$\cdot\left(\frac{q_A q_B}{p_A p_B}\right)^{i+j} - (p_B+q_A)p_A^{r-1}p_B^{r-2}\sum_{\tau=0}^{r-2}\sum_{\lambda=0}^{r-1}\sum_{i=0}^{\min\{\tau,\lambda\}}\sum_{j=0}^{\min\{r-2-\tau,r-1-\lambda\}} \tag{1.24}$$

$$\binom{\lambda}{i}\binom{\tau}{i}\binom{r-1-\lambda}{j}\binom{r-2-\tau}{j}\left(\frac{q_A q_B}{p_A p_B}\right)^{i+j}$$

Since the fourfold series in (1.24) coincide , - we denote them by g(r) - (1.24) reduces to

$$\frac{\partial}{\partial\bar{p}}(U(r,r)+V(r,r)) = (p_A p_B)^{r-2}\,\Delta(2\bar{p}-1)g(r), \tag{1.25}$$

and from this follows immediately that

$$\frac{\partial}{\partial\bar{p}}(U(r,r)+V(r,r)) = \begin{cases} <0, \text{ if } \bar{p}<1/2 \\ =0, \text{ if } \bar{p}=1/2 \\ >0, \text{ if } \bar{p}>1/2 \end{cases} \tag{1.26}$$

P(CS) is consequently minimized by $\bar{p} = 1/2$, i.e. the LFC is given by

$$p_A = \frac{1}{2} + \frac{\Delta^*}{2} \text{ and } p_B = \frac{1}{2} - \frac{\Delta^*}{2}, \text{ i.e. } p_B = q_A \text{ and } p_A = q_B \tag{1.27}$$

1.4 Derivation of the critical r-value

In order to determine the critical r-value, we have to calculate the probability of correct selection when the parameters are in the LFC. Substituting the parameters given above into (1.13), the expression for P(CS) in the LFC has a simpler form as before. We will show that the following expression

$$E(J_{q_B}(\hat{X}_{r,q_A}+1,r-\hat{X}_{r,q_A})) - p_B E(J_{q_B}(\hat{X}_{r,q_A}+1,r-\hat{X}_{r,q_A}-1)) =$$

$$= \sum_{i=0}^{r-1} \sum_{n=0}^{r-1-i} \binom{i+n}{i}\binom{r}{i} q_A^{i+n} p_A^{r+1} - \sum_{i=0}^{r-2} \sum_{n=0}^{r-2-i} \binom{i+n}{i}\binom{r}{i} q_A^{i+n+1} p_A^{r+1} \tag{1.28}$$

is equal to $q_B E(J_{q_B}(\hat{X}_{r-1,q_A}+1, r-\hat{X}_{r-1,q_A}))$ (1.29)

For this purpose we prove that the following equation holds

$$\sum_{i=0}^{r-1} \sum_{n=0}^{r-1-i} \binom{i+n}{i}\binom{r}{i} q_A^{i+n} p_A^{r+1} - q_B E(J_{q_B}(\hat{X}_{r-1,q_A}+1, r-\hat{X}_{r-1,q_A})) = \\ = \sum_{i=0}^{r-2} \sum_{n=0}^{r-2-i} \binom{i+n}{i}\binom{r}{i} q_A^{i+n+1} p_A^{r+1} \tag{1.30}$$

(1.30) is found to be equivalent to

$$\sum_{i=0}^{r-2} \sum_{n=0}^{r-2-i} \binom{i+n+1}{n}\binom{r-1}{i} q_A^{i+n} = \sum_{i=0}^{r-2} \sum_{n=0}^{r-2-i} \binom{i+n}{n}\binom{r}{i} q_A^{i+n}, \tag{1.31}$$

and (1.31) again is equivalent to

$$\sum_{j=0}^{r-2} \sum_{i=0}^{j} \binom{j+1}{j-1}\binom{r-1}{i} q_A^{j} = \sum_{j=0}^{r-2} \sum_{i=0}^{j} \binom{j}{j-i}\binom{r}{i} q_A^{j} \tag{1.32}$$

The last equation is immediately seen to hold because of the well-known identity

$$\sum_{i=0}^{j} \binom{j+1}{j-i}\binom{r-1}{i} = \sum_{i=0}^{j} \binom{j}{j-i}\binom{r}{i} = \binom{j+r}{j}$$

Making now use of (1.30), P(CS) reduces to

$$P(CS) = E(J_{q_B}(\hat{X}_{r-1,q_A}+1, r-\hat{X}_{r-1,q_A})), \tag{1.33}$$

where $q_A = \frac{1}{2} - \frac{\Delta^*}{2}$ and $q_B = \frac{1}{2} + \frac{\Delta^*}{2}$. For preassigned Δ^* and P^*, the critical r-value is obtained by enlarging r as long as P(CS) in the LFC exceeds P^* for the first time.

1.5 Derivation of the expectations

Defining

$$R_B(m,n) := E(N_B | T=(m,n), NT=A); S_B(m,n) := E(N_B | T=(m,n), NT=B), \tag{1.34}$$

where T has the same meaning as in (1.1), we have

$$R_B(m,n)=p_AR_B(m-1,n)+q_AS_B(m,n-1)$$
$$S_B(m,n)=p_BS_B(m,n-1)+q_BR_B(m-1,n)+1 \qquad (1.35)$$
$$R_B(0,n)=S_B(m,0)=0 \text{ for all natural numbers } m,n$$

Noting the initial randomization, $E(N_B)$ is found to be

$$E(N_B) = \frac{1}{2}\left(R_B(r,r)+S_B(r,r)\right) \qquad (1.36)$$

(1.35) is solved by using the generating functions

$$R_B:=\sum_{m=1}^{\infty}\sum_{n=1}^{\infty} R_B(m,n)x^my^n \text{ and } S_B:=\sum_{m=1}^{\infty}\sum_{n=1}^{\infty} S_B(m,n)x^my^n, \qquad (1.37)$$

we next obtain

$$R_B=p_AxR_B+q_AyS_B \text{ and } S_B=p_ByS_B+q_BxR_B+\frac{x}{1-x}\cdot\frac{y}{1-y}, \qquad (1.38)$$

which finally gives

$$R_B = \frac{q_Axy^2}{G(1-x)(1-y)} \; ; \; S_B = \frac{xy(1-p_Ax)}{G(1-x)(1-y)}, \qquad (1.39)$$

where G has the same meaning as in (1.6).
The coefficients of the power series expansion of R_B are found to be

$$R_B(m,n) = \sum_{i=0}^{\min\{n-2,m-1\}} \frac{1}{q_B} J_{q_A}(i+1,m-i)J_{q_B}(i+1,n-i-1) \qquad (1.40)$$

From (1.35) follows that

$$S_B(r,r) = \frac{1}{q_A}R_B(r,r+1) - \frac{p_A}{q_A}R_B(r-1,r+1), \qquad (1.41)$$

and $E(N_B)$ is hence given by

$$\begin{aligned} E(N_B) = {} & \frac{1}{2q_B}\sum_{i=0}^{r-2} J_{q_A}(i+1,r-i)J_{q_B}(i+1,r-i-1) + \\ & + \frac{1}{2q_Aq_B}\sum_{i=0}^{r-1} J_{q_A}(i+1,r-i)J_{q_B}(i+1,r-i) - \qquad (1.42) \\ & - \frac{p_A}{2q_Aq_B}\sum_{i=0}^{r-2} J_{q_A}(i+1,r-i-1)J_{q_B}(i+1,r-i) \end{aligned}$$

An expression for $E(N_A)$ can be easily found by defining

$$\tilde{R}_A(m,n):=E(N_A|T=(n,m),NT=B), \tilde{S}_A(m,n):=E(N_A|T=(n,m),NT=A) \qquad (1.43)$$

and by solving the following system of difference equations and boundary conditions

$$\tilde{R}_A(m,n) = p_B\tilde{R}_A(m-1,n) + q_B\tilde{S}_A(m,n-1)$$
$$\tilde{S}_A(m,n) = p_A\tilde{S}_A(m,n-1) + q_A\tilde{R}_A(m-1,n) +1 \qquad (1.44)$$
$$\tilde{R}_A(0,n) = \tilde{S}_A(m,0) = 0 \quad \text{for all } m,n\in\mathbb{N}$$

$E(N_A)$ is given by

$$E(N_A) = \frac{1}{2}(\tilde{R}_A(r,r) + \tilde{S}_A(r,r)) \qquad (1.45)$$

Noting that (1.44) is obtained from (1.35) by interchanging p_A and p_B as well as q_A and q_B, $E(N_A)$ can be derived directly from (1.42), that is

$$E(N_A)=E(N_B|p_A \leftrightarrows p_B);\ E(N)=E(N_B)+E(N_B|p_A \leftrightarrows p_B) \qquad (1.46)$$

Some comments concerning the use of the table

The P(CS)-values have not been tabulated, because the LFC can be determined exactly, and by evaluating the critical r-value in the LFC, P(CS) is not less than P* for all possible p_A.

]E(·)[have been tabulated throughout.

1.6 Numerical results

Table 1

p_A	0.2	0.3	0.4	0.5	0.6	0.7	0.8	0.9
$P^* = 0.90,\ \Delta^* = 0.1,\ r = 82$								
$E(N_B)$	73	72	71	69	66	62	56	43
$E(N)$	155	154	152	150	147	143	137	124
$P^* = 0.90,\ \Delta^* = 0.2,\ r = 21$								
$E(N_B)$	-	17	16	16	15	13	12	9
$E(N)$	-	37	37	36	35	34	32	29
$P^* = 0.90,\ \Delta^* = 0.3,\ r = 9$								
$E(N_B)$	-	-	7	6	6	5	5	4
$E(N)$	-	-	15	15	14	14	13	12
$P^* = 0.95,\ \Delta^* = 0.1,\ r = 135$								
$E(N_B)$	120	119	116	113	109	102	92	70
$E(N)$	255	253	251	247	243	236	226	204
$P^* = 0.95,\ \Delta^* = 0.2,\ r = 34$								
$E(N_B)$	-	27	26	25	23	21	18	13
$E(N)$	-	61	60	58	57	55	52	47
$P^* = 0.95,\ \Delta^* = 0.3,\ r = 15$								
$E(N_B)$	-	-	11	10	9	8	7	5
$E(N)$	-	-	25	25	24	23	22	20
$P^* = 0.99,\ \Delta^* = 0.1,\ r = 270$								
$E(N_B)$	241	237	232	226	217	204	182	138
$E(N)$	510	506	502	495	486	473	451	407
$P^* = 0.99,\ \Delta^* = 0.2,\ r = 67$								
$E(N_B)$	-	53	51	49	46	41	35	24
$E(N)$	-	119	117	115	112	108	101	91
$P^* = 0.99,\ \Delta^* = 0.3,\ r = 29$								
$E(N_B)$	-	-	20	19	17	16	13	9
$E(N)$	-	-	49	47	46	44	41	37

2 The Selection Model $[2;PW;\max\{S_A,S_B\}=r \text{ or } F_A=F_B=c]$

2.1 Derivation of the P(CS)-value

The first treatment at the outset is determined by randomization because we use the PW-sampling scheme. Sampling terminates whenever either treatment yields r successes or both yield c failures. In either case the treatment with the larger number of successes is selected as best. If the numbers of successes are the same - which can occur only if we stop by reason of failures, i.e. if the procedure has gone through c cycles - the best treatment is determined randomly.

The selection procedure terminates after at most 2(r+c-1) trials. A sequence of maximal length consisting of r B-successes, r-1 A-successes, c-1 B-failures and c A-failures is given below.

$$\underbrace{\bar{A}\;\bar{B}}_{1}\;\underbrace{\bar{A}\;\bar{B}}_{2}\;\ldots\ldots\;\underbrace{\bar{A}\;\bar{B}}_{c-2}\;\underbrace{A\ldots\ldots A}_{r-1}\;\underset{c-1}{\bar{A}}\;\underbrace{B\ldots\ldots B}_{r-1}\;\underset{c-1}{\bar{B}}\;\underset{c}{\bar{A}}\;\underset{r}{B} \qquad (2.1)$$

(Recall that an A,B-failure is denoted by $\bar{A},\bar{B}$, respectively)

In the following, we use the abbreviations

$$\begin{aligned} S_A(c) &:= \text{number of A-successes preceding the c-th A-failure} \\ S_B(c) &:= \text{number of B-successes preceding the c-th B-failure} \\ F_A(r) &:= \text{number of A-failures preceding the r-th A-success} \\ F_B(r) &:= \text{number of B-failures preceding the r-th B-success} \end{aligned} \qquad (2.2)$$

The probability for a correct selection consists of two terms P_1 and P_2, where P_1 denotes the probability that the r-th A-success occurs before the c-th A-failure and that the r-th A-success precedes the r-th B-success. P_2 denotes the probability that the r-th success of each treatment occurs after the c-th failure and that A is selected.

Using the notation of (2.2), P_1 is found to be

$$\begin{aligned} P_1 = {} & P(CS|F_A(r)=F_B(r),0\leq F_A(r)<c)P(F_A(r)=F_B(r),0\leq F_A(r)<c) + \\ & +P(CS|F_A(r)<F_B(r),0\leq F_A(r)<c)P(F_A(r)<F_B(r),0\leq F_A(r)<c) \end{aligned} \qquad (2.3)$$

In case of $F_B(r)<F_A(r)$ no correct selection is possible.

The following diagrams are to illustrate the above formula.

(1) $F_A(r)=F_B(r)=j<c$

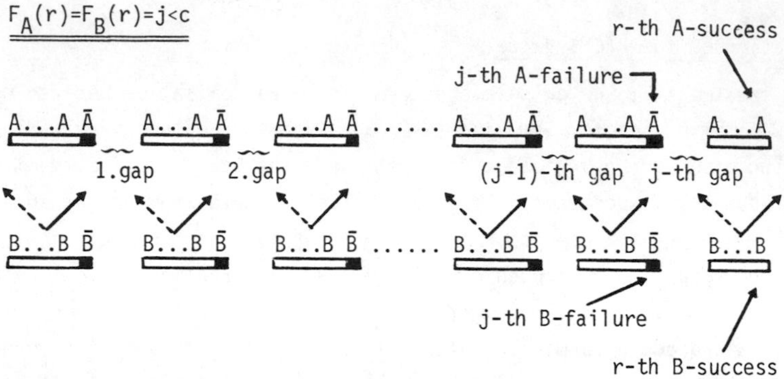

(The symbol A...A Ā has the same meaning as in section 6 of chapter 1, cf. (6.2) and (6.4))

There are two possibilities to combine the sequences.

↗ the r-th A-success precedes the r-th B-success, i.e. a beginning with treatment A implies a correct selection

↖ (dashed) the r-th B-success precedes the r-th A-success, i.e. a beginning with treatment B implies a false selection.

Recalling now that the first treatment at the outset is randomly selected,

we have

$$P(CS|F_A(r)=F_B(r),0\leq F_A(r)<c) = \frac{1}{2}, \qquad (2.4)$$

and the contribution to P_1 is thus given by

$$\frac{1}{2}\, P(F_A(r)=F_B(r),0\leq F_A(r)<c) \qquad (2.5)$$

It follows immediately from the above diagram that no correct selection may take place in case of $F_B(r)<F_A(r)$ because the r-th B-success always precedes the r-th A-success.

(2) $F_A(r)<F_B(r)$ and $F_A(r)=j<c$

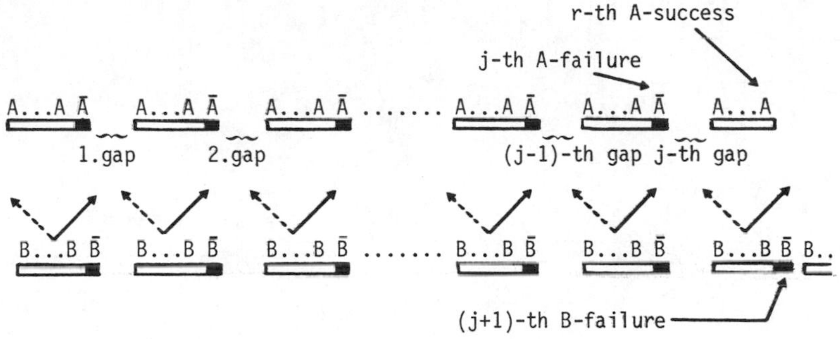

We have again two possibilities to combine the sequences

- the r-th A-success precedes the r-th B-success, i.e. a beginning with treatment A implies a correct selection
- the r-th A-success precedes the r-th B-success because at least $F_A(r)+1$ B-failures occur, i.e. a beginning with treatment B implies again a correct selection.

In either case a correct selection takes place, hence we have

$$P(CS|F_A(r)<F_B(r),0\leq F_A(r)<c) = 1 \tag{2.6}$$

and the contribution to P_1 is thus given by

$$P(F_A(r)<F_B(r),\ 0\leq F_A(r)<c) \tag{2.7}$$

P_1 is now found to be

$$P_1=P(F_A(r)<F_B(r),0\leq F_A(r)<c)+\frac{1}{2}\cdot P(F_A(r)=F_B(r),0\leq F_A(r)<c) \tag{2.8}$$

In determining P_2 we have to note that an equal number of c failures may imply an equal number of successes on the one hand - in this case a correct selection takes place with probability 1/2 only - and different numbers of successes on the other hand. In the latter case a correct selection takes place (and then with probability 1) only if $S_B(c)<S_A(c)$. P_2 is consequently given by

$$P_2=P(S_B(c)<S_A(c),0<S_A(c)<r)+\frac{1}{2}\cdot P(S_B(c)=S_A(c),0\leq S_A(c)<r) \tag{2.9}$$

Recalling that the random variables defined in (2.2) are all together negative binomial random variables, an explicit expression for P_1 and P_2 is easily derived, i.e.

$$P_1=\sum_{j=0}^{c-1}\binom{j+r-1}{j}p_A^r q_A^j\left(\sum_{\tau=j+1}^{\infty}\binom{\tau+r-1}{\tau}p_B^r q_B^\tau+\frac{1}{2}\binom{j+r-1}{j}p_B^r q_B^j\right)=$$

$$=\sum_{j=0}^{c-1}\binom{j+r-1}{j}p_A^r q_A^j\,\frac{1}{2}\,(J_{q_B}(j,r)+J_{q_B}(j+1,r)) \qquad (2.10)$$

$$P_2=\sum_{j=0}^{r-1}\binom{j+c-1}{j}p_A^j q_A^c\sum_{\tau=0}^{j-1}\binom{c+\tau-1}{\tau}p_B^\tau q_B^c+\frac{1}{2}\sum_{j=0}^{r-1}\binom{j+c-1}{j}^2 p_A^j q_A^c p_B^j q_B^c=$$

$$=\sum_{j=0}^{r-1}\binom{j+c-1}{j}p_A^j q_A^c\,\frac{1}{2}(J_{q_B}(c,j)+J_{q_B}(c,j+1)) \qquad (2.11)$$

$$=:\frac{1}{2}\,E^{r-1}(J_{q_B}(c,X_{c,q_A})+J_{q_B}(c,X_{c,q_A}+1)),\ \text{where}$$

$E^{r-1}(Y)$ denotes the expectation of the discrete random variable Y truncated at r-1.

P_1 can be rewritten as

$$P_1=P(F_A(r)<c\leq F_B(r))+P(F_A(r)<F_B(r)<c)+\frac{1}{2}\,P(F_A(r)=F_B(r)<c)=$$

$$=J_{p_A}(r,c)J_{q_B}(c,r)+\sum_{j=0}^{c-1}\binom{j+r-1}{j}p_B^r q_B^j J_{p_A}(r,j)+ \qquad (2.12)$$

$$+\frac{1}{2}\sum_{j=0}^{c-1}\binom{j+r-1}{j}p_B^r q_B^j(J_{p_A}(r,j+1)-J_{p_A}(r,j))$$

Making now use of the above notation, the expression for P(CS) is found to be

$$P(CS)=J_{p_A}(r,c)J_{q_B}(c,r)+\frac{1}{2}\,E^{r-1}(J_{q_B}(c,X_{c,q_A})+J_{q_B}(c,X_{c,q_A}+1))+ \qquad (2.13)$$

$$+\frac{1}{2}\,E^{c-1}(J_{p_A}(r,X_{r,p_B})+J_{p_A}(r,X_{r,p_B}+1))$$

2.2 Derivation of the critical r- and c-values

If r and c are large enough, the random variables given in (2.2) may be expressed by normally distributed random variables. The second term in (2.8) may therefore be neglected. Noting that $E(F_i(r))=rq_i/p_i$ and $Var(F_i(r))=rq_i/p_i^2$; $i\varepsilon\{A,B\}$; we find the following equivalence:

$$F_B(r)>F_A(r) \iff Y^*>\frac{p_B}{p_A}\sqrt{\frac{q_A}{q_B}}\;X^*-\frac{\Delta}{p_A}\sqrt{\frac{r}{q_B}}\;;\quad \Delta:=p_A-p_B,\ \text{where} \qquad (2.14)$$

$$X^* := \frac{F_A(r)-rq_A/p_A}{\sqrt{rq_A}/p_A} \text{ and } Y^* := \frac{F_B(r)-rq_B/p_B}{\sqrt{rq_B}/p_B}$$ are asymptotically standard

normal distributed. P_1 is hence asymptotically given by

$$P_1 \approx P\left(Y^* > \frac{p_B}{p_A}\sqrt{\frac{q_A}{q_B}}\ X^* - \frac{\Delta}{p_A}\sqrt{\frac{r}{q_B}},\ -\sqrt{rq_A} \le X^* \le \frac{cp_A-rq_A}{\sqrt{rq_A}}\right) \approx$$
$$\approx \int_{-\infty}^{(cp_A-rq_A)/\sqrt{rq_A}} \left(1-\Phi\left(\frac{p_B}{p_A}\sqrt{\frac{q_A}{q_B}}\ x^* - \frac{\Delta}{p_A}\sqrt{\frac{r}{q_B}}\right)\right)d\Phi(x^*) \tag{2.15}$$

Replacing x^* by $-x^*$, we get the simpler expression

$$P_1 \approx \int_{(rq_A-cp_A)/\sqrt{rq_A}}^{\infty} \Phi\left(\frac{p_B}{p_A}\sqrt{\frac{q_A}{q_B}}\,x^* + \frac{\Delta}{p_A}\sqrt{\frac{r}{q_B}}\right)d\Phi(x^*) \tag{2.16}$$

The second term of P_2, given by (2.9), may be neglected, too, by using our approximation. Noting that $E(S_i(c))=cp_i/q_i$ and $Var(S_i(c))=cp_i/q_i^2; i\varepsilon\{A,B\}$; and denoting the standardized variables $S_A(c),S_B(c)$ by X,Y, respectively, we finally obtain:

$$P_2 \approx P\left(Y < \frac{q_B}{q_A}\sqrt{\frac{p_A}{p_B}}\ X + \frac{\Delta}{q_A}\sqrt{\frac{c}{p_B}},\ -\sqrt{cp_A} < X < \frac{rq_A-cp_A}{\sqrt{cp_A}}\right) \approx$$
$$\approx \int_{-\infty}^{(rq_A-cp_A)/\sqrt{cp_A}} \Phi\left(\frac{q_B}{q_A}\sqrt{\frac{p_A}{p_B}}\ x + \frac{\Delta}{q_A}\sqrt{\frac{c}{p_B}}\right)d\Phi(x) \tag{2.17}$$

We next consider the two special cases "r fixed, $c\to\infty$" and "c fixed, $r\to\infty$". Holding r fixed and letting $c\to\infty$, we obtain from (2.16):

$$\lim_{c\to\infty} P_1 \approx \int_{-\infty}^{+\infty} \Phi\left(\frac{p_B}{p_A}\sqrt{\frac{q_A}{q_B}}\,x^* + \frac{\Delta}{p_A}\sqrt{\frac{r}{q_B}}\right)d\Phi(x^*), \tag{2.18}$$

and from (2.17) follows

$$0 \le \lim_{c\to\infty} P_2 \le \lim_{c\to\infty} \int_{-\infty}^{(rq_A-cp_A)/\sqrt{cp_A}} d\Phi(x) = \Phi(\lim_{c\to\infty}(rq_A-cp_A)/\sqrt{cp_A}) = 0 \tag{2.19}$$

P(CS) hence reduces to the expression given in (2.18), and that may be rewritten as

$$P(CS) \approx P\left(Y^* < \frac{p_B}{p_A}\sqrt{\frac{q_A}{q_B}}\,X^* + \frac{\Delta}{p_A}\sqrt{\frac{r}{q_B}}\right) \tag{2.20}$$

Noting that $E(Y^* - \frac{p_B}{p_A}\sqrt{\frac{q_A}{q_B}}\,X^*) = 0$ and $Var(Y^* - \frac{p_B}{p_A}\sqrt{\frac{q_A}{q_B}}\,X^*) = 1+p_B^2q_A/p_A^2q_B$, P(CS) is found to be

$$P(CS) \approx \Phi\left(\frac{\Delta\sqrt{r}}{\sqrt{p_A^2 q_B + p_B^2 q_A}}\right) \tag{2.21}$$

Since (2.21) and the P(CS)-value found in (7.4) of chapter 1 coincide, we obtain the same critical r-value as given by (7.8), i.e.

$$r_{max} = \left]\frac{8}{27}\cdot\left(\frac{\lambda(P^*)}{\Delta^*}\right)^2\right[\tag{2.22}$$

This result is intuitively obvious; $c\to\infty$ means that the selection model just being considered reduces to the selection model presented in section 6 of chapter 1.

Holding now c fixed and letting $r\to\infty$, P_1 tends to 0 and

$$\lim_{r\to\infty} P_2 = \int_{-\infty}^{+\infty} \Phi\left(\frac{q_B}{q_A}\sqrt{\frac{p_A}{p_B}}\, x + \frac{\Delta}{q_A}\sqrt{\frac{c}{p_B}}\right) d\Phi(x) \tag{2.23}$$

P(CS) hence reduces to (2.23), i.e.

$$P(CS) \approx \lim_{r\to\infty} P_2 \approx P\left(Y < \frac{q_B}{q_A}\sqrt{\frac{p_A}{p_B}}\, X + \frac{\Delta}{q_A}\sqrt{\frac{c}{p_B}}\right) \tag{2.24}$$

Noting that the random variable $Y - \frac{q_B}{q_A}\sqrt{\frac{p_A}{p_B}}\,X$ has zero mean and variance $1+q_B^2p_A/q_A^2p_B$, we get

$$P(CS) \approx \Phi\left(\frac{\Delta\sqrt{c}}{\sqrt{q_A^2 p_B + q_B^2 p_A}}\right) \tag{2.25}$$

The comparison with (8.6) of chapter 1 gives that the LFC is centered about $\bar{q} = \frac{2}{3} + O(\Delta^{*2})$, and the critical c-value is found to be

$$c_{max} = \left]\frac{8}{27}\left(\frac{\lambda(P^*)}{\Delta^*}\right)^2\right[\tag{2.26}$$

This result is obvious, too; $r\to\infty$ means that the selection model just being considered reduces to that one of section 8.2 of chapter 1.

We will prove afterwards that $E(N_B)$ as well as $E(N)$ are strictly increasing functions of r and c; that is why we conjecture that the best combination of (r,c)-values is given by $r=c=r_{max}=c_{max}$. Nevertheless we have to verify the $(P^*;\Delta^*)$-condition for this special values, i.e. we have to show that $P(CS)\geq P^*$ whenever $\Delta\geq\Delta^*$ and $(r,c)=(r_{max},r_{max})$.

Setting r=c in (2.16) and (2.17), P(CS) is found to be

$$P(CS) \approx P_1+P_2 \approx \int_{\sqrt{r}(q_A-p_A)/\sqrt{q_A}}^{\infty} \Phi\left(\frac{p_B}{p_A}\sqrt{\frac{q_A}{q_B}}\,x + \frac{\Delta}{p_A}\sqrt{\frac{r}{q_B}}\right)d\Phi(x) + \int_{-\infty}^{\sqrt{r}(q_A-p_A)/\sqrt{p_A}} \Phi\left(\frac{q_B}{q_A}\sqrt{\frac{p_A}{p_B}}\,x + \frac{\Delta}{q_A}\sqrt{\frac{r}{p_B}}\right)d\Phi(x) \tag{2.27}$$

If $q_A \neq p_A$ (and r large enough), P(CS) is given by (2.21) or by (2.25); in either case the critical r-value is known to satisfy the $(P^*;\Delta^*)$-condition. Thus we have to investigate only the case $p_A=q_A=1/2$.
(2.27) reduces to

$$P(CS) \approx \int_{\infty}^{0} \Phi\left(q_B\sqrt{\frac{2}{p_B}}\,x + 2\Delta\sqrt{\frac{r}{q_B}}\right)d\Phi(x) + \int_{0}^{\infty} \Phi\left(p_B\sqrt{\frac{2}{q_B}}\,x + 2\Delta\sqrt{\frac{r}{p_B}}\right)d\Phi(x) \tag{2.28}$$

Noting that

$$q_B \gtrless p_B \implies q_B\sqrt{\frac{2}{p_B}}\,x + 2\Delta\sqrt{\frac{r}{p_B}} \gtrless p_B\sqrt{\frac{2}{q_B}}\,x + 2\Delta\sqrt{\frac{r}{q_B}}\,, \tag{2.29}$$

we deduce from (2.28):

$$P(CS) \geq \begin{cases} \int_{-\infty}^{+\infty} \Phi\left(p_B\sqrt{\frac{2}{q_B}}\,x + 2\Delta\sqrt{\frac{r}{q_B}}\right)d\Phi(x), & \text{if } q_B \geq p_B \\ \int_{-\infty}^{+\infty} \Phi\left(q_B\sqrt{\frac{2}{p_B}}\,x + 2\Delta\sqrt{\frac{r}{p_B}}\right)d\Phi(x), & \text{if } q_B < p_B\,, \end{cases} \tag{2.30}$$

and by applying the same transformations used to derive (2.21) and (2.25), we finally obtain:

$$P(CS) \geq \begin{cases} \Phi\left(\dfrac{\Delta\sqrt{r}}{\sqrt{(\frac{1}{2})^2 q_B + \frac{1}{2}p_B^2}}\right), & \text{if } q_B \geq p_B \\ \Phi\left(\dfrac{\Delta\sqrt{r}}{\sqrt{(\frac{1}{2})^2 p_B + \frac{1}{2}q_B^2}}\right), & \text{if } q_B < p_B \end{cases} \tag{2.31}$$

Noting that the first-, second term of (2.31) (Δ replaced by Δ^*) is not less than

$$\min_{p_A-p_B=\Delta^*} \Phi\left(\frac{\Delta^*\sqrt{r}}{\sqrt{p_A^2 q_B + q_A p_B^2}}\right), \quad \min_{p_A-p_B=\Delta^*} \Phi\left(\frac{\Delta^*\sqrt{r}}{\sqrt{q_A^2 p_B + p_A q_B^2}}\right),$$

respectively, we have the desired result, i.e. the $(P^*;\Delta^*)$-condition is satisfied, too, if $r=c=r_{max}$ and $p_A=q_A=1/2$. Let us terminate this section by a short summary:

(a) The best combination of r and c values seems to be given by r=c. This conjecture is supported by the fact, that $E(N_B)$ and $E(N)$ are strictly increasing functions of r <u>and</u> c. r=c implies that P(CS), conceived as a function of $\bar{p}$, is symmetrical around $\bar{p} = \frac{1}{2}$.

(b) The critical r- and c-values are given by
$r_{max}=c_{max}= \left]\frac{8}{27}\cdot\left(\frac{\lambda(P^*)}{\Delta^*}\right)^2\right[$, and there exists two LFC's centered about $\bar{p} = 1/3$ and $\bar{p} = 2/3$.

2.3 Derivation of the expectations

In order to derive an exact expression for $E(N_B)$, we define

$$U(m,n,t):=E(N_B|T=(m,n,t),NT=A);V(m,n,t):=E(N_B|T=(m,n,t),NT=B), \qquad (2.32)$$

where $T=(T_A,T_B,T_C)$ and T_A,T_B denote the additional numbers of A-, B-successes needed to select A,B as best, respectively, and T_c denotes the additional number of failures needed to terminate the sampling procedure. Noting the initial randomization, $E(N_B)$ is found to be

$$E(N_B) = \frac{1}{2}\,(U(r,r,2c) + V(r,r,2c)) \qquad (2.33)$$

In the usual manner, the following system of difference equations and boundary conditions is derived

$$\begin{aligned} &U(m,n,t)=p_AU(m-1,n,t)+q_AV(m,n,t-1)\\ &V(m,n,t)=p_BV(m,n-1,t)+q_BU(m,n,t-1)+1 \qquad (2.34)\\ &U(0,n,t)=V(m,0,t)=U(m,n,0)=V(m,n,0)=0 \quad \text{for all } m,n,t>0 \end{aligned}$$

A solution is obtained by using the generating functions

$$U:=\sum_{m=1}^{\infty}\sum_{n=1}^{\infty}\sum_{t=1}^{\infty}U(m,n,t)x^my^nz^t;\; V:=\sum_{m=1}^{\infty}\sum_{n=1}^{\infty}\sum_{t=1}^{\infty}V(m,n,t)x^my^nz^t \qquad (2.35)$$

We find

$$U = \frac{q_Az}{1-p_Ax}\,V;\; V=p_ByV+q_BzU + \frac{xyz}{(1-x)(1-y)(1-z)}\,, \qquad (2.36)$$

and that implies

$$V = \frac{\frac{x}{1-x}\,\frac{y}{1-y}\,\frac{z}{1-z}}{(1-p_By)\cdot\left(1-\frac{q_Aq_Bz^2}{(1-p_Ax)(1-p_By)}\right)} \tag{2.37}$$

The power series expansion of the numerator of (2.37) is easily seen to be $\sum_{\alpha=1}^{\infty}\sum_{\beta=1}^{\infty}\sum_{\gamma=1}^{\infty} x^\alpha y^\beta z^\gamma$, and that one of the remaining factor is found to be

$$\sum_{j=0}^{\infty}\sum_{\tau=0}^{\infty}\sum_{\lambda=0}^{\infty}\binom{\tau+j-1}{j-1}\binom{\lambda+j}{j}(q_Aq_B)^j p_A^\tau p_B^\lambda x^\tau y^\lambda z^{2j} \tag{2.38}$$

The desired coefficient V(m,n,t) is now obtainable by multiplying the two series, we have

$$V(m,n,t) = \sum_{\tau=0}^{m-1}\sum_{\lambda=0}^{n-1}\sum_{j=0}^{[\frac{t-1}{2}]}\binom{\tau+j-1}{j-1}\binom{\lambda+j}{j}(q_Aq_B)^j p_A^\tau p_B^\lambda \tag{2.39}$$

The coefficient, needed to calculate $E(N_B)$ is hence given by

$$V(r,r,2c) = \frac{1}{q_B}\sum_{j=0}^{c-1} J_{q_A}(j,r)J_{q_B}(j+1,r) \tag{2.40}$$

The corresponding expansion for U is obtained by using the first equation of (2.36). The coefficient of interest is found to be

$$U(m,n,t) = \sum_{\tau=0}^{m-1}\sum_{\lambda=0}^{n-1}\sum_{j=0}^{[\frac{t}{2}-1]}\binom{\tau+j}{j}\binom{\lambda+j}{j} q_A^{j+1} q_B^j p_A^\tau p_B^\lambda \,, \tag{2.41}$$

and U(r,r,2c) is hence given by

$$U(r,r,2c) = \frac{1}{q_B}\sum_{j=0}^{c-1} J_{q_A}(j+1,r)J_{q_B}(j+1,r) \tag{2.42}$$

The desired formula for $E(N_B)$ can now be derived from (2.40) and (2.42), i.e.

$$E(N_B) = \frac{1}{2q_B}\sum_{j=0}^{c-1} J_{q_B}(j+1,r)(J_{q_A}(j,r)+J_{q_A}(j+1,r)) \tag{2.43}$$

To evaluate the exact expression for $E(N_A)$, we next define

$$\widetilde{U}(m,n,t):=E(N_A|T=(n,m,t),NT=B);\widetilde{V}(m,n,t):=E(N_A|T=(n,m,t),NT=A), \tag{2.44}$$

and derive the following system of difference equations and boundary conditions

$$\widetilde{U}(m,n,t)=p_B\widetilde{U}(m-1,n,t)+q_B\widetilde{V}(m,n,t-1)$$

$$\widetilde{V}(m,n,t)=p_A\widetilde{V}(m,n-1,t)+q_A\widetilde{U}(m,n,t-1)+1 \tag{2.45}$$

$$\tilde{U}(0,n,t)=\tilde{V}(m,0,t)=\tilde{U}(m,n,0)=\tilde{V}(m,n,0)=0 \quad \text{for } m,n,t>0 \tag{2.45}$$

(2.45) is obtained from (2.34) by interchanging p_A and p_B as well as q_A and q_B. $E(N_A)$ is thus immediately available from (2.43) by carrying out the interchange of the parameters just mentioned. We find

$$E(N_A) = \frac{1}{2q_A} \sum_{j=0}^{c-1} J_{q_A}(j+1,r)(J_{q_B}(j,r)+J_{q_B}(j+1,r)) \tag{2.46}$$

$E(N)$ is the sum of (2.43) and (2.46), but may be directly derived by solving the following system of difference equations and boundary conditions.

$$C(m,n,t):=E(N|T=(m,n,t),NT=A);D(m,n,t):=E(N|T=(m,n,t),NT=B) \tag{2.47}$$

$$\begin{aligned}&C(m,n,t)=p_A C(m-1,n,t)+q_A D(m,n,t-1)+1\\&D(m,n,t)=p_B D(m,n-1,t)+q_B C(m,n,t-1)+1\\&C(0,n,t)=D(m,0,t)=C(m,n,0)=D(m,n,0)=0 \quad \text{for } m,n,t>0\end{aligned} \tag{2.48}$$

$$E(N) = \frac{1}{2}(C(r,r,2c)+D(r,r,2c)) \tag{2.49}$$

P(CS)-function for $P^{*}= 0.95$; $\Delta^{*}= 0.10$; $r = 81$

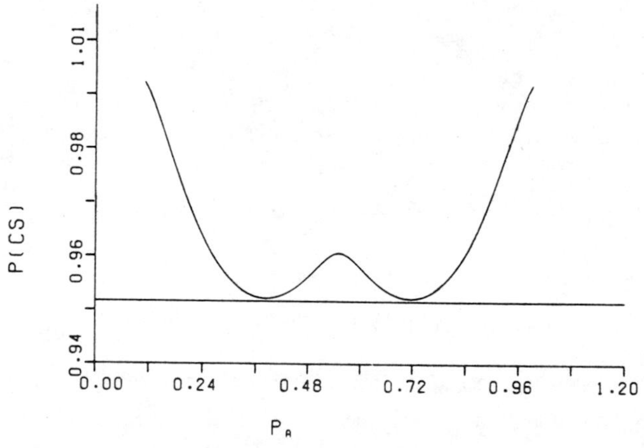

2.4 Numerical results

Table 2

p_A	0.2	0.3	0.4	0.5	0.6	0.7	0.8	0.9
$P^* = 0.90$, $\Delta^* = 0.1$, $r = 49$								
P(CS)	0.934	0.906	0.900	0.909	0.909	0.900	0.906	0.934
$E(N_A)$	62	70	82	91	81	69	61	54
$E(N_B)$	55	62	70	76	66	53	42	29
E(N)	116	132	152	166	146	122	102	83
$P^* = 0.90$, $\Delta^* = 0.2$, $r = 13$								
P(CS)	-	0.922	0.906	0.908	0.913	0.908	0.906	0.922
$E(N_A)$	-	19	21	22	21	18	16	14
$E(N_B)$	-	15	16	17	15	12	9	7
E(N)	-	33	37	38	35	30	25	20
$P^* = 0.90$, $\Delta^* = 0.3$, $r = 6$								
P(CS)	-	-	0.912	0.906	0.909	0.909	0.906	0.912
$E(N_A)$	-	-	10	10	9	9	8	7
$E(N_B)$	-	-	7	7	6	5	4	3
E(N)	-	-	16	16	15	13	11	9
$P^* = 0.90$, $\Delta^* = 0.4$, $r = 4$								
P(CS)	-	-	-	0.922	0.922	0.924	0.922	0.922
$E(N_A)$	-	-	-	6	6	6	5	5
$E(N_B)$	-	-	-	4	4	3	3	2
E(N)	-	-	-	10	9	9	7	6
$P^* = 0.95$, $\Delta^* = 0.1$, $r = 81$								
P(CS)	0.974	0.955	0.951	0.957	0.957	0.951	0.955	0.974
$E(N_A)$	102	116	135	152	135	116	101	90
$E(N_B)$	90	102	116	127	109	88	69	48
E(N)	192	217	251	279	243	203	170	137
$P^* = 0.95$, $\Delta^* = 0.2$, $r = 21$								
P(CS)	-	0.966	0.954	0.955	0.960	0.955	0.954	0.966
$E(N_A)$	-	30	35	37	35	30	26	24
$E(N_B)$	-	24	26	27	24	19	15	10
E(N)	-	54	61	64	58	49	40	33

Table 2 (continued)

p_A	0.2	0.3	0.4	0.5	0.6	0.7	0.8	0.9
$P^* = 0.95$, $\Delta^* = 0.3$, $r = 9$								
P(CS)	-	-	0.954	0.949	0.953	0.949	0.949	0.954
$E(N_A)$	-	-	15	15	15	13	11	10
$E(N_B)$	-	-	10	10	9	8	6	4
E(N)	-	-	24	25	23	20	17	14
$P^* = 0.95$, $\Delta^* = 0.4$, $r = 6$								
P(CS)	-	-	-	0.963	0.963	0.965	0.963	0.963
$E(N_A)$	-	-	-	10	10	9	8	7
$E(N_B)$	-	-	-	6	5	5	4	3
E(N)	-	-	-	15	14	13	11	9
$P^* = 0.99$, $\Delta^* = 0.1$, $r = 161$								
P(CS)	0.997	0.992	0.990	0.992	0.992	0.990	0.992	0.997
$E(N_A)$	202	230	269	308	269	230	202	179
$E(N_B)$	179	202	230	257	216	174	136	92
E(N)	381	432	499	565	484	404	337	271
$P^* = 0.99$, $\Delta^* = 0.2$, $r = 41$								
P(CS)	-	0.995	0.991	0.991	0.994	0.991	0.991	0.995
$E(N_A)$	-	59	69	75	69	59	52	46
$E(N_B)$	-	46	52	54	47	37	27	17
E(N)	-	105	120	129	115	95	78	63
$P^* = 0.99$, $\Delta^* = 0.3$, $r = 18$								
P(CS)	-	-	0.992	0.990	0.992	0.992	0.990	0.992
$E(N_A)$	-	-	30	32	30	26	23	20
$E(N_B)$	-	-	20	20	18	14	10	7
E(N)	-	-	49	52	47	40	33	27
$P^* = 0.99$, $\Delta^* = 0.4$, $r = 11$								
P(CS)	-	-	-	0.993	0.993	0.995	0.993	0.993
$E(N_A)$	-	-	-	19	18	16	14	13
$E(N_B)$	-	-	-	11	10	8	6	4
E(N)	-	-	-	29	27	23	20	16

3 The Selection Model

$[2; VT; \max\{S_A, S_B\} = r \text{ or } \min\{F_A, F_B\} = c]$

3.1 Derivation of the P(CS)-value

The following selection model is in a sense that VT-sampling procedure which corresponds to the PW-sampling procedure investigated in the preceding section.

The proposed procedure uses a cyclic stopping-rule; that means, sampling terminates whenever either treatment yields r successes before its c-th failure - the other treatment may already yield more than c failures - or both yield at least c failures and not yet r successes, i.e. we stop if one treatment yields its c-th failure and the c-th failure of the other treatment has already occurred and both do not yield more than r-1 successes.

In either case the treatment with the largest number of successes is selected. If the number of successes is the same for both treatments, one of them is selected randomly. The maximal number of trials up to a final decision is equal to 2(r+c-1). This may for example occur if up to the (r+c-2)-th trial both treatments have r-1 successes and c-1 failures. Sampling terminates in the following trial, that means, in the (r+c-1)-th trial.

In order to determine P(CS), we must try to find all stopping-sequences that may lead to a correct selection. We have to distinguish six different cases:

(a) A yields r successes and less than c failures, and treatment B yields less than r successes and more failures than treatment A. (3.1)

(b) Both treatments yield r successes and less than c failures. (3.2)

(c) The c-th failure of B occurs before the c-th failure of A, and if A yields c failures it yields less than r successes - this implies that B yields less than r successes -, and if A yields c failures B yields more than c failures. (3.3)

(d) A and B yield c failures in the same trial and less than r successes. (3.4)

(e) The c-th failure of B occurs before the c-th failure of A, and if A yields c failures, B yields c failures, too, and both yield less than r successes. This implies that only successes of B occur after the c-th failure of B up to the termination of the sampling procedure. (3.5)

(f) The c-th failure of A occurs before the c-th failure of B, and if B yields c failures, A yields c failures, too, and both yield less than r successes. This implies that only successes of A occur after the c-th failure of A up to the termination of the sampling procedure. (3.6)

The stopping sequences described in (3.1) and (3.3) lead to a correct selection whereas the stopping sequences in (3.2), (3.4) - (3.6) may lead to a correct selection. In the latter case, the conditional probability of a correct selection is equal to 1/2.

The contributions to P(CS) of (3.1) - (3.6) are as follows:

from (3.1):

$$\sum_{j=0}^{c-1} \binom{r+j-1}{j} p_A^r q_A^j \sum_{\tau=j+1}^{\infty} \binom{r+\tau-1}{\tau} p_B^r q_B^\tau = E^{c-1}(J_{q_B}(X_{r,p_A}+1,r)) \tag{3.7}$$

from (3.2):

$$\frac{1}{2}\sum_{j=0}^{c-1} \binom{r+j-1}{j} p_A^r q_A^j \binom{r+j-1}{j} p_B^r q_B^j = \frac{1}{2} E^{c-1}(J_{q_B}(X_{r,p_A},r)-J_{q_B}(X_{r,p_A}+1,r)) \tag{3.8}$$

from (3.3):

$$\sum_{j=1}^{r-1} \binom{c+j-1}{j} q_A^c p_A^j \sum_{\tau=0}^{j-1} \binom{c+j}{\tau} p_B^\tau q_B^{c+j-\tau} = E^{r-1}(J_{q_B}(c+1,X_{c,q_A})) \tag{3.9}$$

from (3.4):

$$\frac{1}{2}\sum_{j=0}^{r-1} \binom{c+j-1}{j}^2 q_A^c p_A^j q_B^c p_B^j = \frac{1}{2} E^{r-1}(J_{p_B}(X_{c,q_A},c) - J_{p_B}(X_{c,q_A}+1,c)) \tag{3.10}$$

from (3.5)

$$\frac{1}{2}\sum_{j=1}^{r-1} \binom{c+j-1}{j} q_A^c p_A^j \binom{c+j-1}{j-1} q_B^c p_B^j = $$

$$= \frac{p_B}{2q_B} E^{r-1}(J_{p_B}(X_{c,q_A}-1,c+1)-J_{p_B}(X_{c,q_A},c+1)) + \frac{p_B q_A^c}{2q_B} \tag{3.11}$$

Note that we choose the parameters from (0,1), this is no restriction for practical problems as already mentioned in the introduction. On the other hand no difficulties arise in the evaluation of the exact formula for P(CS) if $p_A, p_B \varepsilon \{0,1\}$. These expressions are given in Schriever [160]

from (3.6):

$$\frac{1}{2}\sum_{j=1}^{r-1}\binom{c+j-1}{j} q_B^c p_B^j \binom{c+j-1}{j-1} q_A^c p_A^j = \\ = \frac{p_A}{2q_A} E^{r-2}(J_{p_B}(X_{c+1,q_A}+1,c)-J_{p_B}(X_{c+1,q_A}+2,c)) \tag{3.12}$$

Note that this is the same result as in (3.11).
P(CS) is now available from (3.7) - (3.12).

3.2 Derivation of the critical r and c-values

In using the abbreviations in (2.2) of the preceding section, the "events" (3.1) - (3.6) can be expressed as follows:

$$(3.1) \equiv \{F_A(r) < F_B(r),\ 0 \leq F_A(r) < c\} \tag{3.13}$$

$$(3.2) \equiv \{F_A(r) = F_B(r),\ 0 \leq F_A(r) < c\} \tag{3.14}$$

$$(3.3) \equiv \{S_A(c) > S_B(c+1),\ 0 < S_A(c) < r\} \tag{3.15}$$

$$(3.4) \equiv \{S_A(c) = S_B(c),\ 0 \leq S_A(c) < r\} \tag{3.16}$$

$$(3.5) \equiv \{S_B(c) < S_A(c) < S_B(c+1)+1,\ 0 < S_A(c) < r\} \tag{3.17}$$

$$(3.6) \equiv \{S_A(c) < S_B(c) < S_A(c+1)+1,\ 0 < S_B(c) < r\} \tag{3.18}$$

Recall that for large r and c, the random variables $S_i(c)$ and $F_i(r)$, $i\varepsilon\{A,B\}$, may be expressed by normally distributed random variables; that is why the probability of the events (3.14) and (3.16) tends to 0 and may therefore be neglected. The same is true for the events (3.17) and (3.18).
The event (3.17) is contained in the event $\{S_A(c)+c=X_1+X_2+\ldots+X_{S_A(c)+c}\}$, where the X_i's are independent and identically distributed like the random variable X, i.e. $P(X=1)=p_B$; $P(X=0)=q_B$.
Using the central limit theorem (cf. A2/1), $X_1+X_2+\ldots+X_{S_A(c)+c}$ may be replaced by a normally distributed chance variable (large c), and this implies that the probability of the event (3.17) tends to 0 and may therefore be neglected. The same result may be obtained for the event (3.18). Our first result to determine P(CS) is hence

$$P(CS) \approx P(F_A(r) < F_B(r), 0 \le F_A(r) < c) + P(S_A(c) > S_B(c+1), 0 < S_A(c) < r) \qquad (3.19)$$

Noting that $\sqrt{\frac{c}{c+1}} \approx 1$ if c is large enough, we have

$$S_B(c+1) < S_A(c) \iff \frac{S_B(c+1)-(c+1)p_B/q_B}{\sqrt{(c+1)p_B}/q_B} < \frac{q_B}{q_A}\sqrt{\frac{p_A}{p_B}}\,\frac{S_A(c)-cp_A/q_A}{\sqrt{cp_A}/q_A} + \frac{\Delta}{q_A}\sqrt{\frac{c}{p_B}} \qquad (3.20)$$

Using this equivalence and that one of (2.14), we finally obtain:

$$P(CS) \approx \int_{(rq_A-cp_A)/\sqrt{rq_A}}^{\infty} \Phi\left(\frac{p_B}{p_A}\sqrt{\frac{q_A}{q_B}}\,x + \frac{\Delta}{p_A}\sqrt{\frac{r}{q_B}}\right) d\Phi(x) + \int_{-\infty}^{(rq_A-cp_A)/\sqrt{cp_A}} \Phi\left(\frac{q_B}{q_A}\sqrt{\frac{p_A}{p_B}}\,x + \frac{\Delta}{q_A}\sqrt{\frac{c}{p_B}}\right) d\Phi(x) \qquad (3.21)$$

This expression coincides with the sum of (2.16) and (2.17). The LFC's and the critical r- and c-values are consequently the same as in the preceding section, i.e. we have two LFC's centered about $\bar{p} = 1/3$ and $\bar{p} = 2/3$, and r_{max} and c_{max} are given by

$$r_{max} = c_{max} = \left]\frac{8}{27}\cdot\left(\frac{\lambda(P^*)}{\Delta^*}\right)^2\right[, \qquad (3.22)$$

where $\lambda(P^*)$ denotes - as usual - the $100P^*$-percentile of the standard normal distribution.

3.3 Derivation of the expectations

In order to evaluate the expression for $E(N_A)$, we make use of the representation $N_A = N_{A,CS} + N_{A,FS}$, where $N_{A,CS}$ and $N_{A,FS}$ are given by (6.15) of chapter 1. The contributions of (3.1) - (3.6) to $E(N_{A,CS})$ are as follows:

from (3.1):

$$\sum_{j=0}^{c-1} (r+j)\binom{r+j-1}{j} p_A^r q_A^j J_{q_B}(j+1,r) = \frac{r}{p_A} E^{c-1}(J_{q_B}(X_{r+1,p_A}+1,r)) \qquad (3.23)$$

from (3.2):

$$\frac{1}{2}\sum_{j=0}^{c-1} (r+j)\binom{r+j-1}{j} p_A^r q_A^j \binom{r+j-1}{j} p_B^r q_B^j = \frac{r}{2p_A} E^{c-1}(J_{q_B}(X_{r+1,p_A},r) - J_{q_B}(X_{r+1,p_A}+1,r)) \qquad (3.24)$$

from (3.3):

$$\sum_{j=1}^{r-1} (c+j) \binom{c+j-1}{j} q_A^c p_A^j J_{q_B}(c+1,j) = \frac{c}{q_A} E^{r-1}(J_{q_B}(c+1, X_{c+1,q_A})) \tag{3.25}$$

from (3.4):

$$\frac{1}{2} \sum_{j=0}^{r-1} (c+j) \binom{c+j-1}{j} q_A^c p_A^j (J_{p_B}(j,c) - J_{p_B}(j+1,c)) =$$
$$= \frac{c}{2q_A} E^{r-1}(J_{p_B}(X_{c+1,q_A}, c) - J_{p_B}(X_{c+1,q_A}+1, c)) \tag{3.26}$$

from (3.5):

$$\frac{1}{2} \sum_{j=1}^{r-1} (c+j) \binom{c+j-1}{j} q_A^c p_A^j \binom{c+j-1}{j-1} q_B^c p_B^j =$$
$$= \frac{p_B c}{2q_A q_B} E^{r-1}(J_{p_B}(X_{c+1,q_A}-1, c+1) - J_{p_B}(X_{c+1,q_A}, c+1)) + \frac{p_B c}{2q_A q_B} q_A^{c+1} \tag{3.27}$$

from (3.6):

$$\frac{1}{2} \sum_{j=1}^{r-1} (c+j) \binom{c+j-1}{j} q_B^c p_B^j \binom{c+j-1}{j-1} q_A^c p_A^j =$$
$$= \frac{p_B c}{2q_A q_B} E^{r-1}(J_{p_B}(X_{c+1,q_A}-1, c+1) - J_{p_B}(X_{c+1,q_A}, c+1)) + \frac{p_B c}{2q_A q_B} \cdot q_A^{c+1} \tag{3.28}$$

$E(N_{A,CS})$ is the sum of (3.23) - (3.28).

$E(N_{A,FS})$ can be obtained by using the well-known interchanging technique (cf. (7.9) of chap. 1). We have

$$N_{A,FS} = N_{A,CS}(p_A \leftrightarrows p_B) \tag{3.29}$$

Denoting $E(N_{A,CS})$ by $A(p_A, p_B)$, $E(N_A)$ is found to be

$$E(N_A) = A(p_A, p_B) + A(p_B, p_A) \tag{3.30}$$

3.4 Numerical results

Table 3

p_A	0.2	0.3	0.4	0.5	0.6	0.7	0.8	0.9
$P^* = 0.90$, $\Delta^* = 0.1$, $r = 49$								
P(CS)	0.934	0.906	0.900	0.909	0.909	0.900	0.906	0.934
$E(N_A)$	62	71	82	91	80	69	60	53
$P^* = 0.90$, $\Delta^* = 0.2$, $r = 13$								
P(CS)	-	0.922	0.906	0.908	0.913	0.908	0.906	0.922
$E(N_A)$	-	21	22	23	21	18	16	14
$P^* = 0.90$, $\Delta^* = 0.3$, $r = 6$								
P(CS)	-	-	0.912	0.906	0.909	0.909	0.906	0.912
$E(N_A)$	-	-	12	11	11	9	8	7
$P^* = 0.95$, $\Delta^* = 0.1$, $r = 81$								
P(CS)	0.974	0.955	0.951	0.957	0.957	0.951	0.955	0.974
$E(N_A)$	102	116	136	153	134	115	100	89
$P^* = 0.95$, $\Delta^* = 0.2$, $r = 21$								
P(CS)	-	0.966	0.954	0.955	0.960	0.955	0.954	0.966
$E(N_A)$	-	32	35	37	35	30	26	23
$P^* = 0.95$, $\Delta^* = 0.3$, $r = 9$								
P(CS)	-	-	0.954	0.949	0.953	0.953	0.949	0.954
$E(N_A)$	-	-	17	16	15	13	11	10
$P^* = 0.99$, $\Delta^* = 0.1$, $r = 161$								
P(CS)	0.997	0.992	0.990	0.992	0.992	0.990	0.992	0.997
$E(N_A)$	202	231	269	308	268	230	201	179
$P^* = 0.99$, $\Delta^* = 0.2$, $r = 41$								
P(CS)	-	0.995	0.991	0.991	0.994	0.991	0.991	0.995
$E(N_A)$	-	59	69	75	69	59	52	46
$P^* = 0.99$, $\Delta^* = 0.3$, $r = 18$								
P(CS)	-	-	0.992	0.990	0.992	0.992	0.990	0.992
$E(N_A)$	-	-	32	32	30	26	23	20

4 The Selection Model

$[2;VT;\max\{S_A,S_B\}=r \text{ or } \max\{F_A,F_B\}=c]$

4.1 Derivation of the P(CS)-value

The following selection model uses VT-sampling and a termination-rule similar to that one of the preceding selection procedure, that is, sampling terminates whenever either treatment yields r successes and none of them yields more than c-1 failures, or at least one treatment yields c failures and less than r successes and the other one yields less than r (or exactly r) successes and does not yield more than c failures.

In any case the treatment associated with the larger number of successes is declared to be the best. In case of an equal number of successes, the best treatment is selected randomly. The maximal number of trials up to a final decision is equal to 2(r+c-1), (cf. 3.1). In order to determine P(CS), we must try to find all stopping-sequences that may lead to a correct selection. We have to distinguish five different cases.

(a) A yields r successes and B yields less than r successes and both yield less than c failures. (4.1)

(b) B yields c failures and less than r successes and A yields less than c failures and less than r successes. (4.2)

(c) Both treatments yield r successes and less than c failures. (4.3)

(d) Both treatments yield c failures and less than r successes. (4.4)

(e) A yields r successes and B yields c failures, that means, the r-th A success and the c-th B failure occur in the last trial. (4.5)

The stopping-sequences described in (4.1), (4.2) and (4.5) lead to a correct selection whereas the stopping-sequences of (4.3) and (4.4) may lead to a correct selection only. In this case the conditional probability of a correct selection is equal to 1/2.

The contribution to P(CS) of (4.1) - (4.5) are as follows:

from (4.1):

$$\sum_{j=0}^{c-1} \binom{r+j-1}{j} p_A^r q_A^j \sum_{\tau=\max\{0,r+j-c+1\}}^{r-1} \binom{r+j}{\tau} p_B^{\tau} q_B^{r+j-\tau} \qquad (4.6)$$

from (4.2):

$$\sum_{j=0}^{r-1} \binom{c+j-1}{j} q_B^c p_B^j \sum_{\tau=\max\{0,c+j-r+1\}}^{c-1} \binom{c+j}{\tau} q_A^{\tau} p_A^{c+j-\tau} \qquad (4.7)$$

from (4.3):

$$\frac{1}{2} \sum_{j=0}^{c-1} \binom{r+j-1}{j} p_A^r q_A^j \binom{r+j-1}{j} p_B^r q_B^j = \qquad (4.8)$$
$$= \frac{1}{2} E^{c-1}(J_{q_B}(X_{r,p_A},r) - J_{q_B}(X_{r,p_A}+1,r))$$

from (4.4):

$$\frac{1}{2} \sum_{j=0}^{r-1} \binom{c+j-1}{j} q_A^c p_A^j \binom{c+j-1}{j} q_B^c p_B^j = \qquad (4.9)$$
$$= \frac{1}{2} E^{r-1}(J_{p_B}(X_{c,q_A},c) - J_{p_B}(X_{c,q_A}+1,c))$$

from (4.5):

$$\sum_{j=g(r-c)}^{c-1} \binom{r+j-1}{j} p_A^r q_A^j \binom{r+j-1}{r+j-c} q_B^c p_B^{r+j-c} = \qquad (4.10)$$
$$= \sum_{j=g(r-c)}^{c-1} \binom{r+j-1}{j} p_A^r q_A^j (J_{p_B}(r+j-c,c) - J_{p_B}(r+j-c+1,c)),$$

$$\text{with } g(x) := \begin{cases} 0, & \text{if } x \geq 0 \\ -x, & \text{if } x < 0 \end{cases} \qquad (4.11)$$

P(CS)is the sum of (4.6) - (4.10).

4.2 Derivation of the critical r- and c-values

Making use of the abbreviations given in (2.2) of section 2, the events (4.1) - (4.5) can be expressed as follows:

$$(4.1) \equiv \{F_A(r) < F_B(r), F_A(r) + r < S_B(c) + c,\ 0 \leq F_A(r) < c\} \qquad (4.12)$$

$$(4.2) \equiv \{S_B(c) + c < F_A(r) + r,\ S_B(c) < S_A(c),\ 0 \leq S_B(c) < r\} \qquad (4.13)$$

$$(4.3) \equiv \{F_A(r) = F_B(r),\ 0 \le F_A(r) < c\} \tag{4.14}$$

$$(4.4) \equiv \{S_A(c) = S_B(c),\ 0 \le S_A(c) < r\} \tag{4.15}$$

$$(4.5) \equiv \{F_A(r) + r = S_B(c) + c,\ 0 \le F_A(r) < c,\ 0 \le S_B(c) < r\} \tag{4.16}$$

Using the same arguments as in the preceding section, the probability of the events (4.14) - (4.16) is immediately seen to tend to 0 and may therefore be neglected. Because of the symmetric stopping-rule of the selection procedure just being considered, we conjecture that the best choice among all pairs (r,c) satisfying the $(P^*;\Delta^*)$-condition is a pair with equal components, that means r=c is chosen.

As a first result, we obtain:

$$P(CS) \approx P((4.1)) + P((4.2)) \tag{4.17}$$

Furthermore

$$\begin{aligned}\{F_A(r)<F_B(r),\ 0\le F_A(r)<r\} &= \{F_A(r)<F_B(r),\ 0\le F_A(r)<r,\ F_A(r)<S_B(r)\} \cup \\ &\cup \{F_A(r)<F_B(r),\ 0\le F_A(r)<r,\ F_A(r)\ge S_B(r)\} =: K \cup L\end{aligned} \tag{4.18}$$

K is equal to (4.12):

$$\begin{aligned}&\{S_B(r)<F_A(r),\ S_B(r)<S_A(r),\ 0\le S_B(r)<r\} = \\ &= \{S_B(r)<F_A(r),\ S_B(r)<S_A(r),\ 0\le S_B(r)<r,\ S_A(r)<r\} \cup \\ &\cup \{S_B(r)<F_A(r),\ S_B(r)<S_A(r),\ 0\le S_B(r)<r,\ S_A(r)\ge r\} =: K^* \cup L^*\end{aligned} \tag{4.19}$$

In using in addition the following identities

$$\begin{aligned}K^* &= \{S_B(r) < S_A(r),\ 0 < S_A(r) < r\} \\ L^* &= \{F_A(r) < F_B(r),\ S_B(r) < F_A(r),\ 0 < F_A(r) < r\},\end{aligned} \tag{4.20}$$

the evaluation of P(CS) is now much simpler as before;
we obtain:

$$\begin{aligned}P(CS) &\approx P(K)+P(K^* \cup L^*) \approx P(K^*)+P(K \cup L) = \\ &= P(S_B(r)<S_A(r),\ 0<S_A(r)<r)+P(F_A(r)<F_B(r),\ 0\le F_A(r)<r)\end{aligned} \tag{4.21}$$

The LFC is obtained in the same way as in section 2 and 3, that means

$$r_{max} = c_{max} = \left]\frac{8}{27}\cdot\left(\frac{\lambda(P^*)}{\Delta^*}\right)^2\right[\tag{4.22}$$

where $\lambda(P^*)$ denotes - as usual - the $100P^*$-percentile of the standard normal distribution.

4.3 Derivation of the expectations

To evaluate the exact expression for $E(N_A)$, we use again the representation $N_A=N_{A,CS}+N_{A,FS}$ (cf. (6.15) of chap.1). The contributions of (4.1)-(4.5) to $E(N_{A,CS})$ are as follows:

from (4.1):

$$\sum_{j=0}^{c-1} r \binom{r+j}{j} p_A^r q_A^j \sum_{\tau=\max\{0,r+j-c+1\}}^{r-1} \binom{r+j}{\tau} p_B^\tau q_B^{r+j-\tau} \tag{4.23}$$

from (4.2):

$$\sum_{j=0}^{r-1} c \binom{c+j}{j} q_B^c p_B^j \sum_{\tau=\max\{0,c+j-r+1\}}^{c-1} \binom{c+j}{\tau} q_A^\tau p_A^{c+j-\tau} \tag{4.24}$$

from (4.3):

$$\frac{1}{2}\sum_{j=0}^{c-1} (r+j) \binom{r+j-1}{j} p_A^r q_A^j \binom{r+j-1}{j} p_B^r q_B^j = \tag{4.25}$$

$$= \frac{r}{2p_A} E^{c-1}(J_{q_B}(X_{r+1,p_A},r) - J_{q_B}(X_{r+1,p_A}+1,r))$$

from (4.4):

$$\frac{1}{2}\sum_{j=0}^{r-1} (c+j) \binom{c+j-1}{j} q_A^c p_A^j \binom{c+j-1}{j} q_B^c p_B^j =$$

$$= \frac{c}{2q_A} E^{r-1}(J_{p_B}(X_{c+1,q_A},c) - J_{p_B}(X_{c+1,q_A}+1,c)) \tag{4.26}$$

from (4.5):

$$\sum_{j=g(r-c)}^{c-1} (r+j) \binom{r+j-1}{j} p_A^r q_A^j \binom{r+j-1}{r+j-c} q_B^c p_B^{r+j-c} =$$

$$= \frac{r}{p_A} \sum_{j=g(r-c)}^{c-1} \binom{(r+1)+j-1}{j} p_A^{r+1} q_A^j (J_{p_B}(r+j-c,c)-J_{p_B}(r+j-c+1,c)), \tag{4.27}$$

where $g(x)$ is given by (4.11).

$E(N_{A,CS})$ is now available from (4.23)-(4.27).
Denoting $E(N_{A,CS})$ by $A(p_A,p_B)$ and using the identity $N_{A,FS}=N_{A,CS}(p_A \leftrightarrows p_B)$, we have

$$E(N_A) = A(p_A,p_B) + A(p_B,p_A) \tag{4.28}$$

4.4 Numerical results

Table 4

p_A	0.2	0.3	0.4	0.5	0.6	0.7	0.8	0.9
$P^* = 0.90$, $\Delta^* = 0.1$, $r = 49$								
P(CS)	0.934	0.906	0.900	0.909	0.909	0.900	0.906	0.934
$E(N_A)$	55	61	70	82	82	70	61	55
$P^* = 0.90$, $\Delta^* = 0.2$, $r = 13$								
P(CS)	-	0.922	0.906	0.908	0.913	0.908	0.906	0.922
$E(N_A)$	-	15	17	19	21	19	17	15
$P^* = 0.90$, $\Delta^* = 0.3$, $r = 6$								
P(CS)	-	-	0.912	0.906	0.909	0.909	0.906	0.912
$E(N_A)$	-	-	7	8	9	9	8	7
$P^* = 0.95$, $\Delta^* = 0.1$, $r = 81$								
P(CS)	0.974	0.955	0.951	0.957	0.957	0.951	0.955	0.974
$E(N_A)$	90	102	116	136	136	116	102	90
$P^* = 0.95$, $\Delta^* = 0.2$, $r = 21$								
P(CS)	-	0.966	0.954	0.955	0.960	0.955	0.954	0.966
$E(N_A)$	-	24	27	31	35	31	27	24
$P^* = 0.95$, $\Delta^* = 0.3$, $r = 9$								
P(CS)	-	-	0.954	0.949	0.953	0.953	0.949	0.954
$E(N_A)$	-	-	10	12	14	14	12	10
$P^* = 0.99$, $\Delta^* = 0.1$, $r = 161$								
P(CS)	0.997	0.992	0.990	0.992	0.992	0.990	0.992	0.997
$E(N_A)$	179	202	230	269	269	230	202	179
$P^* = 0.99$, $\Delta^* = 0.2$, $r = 41$								
P(CS)	-	0.995	0.991	0.991	0.994	0.991	0.991	0.995
$E(N_A)$	-	46	52	59	68	59	52	46
$P^* = 0.99$, $\Delta^* = 0.3$, $r = 18$								
P(CS)	-	-	0.992	0.990	0.992	0.992	0.990	0.992
$E(N_A)$	-	-	20	23	27	27	23	20

5 The Selection Model

$[2; PW; |S_A - S_B| = r \text{ or } F_A + F_B = s]$

5.1 Derivation of the P(CS)-value

The following selection procedure uses the PW-sampling scheme and a stopping rule based on the absolute difference of successes and on the sum of failures of both treatments, i.e. sampling terminates whenever the absolute difference of successes equals r or when the sum of failures equals s. In either case, the treatment associated with the larger number of successes is declared as best and ties are, as usual, decided by randomization. r and s must be chosen such that the $(P^*;\Delta^*)$-condition is satisfied.

This selection model can be considered as the truncated version of that one presented in sec. 1 of chap. 1; but by including not only successes but also failures in the termination-rule, the large expected sample sizes in case of small success probabilities are avoided.

To derive the exact expression for P(CS), we next define:

$$\begin{aligned} P_{mn} &:= P(CS|F_A+F_B=s-m,\ S_A-S_B=n,\ NT=A) \\ Q_{mn} &:= P(CS|F_A+F_B=s-m,\ S_A-S_B=n,\ NT=B) \end{aligned} \tag{5.1}$$

From this, the following system of difference equations and boundary conditions is easily derived.

$$\begin{aligned} P_{mn} &= p_A P_{m,n+1} + q_A Q_{m-1,n};\ Q_{mn} = p_B Q_{m,n-1} + q_B P_{m-1,n} \\ P_{mr} &= 1;\ Q_{m,-r} = 0 \text{ for } m \in \{1,\dots,s-1\}, \\ P_{on} = Q_{on} &= \begin{cases} 1, & \text{if } n \in \{1,\dots,r-1\} \\ 1/2, & \text{if } n = 0 \\ 0, & \text{if } n \in \{-1,\dots,-r+1\}, \end{cases} \end{aligned} \tag{5.2}$$

and P(CS) is hence given by:

$$P(CS) = \frac{1}{2}(P_{so} + Q_{so}) \tag{5.3}$$

From (5.2), we deduce:

$q_A q_B\, P_{m-2,n} = -p_B\, P_{m,n-1} + (1+p_A p_B) P_{mn} - p_A P_{m,n+1}$, if $n > -r+1$, and

$q_A q_B\, P_{m-2,-r+1} = P_{m,-r+1} - p_A\, P_{m,-r+2}$, if $n = -r+1$.

The system of difference equations can be written as

$$P(m) = q_A q_B D^{-1} P(m-2) + R, \text{ where}$$
$$P(m) := (P_{m,-r+1}, P_{m,-r+2}, \ldots, P_{mo}, \ldots, P_{m,r-2}, P_{m,r-1})' \tag{5.4}$$

is a column vector, and the matrix D has diagonal elements $(1, 1+p_A p_B, \ldots, 1+p_A p_B)$, $-p_A$ above and $-p_B$ below the leading diagonal, and all other elements of D are equal to zero. The column vector R is the solution of $D \cdot R = (0, \ldots, 0, p_A)'$, i.e. $R = D^{-1} \cdot (0, \ldots, 0, p_A)'$, R is found to be:

$$R = (p_A^{2r-1}, p_A^{2r-2}, \ldots, p_A)' \tag{5.5}$$

Using (5.4), P(s) is immediately seen to be:

$$P(s) = \sum_{j=0}^{c-1} (q_A q_B)^j (D^{-1})^j R + (q_A q_B)^c (D^{-1})^c P(d), \tag{5.6}$$

where $c = [\frac{1}{2} s]$, and d=0 if s is even, and d=1 if s is an odd number. P(0) is obtained from (5.2), i.e.

$$P(0) = (0, \ldots, 0, \tfrac{1}{2}, 1, \ldots, 1)' \tag{5.7}$$

r-th component (pointing to $\tfrac{1}{2}$)

P(1) can be derived by using the recursion formulae (5.2); we obtain:

$$P(1) = (p_A^{r-1}(1-\tfrac{1}{2}q_A), p_A^{r-2}(1-\tfrac{1}{2}q_A), \ldots, p_A(1-\tfrac{1}{2}q_A), (1-\tfrac{1}{2}q_A), 1, \ldots, 1)' \tag{5.8}$$

In the same way as above, we find:

$q_A q_B Q_{m-2,n} = -p_B Q_{m,n-1} + (1+p_A p_B)\, Q_{mn} - p_A Q_{m,n+1}$, if $n < r-1$, and

$q_A q_B Q_{m-2,r-1} = Q_{m,r-1} - p_B Q_{m,r-2} - p_A q_B$, if $n=r-1$

The corresponding recursion formula for Q(m) is hence given by

$$Q(m) = q_A q_B \hat{D}^{-1} Q(m-2) + \hat{R}, \text{ where}$$
$$Q(m) := (Q_{m,-r+1}, Q_{m,-r+2}, \ldots, Q_{mo}, \ldots, Q_{m,r-2}, Q_{m,r-1})' \tag{5.9}$$

is a column vector, and the matrix $\hat{D}$ has diagonal elements $(1+p_A p_B, \ldots, 1+p_A p_B, 1)$, $-p_A$ above and $-p_B$ below the leading diagonal, and all other elements of $\hat{D}$ equal zero. $\hat{R}$ is the solution of $\hat{D}\cdot\hat{R}=(0,\ldots,0,p_Aq_B)'$, i.e. $\hat{R}=\hat{D}^{-1}(0,\ldots,0,p_Aq_B)'$. Q(s) is consequently given by:

$$Q(s) = \sum_{j=o}^{c-1} (q_Aq_B)^j(\hat{D}^{-1})^j\hat{R}+(q_Aq_B)^c(\hat{D}^{-1})^cQ(d), \tag{5.10}$$

where

$$Q(0) = (0,\ldots,0,\tfrac{1}{2},1,\ldots,1)', \text{ and}$$

$$Q(1) = (0,\ldots,0,\underset{\uparrow\atop r\text{-th component}}{\tfrac{1}{2}q_B},\alpha_1,\ldots,\alpha_{r-1}), \text{ where} \tag{5.11}$$

$$\alpha_\nu = q_B(\sum_{j=o}^{\nu-1} p_B^j + \frac{1}{2} p_B^\nu)$$

The desired P(CS)-value is now obtainable from (5.6) and (5.10).

5.2 Derivation of the critical r- and s-values

As pointed out by Fushimi (cf. [63]), we may suppose with a safe conscience that in case of a constant difference $p_A-p_B=\Delta$, the P(CS)-function has two local minima when neither s nor r are too small and s is not too large in comparison with r. For these special cases let $P_1(r,s)$ and $P_2(r,s)$ denote the two minimum values of the P(CS)-function, where $P_1(r,s)$ corresponds to the smaller p_A-value and $P_2(r,s)$ to the larger one. It can be seen from the P(CS)-graph or from a direct computation of the P(CS)-value for various r-,s-, and p_A-values that $P_1(r,s)$ is relative insensitive to a variation of r, provided r is not very small and $P_2(r,s)$ is insensitive to a variation of s if s is large enough. We may conclude from these facts that good approximations of the critical r- and s-values can be obtained by letting $s\to\infty$, $r\to\infty$, respectively. When either r or s tends to ∞, the selection model just being considered reduces to a selection model, the termination rule of which is based only on failures or successes, respectively. If $s=\infty$, sampling terminates whenever $|S_A-S_B|=r$, i.e. the selection model is the same as no. 1 of chap. 1. r_{max} is consequently given by: (cf. (1.18), chap. 1)

$$r_{max} = \left]\frac{\ell n2(1-P^*)}{\ell n(1-\Delta^*)}\right[\tag{5.12}$$

If $r=\infty$, sampling terminates whenever $F_A+F_B=s$. For simplicity, we assume that s=2c. Note that this assumption is not very restrictive! Remembering that the PW sampling scheme is used, sampling terminates if both treatments yield c failures i.e. if $F_A=F_B=c$. c_{max} can be derived in the same way as in sec. 2 of chap. 2. (cf. (2.23)-(2.26) of chap. 2). We obtain

$$c_{max} = \left]\frac{8}{27}\left(\frac{\lambda(P^*)}{\Delta^*}\right)^2\right[\tag{5.13}$$

5.3 Derivation of the expectations

As usual, we next derive the expectation of N_B, the number of patients on the inferior treatment. For this purpose, we define:

$$V_{mn} := E(N_B|F_A+F_B=s-m,\ S_A-S_B=n,\ NT=A)$$
$$W_{mn} := E(N_B|F_A+F_B=s-m,\ S_A-S_B=n,\ NT=B) \tag{5.14}$$

From this we obtain the following recursion formulae and boundary conditions:

$$V_{mn} = p_A V_{m,n+1}+q_A W_{m-1,n};\ W_{mn}=p_B W_{m,n-1}+q_B V_{m-1,n}+1$$
$$V_{mr} = W_{m,-r}=0 \text{ for } m\varepsilon\{1,...,s-1\} \tag{5.15}$$
$$V_{on} = W_{on}=0 \quad \text{for all n with } |n|\varepsilon\{0,1,...,r-1\}$$

$E(N_B)$ is given by:

$$E(N_B) = \frac{1}{2}(V_{so}+W_{so}) \tag{5.16}$$

From (5.15), we deduce:

$$q_Aq_BV_{m-2,n} = -p_BV_{m,n-1}+(1+p_Ap_B)V_{mn}-p_AV_{m,n+1}-q_A \text{ if } n>-r+1, \text{ and}$$

$$q_Aq_BV_{m-2,-r+1} = V_{m,-r+1}-p_AV_{m,-r+2}-q_A, \text{ if } n=-r+1$$

The recursion formula for V(m) is hence:

$$V(m) = q_Aq_BD^{-1}V(m-2)+T, \text{ where}$$
$$V(m) := (V_{m,-r+1},V_{m,-r+2},...,V_{mo},...,V_{m,r-2},V_{m,r-1})' \tag{5.17}$$

is a column vector, and the matrix D is the same as that of (5.6) and T is the solution of $D\cdot T = (q_A,...,q_A)'$, i.e. $T = D^{-1}(q_A,...,q_A)'$. V(s) can be easily seen to be:

$$V(s) = \sum_{j=0}^{c-1}(q_Aq_B)^j(D^{-1})^jT+(q_Aq_B)^c(D^{-1})^cV(d), \tag{5.18}$$

where

$$V(0) = V(1) = (0,...,0)'$$

In the same way, we obtain:

$$q_Aq_BW_{m-2,n}= -p_BW_{m,n-1}+(1+p_Ap_B)W_{mn} -p_AW_{m,n+1}-q_A \text{ if } n<r-1, \text{ and}$$

$$q_Aq_BW_{m-2,r-1} = -p_BW_{m,r-2}+W_{m,r-1}-1 \text{ if } n=r-1$$

The corresponding recursion formula for W(m) is hence given by:

$$W(m) = q_Aq_B\hat{D}^{-1}W(m-2) + \hat{T}, \text{where} \tag{5.19}$$

$$W(m) := (W_{m,-r+1}, W_{m,-r+2}, \ldots, W_{mo}, \ldots, W_{m,r-2}, W_{m,r-1})'$$

is a column vector, and the matrix $\hat{D}$ is the same as that of (5.9), and $\hat{T}$ is the solution of $\hat{D}\cdot\hat{T}=(q_A,\ldots,q_A,1)'$, i.e. $\hat{T}=\hat{D}^{-1}(q_A,\ldots,q_A,1)'$. W(s) is hence given by:

$$W(s) = \sum_{j=0}^{c-1} (q_A q_B)^j (\hat{D}^{-1})^j \hat{T} + (q_A q_B)^c (\hat{D}^{-1})^c W(d), \tag{5.20}$$

where

$$W(0) = (0,\ldots,0)', \text{ and}$$

$$W(1) = (\beta_1, \beta_2, \ldots, \beta_{2r-1})' \text{ with}$$

$$\beta_\nu := (1-p_B^\nu)/q_B, \quad \nu\varepsilon\{1,\ldots,2r-1\}$$

The desired $E(N_B)$-value is now obtainable from (5.18) and (5.20).
The expression for $E(N_A)$ can be derived as follows:

$$\begin{aligned} V^*_{mn} &:= E(N_A | F_A+F_B=s-m,\ S_B-S_A=n,\ NT=A) \\ W^*_{mn} &:= E(N_A | F_A+F_B=s-m,\ S_B-S_A=n,\ NT=B) \end{aligned} \tag{5.21}$$

V^*_{mn} and W^*_{mn} must satisfy the following difference equations and boundary conditions.

$$\begin{aligned} W^*_{mn} &= p_B\, W^*_{m,n+1} + q_B\, V^*_{m-1,n} \\ V^*_{mn} &= p_A\, V^*_{m,n-1} + q_A\, W^*_{m-1,n+1} \\ W^*_{m,r} &= V^*_{m,-r} = 0 \text{ for } m\varepsilon\{1,\ldots,s-1\} \\ W^*_{on} &= V^*_{on} = 0 \quad \text{for all } n \text{ with } |n|\varepsilon\{0,1,\ldots,r-1\} \end{aligned} \tag{5.22}$$

Comparing (5.22) with (5.15), we see immediately that

$$E(N_A) = E(N_B | p_A \leftrightarrows p_B) \tag{5.23}$$

5.4 Numerical results

Table 5

p_A	0.2	0.3	0.4	0.5	0.6	0.7	0.8	0.9
$P^* = 0.90$; $\Delta^* = 0.1$; $r = 17$; $c = 49$								
P(CS)	0.934	0.906	0.900	0.909	0.923	0.930	0.920	0.907
$E(N_B)$	54.24	59.17	62.13	61.78	56.83	46.35	31.91	17.50
E(N)	115.24	126.70	134.33	135.34	126.94	106.76	77.68	48.43
$P^* = 0.90$; $\Delta^* = 0.2$; $r = 8$; $c = 13$								
P(CS)	-	0.922	0.906	0.908	0.919	0.929	0.927	0.916
$E(N_B)$	-	13.90	14.47	14.42	13.49	11.49	8.51	5.20
E(N)	-	31.66	33.50	34.08	32.86	29.29	23.40	16.65
$P^* = 0.90$; $\Delta^* = 0.3$; $r = 5$; $c = 6$								
P(CS)	-	-	0.912	0.906	0.913	0.924	0.928	0.921
$E(N_B)$	-	-	5.92	5.91	5.63	4.98	3.95	2.67
E(N)	-	-	14.55	14.90	14.68	13.66	11.72	9.12
$P^* = 0.90$; $\Delta^* = 0.4$; $r = 4$; $c = 4$								
P(CS)	-	-	-	0.922	0.925	0.935	0.944	0.943
$E(N_B)$	-	-	-	3.57	3.44	3.12	2.59	1.87
E(N)	-	-	-	9.54	9.54	9.15	8.25	6.84
$P^* = 0.95$; $\Delta^* = 0.1$; $r = 23$; $c = 81$								
P(CS)	0.974	0.955	0.951	0.957	0.968	0.971	0.963	0.952
$E(N_B)$	89.55	96.74	99.45	96.03	85.52	67.92	46.26	24.87
E(N)	190.26	207.15	215.11	210.58	191.36	156.93	113.94	70.08
$P^* = 0.95$; $\Delta^* = 0.2$; $r = 11$; $c = 21$								
P(CS)	-	0.966	0.954	0.955	0.964	0.971	0.968	0.959
$E(N_B)$	-	22.42	23.07	22.50	20.40	16.93	12.16	7.12
E(N)	-	51.11	53.50	53.37	50.21	43.62	34.15	23.88
$P^* = 0.95$; $\Delta^* = 0.3$; $r = 7$; $c = 10$								
P(CS)	-	-	0.963	0.958	0.963	0.971	0.972	0.964
$E(N_B)$	-	-	9.75	9.52	8.78	7.47	5.63	3.56
E(N)	-	-	24.07	24.17	23.19	20.93	17.38	13.15
$P^* = 0.95$; $\Delta^* = 0.4$; $r = 5$; $c = 6$								
P(CS)	-	-	-	0.963	0.964	0.970	0.974	0.968
$E(N_B)$	-	-	-	5.10	4.76	4.15	3.28	2.22
E(N)	-	-	-	13.74	13.36	12.41	10.79	8.66

Table 5 (continued)

p_A	0.2	0.3	0.4	0.5	0.6	0.7	0.8	0.9
$P^* = 0.99$; $\Delta^* = 0.1$; $r = 38$; $c = 161$								
P(CS)	0.997	0.992	0.990	0.992	0.996	0.997	0.995	0.992
$E(N_B)$	178.06	191.07	191.73	177.79	151.29	116.64	79.17	41.78
E(N)	378.34	409.24	414.94	390.33	339.21	270.52	195.45	120.42
$P^* = 0.99$; $\Delta^* = 0.2$; $r = 18$; $c = 41$								
P(CS)	-	0.995	0.991	0.991	0.995	0.997	0.996	0.993
$E(N_B)$	-	43.71	44.14	41.49	36.01	28.43	19.83	11.09
E(N)	-	99.77	102.58	98.82	88.88	74.19	57.00	39.42
$P^* = 0.99$; $\Delta^* = 0.3$; $r = 11$; $c = 18$								
P(CS)	-	-	0.992	0.990	0.993	0.996	0.996	0.993
$E(N_B)$	-	-	17.48	16.60	14.69	11.91	8.58	5.09
E(N)	-	-	43.32	42.48	39.31	34.14	27.57	20.53
$P^* = 0.99$; $\Delta^* = 0.4$; $r = 8$; $c = 10$								
P(CS)	-	-	-	0.991	0.991	0.994	0.996	0.995
$E(N_B)$	-	-	-	8.68	7.89	6.60	4.94	3.09
E(N)	-	-	-	23.72	22.66	20.47	17.30	13.60

The numerical results have been calculated directly from (5.6), (5.10), (5.18) and (5.20), that is why no inaccuracy may arise in the computations.
In the original paper of Fushimi [63], the desired values can be calculated after applying a similarity transformation to the matrix D and then evaluating the 2r-1 distinct roots of the equation

$$\frac{1}{2} U_{2r-1}(\omega) - \sqrt{p_A p_B}\, U_{2r-2}(\omega) = 0,$$

where

$$U_n(\omega) := \frac{\sin((n+1)\Theta)}{\sin\Theta} \quad ; \ \omega = \cos\Theta$$

is a Tchebycheff polynomial of the second kind. The numerical results given by us are more accurate than the corresponding results of Fushimi, although the differences are only slight. It should be mentioned that the desired values can be calculated within reasonable bounds of computer-time, even if $\Delta^* = 0.1$ and $P^* = 0.99$

6 The Selection Model

$[k; PW; \max\{S_1,\dots,S_k\}=r \text{ or } \min\{F_1,\dots,F_k\}=c]$

6.1 Introductory remarks

The main disadvantage of the selection models no. 10 and no. 11 of chapter 1 consists in their termination with only probability one. To avoid the occurrence of a never ending sampling or at least the occurrence of a patient horizon, too large for practical purposes, the termination rule has to be based not only on "successes" but also on "failures". The termination rule of selection model no. 2, first considered by Berry, Sobel [35], can be generalized without difficulties to more than 2 treatments. That means, in using the PW-sampling scheme for $k \geq 3$ treatments, as described in 10.1 of chap. 1, sampling terminates whenever one of the k treatments yields its r-th success or all treatments yield at least c failures and less than r successes. Recalling that the occurrence of a failure of the treatment just given to a patient generates a switch to the next treatment in the random order of the treatments, we can say that sampling terminates whenever one treatment yields its r-th success or all treatments yield exactly c failures and less than r successes. In either case the treatment associated with the largest number of successes is selected as best, ties are decided by randomization, which may happen only if the procedure has gone through c cycles. We must determine r and c so that the probability of a correct selection is at least P* whenever the difference between the largest and second largest success parameter is at least Δ^*, i.e.

$$P(CS) \geq P^* \text{ whenever } p_1 - \max_{i>1} p_i =: \Delta \geq \Delta^* \tag{6.1}$$

with $1/k < P^* < 1$ and $\Delta^* \in (0,1)$

The maximal number of trials up to a final decision is equal to $k(r+c-1)$. This may for example occur if k-1 of the k treatments yield c failures and r-1 successes and the k-th treatment yields c-1 failures, too, and r successes.

6.2 Derivation of the P(CS)-value

As usual, let $F_i(r)$ denote the number of A_i-failures preceding the r-th A_i-success, and let $S_i(c)$ denote the number of A_i-successes preceding the c-th A_i-failure; $i\varepsilon\{1,2,\dots,k\}$. The probability of a correct selection is hence the sum of two terms P_1,P_2 which are determined by $(F_i(r))_{i\varepsilon\{1,\dots,k\}}$ and $(S_i(c))_{i\varepsilon\{1,\dots,k\}}$, respectively. P_1 is the probability of selecting A_1 before c cycles and P_2 is the probability of selecting A_1 in exactly c cycles, i.e. in the first case, sampling terminates if one treatment yields r successes and at least one treatment yields less than c failures, and in the second case, sampling terminates if all treatments yield c failures and less than r successes.

P_1 is found to be:

$$P_1 = P(F_1(r) < F_2(r),\dots,F_1(r) < F_k(r),\ 0 \leq F_1(r) < c) +$$
$$+ \sum_{n=1}^{k-1} \sum_{\omega\varepsilon S_n} \sum_{\nu=0}^{k-n-1} \left(\frac{(k-1-n)!(k-\nu-1)!}{k!(k-1-n-\nu)!} P(F_1(r)=F_{i_1}(r)=\dots=F_{i_n}(r), \right. \tag{6.2}$$
$$\left. , F_1(r) < F_{j_1}(r),\dots,F_1(r) < F_{j_{k-n-1}}(r),\ 0 \leq F_1(r) < c \right),$$

where S_n is defined as follows:

$$S_n := \{\omega=(\omega_1,\omega_2) | \omega_1=(i_1,\dots,i_n), \omega_2=(j_1,\dots,j_{k-n-1}),\ i_1<\dots<i_n,$$
$$j_1<\dots<j_{k-n-1},\ i_1,\dots,i_n,j_1,\dots,j_{k-n-1}\varepsilon\{2,\dots,k\} \text{ and} \tag{6.3}$$
$$\{i_1,\dots,i_n\} \cap \{j_1,\dots,j_{k-n-1}\} = \emptyset\}; \text{ (cf. (11.2) of chap. 1)}$$

In the first term of (6.2), the conditional probability of a correct selection is equal to one, and in the second term of (6.2), the conditional probability of a correct selection is equal to $\sum_{\nu=0}^{k-n-1} ((k-1-n)!(k-\nu-1)!/k!(k-1-n-\nu)!)$, which is the probability that in the initial randomization A_1 precedes all A_ℓ with $\ell\varepsilon\{i_1,\dots,i_n\}$. The sampling structure associated with P_1 and P_2 can be illustrated by diagrams similar to those of section 2. Details are omitted, but can be seen in Schriever [159].

The contribution of $(S_i(c))_{i\varepsilon\{1,\dots,k\}}$ to P(CS) is much simpler to evaluate. P_2 is found to be

$$P_2=P(S_2(c)<S_1(c),\dots,S_k(c)<S_1(c),\ 0<S_1(c)<r) +$$
$$+ \sum_{n=1}^{k-1} \sum_{\omega\varepsilon S_n} (\tfrac{1}{n+1} P(S_1(c)=S_{i_1}(c)=\dots=S_{i_n}(c),\ S_{j_1}(c)<S_1(c),\dots \tag{6.4}$$

$$\ldots, S_{j_{k-n-1}}(c) < S_1(c),\ 0 < S_1(c) < r)) \qquad (6.4)$$

The conditional probability of a correct selection equals 1, 1/(n+1) in the first and second term of (6.4), respectively. Recalling that $F_i(r)$ and $S_i(c)$ are all together negative binomial chance variables, P_1 can be rewritten as

$$P_1 = E^{c-1}\Big(\prod_{\ell=2}^{k} J_{q_\ell}(X_{r,p_1}+1,r)\Big) + \sum_{n=1}^{k-1} \sum_{\omega\in S_n} \sum_{\nu=0}^{k-n-1} \Big(\frac{(k-1-n)!(k-\nu-1)!}{k!(k-1-n-\nu)!} \cdot$$
$$\cdot E^{c-1}\Big(\prod_{\ell=1}^{n}(J_{q_{i_\ell}}(X_{r,p_1},r) - J_{q_{i_\ell}}(X_{r,p_1}+1,r)) \cdot \prod_{\ell=1}^{k-n-1} J_{q_{j_\ell}}(X_{r,p_1}+1,r)\Big)\Big) \qquad (6.5)$$

In the same way, P_2 is shown to be

$$P_2 = E^{r-1}\Big(\prod_{\ell=2}^{k} J_{q_\ell}(c,X_{c,q_1})\Big) + \sum_{n=1}^{k-1} \sum_{\omega\in S_n} \frac{1}{n+1} \cdot$$
$$E^{r-1}\Big(\prod_{\ell=1}^{n}(J_{p_{i_\ell}}(X_{c,q_1},c) - J_{p_{i_\ell}}(X_{c,q_1}+1,c)) \cdot \prod_{\ell=1}^{k-n-1} J_{q_{j_\ell}}(c,X_{c,q_1})\Big) \qquad (6.6)$$

P(CS) is the sum of (6.5) and (6.6).

6.3 Derivation of the critical r- and c-values

Applying the central limit theorem to the negative binomial chance variables $F_i(r), S_i(c)$, $i\varepsilon\{1,\ldots,k\}$, we have approximately:

$$P(CS) \approx P(F_1(r) < F_2(r),\ldots,F_1(r) < F_k(r), 0 \le F_1(r) < c) +$$
$$+ P(S_2(c) < S_1(c),\ldots,S_k(c) < S_1(c), 0 < S_1(c) < r) \qquad (6.7)$$

Denoting the standardized variables $F_\lambda(r)$, $S_\lambda(c)$ by $Y_{\lambda r}$, $X_{\lambda c}$ respectively, i.e.

$$Y_{\lambda r} := \frac{F_\lambda(r) - rq_\lambda/p_\lambda}{\sqrt{rq_\lambda}/p_\lambda} \quad \text{and} \quad X_{\lambda c} := \frac{S_\lambda(c) - cp_\lambda/q_\lambda}{\sqrt{cp_\lambda}/q_\lambda}, \qquad (6.8)$$

and denoting in addition the distribution functions of $Y_{\lambda r}$, $X_{\lambda c}$ by $V_{\lambda r}$, $W_{\lambda c}$, respectively, (6.7) can be rewritten as:

$$P(CS) \approx \int_{-\sqrt{rq_1}}^{(cp_1-rq_1)/\sqrt{rq_1}} \prod_{\lambda=2}^{k} \Big(1 - V_{\lambda r}\Big(\frac{p_\lambda}{p_1}\sqrt{\frac{q_1}{q_\lambda}}\,x - \frac{\Delta_{1\lambda}}{p_1}\sqrt{\frac{r}{q_\lambda}}\Big)\Big)dV_{1r}(x) +$$
$$+ \int_{-\sqrt{cp_1}}^{(rq_1-cp_1)/\sqrt{cp_1}} \prod_{\lambda=2}^{k} W_{\lambda c}\Big(x\,\frac{q_\lambda}{q_1}\sqrt{\frac{p_1}{p_\lambda}} + \frac{\Delta_{1\lambda}}{q_1}\sqrt{\frac{c}{p_\lambda}}\Big)dW_{1c}(x), \qquad (6.9)$$

with $\Delta_{1\lambda} := p_1 - p_\lambda$, $\lambda\varepsilon\{2,\ldots,k\}$.

Applying A2/4 to the above formula and then replacing x by -x in the first term of the resulting formula, P(CS) is found to be approximately:

$$P(CS) \approx \int_{(rq_1-cp_1)/\sqrt{rq_1}}^{\infty} \prod_{\lambda=2}^{k} \Phi\left(\frac{p_\lambda}{p_1}\sqrt{\frac{q_1}{q_\lambda}}\,x + \frac{\Delta_{1\lambda}}{p_1}\sqrt{\frac{r}{q_\lambda}}\right) d\Phi(x) + \int_{-\infty}^{(rq_1-cp_1)/\sqrt{cp_1}} \prod_{\lambda=2}^{k} \Phi\left(\frac{q_\lambda}{q_1}\sqrt{\frac{p_1}{p_\lambda}}\,x + \frac{\Delta_{1\lambda}}{q_1}\sqrt{\frac{c}{p_\lambda}}\right) d\Phi(x) \tag{6.10}$$

It is immediately seen from the integral representation of the incomplete beta-function $J_q(s,t)$ that for fixed s and t, $J_q(s,t)$ is an increasing function of q. That is why the incomplete beta functions $J_{q_\ell}(\cdot,\cdot)$ in the first term of (6.5) and (6.6) take their smallest value if we make q_ℓ as small as possible, and that will be the case if we define $p_\ell := p_2^*$ for all $\ell \in \{2,\dots,k\}$, where p_2^* is the second largest success parameter, which will be usually different from p_2, the success parameter associated with treatment A_2 (cf. 10.2, chap.1). In the LFC, (6.10) reduces to:

$$P(CS) \approx \int_{(rq_1-cp_1)/\sqrt{rq_1}}^{\infty} \left(\Phi\left(\frac{p_2^*}{p_1}\sqrt{\frac{q_1}{q_2^*}}\,x + \frac{\Delta}{p_1}\sqrt{\frac{r}{q_2^*}}\right)\right)^{k-1} d\Phi(x) + \int_{-\infty}^{(rq_1-cp_1)/\sqrt{cp_1}} \left(\Phi\left(\frac{q_2^*}{q_1}\sqrt{\frac{p_1}{p_2^*}}\,x + \frac{\Delta}{q_1}\sqrt{\frac{c}{p_2^*}}\right)\right)^{k-1} d\Phi(x), \tag{6.11}$$

with $\Delta := p_1 - p_2^*$.

Letting $c \to \infty$ and holding r fixed, we obtain:

$$P(CS) \approx \int_{-\infty}^{+\infty} \left(\Phi\left(\frac{p_2^*}{p_1}\sqrt{\frac{q_1}{q_2^*}}\,x + \frac{\Delta}{p_1}\sqrt{\frac{r}{q_2^*}}\right)\right)^{k-1} d\Phi(x) \tag{6.12}$$

Letting $r \to \infty$ and holding c fixed, we obtain:

$$P(CS) \approx \int_{-\infty}^{+\infty} \left(\Phi\left(\frac{q_2^*}{q_1}\sqrt{\frac{p_1}{p_2^*}}\,x + \frac{\Delta}{q_1}\sqrt{\frac{c}{p_2^*}}\right)\right)^{k-1} d\Phi(x) \tag{6.13}$$

(6.12), (6.13) give the P(CS)-value of selection procedures, where the termination rule is only based on successes, failures, respectively.

It is immediately seen from (6.11) that in the LFC $\Delta = \Delta^*$ must hold. As pointed out in section 11 of chap. 1, there are two possibilities to determine the critical r-and c-values in the special cases (6.12) and (6.13). In the following, we make only use of the first possibility, because there are only slight differences between the critical r- and c-values obtained by either procedure.

The corresponding results for P(CS) in the LFC are as follows:

6.12 reduces to: (cf.(2.21))

$$P(CS) \approx \left(\Phi\left(\frac{\Delta^* \sqrt{r}}{\sqrt{p_1^2 q_2^* + q_1 p_2^{*2}}}\right)\right)^{k-1} \tag{6.14}$$

6.13 reduces to: (cf.(2.25))

$$P(CS) \approx \left(\Phi\left(\frac{\Delta^* \sqrt{c}}{\sqrt{q_1^2 p_2^* + q_2^{*2} p_1}}\right)\right)^{k-1} \tag{6.15}$$

The critical r- and c-values are obtained in the same way as in section 2, i.e.

$$r_{max} = \left]\frac{8}{27}\left(\frac{\lambda_k(P^*)}{\Delta^*}\right)^2\right[= c_{max}, \text{ where} \tag{6.16}$$

$\lambda_k(P^*)$ is the $100P^{*1/(k-1)}$-percentile of the standard normal distribution. For the special cases (6.12) and (6.13), we have found the LFC's to be centered about $\bar{p}=2/3$, $\bar{p}=1/3$, respectively, where $\bar{p}:=\frac{1}{2}\cdot(p_1+p_2^*)$. For details, see Schriever [159].

We will prove in the following that $E(N_i)$ and $E(N)$ are monotone increasing functions of r <u>and</u> c. That is why we conjecture that the best choice among all pairs (r,c) satisfying the $(P^*;\Delta^*)$-condition (6.1) consists in setting r=c. (6.11) hence reduces to: (Δ replaced by Δ^*)

$$\begin{aligned} P(CS) \approx & \int_{\sqrt{r}(q_1-p_1)/\sqrt{q_1}}^{\infty} \left(\Phi\left(\frac{p_2^*}{p_1}\sqrt{\frac{q_1}{q_2^*}}\,x + \frac{\Delta^*}{p_1}\sqrt{\frac{r}{q_2^*}}\right)\right)^{k-1} d\Phi(x) + \\ & + \int_{-\infty}^{\sqrt{r}(q_1-p_1)/\sqrt{p_1}} \left(\Phi\left(\frac{q_2^*}{q_1}\sqrt{\frac{p_1}{p_2^*}}\,x + \frac{\Delta^*}{q_1}\sqrt{\frac{r}{p_2^*}}\right)\right)^{k-1} d\Phi(x) \end{aligned} \tag{6.17}$$

If $p_1 \neq q_1$, P(CS) is given by (6.14) or (6.15), provided r is large enough, and the critical r-value is the same as in (6.16). Thus we have only to investigate the special case $p_1=q_1=1/2$. We obtain:

$$\begin{aligned} P(CS) \approx & \int_0^{\infty} \left(\Phi\left(p_2^*\sqrt{\frac{2}{q_2^*}}\,x + 2\Delta^*\sqrt{\frac{r}{q_2^*}}\right)\right)^{k-1} d\Phi(x) + \\ & + \int_{-\infty}^{0} \left(\Phi\left(q_2^*\sqrt{\frac{2}{p_2^*}}\,x + 2\Delta^*\sqrt{\frac{r}{p_2^*}}\right)\right)^{k-1} d\Phi(x) \end{aligned} \tag{6.18}$$

Using the fact that

$$q_2^* \gtrless p_2^* \implies q_2^* \sqrt{\frac{2}{p_2^*}}\, x + 2\Delta^* \sqrt{\frac{r}{p_2^*}} \gtrless p_2^* \sqrt{\frac{2}{q_2^*}}\, x + 2\Delta^* \sqrt{\frac{2}{q_2^*}}\,, \tag{6.19}$$

we obtain:

$$P(CS) \geq \begin{cases} \left(\Phi\left(\dfrac{\Delta^*\sqrt{r}}{\sqrt{\left(\frac{1}{2}\right)^2 q_2^* + \frac{1}{2} {p_2^*}^2}}\right)\right)^{k-1}, & \text{if } q_2^* \geq p_2^* \\[2ex] \left(\Phi\left(\dfrac{\Delta^*\sqrt{r}}{\sqrt{\left(\frac{1}{2}\right)^2 p_2^* + \frac{1}{2} {q_2^*}^2}}\right)\right)^{k-1}, & \text{if } p_2^* > q_2^* \end{cases} \tag{6.20}$$

Comparing this result with (6.14) and (6.15), we have that in either case ($q_2^* \geq p_2^*$ and $p_2^* > q_2^*$) $\min P(CS) \geq (\Phi(\Delta^* \sqrt{27 r_{max}/8}))^{k-1}$, and this implies that the pair (r_{max}, r_{max}) satisfies in any case the $(P^*;\Delta^*)$-condition (6.1).

6.4 Derivation of the expectations

Let N_i denote the number of patients on treatment i up to a final decision, $i \in \{1,2,\dots,k\}$ and let $T=(T_1,\dots,T_k,T_{k+1})$ be a random vector, where T_i denotes the number of additional successes of treatment A_i needed to declare treatment A_i as best for $i \in \{1,\dots,k\}$, and T_{k+1} denotes the number of additional failures needed to finish the c-th cycle. Moreover we define

$$R_{ij}(n_1,\dots,n_k,f) := E(N_i \mid T=(n_1,\dots,n_k,f),\ NT=A_j), \tag{6.21}$$

where "$NT=A_j$" means as usual that the next trial is carried out with treatment A_j. From this, we derive the following recursion formulae:

$$\begin{aligned} R_{ij}(n_1,\dots,n_k,f) = {} & p_j R_{ij}(n_1,\dots,n_{j-1},n_j-1,n_{j+1},\dots,n_k,f) + \\ & + q_j R_{i,j+1}(n_1,\dots,n_k,f-1) + \delta_{ij}, \end{aligned} \tag{6.22}$$

with $\delta_{ij} = 1$ for $i=j$ and 0 otherwise; $R_{i,k+1} := R_{i1}$.

The boundary conditions for (6.22) are given by

$$\begin{aligned} & R_{ij}(n_1,\dots,n_{j-1},0,n_{j+1},\dots,n_k,f)=0 \text{ for } n_\lambda>0\ \forall \lambda\neq j \text{ and } f>0 \\ & R_{ij}(n_1,\dots,n_k,0)=0 \text{ for } n_\lambda>0\quad \forall \lambda,i,j \in \{1,\dots,k\} \end{aligned} \tag{6.23}$$

To find a solution of (6.22) satisfying (6.23), we use generating functions R_{ij} defined by:

$$R_{ij} := \sum_{n_1=1}^{\infty} \cdots \sum_{n_k=1}^{\infty} \sum_{f=1}^{\infty} R_{ij}(n_1,\ldots,n_k,f)x_1^{n_1} \cdots x_k^{n_k} y^f \tag{6.24}$$

The desired expectations are given by:

$$\begin{aligned} E(N_i) &= \frac{1}{k} \sum_{j=1}^{k} R_{ij}(r,\ldots,r,kc) \\ E(N) &= \frac{1}{k} \sum_{i=1}^{k} \sum_{j=1}^{k} R_{ij}(r,\ldots,r,kc), \end{aligned} \tag{6.25}$$

where $1/k$ is the probability that the first treatment given to a patient is $A_j, j\varepsilon\{1,\ldots,k\}$. In solving (6.22), we have to distinguish two cases;
In case that $i=j$, we obtain:

$$R_{ii}(1-p_i x_i) = q_i y R_{i,i+1} + \frac{y}{1-y} \prod_{\ell=1}^{k} \frac{x_\ell}{1-x_\ell} \tag{6.26}$$

In case that $i \neq j$, we have

$$R_{ij} = \frac{q_j y}{1-p_j x_j} R_{i,j+1} \tag{6.27}$$

Using (6.27) we obtain from (6.26):

$$R_{ii} = \frac{1}{G^*} \prod_{\substack{\ell=1 \\ \ell\neq i}}^{k} (1-p_\ell x_\ell) \frac{y}{1-y} \prod_{\ell=1}^{k} \frac{x_\ell}{1-x_\ell}, \tag{6.28}$$

where $G^* := (1-p_1x_1)\ldots(1-p_kx_k)-q_1\ldots q_k y^k$, (6.29)

and from (6.27) follows:

$$R_{ij} = R_{ii} \prod_{\ell=j}^{i-1} \frac{q_\ell y}{1-p_\ell x_\ell}, \text{ if } j < i \tag{6.30}$$

$$R_{ij} = R_{ii} \prod_{\ell=1}^{i-1} \frac{q_\ell y}{1-p_\ell x_\ell} \prod_{\ell=j}^{k} \frac{q_\ell y}{1-p_\ell x_\ell}, \text{ if } j > i \tag{6.31}$$

We next determine the power series expansion of (6.28), (6.30) and (6.31). The expansion of $1/G^*$ is found to be

$$\frac{1}{G^*} = \sum_{n_1=0}^{\infty} \cdots \sum_{n_k=0}^{\infty} \sum_{\ell=0}^{\infty} \binom{n_1+\ell}{n_1} \cdots \binom{n_k+\ell}{n_k} p_1^{n_1}\ldots p_k^{n_k} q_1^{\ell}\ldots q_k^{\ell} x_1^{n_1}\ldots x_k^{n_k} y^{k\ell} \tag{6.32}$$

We have to expand the other factors of the product given in (6.28) and then to multiply them with (6.32). The coefficients of the resulting power series are as follows:

$$R_{ii}(n_1,\ldots,n_k,f) = \sum_{\ell=0}^{\sigma(k,f)} \left(\frac{1}{q_i} J_{q_i}(\ell+1,n_i) \prod_{\substack{\lambda=1\\ \lambda\neq i}}^{k} J_{q_\lambda}(\ell,n_\lambda)\right), \tag{6.33}$$

where $\sigma(x,y)$ is defined by:

$$\sigma(x,y) := \begin{cases} [\frac{y}{x}]-1, & \text{if } \gcd(x,y) = x \\ [\frac{y}{x}] & , \text{if } \gcd(x,y) < x , \end{cases} \tag{6.34}$$

and $\gcd(x,y)$ denotes the greatest common divisor of x and y, and [a] denotes the greatest integral number not greater than a. We finally obtain:

$$R_{ii}(r,\ldots,r,kc) = \sum_{\ell=0}^{c-1} \left(\frac{1}{q_i} J_{q_i}(\ell+1,r) \prod_{\substack{\lambda=1\\ \lambda\neq i}}^{k} J_{q_\lambda}(\ell,r)\right) \tag{6.35}$$

In the same way, we obtain for $j < i$:

$$R_{ij}(n_1,\ldots,n_k,f) \sum_{\ell=0}^{\sigma(k,f-(i-j))} \left(\prod_{\substack{\lambda\leq j-1\\ \lambda\geq i+1}} J_{q_\lambda}(\ell,n_\lambda) \frac{1}{q_i} \prod_{\lambda=j}^{i} J_{q_\lambda}(\ell+1,n_\lambda)\right) \tag{6.36}$$

$$R_{ij}(r,\ldots,r,kc) = \sum_{\ell=0}^{c-1} \left(\prod_{\substack{\lambda\leq j-1\\ \lambda\geq i+1}} J_{q_\lambda}(\ell,r)\cdot \frac{1}{q_i} \cdot \prod_{\lambda=j}^{i} J_{q_\lambda}(\ell+1,r)\right) \tag{6.37}$$

The remaining expressions of interest are as follows: $(j > i)$

$$R_{ij}(n_1,\ldots,n_k,f) = \sum_{\ell=0}^{\sigma(k,f-k+(j-i))} \left(\prod_{\substack{\lambda\leq i\\ \lambda\geq j}} J_{q_\lambda}(\ell+1,n_\lambda)\frac{1}{q_i} \prod_{\lambda=i+1}^{j-1} J_{q_\lambda}(\ell,n_\lambda)\right) \tag{6.38}$$

$$R_{ij}(r,\ldots,r,kc) = \sum_{\ell=0}^{c-1} \left(\prod_{\substack{\lambda\leq i\\ \lambda\geq j}} J_{q_\lambda}(\ell+1,r) \frac{1}{q_i} \prod_{\lambda=i+1}^{j-1} J_{q_\lambda}(\ell,r)\right) \tag{6.39}$$

The desired expectations are now obtainable from (6.25), (6.35), (6.37) and (6.39).

6.5 Numerical results

Table 6

k = 3; $P^* = 0.90$									
p_1	0.2	0.3	0.4	0.5	0.6	0.7	0.8	0.9	r
$P(CS \mid \Delta^*=0.1)$	0.952	0.919	0.912	0.923	0.920	0.909	0.916	0.949	79
$E(N \mid \Delta^*=0.1)$	275	311	358	396	342	281	231	179	
$P(CS \mid \Delta^*=0.2)$	-	0.936	0.915	0.917	0.923	0.912	0.909	0.930	20
$E(N \mid \Delta^*=0.2)$	-	73	82	86	78	64	51	40	
$P(CS \mid \Delta^*=0.3)$	-	-	0.926	0.916	0.920	0.916	0.907	0.915	9
$E(N \mid \Delta^*=0.3)$	-	-	34	34	32	27	22	17	
$P(CS \mid \Delta^*=0.4)$	-	-	-	0.916	0.913	0.912	0.903	0.901	5
$E(N \mid \Delta^*=0.4)$	-	-	-	17	16	14	12	9	

k = 3; $P^* = 0.95$									
p_1	0.2	0.3	0.4	0.5	0.6	0.7	0.8	0.9	r
$P(CS \mid \Delta^*=0.1)$	0.981	0.961	0.956	0.963	0.962	0.954	0.959	0.980	114
$E(N \mid \Delta^*=0.1)$	396	448	516	577	495	408	335	258	
$P(CS \mid \Delta^*=0.2)$	-	0.972	0.959	0.960	0.966	0.958	0.956	0.970	29
$E(N \mid \Delta^*=0.2)$	-	106	120	127	113	93	75	57	
$P(CS \mid \Delta^*=0.3)$	-	-	0.966	0.959	0.964	0.962	0.955	0.962	13
$E(N \mid \Delta^*=0.3)$	-	-	49	51	46	39	31	24	
$P(CS \mid \Delta^*=0.4)$	-	-	-	0.969	0.969	0.971	0.965	0.965	8
$E(N \mid \Delta^*=0.4)$	-	-	-	28	26	23	18	15	

k = 3; $P^* = 0.99$									
p_1	0.2	0.3	0.4	0.5	0.6	0.7	0.8	0.9	r
$P(CS \mid \Delta^*=0.1)$	0.998	0.992	0.991	0.993	0.993	0.991	0.992	0.998	197
$E(N \mid \Delta^*=0.1)$	685	774	892	1010	856	706	578	443	
$P(CS \mid \Delta^*=0.2)$	-	0.996	0.992	0.992	0.995	0.992	0.991	0.966	50
$E(N \mid \Delta^*=0.2)$	-	183	208	225	196	159	128	96	
$P(CS \mid \Delta^*=0.3)$	-	-	0.994	0.991	0.993	0.993	0.991	0.993	22
$E(N \mid \Delta^*=0.3)$	-	-	85	88	79	65	52	40	
$P(CS \mid \Delta^*=0.4)$	-	-	-	0.994	0.994	0.995	0.993	0.993	13
$E(N \mid \Delta^*=0.4)$	-	-	-	47	43	36	29	23	

Table 6 (continued)

k = 4; P* = 0.90									
p_1	0.2	0.3	0.4	0.5	0.6	0.7	0.8	0.9	r
P(CS|Δ*=0.1)	0.960	0.926	0.918	0.930	0.926	0.913	0.921	0.957	98
E(N |Δ*=0.1)	450	508	584	649	555	454	369	278	
P(CS|Δ*=0.2)	-	0.946	0.924	0.926	0.933	0.918	0.915	0.938	25
E(N |Δ*=0.2)	-	120	134	141	125	101	80	60	
P(CS|Δ*=0.3)	-	-	0.934	0.923	0.928	0.921	0.909	0.919	11
E(N |Δ*=0.3)	-	-	53	54	49	41	32	24	
P(CS|Δ*=0.4)	-	-	-	0.945	0.943	0.942	0.931	0.929	7
E(N |Δ*=0.4)	-	-	-	31	29	24	20	15	
k = 4; P* = 0.95									
p_1	0.2	0.3	0.4	0.5	0.6	0.7	0.8	0.9	r
P(CS|Δ*=0.1)	0.984	0.964	0.958	0.966	0.965	0.956	0.961	0.983	134
E(N |Δ*=0.1)	615	694	798	894	761	624	506	379	
P(CS|Δ*=0.2)	-	0.976	0.962	0.963	0.969	0.960	0.958	0.973	34
E(N |Δ*=0.2)	-	162	184	195	171	138	109	80	
P(CS|Δ*=0.3)	-	-	0.969	0.961	0.966	0.963	0.955	0.962	15
E(N |Δ*=0.3)	-	-	73	75	68	56	44	33	
P(CS|Δ*=0.4)	-	-	-	0.971	0.970	0.971	0.964	0.963	9
E(N |Δ*=0.4)	-	-	-	40	37	31	25	19	
k = 4; P* = 0.99									
p_1	0.2	0.3	0.4	0.5	0.6	0.7	0.8	0.9	r
P(CS|Δ*=0.1)	0.998	0.993	0.991	0.994	0.993	0.991	0.992	0.998	218
E(N |Δ*=0.1)	1000	1129	1298	1469	1238	1016	823	613	
P(CS|Δ*=0.2)	-	0.996	0.992	0.992	0.995	0.992	0.991	0.996	55
E(N |Δ*=0.2)	-	262	298	321	277	223	176	128	
P(CS|Δ*=0.3)	-	-	0.995	0.993	0.995	0.994	0.992	0.994	25
E(N |Δ*=0.3)	-	-	124	129	114	92	72	53	
P(CS|Δ*=0.4)	-	-	-	0.994	0.994	0.995	0.993	0.993	14
E(N |Δ*=0.4)	-	-	-	65	58	48	38	28	

Table 6 (continued)

k = 5; P^* = 0.90									
p_1	0.2	0.3	0.4	0.5	0.6	0.7	0.8	0.9	r
P(CS|Δ^*=0.1)	0.965	0.931	0.922	0.935	0.930	0.917	0.925	0.961	112
E(N |Δ^*=0.1)	638	720	827	921	783	640	515	381	
P(CS|Δ^*=0.2)	-	0.950	0.927	0.929	0.936	0.918	0.915	0.940	28
E(N |Δ^*=0.2)	-	165	186	195	172	138	108	78	
P(CS|Δ^*=0.3)	-	-	0.945	0.934	0.940	0.931	0.919	0.929	13
E(N |Δ^*=0.3)	-	-	77	79	71	58	45	33	
P(CS|Δ^*=0.4)	-	-	-	0.935	0.931	0.928	0.912	0.909	7
E(N |Δ^*=0.4)	-	-	-	37	34	29	23	17	
k = 5; P^* = 0.95									
p_1	0.2	0.3	0.4	0.5	0.6	0.7	0.8	0.9	r
P(CS|Δ^*=0.1)	0.986	0.965	0.960	0.968	0.966	0.957	0.963	0.985	148
E(N |Δ^*=0.1)	843	952	1093	1225	1038	848	683	503	
P(CS|Δ^*=0.2)	-	0.977	0.962	0.964	0.970	0.959	0.957	0.973	37
E(N |Δ^*=0.2)	-	218	246	261	228	183	143	102	
P(CS|Δ^*=0.3)	-	-	0.973	0.966	0.971	0.967	0.959	0.966	17
E(N |Δ^*=0.3)	-	-	102	105	94	76	59	43	
P(CS|Δ^*=0.4)	-	-	-	0.974	0.973	0.974	0.966	0.965	10
E(N |Δ^*=0.4)	-	-	-	55	50	41	32	24	
k = 5; P^* = 0.99									
p_1	0.2	0.3	0.4	0.5	0.6	0.7	0.8	0.8	r
P(CS|Δ^*=0.1)	0.998	0.993	0.991	0.994	0.994	0.991	0.993	0.998	234
E(N |Δ^*=0.1)	1333	1505	1728	1955	1642	1342	1079	790	
P(CS|Δ^*=0.2)	-	0.996	0.992	0.993	0.995	0.992	0.992	0.996	59
E(N |Δ^*=0.2)	-	347	393	424	364	291	226	160	
P(CS|Δ^*=0.3)	-	-	0.995	0.992	0.994	0.994	0.991	0.994	26
E(N |Δ^*=0.3)	-	-	158	164	144	115	89	63	
P(CS|Δ^*=0.4)	-	-	-	0.994	0.994	0.995	0.993	0.993	15
E(N |Δ^*=0.4)	-	-	-	84	75	61	41	34	

7 The Selection Model

$[k; VT; \max\{S_1,\dots,S_k\}=r \text{ or } \min\{F_1,\dots,F_k\}=c]$

7.1 Derivation of the P(CS)-value

The following selection procedure uses the VT-sampling scheme and the termination rule of the preceding selection model, i.e. sampling terminates whenever at least one treatment yields its r-th success before its c-th failure - the other treatments may already yield more than c failures - or all yield at least c failures and not yet r successes, in other words, we stop if the c-th failure of one treatment occurs and all the other treatments yield already more than c-1 failures and less than r successes.
In either case the treatment associated with the largest number of successes is selected as best, ties are broken by randomization.
The maximal number of trials up to a final decision is equal to k(r+c-1).
To determine the exact formula for P(CS), we must try to find all stopping-sequences that may lead to a correct selection.
We distinguish five different cases:

(a) A_1 yields r successes and less than c failures and $A_2,\dots,A_k$ yield less than r successes, too, and more failures than A_1. (7.1)

(b) $\ell+1$ treatments ($\ell\geq 1$) yield r successes (among them A_1) and less than c failures, and all other treatments yield less than r successes and more failures than A_1. (7.2)

(c) A_1 yields c failures and less than r successes and all other treatments yield more than c failures and less than r successes. This implies that the c-th failure of $A_2,\dots,A_k$ occurs before the c-th failure of treatment A_1, and this is possible only if A_1 yields at least one success. (7.3)

(d) $A_1,A_{i_1},\dots,A_{i_\ell}(1<i_1<i_2<\dots<i_\ell\leq k;\ell\varepsilon\{1,2,\dots,k-1\})$ yield c failures and less than r successes and all other treatments yield more than c failures and less than r successes. $A_1,A_{\tau_1},\dots,A_{\tau_n}$ yield their c-th fai-

lure before the c-th failure of $A_{\lambda_1},\dots,A_{\lambda_{\ell-n}}$, and then only successes of $A_1,A_{\tau_1},\dots,A_{\tau_n}$ occur and $A_{\lambda_1},\dots,A_{\lambda_{\ell-n}}$ yield their c-th failure in the last trial.

$\{i_1,\dots,i_\ell\}=\{\tau_1,\dots,\tau_n\}\cup\{\lambda_1,\dots,\lambda_{\ell-n}\}$ and $\{\tau_1,\dots,\tau_n\}\cap\{\lambda_1,\dots,\lambda_{\ell-n}\}=\emptyset$.

The c-th failure of $A_1,A_{\tau_1},\dots,A_{\tau_n}$ may occur in the same trial but that is not true in general. (7.4)

The maximal value of n is $\ell-1$, because at least one population yields its c-th failure in the last trial: this implies in addition that $A_1,A_{\tau_1},\dots,A_{\tau_n}$ yield at least one success.

(e) $A_1,A_{\tau_1},\dots,A_{\tau_\ell}$ yield c failures and less than r successes and all other treatments yield more than c failures and less than r successes.

$A_1,A_{i_1},\dots,A_{i_\ell}$ yield their c-th failure in the last trial and

$A_{\lambda_1},\dots,A_{\lambda_{\ell-n}}$ yield their c-th failure before the c-th failure of A_1. (7.5)

The maximal value of n is ℓ. In case that ℓ is less than k-1, at least one treatment yields more than c failures and this implies that A_1 yields at least one success. If $\ell=k-1$ and $n<k-1$, A_1 yields at least one success, too. If $\ell=n=k-1$, no success may has been occurred up to the termination of the selection procedure, i.e. the special case $\ell=n=k-1$ must be considered separately.

The stopping-sequences given in (7.1) and (7.3) lead to a correct selection whereas the stopping-sequences of (7.2), (7.4) and (7.5) may lead to a correct selection only. In this case, the conditional probability of a correct selection is equal to $1/(\ell+1)$.

The contributions of (7.1)-(7.5) to P(CS) are as follows:

from (7.1):

$$\sum_{j=0}^{c-1}\binom{j+r-1}{j}p_1^r q_1^j \prod_{\lambda=2}^{k}\Big(\sum_{\tau=j+1}^{r+j}\binom{r+j}{\tau}q_\lambda^\tau p_\lambda^{r+j-\tau}\Big) =$$
$$= \sum_{j=0}^{c-1}\binom{j+r-1}{j}p_1^r q_1^j \prod_{\lambda=2}^{k} J_{q_\lambda}(j+1,r) = E^{c-1}\Big(\prod_{\lambda=2}^{k} J_{q_\lambda}(X_{r,p_1}+1,r)\Big) \tag{7.6}$$

from (7.2):

$$\sum_{\ell=1}^{k-1}\sum_{\omega\varepsilon S_\ell}\sum_{j=0}^{c-1}\left(\frac{1}{\ell+1}\binom{j+r-1}{j}p_1^r q_1^j \prod_{\lambda=1}^{\ell}\left(\binom{j+r-1}{j}p_{i_\lambda}^r q_{i_\lambda}^j\right)\cdot\right.$$

$$\left.\cdot\prod_{\lambda=1}^{k-\ell-1}\left(\sum_{\tau=j+1}^{r+j}\binom{r+j}{\tau}q_{j_\lambda}^\tau p_{j_\lambda}^{r+j-\tau}\right)\right) = \tag{7.7}$$

$$= \sum_{\ell=1}^{k-1}\sum_{\omega\varepsilon S_\ell}\frac{1}{\ell+1}E^{c-1}\left(\prod_{\lambda=1}^{\ell}(J_{q_{i_\lambda}}(X_{r,p_1},r)-J_{q_{i_\lambda}}(X_{r,p_1}+1,r))\prod_{\lambda=1}^{k-\ell-1}J_{q_{j_\lambda}}(X_{r,p_1}+1,r)\right),$$

where S_ℓ is defined as in (11.2) of chap. 1.

from (7.3):

$$\sum_{j=1}^{r-1}\binom{j+c-1}{j}q_1^c p_1^j \prod_{\lambda=2}^{k}\left(\sum_{\tau=0}^{j-1}\binom{c+j}{\tau}p_\lambda^\tau q_\lambda^{c+j-\tau}\right) =$$

$$= \sum_{j=1}^{r-1}\binom{j+c-1}{j}q_1^c p_1^j \prod_{\lambda=2}^{k}J_{q_\lambda}(c+1,j) = E^{r-1}\left(\prod_{\lambda=2}^{k}J_{q_\lambda}(c+1,X_{c,q_1})\right) \tag{7.8}$$

from (7.4):

$$\sum_{\ell=1}^{k-1}\sum_{n=0}^{\ell-1}\sum_{\omega\varepsilon S_\ell(n,\ell-n)}\sum_{j=1}^{r-1}\left(\frac{1}{\ell+1}\binom{j+c-1}{j-1}q_1^c p_1^j\prod_{\nu=1}^{n}\left(\binom{j+c-1}{j-1}q_{\tau_\nu}^c p_{\tau_\nu}^j\right)\cdot\right.$$

$$\left.\cdot\prod_{\nu=1}^{\ell-n}\left(\binom{j+c-1}{j}q_{\lambda_\nu}^c p_{\lambda_\nu}^j\right)\prod_{\nu=1}^{k-\ell-1}\left(\sum_{\alpha=0}^{j-1}\binom{c+j}{\alpha}p_{j_\nu}^\alpha q_{j_\nu}^{c+j-\alpha}\right)\right) =$$

$$= \sum_{\ell=1}^{k-1}\sum_{n=0}^{\ell-1}\sum_{\omega\varepsilon S_\ell(n,\ell-n)}\frac{1}{\ell+1}\cdot\frac{p_1}{q_1}E^{r-2}\left(\prod_{\nu=1}^{n}\left(\frac{p_{\tau_\nu}}{q_{\tau_\nu}}(J_{p_{\tau_\nu}}(X_{c+1,q_1},c+1)-\right.\right. \tag{7.9}$$

$$\left.-J_{p_{\tau_\nu}}(X_{c+1,q_1}+1,c+1))\right)\prod_{\nu=1}^{\ell-n}(J_{p_{\lambda_\nu}}(X_{c+1,q_1}+1,c)-J_{p_{\lambda_\nu}}(X_{c+1,q_1}+2,c))\cdot$$

$$\left.\cdot\prod_{\nu=1}^{k-\ell-1}J_{q_{j_\nu}}(c+1,X_{c+1,q_1}+1)\right),\text{ where}$$

$$\begin{aligned}
&S_\ell(n,\ell-n):=\{\omega=(\omega_1,\omega_2,\omega_3)\,|\,\omega_1=(\tau_1,\dots,\tau_n),\omega_2=(\lambda_1,\dots,\lambda_{\ell-n}),\\
&\omega_3=(j_1,\dots,j_{k-\ell-1});\tau_1<\dots<\tau_n;\lambda_1<\dots<\lambda_{\ell-n};j_1<\dots<j_{k-\ell-1};\\
&\{\tau_1,\dots,\tau_n\}\cup\{\lambda_1,\dots,\lambda_{\ell-n}\}=\{i_1,\dots,i_\ell\},(i_1,\dots,i_\ell):=\omega_1^*;(\omega_1^*,\omega_3)=:\omega\varepsilon S_\ell\},
\end{aligned} \tag{7.10}$$

and ℓ runs from 1 to k-1; in case that ℓ=0, the stopping-sequence, just considered, belongs to the set of sequences described in (c).

from (7.5), if n<k-1:

$$\sum_{\ell=1}^{k-1}\sum_{n=0}^{\min\{\ell,k-2\}}\sum_{\omega\varepsilon S_\ell(n,\ell-n)}\sum_{j=1}^{r-1}\Bigg(\frac{1}{\ell+1}\binom{j+c-1}{j}q_1^c p_1^j\cdot$$

$$\cdot\prod_{\nu=1}^{n}\left(\binom{j+c-1}{j}q_{\tau_\nu}^c p_{\tau_\nu}^j\right)\prod_{\nu=1}^{\ell-n}\left(\binom{j+c-1}{j-1}q_{\lambda_\nu}^c p_{\lambda_\nu}^j\right)\prod_{\nu=1}^{k-\ell-1}\left(\sum_{\alpha=0}^{j-1}\binom{c+j}{\alpha}p_{j_\nu}^{\alpha}q_{j_\nu}^{c+j-\alpha}\right)\Bigg)=$$

$$=\sum_{\ell=1}^{k-1}\sum_{n=0}^{\min\{\ell,k-2\}}\sum_{\omega\varepsilon S_\ell(n,\ell-n)}\sum_{j=1}^{r-1}\Bigg(\frac{1}{\ell+1}\binom{j+c-1}{j}q_1^c p_1^j\cdot\tag{7.11}$$

$$\cdot\prod_{\nu=1}^{n}(J_{p_{\tau_\nu}}(j,c)-J_{p_{\tau_\nu}}(j+1,c))\prod_{\nu=1}^{\ell-n}\frac{p_{\lambda_\nu}}{q_{\lambda_\nu}}(J_{p_{\lambda_\nu}}(j-1,c+1)-J_{p_{\lambda_\nu}}(j,c+1))\cdot$$

$$\cdot\prod_{\nu=1}^{k-\ell-1}J_{q_{j_\nu}}(c+1,j)\Bigg)$$

If ℓ=n=k-1, then all treatments yield their c-th failure in the last trial and the contribution from (e) to P(CS) is as follows:

from (7.5), if n=ℓ=k-1:

$$\sum_{j=0}^{r-1}\frac{1}{k}\prod_{\nu=1}^{k}\left(\binom{j+c-1}{j}q_\nu^c p_\nu^j\right)=\frac{1}{k}E^{r-1}\Big(\prod_{\nu=2}^{k}(J_{p_\nu}(X_{c,q_1},c)-J_{p_\nu}(X_{c,q_1}+1,c))\Big)\tag{7.12}$$

7.2 Derivation of the critical r- and c-values

In using the random variables $F_i(r)$ and $S_i(c)$ defined in 6.2, the "events" (7.1)-(7.5) can be expressed as follows:

$$(7.1)\equiv\{F_1(r)<F_2(r),\ldots,F_1(r)<F_k(r),0\le F_1(r)<c\}\tag{7.13}$$

$$(7.2)\equiv\{F_1(r)=F_{i_1}(r)=\ldots=F_{i_\ell}(r),F_1(r)<F_{j_1}(r),\ldots,F_1(r)<F_{j_{k-\ell-1}}(r),0\le F_1(r)<c\}\tag{7.14}$$

$$(7.3)\equiv\{S_1(c)>S_2(c),\ldots,S_1(c)>S_k(c),0<S_1(c)<r\}\tag{7.15}$$

$$(7.4)\equiv\{S_{\lambda_1}(c)=\ldots=S_{\lambda_{\ell-n}}(c),S_1(c)<S_{\lambda_1}(c)<S_1(c+1)+1,$$

$$S_{\tau_\nu}(c)<S_{\lambda_1}(c)<S_{\tau_\nu}(c+1)+1\ \forall\nu\varepsilon\{1,\ldots,n\};\tag{7.16}$$

$$S_{\lambda_1}(c)\ge S_{j_\nu}(c+1)+1\ \forall\nu\varepsilon\{1,\ldots,k-\ell-1\},0<S_1(c)<r\}$$

$$(7.5) \equiv \{S_1(c)=S_{\tau_1}(c)=\ldots=S_{\tau_n}(c), S_{\lambda_\nu}(c)<S_1(c)<S_{\lambda_\nu}(c+1)+1$$
$$\forall \nu\varepsilon\{1,\ldots,\ell-n\}, S_1(c)\geq S_{j_\nu}(c+1)+1\ \forall\nu\varepsilon\{1,\ldots,k-\ell-1\}, \quad (7.17)$$
$$0<S_1(c)<r\},\ \text{if } n<k-1$$

$$(7.5) \equiv \{S_1(c)=S_2(c)=\ldots=S_k(c), 0\leq S_1(c)<r\},\ \text{if } n=\ell=k-1 \quad (7.18)$$

For large r and c the random variables $F_i(r), S_i(c)$; $i\varepsilon\{1,\ldots,k\}$; can be expressed by normally distributed random variables. Noting this fact and using similar arguments as in 3.2, P(CS) is immediately seen to be approximately:

$$P(CS) \approx P(F_1(r)<F_2(r),\ldots,F_1(r)<F_k(r), 0\leq F_1(r)<c) + $$
$$+ P(S_1(c)>S_2(c),\ldots,S_1(c)>S_k(c), 0<S_1(c)<r) \quad (7.19)$$

This is the same expression as already given in (6.7), and therefore, the analysis to determine r_{max} and c_{max} is the same as in the preceding section, i.e. r_{max} and c_{max} are given by (6.16).

7.3 Derivation of the expectations

In using the VT-sampling scheme, we have to calculate only one expectation, that is $E(N_1)$, and this is carried out by using the representation of N_1 given in (11.25) of chap. 1, i.e. we calculate $E(N_{1,1})$, and all other expectations $E(N_{1,i})$ for $i\geq 2$ can be obtained by using the well-known interchanging technique as demonstrated in sec. 11 of chap. 1.

The contributions of (7.6)-(7.9),(7.11),(7.12) to $E(N_{1,1})$ are as follows:

from (7.6):

$$\sum_{j=0}^{c-1}(r+j)\binom{j+r-1}{j}p_1^r q_1^j \prod_{\lambda=2}^{k} J_{q_\lambda}(j+1,r) = \frac{r}{p_1} E^{c-1}\left(\prod_{\lambda=2}^{k} J_{q_\nu}(X_{r+1,p_1}+1,r)\right) \quad (7.20)$$

from (7.7):

$$\sum_{\ell=1}^{k-1}\sum_{\omega\varepsilon S_\ell}\sum_{j=0}^{c-1}\left(\frac{1}{\ell+1}(r+j)\binom{j+r-1}{j}p_1^r q_1^j \prod_{\lambda=1}^{\ell}(J_{q_{i_\lambda}}(j,r)-J_{q_{i_\lambda}}(j+1,r))\ \cdot\right.$$
$$\cdot \prod_{\lambda=1}^{k-\ell-1} J_{q_{j_\lambda}}(j+1,r)) = \sum_{\ell=1}^{k-1}\cdot\sum_{\omega\varepsilon S_\ell}\frac{r}{p_1(\ell+1)} E^{c-1}\left(\prod_{\lambda=1}^{\ell}(J_{q_{i_\lambda}}(X_{r+1,p_1},r)\ -\right. \quad (7.21)$$
$$- J_{q_{i_\lambda}}(X_{r+1,p_1}+1,r)) \prod_{\lambda=1}^{k-\ell-1} J_{q_{j_\lambda}}(X_{r+1,p_1}+1,r))$$

from (7.8):

$$\sum_{j=1}^{r-1} (c+j)\binom{j+c-1}{j} q_1^c p_1^j \prod_{\lambda=2}^{k} J_{q_\lambda}(c+1,j) = \frac{c}{q_1} E^{r-1}\left(\prod_{\lambda=2}^{k} J_{q_\lambda}(c+1,X_{c+1,q_1})\right) \quad (7.22)$$

from (7.9):

$$\sum_{\ell=1}^{k-1}\sum_{n=0}^{\ell-1}\sum_{\omega\varepsilon S_\ell(n,\ell-n)} \frac{(c+1)p_1}{(\ell+1)q_1^2} E^{r-2}\left(\prod_{\nu=1}^{n}\frac{p_{\tau_\nu}}{q_{\tau_\nu}}(J_{p_{\tau_\nu}}(X_{c+2,q_1},c+1) -\right.$$

$$- J_{p_{\tau_\nu}}(X_{c+2,q_1}+1,c+1)) \prod_{\nu=1}^{\ell-n}(J_{p_{\lambda_\nu}}(X_{c+2,q_1}+1,c)-J_{p_{\lambda_\nu}}(X_{c+2,q_1}+2,c)) \cdot \quad (7.23)$$

$$\left.\cdot \prod_{\nu=1}^{k-\ell-1} J_{q_{j_\nu}}(c+1,X_{c+2,q_1}+1)\right)$$

from (7.11):

$$\sum_{\ell=1}^{k-1}\sum_{n=0}^{\min\{\ell,k-2\}}\sum_{\omega\varepsilon S_\ell(n,\ell-n)}\sum_{j=1}^{r-1}\left(\frac{c}{\ell+1}\binom{c+j}{j} q_1^c p_1^j \prod_{\nu=1}^{n}(J_{p_{\tau_\nu}}(j,c) -\right.$$

$$\left.- J_{p_{\tau_\nu}}(j+1,c)) \prod_{\nu=1}^{\ell-n}\frac{p_{\lambda_\nu}}{q_{\lambda_\nu}}(J_{p_{\lambda_\nu}}(j-1,c+1)-J_{p_{\lambda_\nu}}(j,c+1)) \prod_{\nu=1}^{k-\ell-1} J_{q_{j_\nu}}(c+1,j)\right) \quad (7.24)$$

from (7.12):

$$\sum_{j=0}^{r-1}\frac{1}{k}(c+j)\binom{j+c-1}{j} q_1^c p_1^j \prod_{\nu=2}^{k}(J_{p_\nu}(j,c)-J_{p_\nu}(j+1,c)) =$$

$$= \frac{c}{kq_1} E^{r-1}\left(\prod_{\nu=2}^{k}(J_{p_\nu}(X_{c+1,q_1},c)-J_{p_\nu}(X_{c+1,q_1}+1,c))\right) \quad (7.25)$$

$E(N_{1,1})$ is the sum of (7.20)-(7.25), and $E(N_1)$ is obtained from these formulae by using the interchanging technique.

The P(CS)-function for various values of k, P*,and Δ* is shown by the figures on the following page.

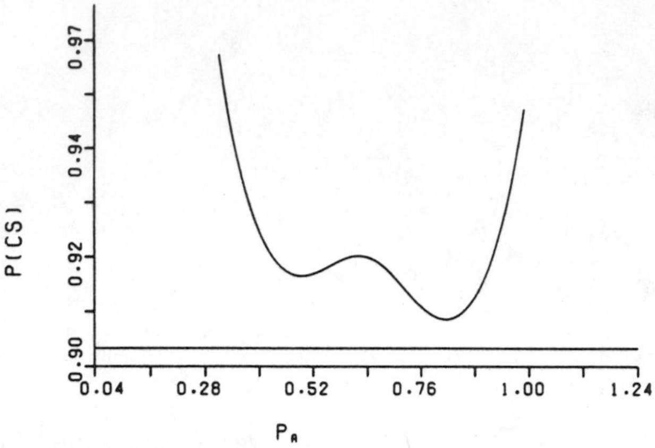

$k = 3$
$P^* = 0.90$
$\Delta^* = 0.30$
$r = 9$

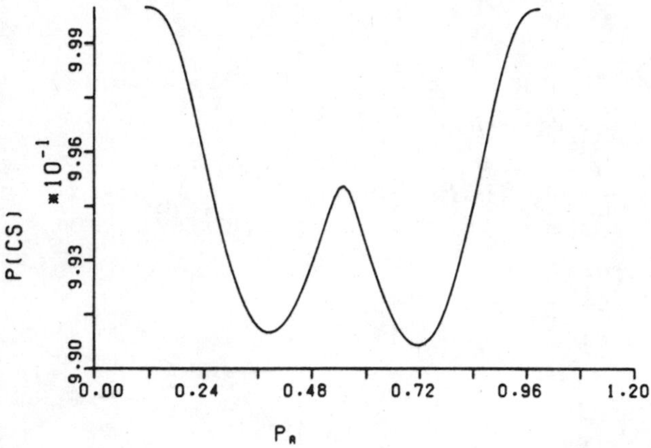

$k = 4$
$P^* = 0.99$
$\Delta^* = 0.10$
$r = 218$

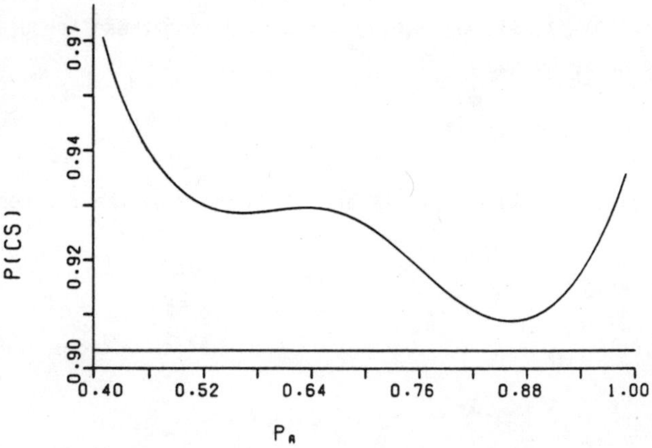

$k = 5$
$P^* = 0.90$
$\Delta^* = 0.40$
$r = 7$

7.4 Numerical results

Table 7

k = 3; P* = 0.90									
p_1	0.2	0.3	0.4	0.5	0.6	0.7	0.8	0.9	r
P(CS|Δ*=0.1)	0.952	0.919	0.912	0.923	0.920	0.909	0.916	0.949	79
E(N |Δ*=0.1)	286	320	372	420	370	315	278	253	
P(CS|Δ*=0.2)	-	0.936	0.915	0.917	0.923	0.912	0.909	0.930	20
E(N |Δ*=0.2)	-	82	93	100	93	80	70	64	
P(CS|Δ*=0.3)	-	-	0.926	0.916	0.920	0.916	0.907	0.915	9
E(N |Δ*=0.3)	-	-	41	42	40	36	32	29	
P(CS|Δ*=0.4)	-	-	-	0.916	0.913	0.912	0.903	0.901	5
E(N |Δ*=0.4)	-	-	-	22	21	20	18	16	
k = 3; P* = 0.95									
p_1	0.2	0.3	0.4	0.5	0.6	0.7	0.8	0.9	r
P(CS|Δ*=0.1)	0.981	0.961	0.956	0.963	0.962	0.954	0.959	0.980	114
E(N |Δ*=0.1)	421	474	552	630	552	470	413	374	
P(CS|Δ*=0.2)	-	0.972	0.959	0.960	0.966	0.958	0.956	0.970	29
E(N |Δ*=0.2)	-	122	140	152	140	120	105	95	
P(CS|Δ*-0.3)	-	-	0.966	0.959	0.964	0.962	0.955	0.962	13
E(N |Δ*=0.3)	-	-	62	64	61	54	47	42	
P(CS|Δ*=0.4)	-	-	-	0.969	0.969	0.971	0.965	0.965	8
E(N |Δ*=0.4)	-	-	-	38	37	33	30	26	
k = 3; P* = 0.99									
p_1	0.2	0.3	0.4	0.5	0.6	0.7	0.8	0.9	r
P(CS|Δ*=0.1)	0.998	0.992	0.991	0.993	0.993	0.991	0.992	0.998	197
E(N |Δ*=0.1)	738	939	978	1128	979	837	734	656	
P(CS|Δ*=0.2)	-	0.996	0.992	0.992	0.995	0.992	0.991	0.996	50
E(N |Δ*-0.2)	-	214	248	275	249	213	186	166	
P(CS|Δ*=0.3)	-	-	0.994	0.991	0.993	0.993	0.991	0.993	22
E(N |Δ*=0.3)	-	-	108	116	108	94	82	73	
P(CS|Δ*=0.4)	-	-	-	0.994	0.994	0.995	0.993	0.993	13
E(N |Δ*=0.4)	-	-	-	66	63	56	49	44	

Table 7 (continued)

k = 4; P^* = 0.90									
p_1	0.2	0.3	0.4	0.5	0.6	0.7	0.8	0.9	r
P(CS|Δ*=0.1)	0.960	0.926	0.918	0.930	0.926	0.913	0.921	0.957	98
E(N |Δ*=0.1)	473	525	609	694	606	514	454	418	
P(CS|Δ*=0.2)	-	0.946	0.924	0.926	0.933	0.918	0.915	0.938	25
E(N |Δ*=0.2)	-	137	155	167	154	131	115	105	
P(CS|Δ*=0.3)	-	-	0.934	0.923	0.928	0.921	0.909	0.919	11
E(N |Δ*=0.3)	-	-	67	69	66	58	50	46	
P(CS|Δ*=0.4)	-	-	-	0.945	0.913	0.942	0.931	0.929	7
E(N |Δ*=0.4)	-	-	-	43	41	37	33	30	
k = 4; P^* = 0.95									
p_1	0.2	0.3	0.4	0.5	0.6	0.7	0.8	0.9	r
P(CS|Δ*=0.1)	0.984	0.964	0.958	0.966	0.965	0.956	0.961	0.983	134
E(N |Δ*=0.1)	661	742	861	989	861	733	645	586	
P(CS|Δ*=0.2)	-	0.976	0.962	0.963	0.969	0.960	0.958	0.973	34
E(N |Δ*=0.2)	-	191	219	239	219	186	163	147	
P(CS|Δ*=0.3)	-	-	0.969	0.961	0.966	0.963	0.955	0.962	15
E(N |Δ*=0.3)	-	-	95	100	94	83	72	65	
P(CS|Δ*=0.4)	-	-	-	0.971	0.970	0.971	0.964	0.963	9
E(N |Δ*=0.4)	-	-	-	58	55	50	44	39	
k = 4; P^* = 0.99									
p_1	0.2	0.3	0.4	0.5	0.6	0.7	0.8	0.9	r
P(CS|Δ*=0.1)	0.998	0.993	0.991	0.994	0.993	0.991	0.992	0.998	218
E(N |Δ*=0.1)	1089	1238	1442	1668	1444	1234	1082	967	
P(CS|Δ*=0.2)	-	0.996	0.992	0.992	0.995	0.992	0.991	0.996	55
E(N |Δ*=0.2)	-	314	364	404	365	312	273	244	
P(CS|Δ*=0.3)	-	-	0.995	0.993	0.995	0.994	0.992	0.994	25
E(N |Δ*=0.3)	-	-	165	177	164	142	124	111	
P(CS|Δ*=0.4)	-	-	-	0.994	0.994	0.995	0.993	0.993	14
E(N |Δ*=0.4)	-	-	-	95	90	80	70	62	

Table 7 (continued)

k = 5; P* = 0.90									
p_1	0.2	0.3	0.4	0.5	0.6	0.7	0.8	0.9	r
P(CS|Δ*=0.1)	0.965	0.931	0.921	0.935	0.930	0.917	0.925	0.961	112
E(N |Δ*=0.1)	678	751	870	997	866	733	648	598	
P(CS|Δ*=0.2)	-	0.950	0.927	0.929	0.936	0.918	0.915	0.940	28
E(N |Δ*=0.2)	-	192	218	235	216	183	160	146	
P(CS|Δ*-0.3)	-	-	0.945	0.934	0.940	0.931	0.919	0.929	13
E(N |Δ*=0.3)	-	-	100	104	98	86	75	68	
P(CS|Δ*=0.4)	-	-	-	0.935	0.931	0.928	0.912	0.909	7
E(N |Δ*=0.4)	-	-	-	53	51	46	40	36	

k = 5; P* = 0.95									
p_1	0.2	0.3	0.4	0.5	0.6	0.7	0.8	0.9	r
P(CS|Δ*=0.1)	0.986	0.965	0.960	0.968	0.966	0.957	0.963	0.985	148
E(N |Δ*=0.1)	913	1024	1189	1370	1189	1010	890	810	
P(CS|Δ*=0.2)	-	0.977	0.962	0.964	0.970	0.959	0.957	0.973	37
E(N |Δ*=0.2)	-	260	298	326	297	253	221	200	
P(CS|Δ*=0.3)	-	-	0.973	0.966	0.971	0.967	0.959	0.966	17
E(N |Δ*=0.3)	-	-	136	143	135	117	102	92	
P(CS|Δ*=0.4)	-	-	-	0.974	0.973	0.974	0.966	0.965	10
E(N |Δ*=0.4)	-	-	-	81	77	69	61	54	

k = 5; P* - 0.99									
p_1	0.2	0.3	0.4	0.5	0.6	0.7	0.8	0.9	r
P(CS|Δ*=0.1)	0.998	0.993	0.991	0.994	0.994	0.991	0.993	0.998	234
E(N |Δ*=0.1)	1461	1661	1935	2242	1937	1656	1452	1298	
P(CS|Δ*=0.2)	-	0.996	0.992	0.993	0.995	0.992	0.992	0.996	59
E(N |Δ*=0.2)	-	421	489	544	489	418	366	327	
P(CS|Δ*=0.3)	-	-	0.995	0.992	0.994	0.994	0.991	0.994	26
E(N |Δ*=0.3)	-	-	214	230	214	185	161	144	
P(CS|Δ*=0.4)	-	-	-	0.994	0.994	0.995	0.993	0.993	15
E(N |Δ*=0.4)	-	-	-	128	121	107	93	83	

8 The Selection Model

$[k; VT; \max\{S_1,\dots,S_k\}=r \text{ or el. } A_i \text{ if } F_i=c]$

8.1 Derivation of the P(CS)-value

The procedure proposed in the following is in a certain sense the extension to more than two treatments of the VT-sampling procedure considered in sec. 4. As indicated, the procedure uses the VT-sampling scheme and a termination-rule allowing to eliminate treatments that turn out to be poor. Sampling terminates whenever one of the competing treatments yields r successes, and each treatment that reaches the c-th failure is eliminated from the sampling procedure. In case that the (k-1)-th treatment is just eliminated, the remaining k-th treatment is declared as best notwithstanding that less than r successes might have been occurred. Ties in the successes are, as usual, decided by randomization. Sampling terminates after at most k(r+c-1) steps.
To determine the exact formula for P(CS), we must try to find all stopping-sequences that may lead to a correct selection.
We have to distinguish seven different cases:

(a) A_1 yields r successes and less than c failures, and all other treatments yield less than r successes and less than c failures when sampling terminates. (8.1)

(b) A_1 yields r successes and less than c failures, and all other treatments yield less than r successes and c failures. $A_{i_1},\dots,A_{i_\ell}$ $(1<i_1<\dots<i_\ell\leq k;$ $\ell\in\{1,2,\dots,k-1\})$ yield their c-th failure in the last trial and the treatments $A_{\lambda_1},\dots,A_{\lambda_{k-\ell-1}}$ $(1<\lambda_1<\dots<\lambda_{k-\ell-1}\leq k)$ are eliminated from the sampling procedure before the last trial. (Note that $\ell\geq 1$ because sampling terminates automatically when only one treatment is left.) (8.2)

(c) A_1 yields r successes and less than c failures and all other treatments yield less than r successes. $A_{i_1},\dots,A_{i_\ell}$, $\ell\geq 0$ yield their c-th failure in the last trial, $A_{\lambda_1},\dots,A_{\lambda_n}$ are eliminated from the sampling procedure before the last trial, and $A_{j_1},\dots,A_{j_{k-\ell-n-1}}$, $0<k-\ell-n-1<k-1$, (8.3)

yield less than c failures when sampling terminates.

(d) $A_1, A_{\tau_1}, \ldots, A_{\tau_m}$ yield r successes and less than c failures, and all other treatments yield less than r successes and less than c failures. (8.4)

(e) $A_1, A_{\tau_1}, \ldots, A_{\tau_m}$ yield r successes and less than c failures, and all other treatments yield less than r successes. $A_{i_1}, \ldots, A_{i_\ell}$ yield their c-th failure in the last trial, $A_{\lambda_1}, \ldots, A_{\lambda_n}$ are eliminated from the sampling procedure before the last trial, and $A_{j_1}, \ldots,$ $A_{j_{k-m-\ell-n-1}}$ yield less than c failures when sampling terminates. (8.5)
$(m \geq 1; \ell \geq 0; n \geq 0; 0 \leq k-m-\ell-1 \leq k-2)$

(f) All treatments yield less than r successes, and A_1 yield less than c failures and all other treatments yield c failures. $A_{i_1}, \ldots, A_{i_\ell}$ yield their c-th failure in the last trial, $\ell \geq 1$, and $A_{\lambda_1}, \ldots, A_{\lambda_{k-\ell-1}}$ are eliminated before the last trial $(0 \leq k-\ell-1 \leq k-2)$. (8.6)

(g) All treatments yield less than r successes and c failures. $A_1, A_{i_1}, \ldots,$ A_{i_ℓ} $(\ell \geq 1)$ yield their c-th failure in the last trial, and $A_{\lambda_1}, \ldots,$ $A_{\lambda_{k-\ell-1}}$ are eliminated before the last trial $(0 \leq k-\ell-1 \leq k-2)$. (8.7)

The stopping-sequences described in (8.1) - (8.3) and (8.6) lead to a correct selection, whereas the stopping-sequences given by (8.4), (8.5) and (8.7) may lead to a correct selection only. In this case, the probability of a correct selection is equal to $1/(\ell+1)$.

The contributions of (8.1) - (8.7) to P(CS) are as follows:

from (8.1):

$$\sum_{j=0}^{c-2} \binom{j+r-1}{j} p_1^r q_1^j \prod_{\nu=2}^{k} \left(\sum_{\tau=\max\{0,r+j-c+1\}}^{r-1} \binom{j+r}{\tau} p_\nu^\tau q_\nu^{j+r-\tau} \right) \tag{8.8}$$

from (8.2):

$$\sum_{\ell=1}^{k-1} \sum_{\omega \varepsilon S_\ell} \sum_{j=\max\{0,c-r\}}^{c-1} \left(\binom{j+r-1}{j} p_1^r q_1^j \prod_{\nu=1}^{\ell} \binom{j+r-1}{c-1} q_{i_\nu}^c p_{i_\nu}^{r+j-c} \right. \tag{8.9}$$

$$\prod_{\nu=1}^{k-\ell-1}\Bigg(\sum_{\tau=0}^{\min\{r-1,r+j-c-1\}}\binom{c+\tau-1}{\tau}q_{\lambda_\nu}^{c}p_{\lambda_\nu}^{\tau}\Bigg)\Bigg),$$

where S_ℓ is defined by (6.3).

from (8.3):

$$\sum_{\ell=0}^{k-2}\sum_{n=0}^{k-2-\ell}\sum_{\substack{\omega\in S(\ell,n,k-\ell-n-1)\\ \omega\notin S(0,0,k-1)}}\sum_{j=\max\{0,c-r\}}^{c-2}\Bigg(\binom{j+r-1}{j}p_1^r q_1^j$$

$$\prod_{\nu=1}^{\ell}\binom{j+r-1}{c-1}q_{i_\nu}^{c}p_{i_\nu}^{r+j-c}\prod_{\nu=0}^{n}\Bigg(\sum_{\tau=0}^{\min\{r-1,r+j-c-1\}}\binom{c+\tau-1}{\tau}p_{\lambda_\nu}^{\tau}q_{\lambda_\nu}^{c}\Bigg) \qquad (8.10)$$

$$\prod_{\nu=1}^{k-\ell-n-1}\Bigg(\sum_{\tau=\max\{0,j+r-c+1\}}^{r-1}\binom{j+r}{\tau}p_{j_\nu}^{\tau}q_{j_\nu}^{j+r-\tau}\Bigg)\Bigg),$$

where

$$\begin{aligned}&S(\ell,n,k-\ell-n-1) := \{\omega=(\omega_1,\omega_2,\omega_3) : \omega_1=(i_1,\ldots,i_\ell),\\ &\omega_2=(\lambda_1,\ldots,\lambda_n),\ \omega_3=(j_1,\ldots,j_{k-\ell-n-1}),\ i_1<\ldots<i_\ell,\\ &\lambda_1<\ldots<\lambda_n,\ j_1<\ldots<j_{k-\ell-n-1},\ \text{and } \{i_1,\ldots,i_\ell,\lambda_1,\ldots,\lambda_n,\\ &j_1,\ldots,j_{k-\ell-n-1}\}=\{2,\ldots,n\},\ \text{and } \omega_s \text{ and } \omega_t \text{ have no common}\\ &\text{components if } s\neq t;\ s,t\in\{1,2,3\}\}\end{aligned} \qquad (8.11)$$

(Note that the case $n=\ell=0$, i.e. $\omega\in S(0,0,k-1)$, is separately considered in (8.8))

from (8.4):

$$\sum_{m=1}^{k-1}\sum_{\omega\in S_m}\sum_{j=0}^{c-2}\Bigg(\frac{1}{m+1}\binom{j+r-1}{j}p_1^r q_1^j\prod_{\nu=1}^{m}\binom{j+r-1}{j}p_{\tau_\nu}^{r}q_{\tau_\nu}^{j}$$

$$\prod_{\nu=1}^{k-m-1}\Bigg(\sum_{i=\max\{0,r+j-c+1\}}^{r-1}\binom{j+r}{i}p_{\lambda_\nu}^{i}q_{\lambda_\nu}^{j+r-i}\Bigg)\Bigg) \qquad (8.12)$$

from (8.5):

$$\sum_{m=1}^{k-1}\sum_{\ell=0}^{k-m-1}\sum_{n=0}^{k-m-\ell-1}\sum_{\substack{\omega\in S(m,\ell,n,k-m-\ell-n-1)\\ \omega\notin S(m,0,0,k-m-1)}}\sum_{j=\max\{0,c-r\}}^{c^*}\Bigg(\frac{1}{m+1}\cdot$$

$$\binom{j+r-1}{j} p_1^r q_1^j \prod_{\nu=1}^{m} \binom{j+r-1}{j} p_{\tau_\nu}^r q_{\tau_\nu}^j \prod_{\nu=1}^{\ell} \binom{j+r-1}{c-1} q_{i_\nu}^c p_{i_\nu}^{r+j-c} \cdot$$

$$\prod_{\nu=1}^{n} \left(\sum_{\alpha=0}^{\min\{r-1,r+j-c-1\}} \binom{c+\alpha-1}{\alpha} q_{\lambda_\nu}^c p_{\lambda_\nu}^\alpha \right) \tag{8.13}$$

$$\prod_{\nu=1}^{k-m-\ell-n-1} \left(\sum_{i=\max\{0,r+j-c+1\}}^{r-1} \binom{j+r}{i} p_{j_\nu}^i q_{j_\nu}^{j+r-i} \right)\Bigg) ,$$

where $S(m,\ell,n,k-m-\ell-n-1)$ has a similar meaning as $S(\ell,n,k-\ell-n-1)$ given by (8.11), and c^* equals $c-1$ if $k-m-\ell-1=0$ and $c-2$ if $k-m-\ell-1>0$.

from (8.6):

$$\sum_{\ell=1}^{k-1} \sum_{\omega\in S_\ell} \sum_{\tau=0}^{r-2} \sum_{\alpha=\tau+1}^{r-1} \Bigg(\binom{c+\iota}{\alpha} p_1^\alpha q_1^{c+\iota-\alpha} \cdot$$

$$\prod_{\nu=1}^{\ell} \binom{c+\tau-1}{\tau} p_{i_\nu}^\tau q_{i_\nu}^c \prod_{\nu=1}^{k-\ell-1} \left(\sum_{i=0}^{\tau-1} \binom{c+i-1}{i} q_{\lambda_\nu}^c p_{\lambda_\nu}^i \right)\Bigg) \tag{8.14}$$

from (8.7):

$$\sum_{\ell=1}^{k-1} \sum_{\omega\in S_\ell} \sum_{j=0}^{r-1} \Bigg(\frac{1}{\ell+1} \binom{j+c-1}{j} q_1^c p_1^j \prod_{\nu=1}^{\ell} \binom{j+c-1}{j} q_{i_\nu}^c p_{i_\nu}^j$$

$$\prod_{\nu=1}^{k-\ell-1} \left(\sum_{\alpha=0}^{j-1} \binom{c+\alpha-1}{\alpha} q_{\lambda_\nu}^c p_{\lambda_\nu}^\alpha \right)\Bigg) \tag{8.15}$$

8.2 Derivation of the critical r- and c-values

In using the random variables $F_i(r)$ and $S_i(c)$, the "events" (8.1) - (8.7) can be rewritten as follows:

$$(8.1) \equiv \{F_1(r) < F_\nu(r), F_1(r)+r<S_\nu(c)+c \;\; \forall\nu\varepsilon\{2,\dots,k\},\; 0\le F_1(r)<c\} \tag{8.16}$$

$$(8.2) \equiv \{F_1(r)<F_\nu(r) \forall\nu\varepsilon\{2,\dots,k\}, r+F_1(r)=c+S_{i_\mu}(c) \;\; \forall\mu\varepsilon\{1,\dots,\ell\}, r+F_1(r)>c+S_{\lambda_\beta}(c) \forall\beta\varepsilon\{1,\dots,k-\ell-1\},\; 0\le F_1(r)<c\} \tag{8.17}$$

$$(8.3) \equiv \{F_1(r)<F_\nu(r)\forall\nu\varepsilon\{2,\dots,k\}, r+F_1(r)=c+S_{i_\nu}(c) \forall\nu\varepsilon\{1,\dots,\ell\}, r+F_1(r)>c+S_{\lambda_\mu}(c)\forall\mu\varepsilon\{1,\dots,n\}, r+F_1(r)<c+S_{j_\mu}(c)\forall\mu\varepsilon\{1,\dots,k-\ell-n-1\}, 0\le F_1(r)<c\} \quad (8.18)$$

$$(8.4) \equiv \{F_1(r)=F_{\tau_\nu}(r)\forall\nu\varepsilon\{1,\dots,m\}, F_1(r)<F_{\lambda_\mu}(r), F_1(r)+r<c+S_{\lambda_\mu}(c)\forall\mu\varepsilon\{1,\dots,k-m-1\}, 0\le F_1(r)<c\} \quad (8.19)$$

$$(8.5) \equiv \{F_1(r)=F_{\tau_\nu}(r)\forall\nu\varepsilon\{1,\dots,m\}, F_1(r)<F_{j_\mu}(r) \forall\mu\varepsilon\{1,\dots,k-m-1\}, F_1(r)+r=S_{i_\beta}(c)+c \forall\beta\varepsilon\{1,\dots,\ell\}, F_1(r)+r>S_{\lambda_\delta}(c)+c\forall\delta\varepsilon\{1,\dots,n\}, F_1(r)+r<S_{j_\alpha}(c)+c\forall\alpha\varepsilon\{1,\dots,k-m-\ell-n-1\}, 0\le F_1(r)<c\} \quad (8.20)$$

$$(8.6) \equiv \{S_i(c)<S_1(c), F_1(r)+r>S_i(c)+c\forall i\varepsilon\{2,\dots,k\}\} \quad (8.21)$$

$$(8.7) \equiv \{S_1(c)=S_{i_\mu}(c)\forall\mu\varepsilon\{1,\dots,\ell\}, S_1(c)>S_{\lambda_\beta}(c) \forall\beta\varepsilon\{1,\dots,k-\ell-1\},\ 0<S_1(c)<r\} \quad (8.22)$$

remark:

The inequality $S_i(c)<r$ holds for all $i\varepsilon\{2,\dots,k\}$ for the stopping-sequences giv by (8.6).
$S_i(c)\geq r$ would imply $F_1(r)+r>S_i(c)+c\geq r+c$, and $F_1(r)>c$ would imply $S_1(c)<r$, and this again would imply $S_i(c)<S_1(c)<r$.
Note that $S_1(c)<r$ must not hold.

Recalling that for large values of r and c the random variables $F_i(r)$ and $S_i(c)$ can be expressed by normally distributed random variables, the probability of t events (8.4), (8.5) and (8.7) is immediately seen to tend to 0. To get a more distinct presentation of the approximate P(CS)-value, we rewrite the event (8.6 as follows (cf. 4.2 of sec. 4):

$$\{S_i(c)<S_1(c), F_1(r)+r>S_i(c)+c\ \forall i\varepsilon\{2,\dots,k\}\} =$$

$= \{S_i(c)<S_1(c), F_1(r)+r>S_i(c)+c \quad \forall i\varepsilon\{2,\ldots,k\},\ 0<S_1(c)<r\} \cup$

$\cup \{S_i(c)<S_1(c), F_1(r)+r>S_i(c)+c \quad \forall i\varepsilon\{2,\ldots,k\},\ S_1(c)\geq r\} =:$

$=: L \cup K$

L and K can again be rewritten as:

$$L = \{S_i(c)<S_1(c) \ \forall i\varepsilon\{2,\ldots,k\},\ 0<S_1(c)<r\} \tag{8.23}$$

$$K = \{F_1(r)<F_i(r), F_1(r)+r>S_i(c)+c \ \forall i\varepsilon\{2,\ldots,k\},\ 0\leq F_1(r)<c\} \tag{8.24}$$

Combining now the events (8.16), (8.17), (8.18) and (8.24), and (8.23), the P(CS)-value is immediately seen to be approximately:

$$\begin{aligned} P(CS) = {} & P(F_1(r)<F_2(r),\ldots,F_1(r)<F_k(r),\ 0\leq F_1(r)<c) + \\ & + P(S_1(c)>S_2(c),\ldots,S_1(c)>S_k(c),\ 0<S_1(c)<r) \end{aligned} \tag{8.25}$$

This term equals that of (6.7) and the critical r and c-values are, hence, determined by using similar arguments as before, i.e. we get

$$r_{max} = c_{max} = \left]\frac{8}{27}\left(\frac{\lambda_k(P^*)}{\Delta^*}\right)^2\right[\text{, where} \tag{8.26}$$

$\lambda_k(P^*)$ is the $100P^{*1/(k-1)}$-percentile of the standard normal distribution.

8.3 Numerical results

Table 8

	0.3	0.4	0.5	0.6	0.7	0.8	0.9
k=3, P*=0.90 Δ*=0.2, r=c=20 P(Cs; sim.) E(NB; sim.) E(N; sim.)	 0.929 23 68	 0.912 26 77	 0.926 29 87	 0.924 32 94	 0.912 29 86	 0.892 26 75	 0.928 23 68
k=3, P*=0.90 Δ*=0.3, r=c=9 P(Cs; sim.) E(NB; sim.) E(N; sim.)	 - - -	 0.928 11 32	 0.909 12 35	 0.919 13 38	 0.913 13 38	 0.916 12 34	 0.914 11 31
k=3, P*=0.90 Δ*=0.4, r=c=5 P(Cs; sim.) E(NB; sim.) E(N; sim.)	 - - -	 - - -	 0.917 6 18	 0.909 7 19	 0.904 7 20	 0.895 7 19	 0.908 6 17
k=3, P*=0.95 Δ*=0.2, r=c=29 P(Cs; sim.) E(NB; sim.) E(N; sim.)	 0.978 33 99	 0.955 37 111	 0.966 42 127	 0.967 46 138	 0.955 42 125	 0.951 37 110	 0.972 33 98
k=3, P*=0.95 Δ*=0.3, r=c=13 P(Cs; sim.) E(NB; sim.) E(N; sim.)	 - - -	 0.963 15 45	 0.964 17 51	 0.970 19 56	 0.957 19 55	 0.956 17 49	 0.966 15 44
k=3, P*=0.95 Δ*=0.4, r=c=8 P(Cs; sim.) E(NB; sim.) E(N; sim.)	 - - -	 - - -	 0.970 10 28	 0.957 11 31	 0.967 11 32	 0.961 11 30	 0.965 10 28
k=3, P*=0.99 Δ*=0.2, r=c=50 P(Cs; sim.) E(NB; sim.) E(N; sim.)	 0.995 57 169	 0.990 64 191	 0.988 72 218	 0.997 80 239	 0.991 72 214	 0.993 63 188	 0.997 56 167
k=3, P*=0.99 Δ*=0.3, r=c=22 P(Cs; sim.) E(NB; sim.) E(N; sim.)	 - - -	 0.996 25 75	 0.987 28 85	 0.993 32 95	 0.988 32 94	 0.990 28 83	 0.996 26 75
k=3, P*=0.99 Δ*=0.4, r=c=13 P(Cs; sim.) E(NB; sim.) E(N; sim.)	 - - -	 - - -	 0.993 15 45	 0.996 17 50	 0.989 18 53	 0.995 17 49	 0.990 15 44

Table 8 (continued)

	0.3	0.4	0.5	0.6	0.7	0.8	0.9
k=4, P^*=0.90 Δ^*=0.2, r=c=25							
P(Cs; sim.)	0.953	0.917	0.920	0.920	0.926	0.919	0.942
E(NB; sim.)	29	32	37	40	37	32	29
E(N; sim.)	113	128	147	157	143	126	112
k=4, P^*=0.90 Δ^*=0.3, r=c=11							
P(Cs; sim.)	-	0.928	0.911	0.925	0.921	0.922	0.917
E(NB; sim.)	-	13	15	16	16	15	13
E(N; sim.)	-	51	57	62	61	55	50
k=4, P^*=0.90 Δ^*=0.4, r=c=7							
P(Cs; sim.)	-	-	0.959	0.937	0.939	0.933	0.922
E(NB; sim.)	-	-	9	10	10	10	9
E(N; sim.)	-	-	33	36	37	35	32
k=4, P^*=0.95 Δ^*=0.2, r=c=34							
P(Cs; sim.)	0.982	0.965	0.965	0.977	0.955	0.947	0.978
E(NB; sim.)	39	43	49	54	49	44	39
E(N; sim.)	154	173	198	216	195	171	152
k=4, P^*=0.95 Δ^*=0.3, r=c=15							
P(Cs; sim.)	-	0.974	0.964	0.960	0.964	0.953	0.957
E(NB; sim.)	-	18	20	22	22	20	18
E(N; sim.)	-	69	77	86	84	76	68
k=4, P^*=0.95 Δ^*=0.4, r=c=9							
P(Cs; sim.)	-	-	0.969	0.967	0.961	0.962	0.956
E(NB; sim.)	-	-	11	12	13	12	11
E(N; sim.)	-	-	42	46	48	45	41
k=4, P^*=0.99 Δ^*=0.2, r=c=55							
P(Cs; sim.)	0.996	0.993	0.991	0.994	0.989	0.991	0.990
E(NB; sim.)	62	70	79	88	79	70	62
E(N; sim.)	248	280	320	353	314	276	245
k=4, P^*=0.99 Δ^*=0.3, r=c=25							
P(Cs; sim.)	-	0.997	0.990	0.991	0.998	0.994	0.995
E(NB; sim.)	-	29	32	36	36	32	29
E(N; sim.)	-	114	129	144	142	126	112
k=4, P^*=0.99 Δ^*=0.4, r=c=14							
P(Cs; sim.)	-	-	0.995	0.992	0.990	0.992	0.995
E(NB; sim.)	-	-	17	18	19	18	17
E(N; sim.)	-	-	64	72	76	71	63

Table 8 (continued)

	0.3	0.4	0.5	0.6	0.7	0.8	0.9
k=5, P*=0.90 Δ*0.2, r=c=28 P(Cs; sim.) E(NB; sim.) E(N; sim.)	 0.950 32 158	 0.925 36 178	 0.921 41 205	 0.927 44 220	 0.904 41 200	 0.911 36 176	 0.950 32 156
k=5, P*=0.90 Δ*=0.3, r=c=13 P(Cs; sim.) E(NB; sim.) E(N; sim.)	 - - -	 0.947 15 74	 0.942 17 84	 0.943 19 92	 0.944 19 90	 0.920 17 81	 0.942 15 73
k=5, P*=0.90 Δ*=0.4, r=c=7 P(Cs; sim.) E(NB; sim.) E(N; sim.)	 - - -	 - - -	 0.942 9 41	 0.937 9 45	 0.925 10 46	 0.921 9 43	 0.911 9 40
k=5, P*=0.95 Δ*=0.2, r=c=37 P(Cs; sim.) E(NB; sim.) E(N; sim.)	 0.976 42 209	 0.964 47 235	 0.970 54 270	 0.972 59 294	 0.973 54 265	 0.958 47 232	 0.960 42 207
k=5, P*=0.95 Δ*=0.3, r=c=17 P(Cs; sim.) E(NB; sim.) E(N; sim.)	 - - -	 0.970 20 97	 0.967 22 109	 0.974 25 121	 0.957 25 120	 0.960 22 107	 0.960 20 95
k=5, P*=0.95 Δ*=0.4, r=c=10 P(Cs; sim.) E(NB; sim.) E(N; sim.)	 - - -	 - - -	 0.973 12 58	 0.971 13 64	 0.979 14 66	 0.965 13 62	 0.966 12 56
k=5, P*=0.99 Δ*=0.2, r=c=59 P(Cs; sim.) E(NB; sim.) E(N; sim.)	 0.996 67 332	 0.992 75 374	 0.992 85 429	 0.994 95 474	 0.995 85 422	 0.990 75 371	 0.995 67 330
k=5, P*=0.99 Δ*0.3, r=c=26 P(Cs; sim.) E(NB; sim.) E(N; sim.)	 - - -	 0.993 30 147	 0.994 33 166	 0.995 38 187	 0.997 37 184	 0.988 33 163	 0.995 30 145
k=5, P*=0.99 Δ*0.4, r=c=15 P(Cs; sim.) E(NB; sim.) E(N; sim.)	 - - -	 - - -	 0.988 18 86	 0.994 20 96	 0.996 21 101	 0.991 20 94	 0.991 18 84

9 Further Elimination Procedures

We conclude this chapter by presenting five new elimination procedures that are designed for identifying the best of $k \geq 3$ alternatives.

9.1 The selection model [k;PW;el.A_i if $S_j-S_i=r$ or if $F_i=c$]

This selection model uses the PW-sampling-scheme and eliminates treatment A_i if there is some treatment A_j for which $S_j-S_i=r$ holds. In case that A_i yields c failures, it is eliminated, too. Sampling terminates in any case after at most $r(kc-1)$ trials. This maximal number of trials is reached if in each cycle each treatment yields a run of r-1 successes followed by a (single) failure. Sampling consequently terminates in the c-th cycle, where the last treatment in the random order of treatments, established at the outset, is selected as best.
Without loss of generality let $A_1A_2 \ldots A_k$ be the random order of treatments, then the abovementioned situation can be illustrated as follows:

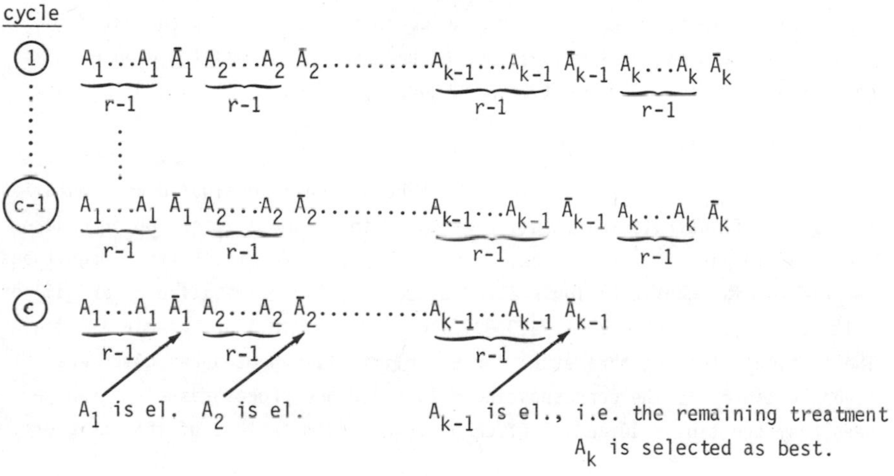

9.2 The selection model [k;VT;el.A_i if $S_j-S_i=r$ or if $F_i=c$]

This selection model uses the VT-sampling scheme and the same termination rule as the preceding selection model. The use of the VT-sampling scheme effects that sampling terminates with probability one only.

9.3 The selection model [k;PW;el.A_i if $S_j-S_i=r$ or stop if $F_1+\ldots+F_k=s$]

This elimination procedure is the extension to more than 2 treatments of the

Fushimi selection model presented in sec. 5. Treatment A_i is eliminated if there is some treatment A_j for which $S_j-S_i=r$ holds. In case that the sum of failures of all treatments equals s, sampling is terminated, and the treatment associated with the largest number of successes is declared as best. Ties in the number of successes are decided by randomization. Sampling terminates in any case after a restricted number of trials.

9.4 The selection model [k;PW;el.A_i if $S_j-S_i=r$ or el.A_i if $\hat{p}_j-\hat{p}_i \geq c/(F_i+F_j)$]

This selection model may be considered as the extension to more than 2 treatments of the Nordbrock selection model presented in sec. 13 of chap. 1. It uses the PW-sampling scheme and eliminates treatment A_i if there exists some treatment A_j for which either $S_j-S_i=r$ or $\hat{p}_j-\hat{p}_i \geq c/(F_i+F_j)$; where $\hat{p}_j$, and $\hat{p}_i$ denote the relative frequencies of successes of treatment A_j, and A_i, respectively. Sampling terminates with probability one only.

9.5 The selection model [k;VT;el.A_i if $S_j-S_i=r$ or el.A_i if $\hat{p}_j-\hat{p}_i \geq d/(F_i+F_j)$]

The fifth elimination procedure considered in this section is the extension to more than 2 treatments of the Nordbrock VT-selection model presented in sec. 14 of chap. 1. Also for this procedure, sampling terminates with probability one only.

All selection models described in 9.1-9.5 have been investigated by simulation. In order to illustrate the relative worth of the eight elimination procedures contained in part 1 of this book, we have tabulated the interesting magnitudes for the special configuration k=3, P*=0.99, and Δ*=0.2. Detailed numerical results for k=3,4,5, P*=0.90,0.95,0.99, and Δ*=0.2,0.3,0.4 are also available. The critical stopping parameters for the elimination procedures E4-E8 are given in table 9; the corresponding values for the procedures E1-E3 can be seen from the tables 10 and 11 of chap. 1, and from table 8 of this chapter.

9.6 Numerical results

Table 9

	E4	E5	E6	E7	E8
k=3;(P*;Δ*)					
(0.90;0.2)	(11,55)	(4,20)	(11,55)	(11,4)	(4,4)
(0.90;0.3)	(7,20)	(3,9)	(7,26)	(7,3)	(3,3)
(0.90;0.4)	(5,12)	(2,5)	(5,14)	(5,2)	(2,2)
(0.95;0.2)	(14,77)	(5,29)	(14,81)	(14,6)	(5,5)
(0.95;0.3)	(9,32)	(3,18)	(9,39)	(9,4)	(3,4)
(0.95;0.4)	(6,16)	(3,8)	(6,20)	(6,3)	(3,3)
(0.99;0.2)	(21,130)	(7,56)	(21,135)	(21,8)	(7,7)
(0.99;0.3)	(13,52)	(5,23)	(13,58)	(13,5)	(5,5)
(0.99;0.4)	(10,28)	(4,13)	(10,36)	(10,4)	(4,4)
k=4;(P*;Δ*)					
(0.90;0.2)	(13,68)	(6,25)	(13,89)	(13,5)	(5,5)
(0.90;0.3)	(8,27)	(4,11)	(8,39)	(8,3)	(3,3)
(0.90;0.4)	(6,14)	(3,6)	(6,22)	(6,3)	(2,2)
(0.95;0.2)	(16,90)	(6,34)	(16,119)	(16,6)	(6,6)
(0.95;0.3)	(10,38)	(4,15)	(10,52)	(10,4)	(4,4)
(0.95;0.4)	(7,18)	(3,9)	(7,31)	(7,3)	(3,3)
(0.99;0.2)	(23,139)	(8,56)	(23,215)	(23,9)	(8,8)
(0.99;0.3)	(15,63)	(5,28)	(15,96)	(15,6)	(5,5)
(0.99;0.4)	(10,28)	(4,14)	(10,53)	(10,5)	(4,4)
k=5;(P*;Δ*)					
(0.90;0.2)	(14,78)	(5,28)	(14,120)	(14,6)	(5,5)
(0.90;0.3)	(9,30)	(3,20)	(9,54)	(9,4)	(3,3)
(0.90;0.4)	(6,14)	(3,7)	(6,30)	(6,3)	(3,2)
(0.95;0.2)	(17,99)	(6,41)	(17,164)	(17,7)	(6,6)
(0.95;0.3)	(11,40)	(4,17)	(11,73)	(11,5)	(4,4)
(0.95;0.4)	(8,20)	(3,10)	(8,42)	(8,3)	(3,3)
(0.99;0.2)	(24,155)	(8,61)	(24,278)	(24,9)	(8,8)
(0.99;0.3)	(15,68)	(5,28)	(15,115)	(15,7)	(5,5)
(0.99;0.4)	(11,32)	(4,15)	(11,68)	(11,5)	(4,4)

Table 10

k=3; P*=0.99; Δ*=0.2									
p_A	0.3	0.4	0.5	0.6	0.7	0.8	0.9	r or s	c or d or s
E1: [3;PW;el.A_i if $S_j-S_i=r$]									
P(CS;sim.)	1.000	1.000	1.000	0.999	0.996	0.999	0.988		
E(NB;sim.)	73	64	54	44	33	22	13	21	-
E(N; sim.)	246	221	190	161	127	93	62		
E2: [3;VT;el.A_i if $S_j-S_i=s$]									
E(CS;sim.)	0.999	0.999	0.996	0.999	0.996	0.999	1.000		
E(NB;sim.)	36	36	36	35	36	36	36	7	-
E(N; sim.)	111	113	112	111	114	114	114		
E3: [3;VT;max{S_1,S_2,S_3}=r or el.A_i if $F_i=c$]									
P(CS;sim.)	0.995	0.990	0.988	0.997	0.991	0.993	0.997		
E(NB;sim.)	57	64	72	80	72	63	56	50	50
E(N; sim.)	169	191	218	239	214	188	167		
E4: [3;PW;el.A_i if $S_j-S_i=r$ or if $F_i=c$]									
P(CS;sim.)	0.992	0.998	1.000	0.999	0.996	0.999	0.988		
E(NB;sim.)	73	64	54	44	33	22	13	21	130
E(N; sim.)	245	221	191	161	128	94	62		
E5: [3;VT;el.A_i if $S_j-S_i=r$ or if $F_i=c$]									
P(CS;sim.)	0.996	0.990	0.998	0.996	0.995	1.000	1.000		
E(NB;sim.)	35	34	33	34	34	34	34	7	56
E(N; sim.)	108	107	107	108	108	110	109		
E6: [3;PW;el.A_i if $S_j-S_i=r$ or stop if $F_1+F_2+F_3=s$]									
P(CS;sim.)	0.991	0.993	0.992	0.997	0.996	0.998	0.994		
E(NB;sim.)	49	50	48	42	34	23	13	21	135
E(N; sim.)	161	170	168	153	129	97	64		
E7: [3;PW;el.A_i if $S_j-S_i=r$ or el.A_i if $\hat{p}_j-\hat{p}_i\geq c/(F_i+F_j)$]									
P(CS;sim.)	1.000	0.993	0.993	0.991	0.994	0.999	0.989		
E(NB;sim.)	24	26	30	34	32	22	13	21	8
E(N; sim.)	84	94	110	128	124	94	62		

Table 10 (continued)

k=3; $P^*=0.99;\Delta^*=0.2$									
p_A	0.3	0.4	0.5	0.6	0.7	0.8	0.9	r or s	c or d or s
E8: [3;VT;el.A_i if $S_j-S_i=r$ or el.A_i if $\hat{p}_j-\hat{p}_i\geq d/(F_i+F_j)$]									
P(CS;sim.)	0.996	0.990	0.989	0.997	0.996	0.999	1.000		
E(NB;sim.)	25	28	32	34	36	36	36	7	7
E(N; sim.)	79	87	102	110	114	114	114		

9.7 Comparison of selection models

When comparing the selection models from sec. 1 - sec. 5, we recognize that the Fushimi selection model, given in sec. 5, is uniformly better than all other selection models. Note that the comparison of these models is more predicatory than the comparison of the models of chap. 1 because the selection models of sec. 2,3,4 of the present chapter have identical P(CS)-values, and for $p_A\leq 0.4$ these P(CS)-values coincide additionally with the corresponding ones of the Fushimi selection model.

The Fushimi selection model is uniformly better than the untruncated model of sec. 1 in chap. 1. Note that for large parameter values P(CS)-values and the expectations of both procedures coincide (equal r-values!), because termination is mainly based on the difference in successes.

For large values of the success parameter p_A, the Fushimi selection model and the selection model of Nordbrock (cf. sec. 13, chap. 1) coincide in their P(CS)-values and expectations, because the termination of sampling is mainly based on the difference in successes, for smaller p_A-values, the Nordbrock selection model yields better results than the Fushimi selection model, i.e. the Nordbrock selection model of sec. 13, chap. 1 is uniformly better (or at least as good) than the Fushimi selection model. Note, however, that the Nordbrock model leads to a terminal decision only with probability one.

The last four sections of this chapter contain selection models constructed for identifying the best of $k\geq 3$ treatments. The selection model of sec. 6 yields better results than the selection model of sec. 7, and the same PW selection model is better than the elimination procedure of sec. 8 if $p_A\geq 0.5$. The selection models of sec. 6 and sec. 7 are the truncated versions of the selection models of sec.10 and sec. 11, chap. 1, respectively; they yield better results than the corresponding models of chap. 1, moreover sampling terminates in any case after a restricted number of trials.

The PW elimination procedure of sec. 15, chap. 1 (E1) seems to be better (with respect to E(N)) than the PW selection model of sec. 6 if $p_A \geq 0.5$. Note, however, that the patient horizon of all models presented in chap. 1 is not restricted. E4 is a truncated version of E1; both elimination procedures yield the same results for $p_A \geq 0.3$, however, E1 is worse than E4 if p_A is very small.
E6 is also a truncated version of E1. E6 is uniformly better (or at least as good) than E4.
E7 is uniformly better than E6; it should be however mentioned that E7 terminates with probability 1 only.
For large values of p_A E1,E4,E6, and E8 yield the same results because elimination is mainly based on the differences in the current numbers of successes.
This brief comparison can be summarized as follows:

The best PW-elimination procedure that terminates in any case after a restricted number of trials is E6, the elimination procedure based on Fushimi's selection model of sec. 5, chap. 2.

The best PW-elimination procedure that terminates with probability one only is E7, the elimination procedure based on the PW-selection model of Nordbrock (cf. sec. 13, chap. 1).

As for the VT-elimination procedure, the corresponding comparisons yield similar results, i.e.

E3 is worse than E2 if $p_A \geq 0.3$; for very small values of p_A E3 is better than E2, moreover E3 terminates after a restricted number of trials.
E2 and E5 yield essentially the same results if $p_A \geq 0.3$. E5 is better than E2 if p_A is very small. Note that E5 is a truncated version of E2.
E8 is uniformly better (or at least as good) than E5; however, E8 terminates with probability 1 only.

We may recapitulate:

The best VT-elimination procedure that terminates in any case after a restricted number of trials is E5.

The best VT-elimination procedure that terminates with probability one only is E8, the elimination procedure based on Nordbrock's VT-selection model (cf. sec. 14, chap. 1).

The results stated above are correct also for the other interesting values of the parameters P^* and Δ^* as well as for k=4, and k=5 treatments.

9.8 Further selection procedures

A further selection procedure appropriate to select the best of k=3 binomial populations has been given by W.L.Eckardt jr. [56]. Properly speaking his selection model consists of two stages; in the first stage sampling is carried out according to the PW-sampling scheme, where at the outset a random order of treatments is established and sampling starts with the leading treatment of that order. The experiment is continued until either one treatment yields r successes or all treatments yield c failures. In the first case the treatment yielding r successes is declared as best whereas in the second case we examine whether $S_i = d\cdot\max\{S_1, S_2, S_3\}$; $i \in \{1,2,3\}$, where $0 \leq d \leq 1$, and treatment A_i is eliminated from the experiment if the above requirement is met by S_i. If the best treatment cannot identified after that examination, sampling is continued (second stage) on the remaining treatments as long as one of them reaches r successes. Simulation results for d=1/2, $\Delta^* = 0.2$ and $P^* = 0.95$ are given in [56].

Another elimination procedure has been proposed by Hoel (cf. D.G.Hoel, "A selection procedure based upon Wald's double dichotomy test", Oak Ridge National Laboratory Technical Report No. TM-3237, Oak Ridge, Tennessee, 1971) Hoel's procedure based on general methods, derived in [91], applied to the double dichotomy problem formulated by A.Wald [198]. The trials are carried out according to the VT-sampling scheme and treatment A_j is eliminated if for some i $S_i - S_j \geq c + n^* d$, where $c > 0$ and $d \leq 0$ are preassigned constants and n^* denotes the number of trials that yield different results for the treatments A_i and A_j. Let

$$\tau_0 = \left(\frac{1-\Delta^*}{1+\Delta^*}\right)^2 \text{ and choose } \tau_1 \text{ such that } \tau_0 < \tau_1 \leq \frac{1}{\tau_0},$$

then the $(P^*;\Delta^*)$-condition is satisfied if c and d are chosen as follows (cf. [91]):

$$c = 2\ln\left(\frac{k-1}{1-P^*}\right)\Big/\ln\left(\frac{\tau_1}{\tau_0}\right) \quad ; \quad d = 2\ln\left(\frac{1+\tau_1}{1+\tau_0}\right)\Big/\ln\left(\frac{\tau_1}{\tau_0}\right) - 1$$

Numerical results for $\tau_1 = 1/\tau_0$, which implies d=0, are given in [98]. For details the reader is referred to the above quoted articles; it should be, however, mentioned that the elimination rule of Hoel's procedure causes more difficulties when actually applied than for instance the selection model of sec.6, chap. 2 where only successes and failures must be counted. This example reflects clearly the dilemma from which we cannot escape, i.e. to look for an overall best procedure that is simply to apply in practice. Comparing the selection models presented in chap. 1 and chap. 2, we recognize immediately that the better selection model is actually associated with the more sophisticated termination rule.

CHAPTER 3

Selection Procedures with Fixed Patient Horizon

1 Historical Remarks

As already mentioned in the introduction, the development of sequential selection models has been strongly influenced by medical formulations of the questions. Although the original idea of using sequential clinical trials seems to be due to A.B.Hill in 1951 (cf. British Med.Bull. 7, 278), the first edition of Armitage's monograph [8] in 1960 and the subsequent review by Anscombe [5] may be considered as the very beginning of an intensive research activity in the area of "controlled clinical trials", where the notion "(controlled) clinical trial" is used as a statistical experiment to select the best of $k \geq 2$ treatments for the same disease. In the following we will give only a brief survey of what has been done in this special field of statistics since 1960, the reader interested in a more detailed survey of adaptive sampling for clinical trials is referred to Hoel et al. [98].

Having in mind that A.Wald's Sequential Analysis [198] had been published only about 10 years before Armitage's book, it seems to be obvious that Armitage's original thinking was based on a fairly straightforward application of sequential analysis to the problem of selecting the better of two medical treatments. A different approach to that problem can be described by the following quotation from his book, "Suppose there are two rival treatments, A and B, for a specific condition and that each of N patients (the horizon) is to receive either A or B. We should like to give the better treatment to as many patients as possible, but do not know initially which is the better. A reasonable proposal would be to do a randomized trial with 2n patients (n on A and n on B) and use, for the remaining N-2n patients, whichever treatment appeared to be better in the trial...". In Armitage's conception the patient horizon N is the number of all patients that are ever to be treated with medicament A or B, if possible with the better of them, and this interpretation of the notion patient horizon is quite different from that we use throughout. Recall that in our conception the patient horizon N is the number of patients involved in the experiment, i.e. the number of experimental units needed to decide upon the relative worth of the treatments in research at the preassigned significance level P^*.

Colton [49] took up Armitage's suggestion and constructed a selection model based on it. He assumes that the individual responses to each treatment are normally distributed with unknown mean - μ_A for treatment A and μ_B for treatment B - and known equal variance σ^2; moreover higher response is assumed to be associated with better effect. The loss function, considered by Colton, is based only on the consequences of treating a patient with one of the competing medicaments; other factors as e.g. economic ones are not incorporated in the loss function. This proceeding harmonizes with the ethical point of view initially mentioned. Minimax, maximin, and Bayesian approaches are considered to derive the optimal size n of the testing-phase.

A very critical and restrictive assumption made by several scientists, working at Armitage's suggestion, is that the patient horizon N must be known, either exactly or at least in statistical terms. It is certainly beyond doubt that this knowledge is not available in general.

Further modifications of the Armitage selection model have been discussed by Chernoff (Proc.Berkeley Symp.Math.Statist.Probability, 5th, 4, 805), Cornfield et al. [51], Canner [46], and Zelen [202]. Canner assumes the response to be only dichotomous. He considers two costs, that are the costs of treating a patient with the inferior treatment and, in addition, the costs of conducting a trial, the so-called experimental costs.Minimax and Bayesian methods are used to determine the optimal size n of the testing-phase.
Zelen applied the PW-sampling scheme to the Colton model; his proceeding will be discussed in detail hereafter, because of its importance for the further development of selection models using adaptive sampling schemes.

2 The Zelen Selection Model

2.1 Definition of the model

M.Zelen [202] has considered the following modification of Colton's model. Like Canner [46], he assumes the response to be dichotomous. In the first stage of experimentation, the treatments are assigned according to the PW-sampling scheme, and the remaining N-n patients receive that treatment which yields the most successes up to the n-th trial.

Recalling that the treatment assignment depends only on the result of the preceding trial, when the PW-sampling scheme is used, the PW-rule can be described by a two-stage Markov chain, where the two stages correspond to the two treatments.

Let the random variables Y_{k-1} be defined as follows:

$$Y_{k-1} := \begin{cases} 1, & \text{if the k-th trial is carried out with treatment A} \\ 0, & \text{if the k-th trial is carried out with treatment B} \end{cases} \tag{2.1}$$

The occupation time of stage A within t trials is then given by

$$Z_t = Y_0 + Y_1 + \dots + Y_{t-1} \tag{2.2}$$

The matrix of transition probabilities is as follows

$$P = \begin{pmatrix} p_{AA} & p_{AB} \\ p_{BA} & p_{BB} \end{pmatrix}, \text{ where} \tag{2.3}$$

$p_{AA} = P(Y_k=1|Y_{k-1}=1) = p_A$ and $p_{BB} = P(Y_k=0|Y_{k-1}=0) = p_B$.

The expected number of A-trials when n trials are carried out ($E(N_A^n)$) is hence given by:

$$\begin{aligned}E(N_A^n) &= E(Z_n|Y_0=1)\,P(Y_0=1) + E(Z_n|Y_0=0)\,P(Y_o=0)\\ &= \frac{1}{2}\,(E(Z_n|Y_0=1) + E(Z_n|Y_o=0))\\ &= \frac{1}{2}\sum_{i=0}^{n-1}\,(E(Y_i|Y_0=1) + E(Y_i|Y_o=0))\\ &= \frac{1}{2}\sum_{i=0}^{n-1}\,(p_{AA}^{(i)} + p_{BA}^{(i)})\end{aligned} \tag{2.4}$$

where the i-th transition probabilities are available from

$$P^i = \frac{1}{p_{AB}+p_{BA}}\begin{pmatrix} p_{BA} & p_{AB}\\ p_{BA} & p_{AB}\end{pmatrix} + \frac{(1-(p_{AB}+p_{BA}))^n}{p_{AB}+p_{BA}}\begin{pmatrix} p_{AB} & -p_{AB}\\ -p_{BA} & p_{BA}\end{pmatrix} \tag{2.5}$$

We finally obtain by using (2.3) and (2.5):

$$E(N_A^n) = \frac{nq_B}{q_A+q_B} + \frac{1}{2}\,\frac{q_A-q_B}{q_A+q_B}\cdot\frac{1-(1-(q_A+q_B))^n}{q_A+q_B} \approx \frac{nq_B}{q_A+q_B} \tag{2.6}$$

The variance of N_A^n is approximately given by

$$Var(N_A^n) \approx \frac{nq_Aq_B(p_A+p_B)}{(q_A+q_B)^3} \tag{2.7}$$

The expected number of B-trials in the first stage of experimentation is obtained in the same way:

$$E(N_B^n) = \frac{nq_A}{q_A+q_B} + \frac{1}{2}\,\frac{q_B-q_A}{q_A+q_B}\cdot\frac{1-(1-(q_A+q_B))^n}{q_A+q_B} \approx \frac{nq_A}{q_A+q_B} \tag{2.8}$$

The expected number of successes in the first stage is immediately seen to be:

$$p_AE(N_A^n) + p_BE(N_B^n) \approx \frac{n(p_Aq_B+p_Bq_A)}{q_A+q_B}\,, \tag{2.9}$$

and the expected proportion of successes among N trials is thus approximately given by (note that N is no random variable but a fixed number):

$$s_r \approx \frac{n(p_Aq_B+p_Bq_A)}{N(q_A+q_B)} + \frac{N-n}{N}\,(p_AP(N_A^n > N_B^n) + p_BP(N_A^n < N_B^n)), \tag{2.10}$$

where the probability $P(N_A^n=N_B^n)$ is assumed to be negligible. Assuming in addition that N_A^n is normally distributed with mean and variance given by (2.6) and (2.7), respectively, $P(N_A^n > N_B^n)$ may be approximately evaluated as follows

$$P(N_A^n > N_B^n) = P(N_A^n > \left[\frac{n}{2}\right]) \approx P(N_A^n > \frac{n}{2}) =$$

$$= P\left(\frac{N_A^n - nq_B/(q_A+q_B)}{\sqrt{nq_Aq_B(p_A+p_B)/(q_A+q_B)^3}} > \frac{n/2 - nq_B/(q_A+q_B)}{\sqrt{nq_Aq_B(p_A+p_B)/(q_A+q_B)^3}}\right) \qquad (2.11)$$

$$= 1 - \Phi\left(\frac{n/2 - nq_B/(q_A+q_B)}{\sqrt{nq_Aq_B(p_A+p_B)/(q_A+q_B)^3}}\right) \quad =: \Phi^*(z)$$

In using the abbreviations

$$p^* := \frac{p_Aq_B+p_Bq_A}{q_A+q_B} \; ; \; p^{**} = p_A\Phi^*(z) + p_B(1-\Phi^*(z)), \qquad (2.12)$$

the expected proportion of successes in the entire experiment s_r can be written as

$$s_r = \frac{n}{N}\, p^* + (1-\frac{n}{N})\, p^{**}, \qquad (2.13)$$

where p^* and p^{**} are the expected proportion of successes in the first and second stage of the experiment. Since the first stage of experimentation is used to procure enough information so as to decide on the relative worth of the competing treatments, it seems to be obvious to determine that value of n which will guarantee the expected proportion of successes in the second stage to be not less than that in the first stage of experimentation, i.e. we have to determine those values of n for which the following condition holds:

$$p^{**} - p^* = (p_A-p_B)(\Phi^*(z) - q_B/(q_A+q_B)) \geq 0 \qquad (2.14)$$

It is immediately obvious that (2.14) is satisfied only if both factors of the righthand side are of the same sign. A straightforward analysis shows that this will be the case if

$$n \geq n_\alpha := z_\alpha^2 \cdot \frac{\alpha(1-\alpha)}{(\alpha-1/2)^2} \cdot \left(\frac{p_A+p_B}{q_A+q_B}\right) \quad \text{for } \alpha \neq 1/2, \qquad (2.15)$$

where $\alpha := q_B/(q_A+q_B)$ and $\Phi^*(z_\alpha) = \alpha$

($\alpha = 1/2$ implies $p_A=p_B$)

2.2 Comparison with a VT-sampling procedure

Another possibility to get information about the relative worth of two rival treatments is to conduct an equal number of trials on both in the first stage of experimentation. Such sort of procedure is usually referred to as VT-sampling procedure or equal assignment procedure. Colton's selection model, for example, is an equal assignment procedure (cf. [49]).

In using an equal assignment procedure with n/2 patient on each treatment in the first stage of experimentation, it is immediately seen that the expected proportion of successes in the first stage is $(p_A+p_B)/2$. Comparing this result with the expected proportion of successes in the first stage of the PW-sampling procedure, given in (2.12), we get:

$$p^* - \frac{p_A+p_B}{2} = \frac{(p_A-p_B)^2}{2(q_A+q_B)} \geq 0 \ , \tag{2.16}$$

i.e. the expected proportion of successes in the first stage for the PW-sampling procedure is always greater than the corresponding proportion for an equal assignment procedure, provided $p_A \neq p_B$.

To evaluate the value of $P(N_A^n > N_B^n)$ we use the fact that in case of an equal assignment procedure N_A^n and N_B^n are stochastically independent, moreover the binomial chance variables N_A^n and N_B^n have asymptotically a normal distribution with mean $\frac{n}{2}\, p_A$, $\frac{n}{2}\, p_B$ and variance $\frac{n}{2}\, p_A q_A$, $\frac{n}{2}\, p_B q_B$, respectively. $N_A^n - N_B^n$ has, therefore, asymptotically a normal distribution with mean $\frac{n}{2}\,(p_A-p_B)$ and variance $\frac{n}{2}\,(p_A q_A + p_B q_B)$. The desired value for $P(N_A^n > N_B^n)$ is thus given by:

$$P(N_A^n - N_B^n > 0) \approx 1 - \Phi\left(\frac{(p_B-p_A)\sqrt{n}}{\sqrt{2(p_A q_A + p_B q_B)}}\right) \tag{2.17}$$

It can be seen from (2.11) and (2.17) that in either case $P(N_A^n > N_B^n)$ tends to 0 (1) as n becomes large and if $p_A > p_B$ $(p_A < p_B)$, i.e. when n is large enough the expected number of successes in the second stage of the experiment is approximately given by $(N-n)\max\{p_A, p_B\}$ for either the PW-sampling or the equal assignment procedure. Assuming n to be an increasing function of the patient horizon N, the PW-sampling procedure is better than the VT-sampling procedure - greater proportion of successes in the first stage and equal proportion of successes in the second stage - , when N is sufficiently large.

We will conclude the comparison of both procedures in anwsering the question how many trials are needed in the first stage when the probabilities of identifying the best treatment, $P(N_A^n > N_B^n)$, are the same. These probabilities, given by (2.11) and (2.17), coincide when

$$\frac{n/2 - nq_B/(q_A+q_B)}{\sqrt{nq_A q_B (p_A+p_B)/(q_A+q_B)^3}} = \frac{(p_B-p_A)/\sqrt{n_e}}{\sqrt{2(p_A q_A + p_B q_B)}} \ , \tag{2.18}$$

where n, n_e denote the number of trials in the first stage of the PW-sampling, equal assignment procedure, respectively.
Writing $n_e = n(1+\delta)$ and substituting it into (2.18), we get:

$$\delta = (p_A+p_B-1)(p_A-p_B)^2/2q_Aq_B(p_A+p_B) \tag{2.19}$$

$\delta=0$, i.e. $p_A=p_B$ or $p_A+p_B=1$, implies that both procedures require the same number of trials in the first stage in order to identify the best treatment with the same probability. $\delta>0$, i.e. $p_A \neq p_B$ and $p_A+p_B>1$, means that the equal assignment procedure requires a larger number of trials to select the best treatment with the same probability. The PW-sampling procedure is, therefore, uniformly better than the equal assignment procedure, if $p_A+p_B>1$ and $p_A \neq p_B$. In case that $p_A+p_B<1$, there are only slight differences between either procedures.

2.3 Determination of the optimal value of n

We will conclude this section with the determination of the optimal n value, i.e. that value of n which maximizes $s_r=s_r(n,N)$, the expected proportion of successes during the entire experiment. Denoting $\max_n s_r(n,N)$ by $s_r(n_0,N)$, the inequalities $s_r(n_0,N) \geq s_r(n_0 \pm 1,N)$ must trivially hold. Let p^{**}_{+1} and p^{**}_{-1} denote the expected number of successes in the second stage if $n=n_0+1$ and $n=n_0-1$, respectively, then we get

$$\begin{aligned} s_r(n_0,N) &= s_r(n_0-1,N) + (p^{**}-p^{**}_{-1})(1-\frac{n_0}{N}) + (p^{*}-p^{**}_{-1})\frac{1}{N} \\ s_r(n_0+1,N) &= s_r(n_0,N) + (p^{**}_{+1}-p^{**})(1-\frac{n_0}{N}) + (p^{*}-p^{**}_{+1})\frac{1}{N}, \end{aligned} \tag{2.20}$$

and the above inequalities, necessary and sufficient for $s_r(n_0,N)$ to be a maximal value of $s_r(n,N)$, can be rewritten as:

$$\begin{aligned} &(1-\frac{n_0}{N})(p^{**}_{-1}-p^{**}) + \frac{1}{N}(p^{**}_{-1}-p^{*}) \leq 0 \\ &(1-\frac{n_0}{N})(p^{**}_{+1}-p^{**}) + \frac{1}{N}(p^{*}-p^{**}_{+1}) \leq 0 \end{aligned} \tag{2.21}$$

A straightforward calculation shows that (2.21) is equivalent to:

$$(1-\frac{1}{N-n_0})\cdot(p^{**}_{+1}-p^{**}) \leq \frac{p^{**}-p^{*}}{N-n_0} \leq (1+\frac{1}{N-n_0})\cdot(p^{**}-p^{**}_{-1}) \tag{2.22}$$

Let z_0 be the corresponding z-value, given by (2.11), when $n=n_0$, then, making use of (2.12), $p^{**}_{\Delta}-p^{**}$ may be expressed as $(\Delta \in \{-1,+1\})$:

$$p^{**}_{\Delta}-p^{**} = (p_A-p_B)\cdot(\Phi^*(z_0\sqrt{\frac{n_0+\Delta}{n_0}}) - \Phi^*(z_0)) \tag{2.23}$$

Using in addition the approximate result

$$\Phi^*(z_0\sqrt{\frac{n_0+\Delta}{n_0}}) - \Phi^*(z_0) \approx \frac{|p_A-p_B|\, z_0\varphi(z_0)\Delta}{2n_0(p_A-p_B)}, \tag{2.24}$$

where φ is the density function of the standard normal distribution, and recalling (2.14) and (2.15), the approximate requirement for n_0 to maximize the function $s_r(n,N)$ is given by

$$\Phi^*(z_0) + z_0\varphi(z_0)\,\frac{N-n_0}{2n_0} = \alpha = \frac{q_B}{q_A+q_B} \tag{2.25}$$

Substituting the Taylor expansion for $\Phi^*(z_\alpha)$ into (2.25), we get:

$$\begin{aligned} \alpha = \Phi^*(z_\alpha) &\approx \Phi^*(z_0) - \varphi(z_0)(z_\alpha - z_0) \\ z_\alpha - z_0 + z_0\,\frac{N-n_0}{2n_0} &= 0, \end{aligned} \tag{2.26}$$

and n_0 can be obtained by solving the following quadratic equation for $\sqrt{n_0}$

$$n_0 - \frac{2}{3}y_\alpha\sqrt{n_0} - \frac{N}{3} = 0, \tag{2.27}$$

where

$$y_\alpha := z_\alpha\sqrt{\frac{\alpha(1-\alpha)(p_A+p_B)}{(\alpha-1/2)^2(q_A+q_B)}}$$

The desired value of n_0 is hence approximately given by

$$n_0 = \frac{N}{3} + \frac{2y_\alpha^2}{9} \pm \frac{2}{9}y_\alpha\sqrt{3N+y_\alpha^2} \tag{2.28}$$

Since both values of n_0 satisfy the first equation of (2.27), each value of n contained in the interval, having limits given by (2.28), is expected to be a good approximation of the optimal value n_0.

It has been shown by Zelen [202] that $s_r(n,N)$ is relatively flat in the neighborhood of n_0, i.e. large values of N effect that the value of $s_r(n,N)$ is fairly insensitive within large neighborhoods of n_0=N/3. (For a detailed analysis the reader is referred to the original paper of Zelen.)

If N is sufficiently large, the maximum value of the expected proportion of successes among N trials is approximately given by

$$s_r(n_0,N) \approx \frac{p_Aq_B+p_Bq_A}{3(q_A+q_B)} + \frac{2}{3}\max\{p_A,p_B\} \tag{2.29}$$

3 The Selection Models

[2;PW;fixed N] and [2;VT;fixed N]

3.1 Introduction

The following selection models, due to Nebenzahl/Sobel [130], use the PW- and VT-sampling scheme with a fixed patient horizon. They resemble to a certain extent that selection model, considered in the preceding section, except that we now require the probability of a correct selection to be at least P^* whereas the Zelen model is, first of all, designed to achieve optimum results for the expected proportion of successes during the entire experiment rather than meeting the probability requirements given by the $(P^*;\Delta^*)$-condition. Moreover, the notion patient horizon is now again to interpret as the whole number of experimental units needed to decide upon the relative worth of the rival treatments at a preassigned significance level P^*. Fixed patient horizon N means that exactly N trials are conducted according to the applied sampling scheme, and after that the decision about the best treatment is taken. We declare that treatment as best which yields the greatest number of successes, ties (in the number of successes) are broken by randomization. Contrary to the selection models that assume an infinite or restricted patient horizon (cf. chap. 1 and chap. 2), the fixed patient horizon selection models have no special termination-rules. Sampling always[1] terminates after exactly N trials, where the constant N is determined so that the (P^*,Δ^*)-condition is satisfied, and the terminal decision is based on the observed number of successes.
When using the VT-sampling scheme the total sample size N is an even integer. In case that the PW-sampling scheme is applied, N may be odd or even. The comparison of both sampling schemes is possible without difficulties for even N, whereas an odd N requires a slight modification of the VT-sampling scheme before a comparison can be carried out. This special case will be discussed later on.

---

1 In case that the same decision can be taken before the N-th trial, sampling may be truncated; cf. e.g. sec. 4, 5, and 6.

3.2 Comparison of the P(CS)-values

As indicated above, we assume N to be even. The main result to be verified in the following is that the PW- and the VT-sampling scheme have equal probabilities of correct selection, i.e.

$$P_N(CS|PW) = P_N(CS|VT) \text{ for all even integer } N; \tag{3.1}$$

the index N indicates that N trials are carried out.

Let $N=2k$, $k \in \mathbb{N}$, then $P_{2k}(CS|VT)$ is easily seen to be

$$P_{2k}(CS|VT) = P_{2k}(S_A>S_B|VT) + \frac{1}{2} P_{2k}(S_A=S_B|VT) = \\ = \sum_{i=1}^{k} \sum_{j=0}^{i-1} \binom{k}{i}\binom{k}{j} p_A^i q_A^{k-i} p_B^j q_B^{k-j} + \frac{1}{2} \sum_{i=0}^{k} \binom{k}{i}^2 (p_A p_B)^i (q_A q_B)^{k-i} \tag{3.2}$$

A direct computation for the PW-sampling scheme for k=1 shows that both P(CS)-values coincide for this special case, i.e.

$$P_2(CS|VT) = \frac{1}{2}(p_A+q_B) = P_2(CS|PW) \tag{3.3}$$

To complete the proof of (3.1), we have to verify the following identity for $n \in \mathbb{N}$

$$P_{2n+2}(CS|VT) - P_{2n}(CS|VT) = P_{2n+2}(CS|PW) - P_{2n}(CS|PW) \tag{3.4}$$

Let A_i, B_i denote the events that treatment A,B are used at the i-th stage of the experiment, respectively, and let $\Delta S := S_A - S_B$ denote the difference in the current numbers of successes. In evaluating the righthand side of (3.4), we have to note that an increase in P(CS|PW) from i to i+1 is associated with the occurrence of A_{i+1}, and a decrease in P(CS|PW) occurs if the (i+1)-th trial is on treatment B, i.e. when B_{i+1} occurs. The increase in P(CS|PW) from i to i+1 is given by:

$$h_i(p_A,p_B) = P_{i+1}(CS,A_{i+1}|PW) - P_i(CS,A_{i+1}|PW) \tag{3.5}$$

To simplify the notation, the symbol "PW" is omitted in the following. Noting that an increase in P(CS) occurs only if the (i+1)-th trial is a success, (3.5) can be rewritten as

$$h_i(p_A,p_B) = \frac{1}{2} p_A P_i(\Delta S=0,A_{i+1}) + \frac{1}{2} p_A P_i(\Delta S=-1, A_{i+1}), \tag{3.6}$$

where $P_i(\Delta S=0,A_{i+1})$ is the probability of all sequences of length i that have an equal number of A- and B-successes, and that terminate with an A-success or a B-failure. The factor 1/2 is due to the randomization at stage i, correspondingly, in the second term of (3.6) $P_i(\Delta S= -1,A_{i+1})$ is the probability of all sequences of length i for which $S_A-S_B= -1$ holds, and that terminate with an A-success or a B-failure. If the (i+1)-th trial is an A-success, ΔS increases by 1, i.e. $\Delta S=0$ at stage i+1, and the best treatment is determined by randomization, reflected by the factor 1/2.

A decrease in P(CS) occurs if the (i+1)-th trial is a successful B-trial. The sequences having $\Delta S=1$, $\Delta S=0$ at stage i hence reduces to sequences with $\Delta S=0$, $\Delta S= -1$, respectively, at stage i+1. Randomization takes place at stage i+1, stage i in the first, second case, respectively. The decrease is consequently given by

$$- \frac{1}{2} p_B P_i(\Delta S=1,B_{i+1}) - \frac{1}{2} p_B P_i(\Delta S=0,B_{i+1}), \tag{3.7}$$

and by applying the well-known interchanging technique, (3.7) is immediately seen to be

$$\begin{aligned} &- \frac{1}{2} p_B P_i(\Delta S= -1,A_{i+1},p_A \leftrightarrows p_B) - \frac{1}{2} p_B P_i(\Delta S=0,B_{i+1},p_A \leftrightarrows p_B) \\ &\quad = -h(p_A,p_B,p_A \leftrightarrows p_B) = -h(p_B,p_A) \end{aligned} \tag{3.8}$$

The difference in the P(CS)-values is hence given by

$$P_{i+1}(CS) - P_i(CS) = h_i(p_A,p_B) - h_i(p_B,p_A) \tag{3.9}$$

The righthand side of (3.4) can now be rewritten as:

$$\begin{aligned} &P_{2n+2}(CS) - P_{2n}(CS) = P_{2n+2}(CS) - P_{2n+1}(CS) + P_{2n+1}(CS) - \\ &P_{2n}(CS) = h_{2n+1}(p_A,p_B) + h_{2n}(p_A,p_B) - (h_{2n+1}(p_B,p_A) + h_{2n}(p_B,p_A)) \\ &\qquad =: t(p_A,p_B) - t(p_B,p_A) \end{aligned} \tag{3.10}$$

To get an explicit expression for $t(p_A,p_B)$, we have to evaluate $P_{2n}(\Delta S=0,A_{2n+1})$, $P_{2n+1}(\Delta S=0, A_{2n+2})$, $P_{2n}(\Delta S= -1,A_{2n+1})$, and $P_{2n+1}(\Delta S= -1,A_{2n+2})$.

The event "$\{\Delta S=0\} \cap A_{2n+1}$" means that the 2n-th trial is an A-success or a B-failure and since $\Delta S=0$, the numbers of A- and B-failures are equal, and that again is possible only if the first trial is carried out on treatment A. The desired probability is, therefore, given by

$$P_{2n}(\Delta S=0,A_{2n+1}) = \frac{1}{2} P_{2n}(\Delta S=0,A_{2n+1} | \text{beg.w.A}) \tag{3.11}$$

In noting that n trials are carried out on either treatment, and that the last

B-trial is a B-failure, whereas the last A-trial may be a success or a failure, an explicit expression for the righthand side of (3.11) is easily obtained by using the following diagram:

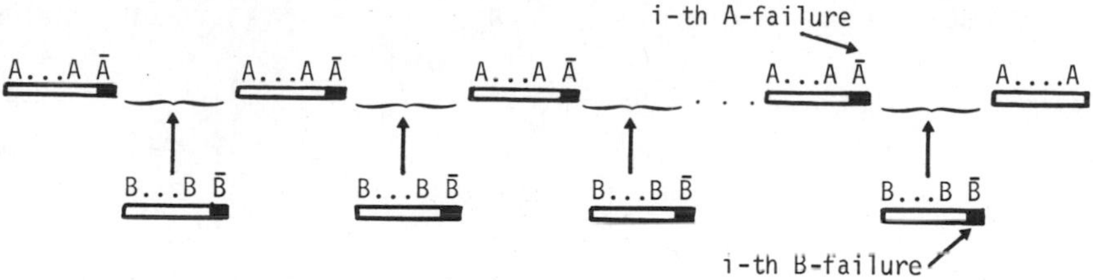

We finally obtain:

$$P_{2n}(\Delta S=0,A_{2n+1}) = \frac{1}{2}\sum_{i=1}^{n}\binom{n}{i} q_A^i p_A^{n-i}\binom{n-1}{i-1} q_B^{i-1} p_B^{n-i} q_B = \tag{3.12}$$

$$= \frac{1}{2}\sum_{i=1}^{n} b(i;n,q_A)b(i-1;n-1,q_B)q_B,$$

where $b(k;n,q) := \binom{n}{k} q^k p^{n-k}$ denotes the binomial probability function.

In order to evaluate $P_{2n+1}(\Delta S=0,A_{2n+2})$, we have to note that the last B-trial is a $\bar{B}$-trial, and that there are n+1 B-trials and n A-trials. The situation can be illustrated by a diagram similar to that given above.

The desired result is as follows

$$P_{2n+1}(\Delta S=0,A_{2n+2}) = \frac{1}{2}\sum_{i=0}^{n}\binom{n}{i} q_A^i p_A^{n-i}\binom{n}{i} q_B^i p_B^{n-i} q_B = \tag{3.13}$$

$$= \frac{1}{2}\sum_{i=0}^{n} b(i;n,q_A)b(i;n,q_B)q_B$$

The remaining probabilities of interest can be evaluated by using similar considerations as before. The desired expressions are thereby easily found to be

$$P_{2n}(\Delta S= -1,A_{2n+1}) = \frac{1}{2}\sum_{i=0}^{n-1}\binom{n-1}{i} q_A^i p_A^{n-1-i}\binom{n}{i} q_B^i p_B^{n-i} q_B = \tag{3.14}$$

$$= \frac{1}{2}\sum_{i=0}^{n-1} b(i;n-1,q_A)b(i;n,q_B)q_B$$

$$P_{2n+1}(\Delta S= -1,A_{2n+2}) = \frac{1}{2}\sum_{i=0}^{n}\binom{n}{i+1} q_A^{i+1} p_A^{n-i-1}\binom{n}{i} q_B^i p_B^{n-i} q_B = \tag{3.15}$$

$$= \frac{1}{2} \sum_{i=0}^{n} b(i+1;n,q_A)b(i;n,q_B)q_B$$

$t(p_A,p_B)$ can now be derived from (3.6),(3.10), and (3.12) - (3.15). We obtain:

$$t(p_A,p_B) = \frac{1}{4} p_A q_B \sum_{i=0}^{n} b(i;n,q_A)(b(i-1;n-1,q_B) + b(i;n,q_B)) + \\ + \frac{1}{4} p_A q_B \sum_{i=0}^{n} b(i;n,q_B)(b(i;n-1,q_A) + b(i+1;n,q_A)) \tag{3.16}$$

The lefthand side of (3.4) is much simpler to evaluate. An increase in $P(CS|VT)$ from 2n to 2n+2 can occur only if the n+1 stage of the experiment results in an A-success and a B-failure. Testing-sequences yielding $\Delta S=0$ or $\Delta S= -1$ at the n-th stage are turned into sequences that yield $\Delta S=1$ or $\Delta S=0$, respectively, at the (n+1)-th stage of the experiment. The increase in $P(CS|VT)$ is consequently given by

$$\frac{1}{2} p_A q_B (P_{2n}(\Delta S=0) + P_{2n}(\Delta S= -1)) =: m(p_A,p_B) \tag{3.17}$$

A decrease in $P(CS|VT)$ from 2n to 2n+2 occur if the (n+1)-th A-trial is a failure and the (n+1)-th B-trial a success. Testing-sequences yielding $\Delta S=0$ or $\Delta S=1$ at the n-th stage are turned into sequences that yield $\Delta S= -1$ or $\Delta S=0$, respectively, at the (n+1)-th stage of the experiment. The decrease is explicitly given by

$$\frac{1}{2} p_B q_A (P_{2n}(\Delta S=0) + P_{2n}(\Delta S=1)) = m(p_B,p_A) \tag{3.18}$$

From (3.17) and (3.16) follows:

$$P_{2n+2}(CS|VT) - P_{2n}(CS|VT) = m(p_A,p_B) - m(p_B,p_A),$$

where (3.19)

$$m(p_A,p_B) = \frac{1}{2} p_A q_B \sum_{i=0}^{n} b(i;n,q_B)(b(i;n,q_A) + b(i+1;n,q_A))$$

To verify the equality, given by (3.4), we have to show that $t(p_A,p_B)-t(p_B,p_A) = = m(p_A,p_B) - m(p_B,p_A)$ holds, and that again can be proved by showing that $s(p_A,p_B) := t(p_A,p_B) - m(p_A,p_B)$ is symmetric with respect to an interchange of p_A and p_B. The last assertion requires only a straightforward analysis to be proved. $s(p_A,p_B)$ is quickly seen to be

$$s(p_A,p_B) = \frac{1}{4} \sum_{i=0}^{n} \binom{n}{i} \left(\binom{n-1}{i-1} (q_A q_B)^i (p_A p_B)^{n-i+1} - \binom{n-1}{i+1} (q_A q_B)^{i+1} (p_A p_B)^{n-i} \right), \tag{3.20}$$

from which the above mentioned symmetry is immediately obvious. The proof of (3.4) and hence that of (3.1) is thereby completed, i.e. $P_N(CS|PW)=P_N(CS|VT)$ for even integers N.

3.3 Comparison of the expectations

We will prove in the following that the PW-procedure is at least as good as the VT-procedure in the sense that the expected number of patients on the poorer treatment under PW will never exceed the corresponding expectation under VT. To see this let $N_B(n)$ denote the number of patients on the poorer treatment on the (n+1)-th and (n+2)-th trials, and let A_n^{**}, B_n^{**} denote the events that the better, poorer treatments, respectively, are used on both the (n+1)-th and the (n+2)-th trials (under PW), and finally let C_n^{**} denote the event that either treatment is used on the (n+1)-th or (n+2)-th trial. It follows:

$$E(N_B(n)|VT) = 1\cdot P(\{(A,\bar{B}),(A,B),(\bar{A},B),(\bar{A},\bar{B})\}) = 1\cdot(\frac{1}{2}\cdot\frac{1}{2}+\frac{1}{2}\cdot\frac{1}{2}+ + \frac{1}{2}\cdot\frac{1}{2}+\frac{1}{2}\cdot\frac{1}{2}) = 1 \tag{3.21}$$

In using the inequality $P(B_n^{**} \mid PW) \leq P(A_n^{**}|PW)$, which will be proved hereafter, we obtain

$$E(N_B(n)|PW) = 0\cdot P(A_n^{**}|PW) + 1P(C_n^{**}|PW) + 2P(B_n^{**}|PW) \leq \leq P(C_n^{**}|PW) + 2(\frac{1}{2} - \frac{1}{2} P(C_n^{**}|PW)) = 1 \tag{3.22}$$

(3.21) and (3.22) jointly imply

$$E(N_B(n)|PW) \leq E(N_B(n)|VT), \tag{3.23}$$

and we consequently have

$$E(N_B|PW) = \sum_{n-o}^{k-1} E(N_B(2n)|PW) \leq \sum_{n-o}^{k-1} E(N_B(2n)|VT) = E(N_B|VT), \tag{3.24}$$

where N=2k is the fixed patient horizon.
The remaining task to do is to prove the inequality $P(B_n^{**}|PW) \leq P(A_n^{**}|PW)$. For this purpose let A_n^*, B_n^* denote the events that the better, poorer treatment, respectively, is used on the (n+1)-th trial. Noting that

$$P(A_n^{**}|PW) = p_A P(A_n^* \mid PW), \quad P(B_n^{**}|PW) = p_B P(B_n^* \mid PW) \tag{3.25}$$

holds, the inequality is proved if we have shown that

$$P(A_n^*|PW) \geq P(B_n^*|PW), \text{ for all even } n \qquad (3.26)$$

Since no ambiguity may arise, the symbol PW is omitted in the following. Noting that $P(A_n^*|PW) + P(B_n^*|PW) = 1$, (3.26) holds if $P(A_n^*|PW) \geq 1/2$ for all even n, and this assertion can be proved by induction.

If n=0, then $P(A_0^*|PW) = P(\text{begin.w.A}) = \frac{1}{2}$, and (3.26) holds true for n=0. Let (3.26) hold for some even n, then, using the recursion formulae,

$$P(A_{n+1}^*) = p_A P(A_n^*) + q_B P(B_n^*);\ P(B_{n+1}^*) = p_B P(B_n^*) + q_A P(A_n^*) \qquad (3.27)$$

$P(A_{n+2}^*)$ is seen to be:

$$P(A_{n+2}^*) = p_A(p_A P(A_n^*) + q_B(1-P(A_n^*))) + q_B(p_B(1-P(A_n^*)) +$$
$$+ q_A P(A_n^*)) = (p_A+p_B-1)^2 P(A_n^*) + (1-p_B)(p_A+p_B) \geq$$
$$\geq \frac{1}{2}(p_A+p_B-1)^2 + (1-p_B)(p_A+p_B) = \frac{1}{2} + \frac{1}{2}(p_A-p_B)(p_A+p_B) \geq \frac{1}{2},$$

and that completes the proof of (3.26), and hence that of (3.24).

3.4 Exact and asymptotic formulae for $E(N_B)$

To derive the explicit formula for $E(N_B)$ when PW-sampling is applied, we make use of the well-known representation $E(N_B) = E(N_{B,CS}) + E(N_{B,FS})$ (cf. (6.15) of chap. 1).

Denoting as usual by S_A and S_B / F_A and F_B the current numbers of successes/failures of treatment A,B, respectively, - in case of a fixed patient horizon N, S_A, S_B, F_A, F_B are the respective quantities within N trials - $E(N_{B,CS})$ is found to be (N=2n):

$$\begin{aligned} E(N_{B,CS}) = & \sum_{i=1}^{n-1} \sum_{j=0}^{\frac{N-2i}{2}-1} (i+j)P(F_A=F_B=i,\ S_A=N-2i-j, S_B=j) \\ & + \sum_{i=1}^{n} \frac{n}{2} P(F_A=F_B=i, S_A=S_B=n-i) \\ & + \sum_{i=0}^{n-1} \sum_{j=0}^{\frac{N-2i}{2}-1} (i+j+1)P(i=F_A=F_B-1, S_A=N-2i-1-j, S_B=j) \end{aligned} \qquad (3.28)$$

$$+\sum_{i=1}^{n}\sum_{j=0}^{\frac{N-2i}{2}}(i+j-1)P(i=F_A=F_B+1,S_A=N-2i+1-j,S_B=j)$$

The individual probabilities can be derived by using diagrams similar to that following (3.11). The exact expression for $E(N_{B,CS})$ is, thereby, easily seen to be

$$\begin{aligned}E(N_{B,CS}) &= \frac{1}{2}\sum_{i=1}^{n-1}\sum_{j=0}^{\frac{N-2i}{2}-1}\Big[(i+j)\Big[\binom{N-i-j}{i}\binom{i+j-1}{i-1}+\binom{i+j}{i}\binom{N-i-j-1}{i-1}\Big]\cdot\\ &\qquad\cdot q_A^i p_A^{N-2i-j} q_B^i p_B^j\Big]\\ &+\frac{n}{2}\sum_{i=1}^{n}\binom{n}{i}\binom{n-1}{i-1}(q_Aq_B)^i(p_Ap_B)^{n-i}\\ &+\frac{1}{2}\sum_{i=0}^{n-1}\sum_{j=0}^{\frac{N-2i}{2}-1}(i+j+1)\binom{N-i-j-1}{i}\binom{i+j}{i}q_A^i p_A^{N-2i-j-1}q_B^{i+1}p_B^j\\ &+\frac{1}{2}\sum_{i=1}^{n}\sum_{j=0}^{\frac{N-2i}{2}}(i+j-1)\binom{N-i-j}{i-1}\binom{i+j-1}{i-1}q_A^i p_A^{N-2i+1-j}q_B^{i-1}p_B^j\end{aligned}\tag{3.29}$$

$E(N_{B,FS})$ is derived in essentially the same way when noting that $E(N_{B,FS}) =$ $= E(N_{A,CS}(p_A \leftrightarrows p_B))$ and

$$\begin{aligned}&E(N_{A,CS}|F_A=F_B=i,S_A=N-2i-j,S_B=j,p_A\leftrightarrows p_B) = N-i-j\\ &E(N_{A,CS}|F_A=F_B=i,S_A=S_B=n-i,p_A\leftrightarrows p_B) = \frac{n}{2}\\ &E(N_{A,CS}|i=F_A=F_B-1,S_A=N-2i-1-j,S_B=j,p_A\leftrightarrows p_B) = N-i-1-j\\ &E(N_{A,CS}|i=F_A=F_B+1,S_A=N-2i+1-j,S_B=j,p_A\leftrightarrows p_B) = N-i-j+1\end{aligned}\tag{3.30}$$

The resulting formula for $E(N_{B,FS})$ is hence given by

$$\begin{aligned}E(N_{B,FS}) &= \frac{1}{2}\sum_{i=1}^{n-1}\sum_{j=0}^{\frac{N-2i}{2}-1}\Big[(N-i-j)\Big[\binom{N-i-j}{i}\binom{i+j-1}{i-1}+\binom{i+j}{i}\binom{N-i-j-1}{i-1}\Big]\cdot\\ &\qquad\cdot q_B^i p_B^{N-2i-j} q_A^i p_A^j\Big] + np_B^{2n}\\ &+\frac{n}{2}\sum_{i=1}^{n}\binom{n}{i}\binom{n-1}{i-1}(q_Aq_B)^i(p_Ap_B)^{n-i}\\ &+\frac{1}{2}\sum_{i=0}^{n-1}\sum_{j=0}^{\frac{N-2i}{2}-1}(N-i-j-1)\binom{N-i-j-1}{i}\binom{i+j}{i}q_B^i p_B^{N-2i-1-j}q_A^{i+1}p_A^j\end{aligned}\tag{3.31}$$

$$+\frac{1}{2}\sum_{i=1}^{n}\sum_{j=0}^{\frac{N-2i}{2}}(N-i-j+1)\binom{N-i-j}{i-1}\binom{i+j-1}{i-1}q_B^i p_B^{N-2i+1-j} q_A^{i-1} p_A^j$$

An approximative expression for $E(N_B|PW)$ can be derived as follows (cf. A1/6):

$$\frac{E(N_B|PW)}{E(N_A|PW)} \approx \frac{E(X_{i,q_B}+i)}{E(X_{i,q_A}+i)} = \frac{ip_B/q_B+i}{ip_A/q_A+i} = \frac{q_A}{q_B} \tag{3.32}$$

Using in addition the fact that $E(N_B|PW) + E(N_A|PW) = N$, we have

$$E(N_B|PW) \approx \frac{q_A}{q_B}(N-E(N_B|PW))$$
$$E(N_B|PW) \approx \frac{q_A}{q_A+q_B} N \tag{3.33}$$

Note that the expression for $E(N_B|PW)$, given by (3.33), is exact if we consider a PW-sampling procedure that terminates after N trials, where an equal number of failures on both treatments occur and the last trial is a faulty one (cf. Zelen [202], (2), p.134).

3.5 Extension of the selection models to odd N

The VT-sampling scheme in its original conception can be applied only if the patient horizon N is even. Therefore, we introduce a slight modification of the VT-sampling scheme, so as to be able to use both procedures irrespective of the patient horizon. To modify the VT-sampling scheme, the treatments are given one at a time to the patients. The first treatment at the outset is determined randomly, and then the treatments are given alternatively. Using this modification of the VT-sampling scheme, the patient horizon N is no longer forced to be even. The termination rule of both procedures is also slightly modified, that is, we declare that treatment as best which yields the most successes after N trials; in case of an equal number of successes the treatment, yielding the least number of failures, is declared as best with a probability θ. The resulting selection models are denoted by $[2;PW(\theta);\text{fixed } N]$ and $[2;VT(\theta);\text{fixed } N]$, respectively. The original PW-procedure is by no means the same as the modified $PW(\theta)$-procedure, both procedures coincide only if $\theta = \frac{1}{2}$, i.e. $PW=PW(\frac{1}{2})$. It will be proved in the following that

$$P(CS|PW(\theta)) \geq P(CS|VT(\theta)) \quad \text{for all } N \text{ and } \theta, \tag{3.34}$$

and that equality for odd N in (3.34) holds if and only if $\theta = 1$.

For $p_A=p_B$, equality holds in (3.34) since both sides equal 1/2. (3.34) is, hence, verified if the following inequality is shown to hold for any θ and $p_A \neq p_B$, and that we have equality if and only if $\theta = 1$.

$$P_{2n+1}(CS|PW(\theta)) - P_{2n}(CS|PW(\theta)) \geq P_{2n+1}(CS|VT(\theta)) - P_{2n}(CS|VT(\theta)) \quad (3.35)$$

for $n \in \{0,1,\ldots,k\}$; $(N=2k+1)$.

Using similar arguments as before, the lefthand side of (3.34), denoted by $L(\theta)$, is seen to be

$$L(\theta) = v_\theta(p_A,p_B) - v_\theta(p_B,p_A) \text{ , where} \quad (3.36)$$

$$v_\theta(p_A,p_B)=(p_A+(1-\theta)q_A - \tfrac{1}{2})P_{2n}(\Delta S=0,A_{2n+1})+\theta p_A P_{2n}(\Delta S= -1,A_{2n+1}) =$$

$$= \frac{1}{2} q_B(p_A+(1-\theta)q_A - \tfrac{1}{2}) \sum_{i=0}^{n} b(i;n,q_A)b(i-1;n-1,q_B) + \quad (3.37)$$

$$+ \frac{1}{2} q_B\theta p_A \sum_{i=1}^{n} b(i;n-1,q_A)b(i;n,q_B)$$

(cf. (3.12) and (3.14)). The formation of the first term in (3.37) can be explained as follows. An increase in $P(CS|PW)$ from 2n to 2n+1 occurs only if the (2n+1)-th trial is an A-trial, and if $\Delta S=0$ or $\Delta S= -1$ after 2n trials. $P_{2n}(\Delta S=0,A_{2n+1})$ is, consequently, the probability of all sequences of length 2n that have an equal number of successes (and that implies an equal number of failures, too), and that terminate with a successful A-trial or a faulty B-trial. The probability of a single sequence is $p_A^i p_B^i q_A^{n-i} q_B^{n-i}$, and the contribution to $P_{2n}(CS|PW)$ is hence $\frac{1}{2} p_A^i p_B^i q_A^{n-i} q_B^{n-i}$. The contribution of that special sequence to $P_{2n+1}(CS|PW)$ is $p_A^i p_B^i q_A^{n-i} q_B^{n-i} p_A$ if the (2n+1)-th trial is a successful A-trial, and $p_A^i p_B^i q_A^{n-i} q_B^{n-i} q_A(1-\theta)$ if the (2n+1)-th trial is a faulty B-trial. The increase of P(CS) due to that sequence is hence

$$(p_A+q_A(1-\theta) - \tfrac{1}{2})\; p_A^i p_B^i q_A^{n-i} q_B^{n-i} \;. \quad (3.38)$$

The righthand side of (3.35), denoted by $R(\theta)$, is easily shown to be

$$R(\theta) = w_\theta(p_A,p_B) - w_\theta(p_B,p_A), \text{ where} \quad (3.39)$$

$$w_\theta(p_A,p_B) = (p_A+(1-\theta)q_A - \tfrac{1}{2})\; P_{2n}(\Delta S=0,A_{2n+1}) +$$

$$+ (1-\theta)p_A P_{2n}(\Delta S= -1,A_{2n+1}) = \quad (3.40)$$

$$= \frac{1}{2} (p_A+(1-\theta)q_A - \tfrac{1}{2}) \sum_{i=0}^{n} b(i;n,q_A)b(i;n,q_B) +$$

$$+ \frac{1}{2}(1-\theta)p_A \sum_{i=0}^{n} b(i+1;n,q_A)b(i;n,q_B)$$

To prove (3.34) we use a similar proceeding as before, i.e. we evaluate the difference

$$d_\theta(p_A,p_B) := v_\theta(p_A,p_B) - w_\theta(p_A,p_B) \tag{3.41}$$

A straightforward analysis shows that $d_\theta(p_A,p_B)$ is given by

$$d_\theta(p_A,p_B) = k(p_A,p_B) + (1-\theta)\ell(p_A,p_B), \text{ where}$$

$$\begin{aligned} k(p_A,p_B) = \frac{1}{2}\sum_{i=0}^{n-1} \binom{n}{i}\binom{n-1}{i}(q_Aq_B)^i(p_Ap_B)^{n-i}(\tfrac{1}{2} - (p_A+p_B)) + \\ + \frac{1}{2}\sum_{i=1}^{n}\binom{n-1}{i-1}\binom{n}{i-1}(q_Aq_B)^{i-1}(p_Ap_B)^{n-i+1} \end{aligned} \tag{3.42}$$

$$\ell(p_A,p_B) = -\frac{1}{2}\sum_{i=0}^{n-1}\binom{n-1}{i} b(i;n,q_Aq_B)q_A - \frac{1}{2}\sum_{i=1}^{n-1}\binom{n-1}{i-1} b(i-1;n,q_Aq_B)(\frac{nq_A}{i} + q_B) \tag{3.43}$$

It follows directly from (3.42) that $k(p_A,p_B)$ is symmetric with respect to an interchange of p_A and p_B, i.e. $k(p_A,p_B)=k(p_B,p_A)$. From (3.43) follows that

$$\ell(p_A,p_B)-\ell(p_B,p_A)=\frac{1}{2}(q_B-q_A)\sum_{i=0}^{n-1}\binom{n-1}{i}(b(i;n,q_Aq_B)+b(i-1;n,q_Aq_B)), \tag{3.44}$$

which is positive since $q_B > q_A$.

Using (3.44) and the symmetry of $k(p_A,p_B)$, (3.35) reduces to

$$L(\theta) - R(\theta) = (1-\theta)(\ell(p_A,p_B) - \ell(p_B,p_A)) \geq 0 , \tag{3.45}$$

and this implies that equality in (3.34), provided $p_A \neq p_B$, holds if and only if $\theta = 1$.

3.6 Numerical results

The only parameter that can be manipulated to force P(CS) to be greater than P* if the success parametersare in the LFC is the patient horizon N. Therefore, we must look for the LFC and then enlarge N as long as P(CS) (in the LFC) exceeds the preassigned P* for the first time. Making use of (3.1), the desired N-values can be seen from Huyett/Sobel [106].

Table 1

$P^*=0.90$; p_A	0.2	0.3	0.4	0.5	0.6	0.7	0.8	0.9	N
$P(CS\|\Delta^*=0.1)$	0.966	0.932	0.912	0.903	0.903	0.912	0.932	0.966	
$E(N_B\|\Delta^*=0.1)$	78.1	77.5	76.6	75.5	73.8	71.2	66.6	55.9	166
$\hat{E}(N_B\|\Delta^*=0.1)$	78.1	77.5	76.6	75.5	73.8	71.1	66.4	55.3	
$P(CS\|\Delta^*=0.2)$	-	0.951	0.923	0.908	0.903	0.908	0.923	0.951	
$E(N_B\|\Delta^*=0.2)$	-	18.4	18.1	17.6	16.9	15.9	14.3	11.1	42
$\hat{E}(N_B\|\Delta^*=0.2)$	-	18.4	18.0	17.5	16.8	15.7	14.0	10.5	
$P(CS\|\Delta^*=0.3)$	-	-	0.935	0.912	0.902	0.902	0.912	0.935	
$E(N_B\|\Delta^*=0.3)$	-	-	7.3	7.0	6.7	6.2	5.4	4.2	18
$\hat{E}(N_B\|\Delta^*=0.3)$	-	-	7.2	6.9	6.5	6.0	5.1	3.6	
$P(CS\|\Delta^*=0.4)$	-	-	-	0.924	0.906	0.901	0.906	0.924	
$E(N_B\|\Delta^*=0.4)$	-	-	-	3.7	3.5	3.2	2.8	2.2	10
$\hat{E}(N_B\|\Delta^*=0.4)$	-	-	-	3.6	3.3	3.0	2.5	1.7	
$P^*=0.95$;									
$P(CS\|\Delta^*=0.1)$	0.990	0.972	0.958	0.951	0.951	0.958	0.972	0.990	
$E(N_B\|\Delta^*=0.1)$	127.1	126.0	124.6	122.8	120.1	115.8	108.2	90.6	270
$\hat{E}(N_B\|\Delta^*=0.1)$	127.1	126.0	124.6	122.7	120.0	115.7	108.0	90.0	
$P(CS\|\Delta^*=0.2)$	-	0.983	0.966	0.955	0.952	0.955	0.966	0.983	
$E(N_B\|\Delta^*=0.2)$	-	29.8	29.2	28.4	27.3	25.7	22.9	17.6	68
$\hat{E}(N_B\|\Delta^*=0.2)$	-	29.7	29.1	28.3	27.2	25.5	22.7	17.0	
$P(CS\|\Delta^*=0.3)$	-	-	0.976	0.961	0.953	0.953	0.961	0.976	
$E(N_B\|\Delta^*=0.3)$	-	-	12.1	11.6	11.0	10.2	8.9	6.6	30
$\hat{E}(N_B\|\Delta^*=0.3)$	-	-	12.0	11.5	10.9	10.0	8.6	6.0	
$P(CS\|\Delta^*=0.4)$	-	-	-	0.975	0.963	0.960	0.963	0.975	
$E(N_B\|\Delta^*=0.4)$	-	-	-	6.5	6.1	5.6	4.8	3.6	18
$\hat{E}(N_B\|\Delta^*=0.4)$	-	-	-	6.4	6.0	5.4	4.5	3.0	
$P^*=0.99$;									
$P(CS\|\Delta^*=0.2)$	-	0.999	0.995	0.991	0.990	0.991	0.995	0.999	
$E(N_B\|\Delta^*=0.2)$	-	58.7	57.5	55.9	53.7	50.4	44.9	34.1	134
$\hat{E}(N_B\|\Delta^*=0.2)$	-	58.6	57.4	55.8	53.6	50.2	44.7	33.5	

$\hat{E}(N_B|\cdot)$ is the approximative expectation according to (3.33).

Table 1 (continued)

$P^*=0.99$; p_A	0.2	0.3	0.4	0.5	0.6	0.7	0.8	0.9	N
$P(CS\|\Delta^*=0.3)$	-	-	0.997	0.993	0.990	0.990	0.993	0.997	
$E(N_B\|\Delta^*=0.3)$	-	-	23.3	22.4	21.2	19.5	16.9	12.2	58
$\hat{E}(N_B\|\Delta^*=0.3)$	-	-	23.2	22.3	21.1	19.3	16.6	11.6	
$P(CS\|\Delta^*=0.4)$	-	-	-	0.996	0.992	0.990	0.992	0.996	
$E(N_B\|\Delta^*=0.4)$	-	-	-	11.5	10.8	9.8	8.3	5.9	32
$\hat{E}(N_B\|\Delta^*=0.4)$	-	-	-	11.4	10.7	9.6	8.0	5.3	

3.7 Equivalence to Hoel's selection model

The fixed sample size selection model with PW-sampling, due to Nebenzahl/Sobel [130] is to a certain extent equivalent to the Hoel selection procedure [92] given in sec. 1 of chap. 2. Strictly speaking, the fixed sample size selection model and a slight modification of Hoel's selection model are equivalent in the sense that they reach the same terminal decision (cf. M.Pradhan/Y.S.Sathe, JASA 69, no. 346, 475-476).

Let $S_{i,n}, F_{i,n}$ denote the numbers of successes, failures of treatment i; $i\varepsilon\{A,B\}$, in n trials, respectively, and let $R_{i,n}=S_{i,n}+F_{j,n}$; $i,j\varepsilon\{1,2\}$ and $i\neq j$; then sampling terminates according to Hoel's stopping-rule if either $R_{1,n}$ or $R_{2,n}$ equal r for some n, and treatment no. 1 or no. 2, respectively, is declared as best. We know that Hoel's selection procedure terminates after at most 2r-1 trials.

Let the fixed patient horizon N be an odd number, i.e. N=2r-1 for some r, and let the first trial be carried out with treatment no. i, then Hoel's termination rule is modified by randomizing if $R_{j,2r-1}=r$. (Recall that the original termination-rule of Hoel's model prescribes to select treatment j if $R_{j,2r-1}=r$). The fixed sample size selection model is equivalent to this modification of Hoel's model, i.e. we have

Theorem 3.1

Assuming that the first trial is on treatment i, then the following equivalences hold

(a) $S_{i,2r-1} > S_{j,2r-1} \iff R_{i,n} = r$ for some n with $r \leq n \leq 2r-1$

(b) $S_{j,2r-1} > S_{i,2r-1} \iff R_{j,n} = r$ for some n with $r \leq n \leq 2r-2$

(c) $S_{i,2r-1} = S_{j,2r-1} \iff R_{j,2r-1} = r$

Proof:

(a) Let $R_{i,n}=S_{i,n}+F_{j,n}=r$ for some r with $r\leq n\leq 2r-1$, then $R_{i,n}-R_{j,n}=S_{i,n}-S_{j,n}=r-$ $-(n-r)=2r-n$, because the last failure that occurs is on treatment j. Since the (n+1)-th trial is on treatment i, there may occur at most 2r-n-2 successful trials on treatment j in the remaining 2r-n-1 trials, hence $S_{i,2r-1}-$ $-S_{j,2r-1}>0$.
If on the other hand $S_{i,2r-1}-S_{j,2r-1}>0$, then either $F_{j,2r-1}=F_{i,2r-1}$ or $F_{j,2r-1}=F_{i,2r-1}-1$ because the first trial is on treatment i. In either case

$$R_{i,2r-1}=S_{i,2r-1}+F_{j,2r-1}\geq S_{j,2r-1}+1+F_{i,2r-1}-1=R_{j,2r-1}$$

Recalling that $R_{i,2r-1}+R_{j,2r-1}=2r-1$, then the above inequality implies that $R_{i,2r-1}\geq r$, and that means that there exists some n; $r\leq n\leq 2r-1$; with $R_{i,n}=r$.

The proof of (b) is analogously when noting that a beginning with treatment i and $R_{j,n}=r$ jointly imply that $F_{i,n}-F_{j,n}=1$.

(c) $R_{j,2r-1}=r$ means that the last failure is on treatment i, hence $F_{i,2r-1}=$ $=F_{j,2r-1}+1$, and that implies $S_{i,2r-1}-S_{j,2r-1}=r-1-F_{j,2r-1}-r+F_{j,2r-1}+1=$ $=0$, i.e. $S_{i,2r-1}=S_{j,2r-1}$.

On the other hand $S_{i,2r-1}=S_{j,2r-1}$ and $F_{i,2r-1}=F_{j,2r-1}+1$ imply immediately that $R_{j,2r-1}=r$. ■

In case that the fixed patient horizon N is even, i.e. N=2r, Hoel's termination-rule is modified as follows: sampling terminates whenever $R_{i,n}=r$ $(r\leq n\leq 2r-1)$ or $R_{j,n}=r+1$ $(r\leq n\leq 2r)$ and treatment i,j, respectively, is declared as best; in case that $R_{i,2r}=r$, the best treatment is determined by randomization. The first treatment given at the outset is again denoted by i.
The equivalence of the fixed sample size procedure and the modified selection model of Hoel is given by the following theorem.

Theorem 3.2

Assuming that the first trial is on treatment i, then the following equivalences hold

(a) $S_{i,2r}>S_{j,2r} \iff R_{i,n}=r$ for some n with $r\leq n\leq 2r-1$

(b) $S_{j,2r}>S_{i,2r} \iff R_{j,n}=r+1$ for some n with $r\leq n\leq 2r$

(c) $S_{i,2r}=S_{j,2r} \iff R_{i,2r}=r$

Proof:

(a) $R_{i,n}=r$ for some n with $r\leq n\leq 2r-1$ and the beginning with treatment i jointly imply that $F_{i,n}=F_{j,n}$, and therefore $S_{i,n}-S_{j,n}=2r-n$. Recalling that the (n+1)-th trial is on treatment i, it follows immediately that $S_{i,2r}-S_{j,2r}>0$. On the other hand, $S_{i,2r}>S_{j,2r}$ implies that $R_{i,2r}\geq R_{j,2r}$, and that means $R_{i,2r}\geq r$. Therefore, $R_{i,n}=r$ for some n with $r\leq n\leq 2r-1$. ($R_{i,2r}=r$ would imply $S_{i,2r}=S_{j,2r}$ cf.(c).)

(b) Recalling that the first trial at the outset is on treatment i, then $R_{j,n}=$ $=r+1$ for some n with $r\leq n\leq 2r$ implies $F_{i,n}=F_{j,n}+1$, and that gives $S_{j,n}-S_{i,n}=$ $=2r-n+1$, hence $S_{j,2r}-S_{i,2r}>0$.
On the other hand $S_{j,2r}>S_{i,2r}$ and $F_{i,2r}=F_{j,2r}$ or $F_{i,2r}=F_{j,2r}+1$ imply

$$R_{j,2r} = S_{j,2r} + F_{i,2r} \geq S_{i,2r} + 1 + F_{j,2r} = R_{i,2r} + 1, \text{ i.e.}$$

$R_{j,2r}\geq R_{i,2r}+1$, and that implies $R_{j,2r}\geq r+1$, i.e. there exists some n ($r\leq n\leq 2r$) with $R_{j,n}=r+1$.

(c) $R_{i,2r}$ implies $F_{i,2r}=F_{j,2r}$ and that means $S_{i,2r}=S_{j,2r}$.
On the other hand, $S_{i,2r}=S_{j,2r}$ implies $F_{i,2r}=F_{j,2r}$, and hence $R_{i,2r}=r$.

■

We have seen that Hoel's selection model and the fixed sample size model of Nebenzahl/Sobel provide essentially the same; the numerical results, however, show that Hoel's model is better than the fixed sample size model in the sense that a decision is reached in fewer trials (cf. table 1 of chap. 2 and the numerical results presented above in 3.6).

4 The Selection Models

[2;PW;fixed N] and [2;VT;fixed N]

with Curtailment

4.1 Introductory remarks

In the following we consider essentially the same selection models as in the preceding section. Therefore, it is obvious that the P(CS)-values and the patient horizon N can be adopted from section 3. The fundamental new idea, to be proved in this section, is that it is sometimes possible to curtail the sampling procedures, i.e. in some situations sampling may be terminated before the N-th trial, and the same decision upon the relative worth of the competing treatments can be taken that would have been reached if sampling had been continued till the N-th trial. In other words, some testing-sequences are such that the full information, needed to take the terminal decision, is available at a stage t earlier than N, and all trials after t have no influence on that decision. In case that such special testing-sequences occur it is obvious to stop sampling at stage t and to decide which treatment is to be declared as best. The curtailment effects that the expected number of patients on the poorer treatment and the whole number of patients involved in the experiment of the selection models with curtailment are smaller than the corresponding values of the selection models without curtailment.

4.2 The PW-sampling procedure with curtailment

In order to prove that sampling may be sometimes curtailed we need the following lemma.

Lemma 4.1

Assuming that the first trial is on treatment A, then the following implications hold

(a) $S_{A,r} - S_{B,r} > 2n-r \implies S_{A,r-1} - S_{B,r-1} \geq 2n-r+1$

(b) $S_{A,r} - S_{B,r} \leq 2n-r \implies S_{A,r-1} - S_{B,r-1} < 2n-r+1$,

where $S_{i,r}$ denotes the number of successes of treatment i, $i\varepsilon\{A,B\}$, in r trials.

Proof of (a):

Let $S_{A,r}-S_{B,r} > 2n-r$ and suppose that $S_{A,r-1}-S_{B,r-1} < 2n-r+1$, i.e. $S_{A,r-1}-S_{B,r-1} \leq 2n-r$. Noting that the contribution of the r-th trial to the difference of successes is either 0, 1 or -1, we get $S_{A,r-1}-S_{B,r-1}=2n-r$, and the r-th trial must be a successful A-trial. This again implies that among r-1 trials there is an equal number of faulty A- and B-trials, say i. Therefore, $S_{A,r-1}-S_{B,r-1}=2n-r$ and $S_{A,r-1}+S_{B,r-1}=r-1-2i$, from which $S_{A,r-1}=n-i-1/2$ follows, and that is evidently a contradiction.

Part (b) of lemma 4.1 is proved similarly.

■

We are now able to define and prove when sampling may be curtailed.

Theorem 4.2

Assume that the first trial is on treatment A, and let D_r denote the event that $S_{A,r}-S_{B,r}=2n-r$, $r\varepsilon\{n,n+1,\ldots,2n-1\}$, then the following equivalence hold

$$S_{A,2n} - S_{B,2n} > 0 \iff \text{exactly one } D_r \text{ occurs.}$$

Proof:

Let $S_{A,2n}-S_{B,2n} > 0$, then by part (b) of lemma 4.1,

$$S_{A,2n-1} - S_{B,2n-1} \geq 2n-(2n-1) = 1, \text{ i.e.}$$

$$\text{either } S_{A,2n-1} - S_{B,2n-1} = 1$$

$$\text{or } S_{A,2n-1} - S_{B,2n-1} > 1$$

In the first case, the occurrence of at least one D_r is proved, in the second cas lemma 4.1(b) is applied again. We finally obtain that

$$\text{either } S_{A,n+1} - S_{B,n+1} = 2n-(n+1) = n-1$$

$$\text{or } S_{A,n+1} - S_{B,n+1} > n-1$$

If D_{n+1} does not occur, we get $S_{A,n}-S_{B,n} \geq 2n-n = n$. Since $S_{A,n}-S_{B,n} \ngtr n$, it is obvious that the event D_n must occur.

It ist left to prove that exactly one D_r occurs.
Suppose D_r and D_t occur, and let $t < r$, i.e. $r = t+k$ where k is some natural number. In using the negation of lemma 4.1(b), we obtain

$$S_{A,t} - S_{B,t} \geq 2n-t \implies S_{A,t+1} - S_{B,t+1} > 2n-(t+1)$$

$$S_{A,t+1} - S_{B,t+1} \geq 2n-(t+1) \implies S_{A,t+2} - S_{B,t+2} > 2n-(t+2)$$

$$\vdots \qquad\qquad\qquad \vdots$$

$$S_{A,t+k-1} - S_{B,t+k-1} \geq 2n-(t+k-1) \implies S_{A,t+k} - S_{B,t+k} > 2n-(t+k),$$

and that is obviously a contradiction.
The case $t > r$ is proved analogously.
If on the other hand exactly one D_t occurs, then also $S_{A,2n}-S_{B,2n} > 0$; this is immediately seen by using the same considerations as above with $k=2n-t$.

■

<u>Theorem 4.3</u>

Assume that the first trial is on treatment A, and let D_r^* denote the event that $S_{B,r}-S_{A,r}=2n-r+1$, $r\varepsilon\{n+1,n+2,\ldots,2n\}$, then the following equivalence holds.

$$S_{B,2n}-S_{A,2n} > 0 \iff \text{exactly one } D_r^* \text{ occurs.}$$

<u>Proof:</u>

Supposing that D_r^* occurs, then $S_{B,r}-S_{A,r}=2n-r+1$. In the remaining $2n-r$ trials at most $2n-r$ A-successes may occur, and therefore $S_{B,2n}-S_{A,2n} \geq 2n-r+1-(2n-r)=1$.
On the other hand, if $S_{B,2n}-S_{A,2n} > 0$, then at least one D_r^* must occur; this is immediately obvious from the diagram below.
Suppose that D_r^* and D_t^* occur, and let $t > r$ without loss of generality, i.e.

$$S_{B,r} - S_{A,r} = 2n-r+1 \text{ and } S_{B,t} - S_{A,t} = 2n-t+1,$$

hence

$$\underbrace{(S_{B,t}-S_{B,r})}_{\geq 0} - \underbrace{(S_{A,t}-S_{A,r})}_{\leq t-r} = -(t-r)$$

Equality can hold only if $S_{B,t}-S_{B,r}=0$ and $S_{A,t}-S_{A,r}=t-r$; therefore, after the r-th trial t-r successful A-trials must occur, and that forces the r-th trial to be a successful A-trial or a faulty B-trial.

In the first case, an equal number of faulty A- and B-trials, say i, among r-1 trials occurs. Therefore, $S_{B,r-1}+S_{A,r-1}=r-1-2i$ and consequently $S_{B,r}+S_{A,r}=r-2i$. Noting in addition that $S_{B,r}-S_{A,r}=2n-r+1$, we obtain $S_{B,r}=n-i+1/2$, and that is obviously a contradiction.

In the second case i+1 faulty A-trials and i faulty B-trials among r-1 trials occur, i.e.

$$S_{B,r-1} + S_{A,r-1} = r-1-2i-1 \text{ , hence}$$

$$S_{B,r} + S_{A,r} = r-2i-2 \text{ , and}$$

$$S_{B,r} - S_{A,r} = 2n-r+1 \text{ , and that implies}$$

$S_{B,r}=n-i-1/2$; this completes the proof of theorem 4.3.

■

The theorems 4.2 and 4.3 prescribe at what stage sampling may be curtailed. In using the PW-sampling scheme, the first treatment at the outset is determined randomly and denoted by 1, the other treatment is denoted by 2. (Recall that we do not know whether treatment 1 equals A or B.) Hence, the following rule for curtailment is implied by theorem 4.2 and theorem 4.3:

(a) If $S_{1,r}-S_{2,r}=2n-r$ for some $r\varepsilon\{n,n+1,...,2n-1\}$, then sampling may be curtailed after the r-th trial and $S_{1,2n}>S_{2,2n}$ may be concluded, i.e. treatment no. 1 will be declared as best. (4.1)

(b) If $S_{2,r}-S_{1,r}=2n-r+1$ for some $r\varepsilon\{n+1,n+2,...,2n\}$, then sampling may be curtailed after the r-th trial and $S_{2,2n}>S_{1,2n}$ may be concluded, i.e. treatment no. 2 will be declared as best.

The performance of the PW-selection procedure can be illustrated by the following diagram. We consider the special case n=10.

■ points of curtailment according to (4.1)a

○ points of curtailment according to (4.1)b

↗ success of treatment no. 1

→ failure of treatment no. 1

↘ success of treatment no. 2

→ failure of treatment no. 2

Note that the first arrow in the diagram is always of the kind ↗ or →, since the first treatment at the outset is denoted by 1.

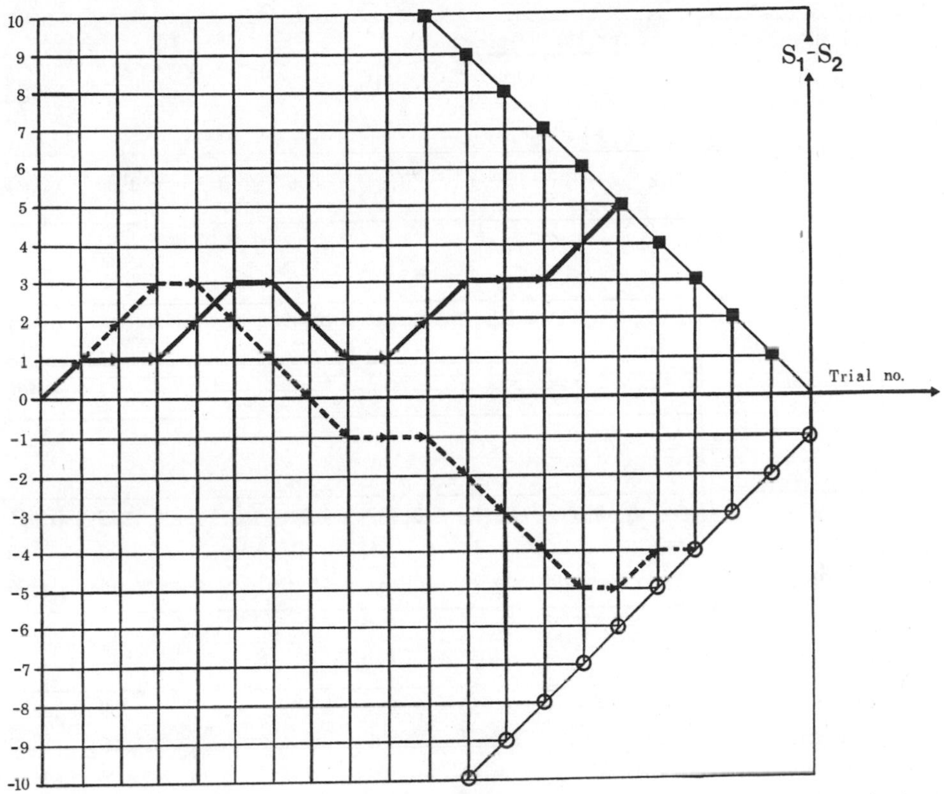

4.3 The VT-sampling procedure with curtailment

In using the VT-sampling scheme, a criterion for curtailment is much simpler to obtain. Since we do not know which treatment is the best, we denote one of the rival treatments by 1 and the other by 2. If we denote by $F_{i,r}$ the number of failures of treatment i in r trials on treatment i, then the following rule for curtailment holds.

If $S_{i,r}+F_{j,r}\varepsilon\{n+1,n+2\}$, then sampling may be curtailed after the r-th trial and $S_{i,n}>S_{j,n}$ may be concluded, $i\varepsilon\{1,2\}$, i.e. treatment no. i will be declared as best. (4.2)

(4.2) is immediately obvious. Let $S_{i,r}+F_{j,r}=n+1$, then $S_{i,r}=k$, $F_{j,r}=n+1-k$, $F_{i,r}=r-k$, and $S_{j,r}=r-n-1+k$ for some k. Therefore,

$$S_{i,r} - S_{j,r} = k-(r-n-1+k) = n-r+1$$

Since maximal n-r successes of j may occur in the remaining n-r trials on treatment j, $S_{i,n}-S_{j,n}\geq 1$ in any case. The proof of (4.2) in case that $S_{i,r}+F_{j,r}=n+2$ is essentially the same.

4.4 Numerical results for the PW-procedure

Table 2

p_A	0.2	0.3	0.4	0.5	0.6	0.7	0.8	0.9
$P^*=0.90, \Delta^*=0.1$ N=166								
P(CS;sim.)	0.969	0.930	0.908	0.904	0.903	0.915	0.919	0.977
E(NB;sim.)	77	76	74	72	70	66	59	45
E(N; sim.)	164	163	161	158	156	151	144	129
$P^*=0.90, \Delta^*=0.2$ N=42								
P(CS;sim.)	-	0.955	0.924	0.907	0.910	0.910	0.919	0.955
E(NB;sim.)	-	18	17	16	15	14	12	8
E(N; sim.)	-	40	40	39	37	36	34	30
$P^*=0.90, \Delta^*=0.3$ N=18								
P(CS;sim.)	-	-	0.938	0.908	0.908	0.905	0.909	0.936
E(NB;sim.)	-	-	7	6	6	5	5	3
E(N; sim.)	-	-	17	16	16	15	14	12
$P^*=0.90, \Delta^*=0.4$ N=10								
P(CS;sim.)	-	-	-	0.930	0.913	0.908	0.903	0.931
E(NB;sim.)	-	-	-	3	3	3	2	2
E(N; sim.)	-	-	-	9	8	8	7	7
$P^*=0.95, \Delta^*=0.1$ N=270								
P(CS;sim.)	0.990	0.962	0.951	0.945	0.949	0.958	0.968	0.991
E(NB;sim.)	126	124	121	118	114	106	94	72
E(N; sim.)	267	265	262	259	254	247	233	210
$P^*=0.95, \Delta^*=0.2$ N=68								
P(CS;sim.)	-	0.985	0.974	0.949	0.952	0.948	0.962	0.986
E(NB;sim.)	-	29	28	27	25	22	19	13
E(N; sim.)	-	65	65	63	61	59	55	49

Table 2 (continued)

p_A	0.2	0.3	0.4	0.5	0.6	0.7	0.8	0.9
$P^*=0.95, \Delta^*=0.3$ N=30								
P(CS;sim.)	-	-	0.983	0.950	0.959	0.955	0.959	0.983
E(NB;sim.)	-	-	11	11	10	9	7	5
E(N; sim.)	-	-	28	27	26	25	23	21
$P^*=0.95, \Delta^*=0.4$ N-18								
P(CS;sim.)	-	-	-	0.984	0.965	0.960	0.955	0.974
E(NB;sim.)	-	-	-	6	5	5	4	3
E(N; sim.)	-	-	-	16	15	14	14	12
$P^*=0.99, \Delta^*=0.1$ N=540								
P(CS;sim.)	1.000	0.998	0.996	0.989	0.985	0.990	0.996	0.999
E(NB;sim.)	251	247	241	235	226	212	186	141
E(N; sim.)	534	529	523	517	507	494	465	421
$P^*=0.99, \Delta^*=0.2$ N=134								
P(CS;sim.)	-	1.000	0.997	0.992	0.986	0.990	0.991	0.998
E(NB;sim.)	-	57	55	52	48	44	37	25
E(N; sim.)	-	129	127	125	120	116	107	95
$P^*=0.99, \Delta^*=0.3$ N=58								
P(CS;sim.)	-	-	0.998	0.987	0.990	0.994	0.990	0.997
E(NB;sim.)	-	-	22	21	19	16	14	9
E(N; sim.)	-	-	54	53	51	47	45	39
$P^*=0.99, \Delta^*=0.4$ N=32								
P(CS;sim.)	-	-	-	0.997	0.995	0.989	0.990	0.997
E(NB;sim.)	-	-	-	10	9	8	6	4
E(N; sim.)	-	-	-	28	27	26	24	21

4.5 Numerical results for the VT-procedure

Table 3

p_A	0.2	0.3	0.4	0.5	0.6	0.7	0.8	0.9
$P^*=0.90, \Delta^*=0.1$ N=166								
P(CS;sim.)	0.961	0.937	0.924	0.891	0.906	0.921	0.932	0.969
E(NB;sim.)	76	76	76	76	76	76	76	76
E(N; sim.)	153	153	152	153	152	153	153	153
$P^*=0.90, \Delta^*=0.2$ N=42								
P(CS;sim.)	-	0.953	0.927	0.916	0.904	0.894	0.928	0.952
E(NB;sim.)	-	18	18	18	18	18	18	18
E(N; sim.)	-	37	37	37	37	37	37	37
$P^*=0.90, \Delta^*=0.3$ N=18								
P(CS;sim.)	-	-	0.950	0.905	0.907	0.895	0.914	0.929
E(NB;sim.)	-	-	8	8	8	8	8	8
E(N; sim.)	-	-	16	15	15	15	16	15
$P^*=0.90, \Delta^*=0.4$ N=10								
P(CS;sim.)	-	-	-	0.928	0.894	0.890	0.906	0.923
E(NB;sim.)	-	-	-	4	4	4	4	4
E(N; sim.)	-	-	-	9	9	9	9	9
$P^*=0.95, \Delta^*=0.1$ N=270								
P(CS;sim.)	0.991	0.973	0.958	0.943	0.956	0.954	0.969	0.989
E(NB;sim.)	123	124	123	124	124	124	124	124
E(N; sim.)	247	247	247	247	247	248	247	247
$P^*=0.95, \Delta^*=0.2$ N=68								
P(CS;sim.)	-	0.984	0.966	0.956	0.955	0.953	0.964	0.974
E(NB;sim.)	-	29	29	29	29	29	29	29
E(N; sim.)	-	59	59	59	59	59	59	59

Table 3 (continued)

p_A	0.2	0.3	0.4	0.5	0.6	0.7	0.8	0.9
P*=0.95,Δ*=0.3 N=30								
P(CS;sim.)	-	-	0.984	0.955	0.944	0.949	0.965	0.978
E(NB;sim.)	-	-	12	12	12	13	12	12
E(N; sim.)	-	-	25	25	25	25	25	25
P*=0.95,Δ*=0.4 N=18								
P(CS;sim.)	-	-	-	0.967	0.966	0.968	0.961	0.972
E(NB;sim.)	-	-	-	7	7	7	7	7
E(N; sim.)	-	-	-	15	15	15	15	15
P*=0.99,Δ*=0.1 N=540								
P(CS;sim.)	0.999	0.999	0.992	0.991	0.991	0.994	0.997	1.000
E(NB;sim.)	247	246	247	247	247	247	246	247
E(N; sim.)	493	492	494	493	494	493	493	493
P*=0.99,Δ*=0.2 N=134								
P(CS;sim.)	-	0.999	0.994	0.991	0.984	0.990	0.993	0.999
E(NB;sim.)	-	57	57	57	57	57	57	57
E(N; sim.)	-	114	114	114	114	114	114	114
P*=0.99,Δ*=0.3 N=58								
P(CS;sim.)	-	-	0.997	0.989	0.991	0.994	0.993	0.995
E(NB;sim.)	-	-	23	23	23	23	23	23
E(N; sim.)	-	-	47	47	46	47	47	46
P*=0.99,Δ*=0.4 N=32								
P(CS;sim.)	-	-	-	0.996	0.995	0.991	0.991	0.998
E(NB;sim.)	-	-	-	12	12	12	12	12
E(N; sim.)	-	-	-	25	25	25	25	25

5 The Selection Model [2;VT;fixed N] with Truncation Based on $|S_A - S_B|$

5.1 Description of the model

The selection model to be considered in the following uses the VT-sampling scheme together with a fixed patient horizon N=2n. If at any stage $k < n$ the absolute difference in the current numbers of successes equals s, then sampling is terminated and the treatment associated with the larger number of successes is declared as best. s is chosen such that $P(CS) \geq P^*$ whenever $p_A - p_B \geq \Delta^*$. In case that no truncation arises within n test-stages, it is supposed that there are no significant differences between the treatments.

5.2 Derivation of the P(CS)-value

Let $P_k(m)$ be the probability of a correct selection at or before the m-th stage (then, 2m-trials have been conducted) of the experiment, given that initially $S_A - S_B + s = k$, i.e.

$$P_k(m) := P(CS \text{ at or before stage } m \mid S_A - S_B + s = k) \tag{5.1}$$

An s has been added to the difference of successes in order to simplify some of the following analysis. Sampling, hence, terminates when $S_A - S_B + s = 2s$ or 0 or when $|S_{A,k} - S_{B,k}| \leq k+s-n-1$, where $S_{A,k}$ denotes the number of A-successes within k A-trials. In the latter case sampling can be terminated and the treatments can be considered to be of equal value, because the absolute difference of successes cannot exceed s-1 in the remaining n-k stages of the experiment. $P_k(m)$ satisfies the following system of difference equations and boundary conditions

$$P_k(m+1) = p_A q_B P_{k+1}(m) + (p_A p_B + q_A q_B)\, P_k(m) + q_A p_B P_{k-1}(m)$$
$$P_o(m) = 0;\; P_{2s}(m) = 1;\; P_k(0) = 0,\; k \neq 2s;\; P_{2s}(0) = 1 \qquad (5.2)$$

Defining new variables by

$$Q_k(m) := P_k(\infty) - P_k(m), \qquad (5.3)$$

and noting that $P_k(\infty)=P_{k-s}$, where P_{k-s} is given by (2.1) of chap. 1, we see that $Q_k(m)$ satisfies the same system of difference equations as $P_k(m)$, i.e.

$$Q_k(m+1) = \alpha Q_{k+1}(m) + \beta Q_k(m) + \gamma Q_{k-1}(m),$$
$$\text{where } \alpha := p_A q_B;\; \beta := p_A p_B + q_A q_B;\; \gamma := q_A p_B \qquad (5.4)$$

The new boundary conditions are

$$Q_o(m) = Q_{2s}(m) = 0;\; Q_{2s}(0) = 0;\; Q_k(0) = P_k(\infty);\; k \neq 2s \qquad (5.5)$$

The set of difference equations given by (5.4) can be rewritten in matrix form, i.e.

$$Q(m+1) = AQ(m) \text{ or } Q(m) = A^m Q(0),$$

where

$$Q(m) = \begin{pmatrix} Q_1(m) \\ Q_2(m) \\ \vdots \\ \vdots \\ Q_{2s-1}(m) \end{pmatrix} \quad ; \qquad A := \begin{pmatrix} \beta & \alpha & 0 & \cdots & \cdots & 0 \\ \gamma & \beta & \alpha & 0 & \cdots & 0 \\ 0 & \ddots & \ddots & \ddots & \ddots & \vdots \\ \vdots & & \ddots & \ddots & \ddots & 0 \\ \vdots & & & \ddots & \ddots & \alpha \\ 0 & \cdots & \cdots & 0 & \gamma & \beta \end{pmatrix} \qquad (5.6)$$

In order to determine the elements of A^m, we first transform A into a symmetric matrix B. This is done by using the matrix T whose elements are given by

$$t_{j\ell} = \left(\gamma^{j-1} \alpha^{2s-j-1}\right)^{1/2} \delta_{j\ell} \qquad (5.7)$$

We obtain

$$B = T^{-1}AT \quad = \begin{pmatrix} \beta & \xi & 0 & \cdots & \cdots & 0 \\ \xi & \beta & \xi & 0 & \cdots & 0 \\ 0 & \ddots & \ddots & \ddots & & \vdots \\ \vdots & & \ddots & \ddots & \ddots & 0 \\ \vdots & & & \ddots & \ddots & \xi \\ 0 & \cdots & \cdots & 0 & \xi & \beta \end{pmatrix} , \tag{5.8}$$

where $\xi = (\alpha\gamma)^{1/2} = (p_A p_B q_A q_B)^{1/2}$

The spectral properties of B are known (cf.e.g. [157]).
The 2s-1 real eigenvalues of B, and consequently of A, are as follows

$$\lambda_j = \beta + 2\xi\cos\left(\frac{\Pi j}{2s}\right) ; \; j=1,2,\ldots,2s-1 \tag{5.9}$$

λ_1 is the largest eigenvalue and by using that $|\lambda_j| < \beta + 2\xi \leq 1$, all eigenvalues are seen to be less than one in absolute value. The corresponding eigenvectors $\mathfrak{u}_j$ have components

$$u_{jr} = \frac{1}{\sqrt{s}} \sin\left(\frac{\Pi jr}{2s}\right) \quad ; \; r=1,2,\ldots,2s-1 \tag{5.10}$$

Recalling the spectral properties of a symmetric matrix, B is seen to be similar to a diagonal matrix D that has the eigenvalues λ_j as elements in its diagonal. Moreover

$$B = UDU^T \quad , \tag{5.11}$$

where the matrix U has elements u_{rj}; $j,r\varepsilon\{1,\ldots,2s-1\}$, given by (5.10), i.e. the j-th column of U is just the j-th eigenvector. U^T denotes the transposed of U. The elements of B^m, denoted by $b_{jk}^{(m)}$, are obtained by using the identity $B^m = UD^mU^T$. We have

$$b_{jk}^{(m)} = \frac{1}{s} \sum_{r=1}^{2s-1} \lambda_r^m \sin\left(\frac{\Pi jr}{2s}\right) \cdot \sin\left(\frac{\Pi rk}{2s}\right) \tag{5.12}$$

The elements of A^m are now available by using (5.7) and (5.8), we obtain

$$a_{jk}^{(m)} = t_{jj} b_{jk}^{(m)} t_{kk}^{-1} = \left(\frac{\gamma}{\alpha}\right)^{(j-k)/2} b_{jk}^{(m)}, \tag{5.13}$$

and this expression together with (5.6) finally gives

$$Q_k(m) = \sum_{j=1}^{2s-1} a_{kj}^{(m)} Q_j(0), \text{ i.e.}$$

$$Q_k(m) = \frac{1}{s} \left(\frac{\gamma}{\alpha}\right)^{k/2} \sum_{r=1}^{2s-1} \lambda_r^m \sin\left(\frac{\Pi rk}{2s}\right) \sum_{j=1}^{2s-1} \left(\frac{\alpha}{\gamma}\right)^{j/2} \sin\left(\frac{\Pi jr}{2s}\right) Q_j(0), \tag{5.14}$$

where $Q_j(0)=P_j(\infty)=P_{j-s}=\dfrac{1-\delta^j}{1-\delta^{2s}}$; $\delta=\frac{\gamma}{\alpha}$; is known from (2.4) of chap. 1.

Writing the sine term in exponential form, i.e. $\sin\lambda=(e^{i\lambda}-e^{-i\lambda})/2i$, where $i^2=-1$, the sum over j can be carried out. A straightforward analysis gives the desired result, i.e.

$$Q_k(m) = \frac{1}{s}\,\delta^{\frac{k}{2}-s+\frac{1}{2}} \sum_{r=1}^{2s-1} \frac{\lambda_r^m(-1)^{r+1}\sin\left(\frac{\pi rk}{2s}\right)\sin\left(\frac{\pi r}{2s}\right)}{1-2\sqrt{\delta}\cos\left(\frac{\pi r}{2s}\right)+\delta}, \tag{5.15}$$

and the probability of a correct selection is hence given by

$$P(CS) = P_s(n) = P_s(\infty) - Q_s(n) = P_0 - Q_s(n), \text{ i.e.}$$

$$P(CS) = \frac{1}{1+\delta^s} - \frac{1}{s}\,\delta^{(1-s)/2} \sum_{r=1}^{2s-1} \frac{\lambda_r^n(-1)^{r+1}\sin\left(\frac{\pi r}{2s}\right)\sin\left(\frac{\pi r}{2}\right)}{1-2\sqrt{\delta}\cos\left(\frac{\pi r}{2s}\right)+\delta}, \tag{5.16}$$

where P_0 is given by (2.5) of chap. 1.

5.3 Derivation of the probability of declaring the two treatments equal

Let $W_k(m)$ denote the probability that none of the competing treatments is declared as best after the m-th stage of the experiment, given that initially $S_A-S_B+s=k$, i.e.

$$W_k(m) := P(\text{no decision after the m-th stage } | S_A-S_B+s=k) \tag{5.17}$$

It is immediately obvious that the $W_k(m)$ satisfy the same system of difference equations as the $P_k(m)$, i.e.

$$W_k(m+1) = \alpha W_{k+1}(m) + \beta W_k(m) + \gamma W_{k-1}(m), \tag{5.18}$$

and the corresponding boundary conditions are as follows

$$W_k(0) = 1;\ k \neq 0,2s;\ W_0(m) = W_{2s}(m) = 0 \tag{5.19}$$

Therefore, the same analysis as before is needed in order to derive the explicit expression for $W_k(m)$. We get

$$W_k(m) = \frac{1}{s}\,\delta^{k/2} \sum_{r=1}^{2s-1} \lambda_r^m \sin\left(\frac{\pi kr}{2s}\right) \sum_{j=1}^{2s-1} \delta^{-j/2}\sin\left(\frac{\pi jr}{2s}\right)\cdot W_j(0), \tag{5.20}$$

and after evaluating the sum over j, we finally obtain

$$W_k(m) = \frac{1}{s}\,\delta^{(k+1)/2}\sum_{r=1}^{2s-1}\lambda_r^m \sin\left(\frac{\Pi kr}{2s}\right)\cdot\sin\left(\frac{\Pi r}{2s}\right)\frac{(1+(-1)^{r+1}\delta^{-s})}{1-2\sqrt{\delta}\cos\left(\frac{\Pi r}{2s}\right)+\delta} \tag{5.21}$$

The probability of declaring the two treatments equal is consequently given by

$$W_s(n) = \frac{1}{s}\,\delta^{(s+1)/2}\sum_{r=1}^{2s-1}\lambda_r^n \sin\left(\frac{\Pi r}{2}\right)\sin\left(\frac{\Pi r}{2s}\right)\cdot\frac{(1+(-1)^{r+1}\delta^{-s})}{1-2\sqrt{\delta}\cos\left(\frac{\Pi r}{2s}\right)+\delta} \tag{5.22}$$

5.4 Derivation of an upper bound for $E(N_B)$

As above-mentioned sampling can be terminated before stage n if it is obvious that neither of the treatments can be declared as best in the remaining trials. To get an upper bound to the expected number of patients on the poorer treatment, we assume that sampling will be terminated at the n-th stage if no decision is made.

Let $U_k(m)$ be the expected number of patients on the poorer treatment after m stages given that initially $S_A-S_B+s=k$, i.e.

$$U_k(m) := E(N_B \text{ after m stages } | S_A-S_B+s=k) \tag{5.23}$$

The $U_k(m)$ satisfy the following system of difference equations and boundary conditions

$$\begin{aligned} U_k(m+1) &= \alpha U_{k+1}(m) + \beta U_k(m) + \gamma U_{k-1}(m) + 1 \\ U_0(m) &= U_{2s}(m) = 0;\ U_k(0) = 0 \end{aligned} \tag{5.24}$$

Denoting by U(m) a column vector whose k-th element is $U_k(m)$, and by V a column vector whose elements are all equal to 1, then (5.24) can be written as

$$U(m+1) = AU(m) + V\ , \tag{5.25}$$

where the matrix A is that of (5.6). Noting that U(0) has all elements equal to zero, we get

$$U(m) = (E + A + A^2 + \dots + A^{m-1})\ V := T(m)V, \tag{5.26}$$

where E is the unit matrix. The elements of T(m) are obtained by using (5.12) and (5.13), we get

$$(T(m))_{jk} = \delta^{(j-k)/2}\,\frac{1}{s}\sum_{r=1}^{2s-1}\sum_{\tau=0}^{m-1}\lambda_r^{\tau}\sin\left(\frac{\Pi jr}{2s}\right)\ \sin\left(\frac{\Pi rk}{2s}\right) = \tag{5.27}$$

$$= \delta^{(j-k)/2}\,\frac{1}{s}\sum_{r=1}^{2s-1}\frac{1-\lambda_r^m}{1-\lambda_r}\sin\left(\frac{\Pi jr}{2s}\right)\sin\left(\frac{\Pi rk}{2s}\right)$$

$U_k(m)$ is hence given by

$$U_k(m) = \sum_{j=1}^{2s-1}(T(m))_{kj}\ , \text{ i.e.} \tag{5.28}$$

$$U_k(m) = \frac{\delta^{k/2}}{s}\sum_{r=1}^{2s-1}\frac{1-\lambda_r^m}{1-\lambda_r}\sin\left(\frac{\Pi rk}{2s}\right)\sum_{j=1}^{2s-1}\delta^{-j/2}\sin\left(\frac{\Pi jr}{2s}\right)$$

$$= \frac{\delta^{(k+1)/2}}{s}\sum_{r=1}^{2s-1}\frac{1-\lambda_r^m}{1-\lambda_r}\;\frac{(1+(-1)^{r+1}\delta^{-s})\sin\left(\frac{\Pi rk}{2s}\right)\sin\left(\frac{\Pi r}{2s}\right)}{1-2\sqrt{\delta}\cos\left(\frac{\Pi r}{2s}\right)+\delta} \tag{5.29}$$

The desired expectation is, therefore, given by

$$E(N_B) = U_s(n) = \frac{\delta^{(s+1)/2}}{s}\sum_{r=1}^{2s-1}\frac{1-\lambda_r^n}{1-\lambda_r}\;\frac{(1+(-1)^{r+1}\delta^{-s})\sin\left(\frac{\Pi r}{2}\right)\sin\left(\frac{\Pi r}{2s}\right)}{1-2\sqrt{\delta}\cos\left(\frac{\Pi r}{2s}\right)+\delta} \tag{5.30}$$

5.5 Derivation of the truncation point s and of the patient horizon N

As usual we assume that P* and Δ* are preassigned constants known to the experimenter. Thus, we have two parameters, that are s an N, which may be varied in such a way that certain probability requirements are met. Strictly speaking, we consider two different situations. At first we suppose that N is fixed and cannot be altered. In such cases it may happen that P(CS) is always smaller than P*. The only thing we can do is to choose that s which maximizes the P(CS)-value when the parameters p_A and p_B are in the LFC. The respective maximum of P(CS) will be denoted by P_{max}. The second possibility consists in determining the smallest value of N such that P(CS) is at least P* whenever p_A-p_B is at least Δ*. N is consequently the smallest number of experimental units that must be available so that the experimenter can be certain of conducting the selection procedure at the preassigned significance level P*.

Let N be fixed. Making use of (2.6) of chap. 1, the P(CS)-function, given by (5.16), is easily seen to be a function of $\bar{p}(1-\bar{p})$, i.e.

$$P(CS) =: f(\bar{p}) = g(\bar{p}(1-\bar{p})) \tag{5.31}$$

Therefore, $f(1-\bar{p})=g((1-\bar{p})(1-(1-\bar{p})))=g(\bar{p}(1-\bar{p}))=f(\bar{p})$, i.e. P(CS) is symmetric around

$\bar{p}=1/2$. Recalling that $\bar{p}\varepsilon(\frac{\Delta^*}{2},1-\frac{\Delta^*}{2})$, we look for minima at the points $\bar{p}=\Delta^*/2$, $\bar{p}=1-\Delta^*/2$ and $\bar{p}=1/2$. Since the P(CS)-function is one at $\bar{p}=\Delta^*/2$ and $\bar{p}=1-\Delta^*/2$ it is conjectured that $\bar{p}=1/2$ is the unique minimum of the P(CS)-function. This conjecture is strongly supported by numerous numerical calculations. In order to maximize the P(CS)-function when N is fixed, we must choose that value of s which maximizes the P(CS)-function at $\bar{p}=1/2$. This value will be obtained numerically.

In a second step we will determine the minimal N needed to guarantee that the $(P^*;\Delta^*)$-condition is satisfied. Since the λ_j are all less than one in absolute value (cf. (5.9)), we have approximately from (5.16):

$$P(CS) = \frac{1}{1+\delta^s} - \frac{1}{s}\,\delta^{(1-s)/2}\,\frac{\lambda_1^n \sin\left(\frac{\Pi}{2s}\right)}{1+\delta-2\sqrt{\delta}\cos\left(\frac{\Pi}{2s}\right)} + O(\lambda_3^n) \tag{5.32}$$

Recall that P(CS) is minimized at $\bar{p}=1/2$, therefore, s cannot be less than the critical s_{max} given by (2.10) of chap. 1, because the second term in (5.32) is positive provided $s\geq 1$, i.e.

$$s \geq \left]\frac{\ell n(1-P^*)}{2\ell n\left(\frac{1-\Delta^*}{1+\Delta^*}\right)}\right[= s_{max} \tag{5.33}$$

Neglecting terms of order $O(\lambda_3^n)$ and letting $\bar{p}=1/2$, then we find from (5.32)

$$P^* = P(CS) = \frac{1}{1+\mu^{2s}} - \frac{1}{s}\,\mu^{(1-s)}.\,\frac{\lambda_1^n(1/2)\sin\left(\frac{\Pi}{2s}\right)}{1+\mu^2-2\mu\cos\left(\frac{\Pi}{2s}\right)} \tag{5.34}$$

where $\mu := \frac{1-\Delta^*}{1+\Delta^*}$ and $\lambda_1^n(1/2)$ is the special value of λ_1^n at $\bar{p}=1/2$.

Using the abbreviations

$$P_s(\infty,\tfrac{1}{2}) = \frac{1}{1+\mu^{2s}} \text{ and } a(s) = \frac{1}{s}\,\mu^{(1-s)}.\,\frac{\sin\left(\frac{\Pi}{2s}\right)}{1+\mu^2-2\mu\cos\left(\frac{\Pi}{2s}\right)} \tag{5.35}$$

we get

$$P^* - P_s(\infty,\tfrac{1}{2}) = -\lambda_1^n(1/2)a(s)$$

$$n(s,P^*)\ell n(\lambda_1(1/2)) = \ell n(P_s(\infty,\tfrac{1}{2}) - P^*) - \ell n(a(s))$$

$$n(s,P^*) = \left]\frac{\ell n(P_s(\infty,\frac{1}{2})-P^*)-\ell n(a(s))}{\ell n(\lambda_1(1/2))}\right[\tag{5.36}$$

The desired n-value is hence given by the minimum over s of $n(s,P^*)$, where s must be at least s_{max}, given by (5.33), i.e.

$$n = \min_{s \geq s_{max}} n(s,P^*) \; ; \; N = 2n \tag{5.37}$$

5.6 Numerical results

Table 4

p_A	0.2	0.3	0.4	0.5	0.6	0.7	0.8	0.9
$P^*=0.75$; $\Delta^*=0.1$; $n=50$; $s=3$								
$P(CS\|p_B=p_A-\Delta^*)$	0.830	0.791	0.765	0.752	0.752	0.765	0.791	0.830
$P(CS\|p_A=p_B)$	0.430	0.466	0.478	0.481	0.478	0.466	0.430	0.316
$W_s(n\|p_B=p_A-\Delta^*)$	0.097	0.052	0.032	0.025	0.025	0.032	0.052	0.097
$W_s(n\|p_A=p_B)$	0.139	0.069	0.045	0.039	0.045	0.069	0.139	0.367
$U_s(n\|p_B=p_A-\Delta^*)$	23.3	19.2	16.9	15.9	15.9	16.9	19.2	23.3
$U_s(n\|p_A=p_B)$	24.9	20.2	18.0	17.4	18.0	20.2	24.9	34.8
$P^*=0.75$; $\Delta^*=0.2$; $n=13$; $s=2$								
$P(CS\|p_B=p_A-\Delta^*)$	-	0.770	0.764	0.758	0.756	0.758	0.764	0.770
$P(CS\|p_A=p_B)$	-	0.390	0.416	0.423	0.416	0.390	0.332	0.202
$W_s(n\|p_B=p_A-\Delta^*)$	-	0.178	0.128	0.103	0.095	0.103	0.128	0.178
$W_s(n\|p_A=p_B)$	-	0.219	0.168	0.154	0.168	0.219	0.336	0.595
$U_s(n\|p_B=p_A-\Delta^*)$	-	7.5	6.7	6.3	6.1	6.3	6.7	7.5
$U_s(n\|p_A=p_B)$	-	7.7	7.1	6.9	7.1	7.7	8.9	10.9
$P^*=0.75$; $\Delta^*=0.3$; $n=5$; $s=1$								
$P(CS\|p_B=p_A-\Delta^*)$	-	-	0.801	0.775	0.762	0.762	0.775	0.801
$P(CS\|p_A=p_B)$	-	-	0.481	0.484	0.481	0.467	0.427	0.315
$W_s(n\|p_B=p_A-\Delta^*)$	-	-	0.066	0.031	0.021	0.021	0.031	0.066
$W_s(n\|p_A=p_B)$	-	-	0.038	0.031	0.038	0.066	0.145	0.371
$U_s(n\|p_B=p_A-\Delta^*)$	-	-	2.2	1.9	1.8	1.8	1.9	2.2
$U_s(n\|p_A=p_B)$	-	-	2.0	1.9	2.0	2.2	2.7	3.5

remark:
The possibility of an earlier truncation of the procedure, mentioned at the beginning of our considerations, has no influence on the P(CS)-values and on the expectations.

Table 4 (continued)

p_A	0.2	0.3	0.4	0.5	0.6	0.7	0.8	0.9
$P^*=0.75;\ \Delta^*=0.4;\ n=3;\ s=1$								
$P(CS\|p_B=p_A-\Delta^*)$	-	-	-	0.787	0.784	0.782	0.784	0.787
$P(CS\|p_A=p_B)$	-	-	-	0.437	0.430	0.402	0.343	0.224
$W_s(n\|p_B=p_A-\Delta^*)$	-	-	-	0.125	0.085	0.074	0.085	0.125
$W_s(n\|p_A=p_B)$	-	-	-	0.125	0.141	0.195	0.314	0.551
$U_s(n\|p_B=p_A-\Delta^*)$	-	-	-	1.7	1.6	1.6	1.6	1.7
$U_s(n\|p_A=p_B)$	-	-	-	1.7	1.8	1.9	2.1	2.5
$P^*=0.90;\ \Delta^*=0.1;\ n=155;\ s=7$								
$P(CS\|p_B=p_A-\Delta^*)$	0.955	0.928	0.911	0.902	0.902	0.911	0.928	0.955
$P(CS\|p_A=p_B)$	0.318	0.377	0.403	0.410	0.403	0.377	0.318	0.186
$W_s(n\|p_B=p_A-\Delta^*)$	0.042	0.050	0.048	0.046	0.046	0.048	0.050	0.042
$W_s(n\|p_A=p_B)$	0.364	0.246	0.194	0.179	0.194	0.246	0.364	0.628
$U_s(n\|p_B=p_A-\Delta^*)$	67.9	64.6	61.8	60.2	60.2	61.8	64.6	67.9
$U_s(n\|p_A=p_B)$	107.8	93.3	85.9	83.7	85.9	93.3	107.8	132.9
$P^*=0.90;\ \Delta^*=0.2;\ n=39;\ s=3$								
$P(CS\|p_B=p_A-\Delta^*)$	-	0.957	0.927	0.908	0.902	0.908	0.927	0.957
$P(CS\|p_A=p_B)$	-	0.435	0.453	0.458	0.453	0.435	0.387	0.260
$W_s(n\|p_B=p_A-\Delta^*)$	-	0.026	0.024	0.020	0.019	0.020	0.024	0.026
$W_s(n\|p_A=p_B)$	-	0.130	0.093	0.083	0.093	0.130	0.255	0.480
$U_s(n\|p_B=p_A-\Delta^*)$	-	14.2	13.3	12.6	12.4	12.6	13.3	14.2
$U_s(n\|p_A=p_B)$	-	19.1	17.3	16.7	17.3	19.1	22.9	30.1
$P^*=0.90;\ \Delta^*=0.3;\ n=16;\ s=2$								
$P(CS\|p_B=p_A-\Delta^*)$	-	-	0.941	0.915	0.902	0.902	0.915	0.941
$P(CS\|p_A=p_B)$	-	-	0.447	0.452	0.447	0.426	0.375	0.247
$W_s(n\|p_B=p_A-\Delta^*)$	-	-	0.033	0.028	0.024	0.024	0.028	0.033
$W_s(n\|p_A=p_B)$	-	-	0.107	0.096	0.107	0.148	0.250	0.507
$U_s(n\|p_B=p_A-\Delta^*)$	-	-	6.1	5.7	5.5	5.5	5.7	6.1
$U_s(n\|p_A=p_B)$	-	-	7.6	7.3	7.6	8.3	9.8	12.6

Table 4 (continued)

p_A	0.2	0.3	0.4	0.5	0.6	0.7	0.8	0.9
P*=0.90; Δ*=0.4; n=10; s=2								
$P(CS\|p_B=p_A-\Delta^*)$	-	-	-	0.932	0.920	0.916	0.920	0.932
$P(CS\|p_A=p_B)$	-	-	-	0.376	0.367	0.338	0.274	0.151
$W_s(n\|p_B=p_A-\Delta^*)$	-	-	-	0.056	0.054	0.053	0.054	0.056
$W_s(n\|p_A=p_B)$	-	-	-	0.248	0.265	0.325	0.451	0.697
$U_s(n\|p_B=p_A-\Delta^*)$	-	-	-	4.7	4.5	4.5	4.5	4.7
$U_s(n\|p_A=p_B)$	-	-	-	6.3	6.4	6.9	7.7	8.9
P*=0.95; Δ*=0.1; n=241; s=9								
$P(CS\|p_B=p_A-\Delta^*)$	0.988	0.971	0.958	0.951	0.951	0.958	0.971	0.988
$P(CS\|p_A=p_B)$	0.304	0.364	0.391	0.399	0.391	0.364	0.304	0.172
$W_s(n\|p_B=p_A-\Delta^*)$	0.012	0.021	0.024	0.024	0.024	0.024	0.021	0.012
$W_s(n\|p_A=p_B)$	0.392	0.272	0.218	0.202	0.218	0.272	0.392	0.655
$U_s(n\|p_B=p_A-\Delta^*)$	89.4	87.5	85.3	84.1	84.1	85.3	87.5	89.4
$U_s(n\|p_A=p_B)$	172.4	150.3	138.9	135.4	138.9	150.3	172.4	210.0
P*=0.95; Δ*=0.2; n=61; s=4								
$P(CS\|p_B=p_A-\Delta^*)$	-	0.986	0.968	0.955	0.951	0.955	0.968	0.986
$P(CS\|p_A=p_B)$	-	0.413	0.435	0.441	0.435	0.413	0.360	0.229
$W_s(n\|p_B=p_A-\Delta^*)$	-	0.009	0.012	0.012	0.012	0.012	0.012	0.009
$W_s(n\|p_A=p_B)$	-	0.173	0.130	0.118	0.130	0.173	0.279	0.541
$U_s(n\|p_B=p_A-\Delta^*)$	-	19.7	19.1	18.5	18.3	18.5	19.1	19.7
$U_s(n\|p_A=p_B)$	-	32.7	29.8	28.9	29.8	32.7	38.5	49.3
P*=0.95; Δ*=0.3; n=26; s=3								
$P(CS\|p_B=p_A-\Delta^*)$	-	-	0.976	0.962	0.954	0.954	0.962	0.976
$P(CS\|p_A=p_B)$	-	-	0.390	0.397	0.390	0.362	0.301	0.171
$W_s(n\|p_B=p_A-\Delta^*)$	-	-	0.020	0.023	0.023	0.023	0.023	0.020
$W_s(n\|p_A=p_B)$	-	-	0.221	0.205	0.221	0.276	0.398	0.658
$U_s(n\|p_B=p_A-\Delta^*)$	-	-	9.8	9.5	9.4	9.4	9.5	9.8
$U_s(n\|p_A=p_B)$	-	-	15.3	14.9	15.3	16.3	18.8	22.7

Table 4 (continued)

p_A	0.2	0.3	0.4	0.5	0.6	0.7	0.8	0.9
$P^*=0.95;\ \Delta^*=0.4;\ n=14;\ s=2$								
$P(CS\|p_B=p_A-\Delta^*)$	-	-	-	0.974	0.959	0.954	0.959	0.974
$P(CS\|p_A=p_B)$	-	-	-	0.434	0.428	0.404	0.348	0.218
$W_s(n\|p_B=p_A-\Delta^*)$	-	-	-	0.014	0.014	0.014	0.014	0.014
$W_s(n\|p_A=p_B)$	-	-	-	0.132	0.145	0.192	0.304	0.564
$U_s(n\|p_B=p_A-\Delta^*)$	-	-	-	4.8	4.7	4.6	4.7	4.8
$U_s(n\|p_A=p_B)$	-	-	-	7.1	7.3	8.0	9.2	11.5
$P^*=0.99;\ \Delta^*=0.1;\ n=453;\ s=14$								
$P(CS\|p_B=p_A-\Delta^*)$	0.999	0.996	0.993	0.990	0.990	0.993	0.996	0.999
$P(CS\|p_A=p_B)$	0.245	0.308	0.338	0.347	0.338	0.308	0.245	0.121
$W_s(n\|p_B=p_A-\Delta^*)$	0.001	0.003	0.005	0.006	0.006	0.005	0.003	0.001
$W_s(n\|p_A=p_B)$	0.511	0.384	0.323	0.305	0.323	0.384	0.511	0.757
$U_s(n\|p_B=p_A-\Delta^*)$	140.0	139.6	139.0	138.6	138.6	139.0	139.6	140.0
$U_s(n\|p_A=p_B)$	358.7	321.3	301.2	294.8	301.2	321.3	358.7	416.5
$P^*=0.99;\ \Delta^*=0.2;\ n=112;\ s=7$								
$P(CS\|p_B=p_A-\Delta^*)$	-	0.998	0.995	0.992	0.990	0.992	0.995	0.998
$P(CS\|p_A=p_B)$	-	0.306	0.337	0.346	0.337	0.306	0.243	0.120
$W_s(n\|p_B=p_A-\Delta^*)$	-	0.002	0.004	0.006	0.006	0.006	0.004	0.002
$W_s(n\|p_A=p_B)$	-	0.387	0.327	0.309	0.327	0.387	0.514	0.759
$U_s(n\|p_B=p_A-\Delta^*)$	-	35.0	34.9	34.7	34.6	34.7	34.9	35.0
$U_s(n\|p_A=p_B)$	-	79.9	74.9	73.4	74.9	79.9	89.0	103.0
$P^*=0.99;\ \Delta^*=0.3;\ n=50;\ s=5$								
$P(CS\|p_B=p_A-\Delta^*)$	-	-	0.997	0.993	0.991	0.991	0.993	0.997
$P(CS\|p_A=p_B)$	-	-	0.308	0.317	0.308	0.277	0.213	0.099
$W_s(n\|p_B=p_A-\Delta^*)$	-	-	0.003	0.006	0.007	0.007	0.006	0.003
$W_s(n\|p_A=p_B)$	-	-	0.385	0.366	0.385	0.447	0.573	0.803
$U_s(n\|p_B=p_A-\Delta^*)$	-	-	16.6	16.6	16.5	16.5	16.6	16.6
$U_s(n\|p_A=p_B)$	-	-	35.7	35.0	35.7	37.8	41.5	46.9

Table 4 (continued)

p_A	0.2	0.3	0.4	0.5	0.6	0.7	0.8	0.9
$P^*=0.99;\ \Delta^*=0.4;\ n=26;\ s=3$								
$P(CS\|p_B=p_A-\Delta^*)$	-	-	-	0.996	0.992	0.990	0.992	0.996
$P(CS\|p_A=p_B)$	-	-	-	0.397	0.390	0.362	0.301	0.171
$W_s(n\|p_B=p_A-\Delta^*)$	-	-	-	0.002	0.003	0.004	0.003	0.002
$W_s(n\|p_A=p_B)$	-	-	-	0.205	0.221	0.276	0.398	0.658
$U_s(n\|p_B=p_A-\Delta^*)$	-	-	-	7.5	7.4	7.4	7.4	7.5
$U_s(n\|p_A=p_B)$	-	-	-	14.9	15.3	16.5	18.8	22.7

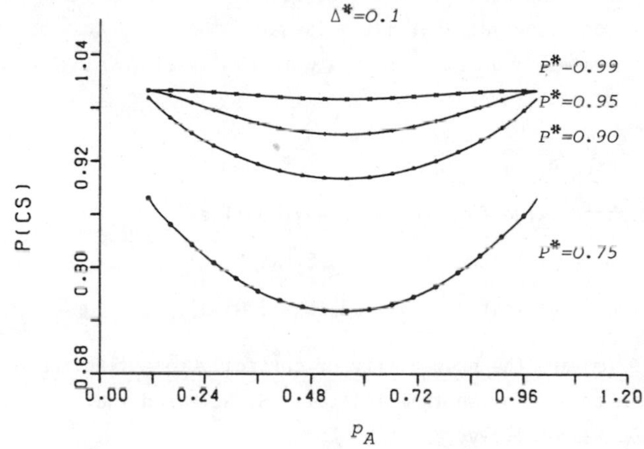

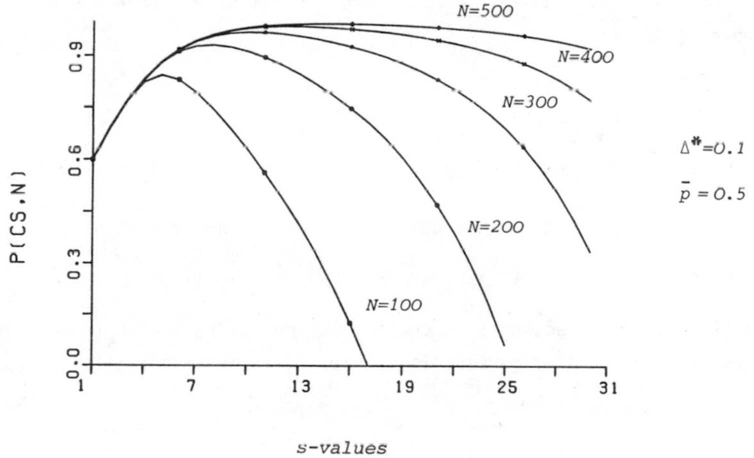

6 The Selection Model $[2;PW;\text{fixed } N]$ with Truncation Based on $|S_A-S_B|$

6.1 Description of the model

In this section we consider a fixed patient horizon selection model that uses the PW-sampling scheme. If at any stage $k<N$ of the experiment the absolute difference in the current numbers of successes equals r, then sampling is terminated and the treatment associated with the larger number of successes is declared as best. In case that no truncation occurs within N trials, both treatments are considered to be of equal worth.

6.2 Derivation of the P(CS)-value

To derive the desired value for P(CS), we next define

$$\begin{aligned} P_k(m) &:= P(\text{CS in at most m trials } |S_A-S_B=k,\ NT=A) \\ Q_k(m) &:= P(\text{CS in at most m trials } |S_A-S_B=k,\ NT=B), \end{aligned} \tag{6.1}$$

i.e. $P_k(m)$ and $Q_k(m)$ are the probability of correctly selecting the best treatment in at most m trials, given that initially $S_A-S_B=k$ and that the next trial is on treatment A,B, respectively.

The $P_k(m)$ and $Q_k(m)$ satisfy the following system of difference equations and boundary conditions

$$\begin{aligned} P_k(m+1) &= p_A\, P_{k+1}(m) + q_A Q_k(m) \\ Q_k(m+1) &= p_B\, Q_{k-1}(m) + q_B Q_k(m) \\ P_r(m) &= 1; Q_{-r}(m) = 0;\ P_k(0) = Q_k(0) = 0 \text{ for } k \neq r;\ P_r(0) = 1. \end{aligned} \tag{6.2}$$

Similar to our proceeding in section 5, we define new variables by

$$P_k^*(m) := P_k(\infty) - P_k(m), \text{ and } Q_k^*(m) = Q_k(\infty) - Q_k(m) \tag{6.3}$$

Recalling (1.3) of section 1 in chapter 1, and noting that $P_k(\infty)=P_k$, and $Q_k(\infty)=$ $=Q_k$ as defined in (1.2) in sec. 1 of chap. 1, it is immediately obvious that the variables $P_k^*(m)$ and $Q_k^*(m)$ satisfy the same system of difference equations as the variables $P_k(m)$ and $Q_k(m)$, i.e.

$$\begin{aligned} P_k^*(m+1) &= p_A P_{k+1}^*(m) + q_A Q_k^*(m) \\ Q_k^*(m+1) &= p_B Q_{k-1}^*(m) + q_B P_k^*(m) \end{aligned} \tag{6.4}$$

The new boundary conditions are as follows

$$\begin{aligned} &P_r^*(m) = 0;\ Q_{-r}^*(m) = 0;\ P_k^*(0) = (q_B - q_A\lambda^{k+r})/(q_B - q_A\lambda^{2r}), \text{ and} \\ &Q_k^*(0) = q_B(1-\lambda^{k+r})/(q_B - q_A\lambda^{2r}) \text{ for } k \neq r \text{ (cf. (1.7) and (1.8) of chap.1);} \\ &P_r^*(0) = 0 \end{aligned} \tag{6.5}$$

The probability of a correct selection is hence given by

$$\begin{aligned} P(CS) &= \tfrac{1}{2}(P_0(N) + Q_0(N)) = \tfrac{1}{2}(P_0(\infty)+Q_0(\infty)-P_0^*(N)-Q_0^*(N)) \\ &= \frac{1}{2}\,\frac{2q_B-(q_A+q_B)\lambda^r}{q_B-q_A\lambda^{2r}} - \frac{1}{2}\,(P_0^*(N)+Q_0^*(N)) \end{aligned} \tag{6.6}$$

As already mentioned in 5.5 of the preceding section, there may be two different situations. If N is fixed and P^*,Δ^* are preassigned constants, then it may occur that P(CS) can never exceed P^* in the LFC. In such situations we determine that r which maximizes P(CS) in the LFC, i.e. we determine that r which guarantees P^* to be $\max_r \min_{p_A} P(CS)$. If N is large enough so that for preassigned P^* and Δ^* P(CS) can exceed P^*, then we look for the smallest r that guarantees P(CS) to be at least P^* in the LFC, i.e. we look for

$$r^* = \min\{r : \min_{p_A} P(CS; p_A, \Delta^*, r) \geq P^*\} \tag{6.7}$$

For practical applications it is useful to know that value of N which guarantees that the best treatment is identified at the preassigned significance level P^*, whenever $p_A - p_B \geq \Delta^*$. The desired values can be obtained by solving (6.4) by an eigenfunction expansion as shown in the preceding section. It is, however, much simpler to make direct use of the recursion formulae (6.4) so as to generate the desired values $P_0^*(N)$ and $Q_0^*(N)$, needed to compute P(CS) for given p_A, Δ^*, and P^*.

6.3 Derivation of the probability of declaring the two treatments equal

Let $R_k(m), T_k(m)$ denote the probability that none of the competing treatments is declared as best after the m-th trial, given that $S_A - S_B = k$, and that the next trial is on treatment A,B, respectively, i.e.

$$R_k(m) := P(\text{no decision after the m-th trial} \mid S_A - S_B = k,\ NT=A)$$
$$T_k(m) := P(\text{no decision after the m-th trial} \mid S_A - S_B = k,\ NT=B) \tag{6.8}$$

The $R_k(m)$ and $T_k(m)$ satisfy the following system of difference equations and boundary conditions.

$$R_k(m+1) = p_A R_{k+1}(m) + q_A T_k(m)$$
$$T_k(m+1) = p_B T_{k-1}(m) + q_B R_k(m) \tag{6.9}$$
$$R_k(0) = T_k(0) = 1;\ R_r(m) = T_{-r}(m) = 0$$

The probability of declaring the two treatments equal is given by

$$P_{nd} := P(\text{"no difference"}) = \frac{1}{2}\,(R_0(N) + T_0(N)) \tag{6.10}$$

P_{nd} is directly computed by using (6.9).

6.4 Derivation of $E(N_B)$

To derive the expected number of patients on the inferior treatment we define

$$U_k(m) := E(N_B \text{ after m trials} \mid S_A - S_B = k,\ NT=A)$$
$$V_k(m) := E(N_B \text{ after m trials} \mid S_A - S_B = k,\ NT=B) \tag{6.11}$$

The $U_k(m)$ and $V_k(m)$ satisfy the following system of difference equations and boundary conditions

$$U_k(m+1) = p_A U_{k+1}(m) + q_A V_k(m)$$
$$V_k(m+1) = p_B V_{k-1}(m) + q_B U_k(m) + 1 \tag{6.12}$$
$$U_k(0) = V_k(0) = 0;\ U_r(m) = 0;\ V_{-r}(m) = 0$$

The desired expectation is hence given by

$$E(N_B) = \frac{1}{2}\,(U_0(N) + V_0(N)) \tag{6.13}$$

(6.12) is used to generate $U_0(N)$ and $V_0(N)$.

Using the well-known interchanging technique, $E(N_A)$ is found to be

$$E(N_A) = E(N_B \mid p_A \leftrightarrows p_B) = \frac{1}{2}(U_0(N \mid p_A \leftrightarrows p_B) + V_0(N \mid p_A \leftrightarrows p_B)) \tag{6.14}$$

In practice sampling will be truncated if no decision in favor of either treatment is possible. Strictly speaking, if $|S_{A,N-r+k}-S_{B,N-r+k}|<k$, then $|S_{A,N}-S_{B,N}|<r$ in any case, and sampling can be truncated after the (N-r+k)-th trial, and the treatments are declared to be of equal worth. The following diagram is to illustrate the performance of the selection procedure.

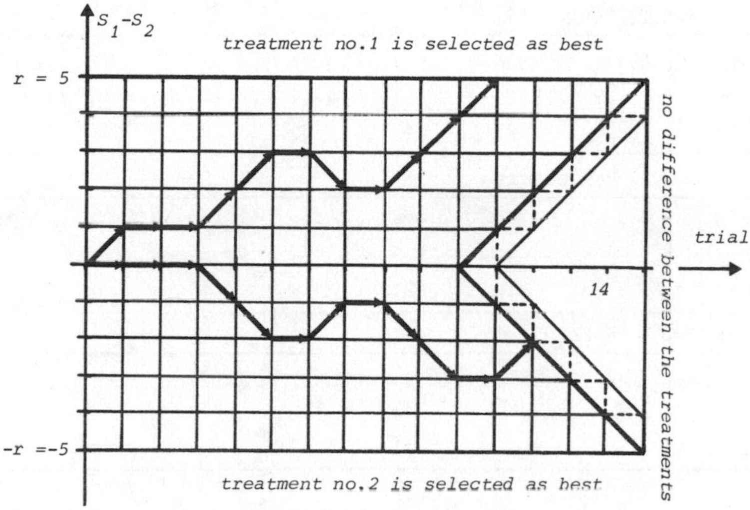

If the straight line $S_1-S_2=r$, $S_1-S_2=-r$ is reached, then treatment no. 1, no. 2 is declared as best, respectively. In case that the straight lines $S_1-S_2=n-(N-r)$ or $S_1-S_2=-n+(N-r)$ are crossed, the treatments are declared to be of equal worth. (Note that the first treatment at the outset is again denoted by no. 1.)

The numerical results to be presented in the following have been obtained by simulation. It was not possible to get the exact results within reasonable bounds of computer time by using the corresponding recursion formulae. Simulation studies for $\Delta^*=0.1$ have been omitted because of the same reason.

6.5 Numerical results

Table 5

p_A	0.3	0.4	0.5	0.6	0.7	0.8	0.9
$P^*=0.90$; $\Delta^*=0.2$; N=106; r=7							
P(CS;sim.)	0.938	0.936	0.933	0.937	0.904	0.913	0.903
E(NB;sim.)	25	21	17	14	10	7	5
E(N; sim.)	56	49	41	35	26	20	14
$P^*=0.90$; $\Delta^*=0.3$; N=47; r=5							
P(CS;sim.)	-	0.925	0.929	0.941	0.931	0.915	0.913
E(NB;sim.)	-	10	9	7	6	4	3
E(N; sim.)	-	25	21	19	15	12	9
$P^*=0.90$; $\Delta^*=0.4$; N=22; r=3							
P(CS;sim.)	-	-	0.914	0.927	0.924	0.913	0.900
E(NB;sim.)	-	-	4	4	3	2	1
E(N; sim.)	-	-	11	9	8	6	5
$P^*=0.95$; $\Delta^*=0.2$; N=156; r=10							
P(CS;sim.)	0.968	0.962	0.969	0.974	0.968	0.961	0.962
E(NB;sim.)	34	29	27	21	16	11	6
E(N; sim.)	77	68	63	51	42	30	21
$P^*=0.95$; $\Delta^*=0.3$; N=66; r=6							
P(CS;sim.)	-	0.964	0.967	0.974	0.961	0.958	0.957
E(NB;sim.)	-	13	10	9	7	5	3
E(N; sim.)	-	31	26	23	19	15	11
$P^*=0.95$; $\Delta^*=0.4$; N=37; r=5							
P(CS;sim.)	-	-	0.970	0.957	0.972	0.974	0.955
E(NB;sim.)	-	-	7	6	5	3	2
E(N; sim.)	-	-	18	16	13	11	9
$P^*=0.99$; $\Delta^*=0.2$; N=277; r=17							
P(CS;sim.)	0.994	0.993	0.990	0.997	0.999	0.993	0.992
E(NB;sim.)	60	52	44	35	27	19	11
E(N; sim.)	136	121	104	87	70	54	38

Table 5 (continued)

p_A	0.3	0.4	0.5	0.6	0.7	0.8	0.9
P*=0.99; Δ*=0.3; N=126; r=10							
P(CS;sim.)	-	0.996	0.996	0.999	0.997	0.993	0.993
E(NB;sim.)	-	21	18	15	11	8	5
E(N; sim.)	-	52	45	39	31	25	19
P*=0.99; Δ*=0.4; N=65; r=7							
P(CS;sim.)	-	-	0.996	0.997	0.996	0.992	0.998
E(NB;sim.)	-	-	9	8	6	4	3
E(N; sim.)	-	-	25	22	19	15	12

6.6 Comparison of the selection models

As already mentioned at the end of sec.3, the fixed sample size selection-model of Nebenzahl/Sobel and that of Hoel (cf. sec.1, chap.2) provide essentially the same, however, when using Hoel's model, a decision can be reached in fewer trials.

The selection model of Nebenzahl/Sobel is uniformly worse than the corresponding truncated selection model of Pradhan/Sathe (cf. sec.4), and the Pradhan/Sathe PW-selection model and the Hoel selection model provide essentially the same, but the Hoel selection model yields slightly smaller expectations.

The Kiefer/Weiss PW-selection model (cf. sec.6) is better than the Kiefer/Weiss VT-selection model (cf. sec.5) with respect to $E(N_B)$ $(E(N))$ if $p_A \geq 0.7$ (0.8).

The Kiefer/Weiss PW-selection model is better than the Pradhan/Sathe PW-selection model when the success parameter p_A is large, and the Kiefer/Weiss VT-selection model is worse than the Pradhan/Sathe PW-selection model if p_A is large enough ($\geq$ 0.8), where the comparison is dependent on the actual values of Δ*, and P*. Details can be seen from the corresponding tables.

7 Selection Models Based on the Randomized Play-the-Winner Rule

7.1 Introductory remarks

The play-the-winner sampling rule requires an instantaneous response, i.e. the experimenter must be able to decide within a reasonable time period after the application of each treatment whether the result of a single trial is a success or a failure. Since the outcome of a trial prescribes which treatment has to be applied in the next trial, delayed response must be avoided. We have already mentioned in the introduction that the so-called randomized play-the-winner design is applicable also in case if delayed response arises. In the following we consider a generalized randomized play-the-winner design, due to Wei and Durham [55], that contains as a special case the randomized play-the-winner design presented in the introduction. The treatment assignment according to the RPW-rule runs as follows: We have balls of two different types which are marked A or B, and u balls of each type are placed into a box. When a patient enters the experiment a ball is drawn at random and replaced. The treatment assignment is conducted according to the balls drawn from the box, i.e. for instance that the first treatment at the outset is determined randomly, because an equal number of A and B balls are in the box before the experiment starts. Whenever the response of a patient is available, the structure of the box is changed in the following way: If the trial has been carried out on treatment i and if the response is a success, then an additional β balls of type i and an additional α balls of type j are placed into the box. If the response is a failure, then an additional α balls of type i and an additional β balls of type j are placed into the box, where $\beta \geq \alpha \geq 0$; $i,j \in \{A,B\}$, and $i \neq j$. When the box is empty, the next treatment assignment is determined randomly. The sampling-rule just described is denoted by RPW(u,α,β). Selection models using RPW(u,α,β) together with delayed response have not yet been investigated, neither analytically nor by simulation. In the following we will assume that the response of each patient is instantaneous.

7.2 The expected number of patients on the better treatment within n trials

Let $S_{i,n}, F_{i,n}$, and $R_{i,n}$; $i \in \{A,B\}$; be defined as in section 3, then the following identities are immediately obvious.

$$\begin{aligned} R_{A,n} &= u+\beta(S_{A,n}+F_{B,n}) + \alpha(S_{B,n}+F_{A,n}) \\ R_{B,n} &= u+\beta(S_{B,n}+F_{A,n}) + \alpha(S_{A,n}+F_{B,n}) \\ R_{A,n} + R_{B,n} &= 2u + n(\alpha+\beta) =: c_n \end{aligned} \tag{7.1}$$

The random variables $R_{A,n}$ form a stochastic process with transition probabilities as follows:

$$\begin{aligned} P(R_{A,n+1}=R_{A,n}+\beta|R_{A,n}) &= (p_A R_{A,n}+q_B R_{B,n})/c_n \\ P(R_{A,n+1}=R_{A,n}+\alpha|R_{A,n}) &= (q_A R_{A,n}+p_B R_{B,n})/c_n \end{aligned} \tag{7.2}$$

(7.2) is immediately seen to hold when noting that $R_{i,n}/c_n$ is the relative frequency of balls of type i after n trials, and that β or α additional A balls are added to the box if the (n+1)-th trial results in a successful A-trial/faulty B-trial or a faulty A-trial/successful B-trial, respectively.

Using conditional expectations, we find for $n \in \mathbb{N}$:

$$\begin{aligned} E(R_{A,n+1}) &= E(E(R_{A,n+1}|R_{A,n})) = E((R_{A,n}+\beta)(p_A R_{A,n}+q_B R_{B,n})/c_n + \\ &\quad + (R_{A,n}+\alpha)(q_A R_{A,n}+p_B R_{B,n})/c_n) = \\ &= (1+(p_A-q_B)(\beta-\alpha)/c_n)E(R_{A,n}) + \alpha p_B+\beta q_B \end{aligned} \tag{7.3}$$

The corresponding recursion formula for $E(R^2_{A,n})$ is obtained in the same way, we get

$$\begin{aligned} E(R^2_{A,n+1}) &= (1+2(p_A-q_B)(\beta-\alpha)/c_n)E(R^2_{A,n}) + 2(\alpha p_B+\beta q_B)E(R_{A,n}) + \\ &\quad + ((\beta^2-\alpha^2)(p_A-q_B)/c_n)E(R_{A,n}) + \beta^2 q_B+\alpha^2 p_B = \\ &= (1+2(p_A-q_B)(\beta-\alpha)/c_n)E(R^2_{A,n}) + 2(\alpha p_B+\beta q_B)E(R_{A,n}) + o(n) \end{aligned} \tag{7.4}$$

The expected number of patients assigned to treatment A after n assignments, denoted by $N_{A,n}$, can be derived by using (7.3) and the following identity.

$$R_{A,n} = u+p_A\beta N_{A,n}+q_A\alpha N_{A,n}+p_B\alpha N_{B,n}+q_B\beta N_{B,n} \tag{7.5}$$

Noting that $N_{A,n}+N_{B,n}=n$, we finally obtain

$$E(N_{A,n}) = (E(R_{A,n})-u-n(\beta q_B+\alpha p_B))/((p_A-q_B)(\beta-\alpha)), \tag{7.6}$$

where $p_A \neq q_B$ and $\beta \neq \alpha$.

Numerical results can be obtained by using (7.6) and the recursion formula for $E(R_{A,n})$, where $E(R_{A,1})=(p_A+q_B)/2$. The corresponding expectations for the PW-rule

can be computed by using (2.6) of section 2.
The limiting behaviour of $E(R_{A,n})$, $E(N_{A,n})$, and $E(R^2_{A,n})$ is investigated by making use of Lemma 6.6 of Freedman [62] (cf. A3/4).

With $x_n:=E(R_{A,n})$, $a_n:=1+\dfrac{(p_A-q_B)(\beta-\alpha)}{2u+(\alpha+\beta)n}$; $a:=(p_A-q_B)(\beta-\alpha)$; $b:=2u$; $c:=(\alpha+\beta)$; $b_n:=(\alpha p_B+\beta q_B)=:B$, and $d=0$, we find:

$$\lim_{n\to\infty}\frac{E(R_{A,n})}{n}=(\alpha p_B+\beta q_B)/(1-(p_A-q_B)(\beta-\alpha)/(\alpha+\beta)) \tag{7.7}$$

Applying A3/4 again with $x_n:=E(R^2_{A,n})$, $a_n:=1+\dfrac{2(p_A-q_B)(\beta-\alpha)}{2u+(\alpha+\beta)n}$, $a:=2(p_A-q_B)(\beta-\alpha)$, $b:=2u$, $c:=\alpha+\beta$, $b_n=2(\alpha p_B+\beta q_B)E(R_{A,n})+o(n)$, where $b_n\approx 2(\alpha p_B+\beta q_B)^2 n/(1-(p_A-q_B)(\beta-\alpha)/(\alpha+\beta))=:Bn$ according to (7.7), then it follows:

$$\lim_{n\to\infty}\frac{E(R^2_{A,n})}{n^2}=(\alpha p_B+\beta q_B)^2/(1-(p_A-q_B)(\beta-\alpha)/(\alpha+\beta))^2 \tag{7.8}$$

Applying (7.7) to the identity given in (7.6), we finally get

$$\lim_{n\to\infty}\frac{E(N_{A,n})}{n}=(\alpha p_B+\beta q_B)/(\alpha(p_A+p_B)+\beta(q_A+q_B)) \tag{7.9}$$

(7.9) is an increasing function of β/α and tends to $q_B/(q_A+q_B)\geq 1/2$ as β/α become large, i.e. if β is large with respect to α, then more patients are treated with the better medicament. The use of the RPW-rule causes hence a bias in favor of testing the better treatment if β/α is large enough. For the special rule RPW($u,0,\beta$) (7.9) equals $q_B/(q_A+q_B)$, and that is the asymptotic proportion of trials carried out with treatment A when the PW-sampling scheme is used, as can be seen from (2.6) of sec. 2. That means, that RPW($u,0,\beta$)- and PW-sampling are of equal value for assigning more patients to the best treatment if the patient horizon is large.
The randomized play-the-winner rule can be combined with several termination-rules. In the following we give only exact results for RPW(0,0,1) and the termination rule of Hoel's selection model, presented in sec. 1 of chap. 2.

7.3 Derivation of the P(CS)-values

For the special case $\alpha=0,\beta=1$, (7.7) and (7.8) reduces to

$$\lim_{n\to\infty}\frac{E(R_{A,n})}{n}=\frac{q_B}{q_A+q_B}\quad\text{and}\quad\lim_{n\to\infty}\frac{E(R^2_{A,n})}{n^2}=\left(\frac{q_B}{q_A+q_B}\right)^2 \tag{7.10}$$

Therefore, $\lim_{n\to\infty} Var(R_{A,n}/n)=0$, and that means:

$$R_A/n \to q_B/(q_A+q_B), \text{ in probability, as } n\to\infty. \tag{7.11}$$

In using Hoel's termination-rule, a correct selection occurs if and only if $R_{A,n}\geq r$ for some $n\leq 2r-1$, and since $R_{A,n}$ is an increasing "function" of n, we have that a correct selection occurs if and only if $R_{A,2r-1}\geq r$, i.e.

$$P(CS) = P(R_{A,2r-1}\geq r) \tag{7.12}$$

Noting that $r/(2r-1)\to 1/2$ as r becomes large, and that $q_B/(q_A+q_B)>1/2$ if $p_A>p_B$, then the following holds for all $\varepsilon>0$ and large r

$$P\left(\frac{R_{A,r}}{r} < \frac{q_B}{q_A+q_B} + \varepsilon\right) \geq P\left(\frac{R_{A,2r-1}}{2r-1} < \frac{r}{2r-1} + \varepsilon\right) \tag{7.13}$$

Recalling that the lefthand side tends to 0 as $r\to\infty$, we have

$$\lim_{r\to\infty} P(CS) = \lim_{r\to\infty} P(R_{A,2r-1}\geq r) = 1 \tag{7.14}$$

Therefore, we can choose r such that P(CS) is not less than a preassigned P^* for all possible parameter configurations (p_A,p_B) with $p_A-p_B=\Delta^*$.

To derive the exact values for a correct selection, we define

$$U(i,j) := P(CS|r\text{-}i \text{ A-balls and } r\text{-}j \text{ B-balls}), \text{ i.e.} \tag{7.15}$$

U(i,j) denotes the probability of a correct selection if r-i A-balls and r-j B-balls are in the box.

The U(i,j) satisfy the following system of difference equations and boundary conditions.

$$U(i,j) = \frac{(r-i)p_A+(r-j)q_B}{2r-i-j}\,U(i-1,j) + \frac{(r-i)q_A+(r-j)p_B}{2r-i-j}\,U(i,j-1)$$
$$U(i,0) = 0;\ U(0,j) = 1, \text{ and } i+j\neq 2r,\ i,j\in\{1,\dots,r\}. \tag{7.16}$$

Recalling that the first assignment is determined randomly, and that an additional A-ball or B-ball is placed into the box if the trial results in a successful A-trial/faulty B-trial or successful B-trial/faulty A-trial, then P(CS) is found to be

$$P(CS) = \tfrac{1}{2}(p_A+q_B)\,U(r-1,r) + \tfrac{1}{2}(q_A+p_B)\,U(r,r-1) \tag{7.17}$$

7.4 Derivation of the expectations

To derive the exact formula for $E(N_B)$ we define

$$V(i,j) := E(N_B | r\text{-}i \text{ A-balls and } r\text{-}j \text{ B-balls}) =: E(N_B | r-i, r-j) \tag{7.18}$$

From this we deduce:

$$\begin{aligned} V(i,j) &= E(N_B | r-i, r-j, NT=A)\,\frac{r-i}{2r-i-j} + E(N_B | r-i, r-j, NT=B)\,\frac{r-j}{2r-i-j} = \\ &= (E(N_B | r-(i-1), r-j)p_A + E(N_B | r-i, r-(j-1))q_A)\,\frac{r-i}{2r-i-j} + \\ &+ (E(N_B | r-i, r-(j-1))p_B + E(N_B | r-(i-1), j-1)q_B + 1)\,\frac{r-j}{2r-i-j} = \\ &= \frac{(r-i)p_A + (r-j)q_B}{2r-i-j}\,V(i-1,j) + \frac{(r-i)q_A + (r-j)p_B}{2r-i-j}\,V(i,j-1) + \frac{r-j}{2r-i-j} \end{aligned} \tag{7.19}$$

The boundary conditions are as follows:

$$V(i,o) = V(o,j) = 0;\ i,j \in \{1,\dots,r\} \text{ and } (i,j) \neq (r,r) \tag{7.20}$$

The exact formula for $E(N_B)$ is found to be

$$E(N_B) = \tfrac{1}{2}(p_A + q_B)V(r-1,r) + \tfrac{1}{2}(q_A + p_B)V(r,r-1) + \tfrac{1}{2} \tag{7.21}$$

The desired results for $E(N)$ are obtained in essentially the same way, i.e. defining

$$W(i,j) = E(N | r-i, r-j), \tag{7.22}$$

we see that the $W(i,j)$ satisfy the following system of difference equations and boundary conditions

$$\begin{aligned} W(i,j) &= \frac{(r-i)p_A + (r-j)q_B}{2r-i-j}\,W(i-1,j) + \frac{(r-i)q_A + (r-j)p_B}{2r-i-j}\,W(i,j-1) + 1 \\ W(i,o) &= W(o,j) = 0;\ i,j \in \{1,\dots,r\} \text{ and } (i,j) \neq (r,r) \end{aligned} \tag{7.23}$$

Using similar arguments as above, $E(N)$ is found to be

$$E(N) = \tfrac{1}{2}\,(p_A + q_B)\,W(r-1,r) + \tfrac{1}{2}\,(q_A + p_B)\,W(r,r-1) + 1 \tag{7.24}$$

7.5 Numerical results

Table 6

p_A	0.3	0.4	0.5	0.6	0.7	0.8	0.9
P^*=0.90; Δ^*=0.2; r=39							
P(CS;sim.)	0.941	0.961	0.958	0.976	0.965	0.946	0.908
E(NB;sim.)	30	29	28	26	24	21	18
E(N; sim.)	69	68	67	64	62	59	54
P^*=0.90; Δ^*=0.3; r=12							
P(CS;sim.)	-	0.928	0.936	0.935	0.931	0.920	0.914
E(NB;sim.)	-	8	8	7	6	6	5
E(N; sim.)	-	20	19	18	18	17	16
P^*=0.90; Δ^*=0.4; r=6							
P(CS;sim.)	-	-	0.916	0.916	0.921	0.920	0.922
E(NB;sim.)	-	-	3	3	3	3	2
E(N; sim.)	-	-	9	9	8	8	7
P^*=0.95; Δ^*=0.3; r=22							
P(CS;sim.)	-	0.978	0.990	0.975	0.981	0.973	0.958
E(NB;sim.)	-	15	14	13	12	10	9
E(N; sim.)	-	36	35	34	33	31	29
P^*=0.95; Δ^*=0.4; r=10							
P(CS;sim.)	-	-	0.971	0.967	0.963	0.960	0.961
E(NB;sim.)	-	-	6	5	5	4	3
E(N; sim.)	-	-	15	15	14	13	12
P^*=0.99; Δ^*=0.3; r=46							
P(CS;sim.)	-	0.995	0.997	0.998	0.998	0.996	0.990
E(NB;sim.)	-	31	29	26	23	20	16
E(N; sim.)	-	76	75	72	69	65	60
P^*=0.99; Δ^*=0.4; r=18							
P(CS;sim.)	-	-	0.992	0.994	0.994	0.995	0.990
E(NB;sim.)	-	-	10	9	8	7	6
E(N; sim.)	-	-	28	27	26	24	22

The RPW-rule have been investigated for u=0, u=10, and u=100 together with the termination rules of Hoel, Berry/Sobel, Fushimi, and Nordbrock.
All selection models proved to be uniformly worse than the corresponding PW-selection models. Moreover the RPW-models become more and more worse if u increases.
Note that the RPW-model presented here has no fixed patient horizon, however, the results indicate that the corresponding fixed patient horizon selection model is also uniformly worse than a PW-selection model.

Table 7

	$\Delta^*=0.1$; N=20; $\alpha=0$; $\beta=1$							
	$E(N_A^{20}\|PW)$	$E(N_A^{20}\|RPW)$						
p_A		u=0	u=2	u=4	u=6	u=8	u=10	u=100
0.2	10.57	10.59	10.46	10.38	10.33	10.29	10.26	10.04
0.3	10.64	10.66	10.50	10.41	10.35	10.30	10.27	10.04
0.4	10.74	10.76	10.54	10.44	10.37	10.32	10.28	10.04
0.5	10.87	10.88	10.59	10.46	10.39	10.33	10.29	10.04
0.6	11.05	11.03	10.65	10.50	10.41	10.34	10.30	10.04
0.7	11.33	11.23	10.71	10.53	10.43	10.36	10.31	10.04
0.8	11.80	11.50	10.79	10.57	10.46	10.38	10.33	10.05
0.9	12.78	11.85	10.88	10.62	10.48	10.40	10.34	10.05

($E(N_A^{20}|PW)$ denotes the expected number of patients on treatment A within 20 trials.)

The numerical results, given in table 7, clearly demonstrate the effect of u, the number of balls of each type contained in the box before sampling starts. If u is too large with respect to the number of trials needed to take the terminal decision, then nearly 50 % of the patients are treated with either treatment. This fact is immediately obvious because a large value of u effects that the administration of the treatments is essentially conducted by randomization (nearly equal probabilities for the treatments to be administered in the next trial!). The best choice of u, i.e. the greatest bias in favor of testing the better treatment, is obtained for u=0. The corresponding values for the PW-sampling scheme indicate that in this special case both sampling rules provide essentially the same.

8 Supplementary Investigations — Topics Requiring Further Research

Two more sampling rules already mentioned in the introduction have been investigated by simulation studies. The FL-sampling scheme has been applied together with the termination rules proposed by Hoel (cf. sec.1, chap.2), Berry/Sobel (cf. sec.2, chap.2), and Fushimi (cf. sec.5, chap. 2). It turned out that the FL-sampling scheme provides essentially the same results, i.e. we did not recognize any substantial difference between the PW- and the FL-sampling scheme. The PW-2F-sampling scheme has been applied together with the termination rules proposed by Hoel, Berry/Sobel, Fushimi, and Nordbrock (cf. sec.13, chap.1). We found that PW-2F provides no better results than the original PW-sampling, quite on the contrary, the P(CS)-values turned out to be slightly smaller than the corresponding values of the PW-procedures, and the expectations proved to be of essentially the same magnitude, except of the Berry/Sobel termination rule; for this selection model the PW-2F sampling scheme yields the threefold of the expectations of the corresponding PW selection model. For $m \geq 3$ the PW-mF-rule yields even worse results.

Almost all selection models considered in part 1 require an instantaneous response. The randomized play-the-winner rule presented in sec.7 of chap.3 is indeed applicable if delayed observations arise, but we did not investigate this case. The few selection models designed for delayed response, known so far, have termination rules mainly based on likelihood ratios, and the times to response are tacitly assumed to be independent of the response itself. Moreover the times to response are assumed to have negative exponential or geometric distributions in the first instance. The investigations carried out so far base on simulation studies.

The reader interested in this subject is referred to Hoel/Sobel/Weiss [97], and [98], Flehinger/Louis [60], and [61], and Flehinger/Louis/Miller [59]. The models presented there are occasionally fairly complicated, which may turn out to be disadvantageous for practical applications. This fact is by no means surprising because the development of the theory of sequential selection models allowing delayed response is still in leading strings.

There remains much to be done in this special field of statistics. For instance the response may be judged according to more than one criterion, and the relative worth of competing treatments reflected by the obtained responses will be determined by using a special preference order on the space of all possible response vectors. However, the more sophisticated the proposed selection model, the more difficulties arise in treating such a model analytically. Therefore, it is not surprising that most results concerning such models, known so far, base on simulation studies. The natural extension of dichotomous response models are continuous response models, i.e. the response obtained after administration of some treatment is measured by a continuously distributed random variable. Selection models allowing continuous response will be considered in the following part 2.

Part 2

Continuous Response Selection Models

Introduction

The second part of the textbook is reserved to procedures that are suitable for selecting the best from several - in the following $k \geq 2$ - alternatives in some well-defined sense of best, where the worth of the alternatives can be quantified by continuously distributed random variables, i.e. part 2 is reserved to the so-called continuous-response-selection models.

One -and Two - Factor- Designs

Whereas in part one all statistical experiments, used to quantify the alternatives, require constant success parameters during the whole experiment, i.e. random variables representing the statistical experiment must all have the same distribution, that is a binomial distribution with index one and success parameter p_i, if treatment A_i is used, in this part we also consider statistical experiments, where the random variables representing some alternative have not necessarily all identical distributions. Therefore, in this part, the notions "population" and "treatment" are of different meaning.
In order to explain the special meaning of the two notions in this part, we consider the following example: A farmer wants to select from k varieties of wheat that which yields the highest average amount. In doing this,the following experimental design will be possible: There exists one homogeneous field which can be partitioned into k homogeneous parcels. For some year the k varieties of wheat will be sown to the k parcels and the yearly wheat-harvests of the k varieties will be weighted and noted. If the experimental conditions remain constant during the years of investigation, then, apart from a lot of small influences, only one, the interesting factor "variety of wheat", influences the results of the experiments. But if on the other hand the conditions vary extremely from year to year - maybe it is not possible to use the same field every year, or there are substantial differences in the weather conditions during the years of investiga-

tion, - then apart from the combined small influences, the magnitude of the wheat-harvests is affected by several factors. Corresponding to the two different models, we distinguish between one-factor- and multi-factor-designs.

In the first case, we speak of a population-model in the following. Moreover, it is assumed that the random experiment, yielding the score for some alternative, may be repeated infinitely often under the same conditions. Instead of k alternatives, we speak of k populations. In the preceding example, the parcels planted with the k varieties of wheat in some year are the elements of the populations. The observable variable, the random variable, is the wheat-harvest of any parcel.

Among the multi-factor-models, we only consider the two-factor-model which is often called the "randomized-block-design" in literature. One block (the field in some year) consists of k fields (parcels), to which k treatments (the varieties of wheat) are attached randomly. The effects of the treatment and the block on the variable "wheat-harvest" , are called treatment-effect and block-effect, respectively. In the preceding example the random variable "wheat-harvest" is affected by the interesting factor "variety of wheat" and for instance by the non-interesting factor "condition of the soil in some special year". To distinguish from the one-factor-model, in case of two-factor-models, we speak of treatments instead of populations.

According to the distinction between the two experimental designs, the k alternatives are called populations or treatments, respectively, and the random variable $X_i, 1 \leq i \leq k$, representing the alternative $\Pi_i, 1 \leq i \leq k$, is hence either a linear function of the population mean and some random residual or a (linear) function of the interesting effect of the treatment Π_i, the block-effect, and some random residual.

In the following we assume that each of the k alternatives $\Pi_i, 1 \leq i \leq k$, can be represented by the realization $\vec{x}_i := (x_{i1}, \ldots, x_{in})$ of a random vector $\vec{X}_i := (X_{i1}, \ldots, X_{in})$ of dimension n. In case of a population-model, the n components of the random vector have all the same distribution function $F_i, 1 \leq i \leq k$, whereas in a two-factor-design the distribution functions $F_{i1}, \ldots, F_{in}$ of the components may be different. In either case the random vector $\vec{X}_i$ is called a random sample and its dimension n is, as usual, referred to as sample size.

The Aim of Selection

After the agreement, whether and which one of the two experimental designs described above is appropriate, the meaning of "best" can be specified. As a matter of principle, this specification ought to be done prior to the beginning of samp-

ling. In case of a population-model the aim may be to select that population which is associated with the largest (smallest) mean or the smallest scale-parameter. If a two-factor-design is to be chosen, then the treatment having the greatest (smallest) positive effect on the observable variable is denoted as best.
Since the selection of the population with the smallest mean or the treatment with the smallest effect can be done in exactly the same way as selecting the alternative with the greatest parameter, we also denote the alternative with the greatest parameter as best.
Throughout this part of the textbook we assume that the random variables X_{ij}, $1 \le i \le k, 1 \le j \le n$, have distribution functions $F_{ij}(x)$, $1 \le i \le k$, $1 \le j \le n$, that differ only in their location-parameters $\Delta_{ij}, 1 \le i \le k, 1 \le j \le n$, and/or scale parameters $\sigma_{ij}, 1 \le i \le k$, $1 \le j \le n$, i.e. $F_{ij}(x)=F(x,\Delta_{ij},\sigma_{ij})$, where F denotes some continuous distribution function. If an one-factor-design is given, then Δ_{ij} is equal to Δ_i for all $j=1,\ldots,n$, whereas in a two-factor-design Δ_{ij} is a linear function of the treatment-effect Δ_i and the block-effect Δ_j.

Making the additional assumption that F is the standard normal distribution function, beginning with Bechhofer [28], a great number of statistical selection procedures with various fixings of an aim have been proposed and developed to be applicable in practice. Extensive bibliographies are given by Bechhofer, Kiefer and Sobel [31] and by Gibbons, Olkin and Sobel [70]. Besides the consideration of selection procedures applicable in case of other special distribution functions (e.g. gamma-distribution), during the last ten years the investigation of procedures without special assumptions concerning the distribution function, i.e. the investigation of nonparametric or distribution-free selection procedures, has reached much interest.

Procedures Based on Rank-Order Statistics

Like the first proposal of a distribution-free selection procedure in continuous response models, made by E.L.Lehmann [116], nearly all procedures proposed so far are based on so-called ranks, where in case of k random vectors it must be distinguished between joint ranks and the ranks of the components within each of the k random vectors, which we call single ranks.
Prior to the definition of joint and single ranks, we generally define the rank-vector $R:=(R_1,\ldots,R_m)$ associated with some random vector $\vec{X}:=(X_1,\ldots,X_m)$. The rank $R_i, 1 \le i \le m$, of the random variable $X_i, 1 \le i \le m$, is defined as the number of random

variables $X_j, 1\le j\le m$, satisfying the condition $X_j \le X_i$. Therefore the components $R_i, 1\le i\le m$, of the rank vector R may take only values from the set $\{1,2,\dots,m\}$. Because of the assumption that the components $X_i, 1\le i\le m$, are independent and continuously distributed, we know that $X_i \ne X_j$ with probability 1, for all $1\le i\ne j\le m$. Therefore the rank-vector R is an element of $\mathfrak{S}_m$ with probability one, where $\mathfrak{S}_m$ is the symmetric (permutations-)group of order m.
Suppose there are given the realizations of the three random vectors $\vec{X}_1 = (X_{11},X_{12},X_{13})$, $\vec{X}_2=(X_{21},X_{22},X_{23})$ and $\vec{X}_3=(X_{31},X_{32},X_{33})$. The corresponding single rank-vectors are elements of $\mathfrak{S}_3$ with probability one. To determine the joint rank vector associated with the three vectors $\vec{X}_1,\vec{X}_2$ and $\vec{X}_3$, we must combine the three vectors to the joint vector $\vec{X}:=(\vec{X}_1,\vec{X}_2,\vec{X}_3)$, having dimension 9. The joint rank $R_{ij}, 1\le i,j\le 3$, of the component $X_{ij}, 1\le i,j\le 3$, is defined as the number of components $X_{\ell r}, 1\le \ell,r\le 3$, satisfying $X_{\ell r}\le X_{ij}, 1\le \ell,r,i,j\le 3$, and the joint rank vector $R=(R_{11},R_{12},\dots,R_{33})$ is an element of $\mathfrak{S}_9$ with probability 1.
All selection procedures to be considered in this part are based on statistics that are functions of the joint rank-vectors or the single rank-vectors associated with the k alternatives under investigation.

The Selection Models

The procedure proposed by E.L.Lehmann, and also some modifications of it, need the assumption that the populations - the mentioned procedures are only applicable in population-models - have symmetric distributions and differ only in their location-parameters. The best population is that, which has the greatest location-parameter. If the procedure leads to the selection of the best population, then we speak of a correct selection (CS).
One type of selection procedures are based on the indifference-zone-model, first proposed by Bechhofer [28].
If

$$\Delta_{(1)} \le \Delta_{(2)} \le \dots \le \Delta_{(k-1)} \le \Delta_{(k)}$$

are the ordered location-parameters of the given k populations, then it is assumed that the two greatest parameters satisfy the inequality $\Delta_{(k)} > \Delta_{(k-1)}$. The vector $\Delta=(\Delta_1,\dots,\Delta_k)$ is called the parameter-configuration. The aim of the procedures is to select the population associated with $\Delta_{(k)}$. It is sometimes possible that a population is selected, which is not associated with the greatest location-parameter. In this case, we speak of an incorrect selection.

Because of the aim to find a selection procedure which maximizes the probability of CS, formally P(CS), whenever $\Delta_{(k)} > \Delta_{(k-1)}$, P(CS) is a measure of quality of the selection procedures. Starting with the opinion, that a discrimination between two populations with respect to their worth, is meaningful only, if the difference between the corresponding location-parameters is greater than some minimal value Δ^*, in the indifference-zone-model the set of all possible parameter-configurations is divided into a preference-zone (PZ), which contains all configurations Δ, satisfying $\Delta_{(k)}-\Delta_{(k-1)}\geq\Delta^*$, and an indifference-zone (IZ), which is the set of all configurations Δ, satisfying the inequality $\Delta_{(k)}-\Delta_{(k-1)}<\Delta^*$. Thereby Δ^* denotes some real number, determined with respect to economic aspects and in dependence on the given problem, which gives the limit beyond that the selection of the population associated with $\Delta_{(k)}$ is strongly preferred. The above mentioned partition of the parameter-space can be graphically described as follows.

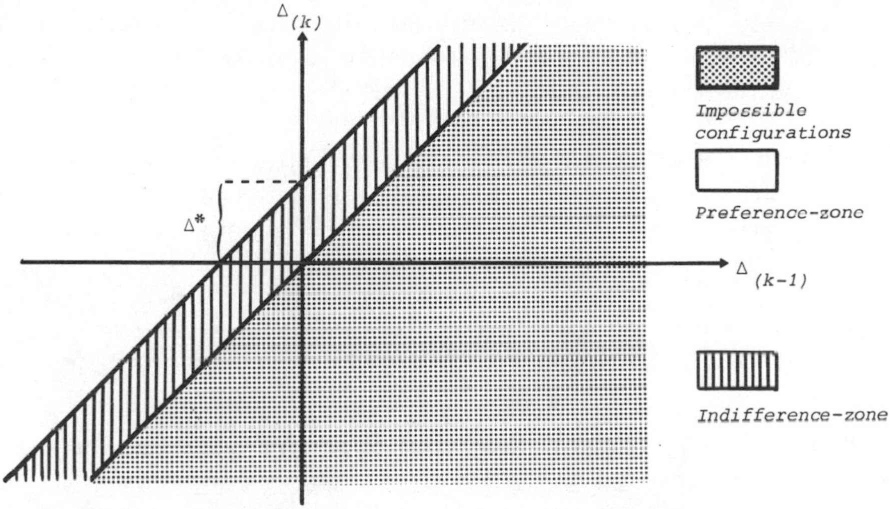

For the parameter-configurations within the PZ the probability $P(CS|\Delta)$ should be as great as possible.
As already indicated, the probability P(CS) takes different values in dependence on the configuration $\Delta\varepsilon PZ$. Therefore, we are especially interested in configurations of PZ, that yield the minimal value of P(CS). Such configurations are denoted as least favorable configurations (LFC). If Δ_{LFC} is such a LFC, then it follows:

$$P(CS|\Delta) \geq P(CS|\Delta_{LFC}), \text{ for all } \Delta\varepsilon PZ.$$

There are two groups of procedures based on the indifference zone model. The first includes procedures with fixed sample size. For given constant Δ^* and mini-

mal probability P^*, $1/k \leq P^* \leq 1$, the sample size n must be determined such that $P(CS|\Delta) \geq P^*$, for all $\Delta \varepsilon PZ$. To guarantee that this probability requirement is always met, n must be chosen such that $P(CS|\Delta_{LFC}) = P^*$. Therefore, to determine n, at least one LFC must be known.
Rizvi and Woodworth [154] have proved, that in case of indifference-zone-procedures based on joint ranks, no LFC can be given explicitly in general. Additionally, as also in case of modifications, which are based on single ranks and where the LFC can be determined in general (see for example in Randles [149]), the difficulty arises that the probability $P(CS|\Delta_{LFC})$ depends upon the unknown distribution of the populations in research. For this reason, indifference-zone-procedures with fixed sample size are not considered in this textbook.

The second group of indifference-zone-procedures consists of sequential selection procedures, where the sample size n is the realization of a random variable, the so-called stopping-variable N. On principle, these procedures are applicable in one- and two-factor-designs.

Another type of selection procedures are the so-called subset-selection procedures They are based on a selection-model, proposed by Gupta [78], where the set of possible configurations is not divided into two parts. The most substantial difference in comparison with the first type of procedures is that not a fixed number of alternatives but a subset including a random number of alternatives is selected. That means, in applying a subset-selection procedure the number T of selected alternatives is a random variable.

Nonparametric Subset-Selection Procedures

Depending on the experimental design, three types of subset-selection procedures will be considered.
In the first case, a population-model is assumed, where the distribution functions of the considered populations are all identical except of a single parameter θ_i, $1 \leq i \leq k$, (location- or scale-parameter) which allows an ordering of the populations in the following way:

$$\theta_\ell < \theta_j \Longrightarrow F(x,\theta_\ell) \not\equiv F(x,\theta_j) \text{ and } F(x,\theta_\ell) \geq F(x,\theta_j), \quad \forall x, 1 \leq \ell, j \leq k.$$

This condition is to be illustrated by the figures below showing the normal distribution function in case of (a) two different location-parameters and (b) two different scale-parameters. Obviously, in case (a) the condition is satisfied, whereas in case (b) it fails to hold.

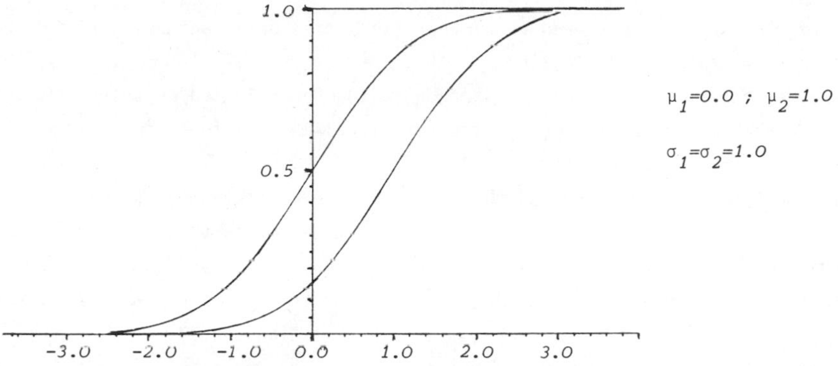

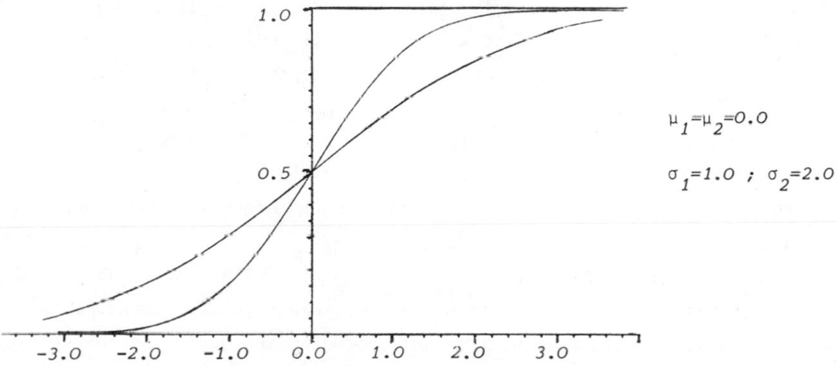

In the considered design, the best population is that, which is associated with the greatest parameter $\theta_{(k)}$, where $\theta_{(1)} \leq \theta_{(2)} \leq \ldots \leq \theta_{(k)}$ are the ordered parameters. We speak of a correct selection (CS) if the selected subset includes the best population. The probability condition in case of subset selection procedures runs as follows: the probability P(CS) is not smaller than the preassigned minimal probability P^*, for each possible configuration, $1/k \leq P^* < 1$. This requirement is met, if the minimal probability of CS is equal to P^*. Therefore, we have to determine the LFC for subset-selection procedures, too.

McDonald [121], and Gupta and McDonald [80] have proposed three classes of decision rules, which can be used for the implementation of non-parametric subset-selection procedures. For two of the mentioned classes, a LFC is given, if the

two greatest parameters are equal, whereas for the third class a LFC is given if all k parameters coincide.
Each of the three classes uses rank-order statistics based on the joint ranks associated with the k populations.
Rank-order statistics are sums of scores, which are evaluated in using some pre-assigned scorefunction. In general the scorefunction is a function of the inverse of some continuous distribution function.
Hâjek and Sidâk [86] have proved that the efficiency of the well-known tests based on rank-order statistics is influenced by the choice of the special score-function. Therefore, in this textbook the question is discussed, whether and how much the form of the applied statistic influences the efficiency of subset-selection procedures. Efficiency in case of subset-selection procedures means the property to select a subset of minimal size in using samples of fixed size and satisfying the probability condition inf $P(CS) \geq P^*$.
Let V1 and V2 be two comparable selection procedures, both satisfying the condition inf $P(CS) \geq P^*$, then V1 is said to be more efficient than V2, if $E(T1) \leq E(T2)$, where T1 and T2 are the random numbers of selected populations when the procedures V1 and V2 are used, respectively.
The second design, for which we consider subset-selection procedures, is also a population model. But it is now not required that the distribution functions can be ordered as in the first case, i.e. in the present model location- and scale-parameters of the k populations may be different.
To solve this general Behrens-Fisher-problem two generalized procedures based on rank-order statistics, too, are derived. The efficiency of this generalized class of subset-selection procedures is also influenced by the choice of the special scorefunction.

Finally we will consider nonparametric subset-selection procedures that are applicable in randomized-block-designs.
Extended Monte-Carlo-studies give rise to the assumption that also in case of randomized-block-designs the effiency of subset-selection procedures based on rank-order statistics is influenced by the choice of the scorefunction.

Although the subset-selection procedures are not appropriate to select exactly one, i.e. the best, alternative, they may be very useful instruments to reach a decision. In practical selection problems the alternatives must be often compared with respect to more than one point of view. If for example from a number of varieties of grapes the best should be selected, then several aspects must be taken into account in order to quantify the relative worth of the alternatives.

Depending on the preference order, the interesting attributes of quality may be considered in sequence. If the highest amount of wine has priority, then a pre-selection with respect to this criterion can be done by using a subset-selection procedure. Afterwards the varieties of grapes included in the selected subset can be compared with respect to further criteria as for example capability of resisting against frost and destructive insects, attributes of taste etc., for selecting the over-all best variety of grapes.

Nonparametric Sequential Selection Procedures

As already mentioned above the construction of sequential selection procedures is possible when using the indifference-zone-model.
In his thesis, Geertsema [66] has proposed methods for constructing sequential confidence intervals of fixed length for the location-parameter of a symmetrically distributed population and in applying this idea, he [67] has formulated a non-parametric sequential selection procedure.
Later on, Swanepoel and Geertsema [191] have given a general class of nonparametric sequential selection procedures which are applicable to population-models with symmetrical distributions. Thereby, they have used the theory, developed by Sen and Ghosh [167], for constructing sequential confidence intervals of fixed length by using general one-sample rank-order statistics, i.e. rank-order statistics which are functions of single ranks.

All selection procedures included in the mentioned class are based on the following idea: From k symmetrically distributed populations, on the premises of equal scale-parameters, that population ought to be selected which is associated with the greatest location-parameter. Because of the assumption of an indifference-zone-model only such parameter-configurations are of interest which satisfy: $\Delta_{(k)}-\Delta_{(k-1)}\geq\Delta^*$. In using an appropriate estimator $\hat{\Delta}_{(i)}$ for each location-parameter $\Delta_{(i)}, 1\leq i\leq k$, a confidence-interval of the fixed length Δ^* is constructed in such a way that each pair of confidence intervals contains the two parameters, they are constructed for, with probability P*. From this and $\Delta_{(k)}-\Delta_{(k-1)}\geq\Delta^*$ it follows that the inequality $\hat{\Delta}_{(k)}\geq\hat{\Delta}_{(k-1)}$ is satisfied with probability P*. That means, $P(CS|\Delta_{LFC})\geq P^*$ holds true, if the population associated with the greatest estimate is selected. The following figure may illustrate the situation described above, if a LFC is given.

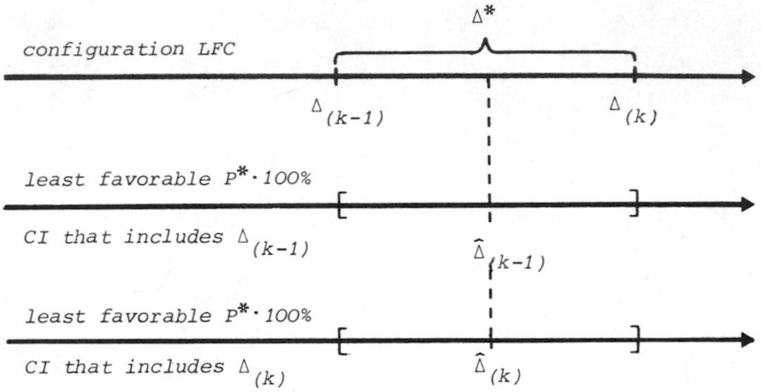

The sequential selection procedure consists essentially in the sequential construction of confidence-intervals of fixed length Δ^*, where at each stage of sampling with $n \geq n_0$—n_0 denotes the minimal sample size at which the procedure starts — it must be proved whether the aim: "confidence-intervals of fixed maximum length Δ^*" is reached or not. If the aim is reached then the selection procedure terminates, otherwise the procedure is continued at the next stage of sampling. This rule is generally called the stopping-rule. The possible realizations of the stopping variable N are the integers not less than n_0.

Of course, this type of selection procedure is only convenient if it actually stops at any stage of sampling. For the mentioned procedures it can be proved that $P(N<\infty)=1$, i.e. the sequential selection procedures terminate with probability 1. The used estimators $\hat{\Delta}_{(i)}, 1 \leq i \leq k$, are functions of general one-sample rank-order statistics. For this reason the influence of the chosen scorefunction on the efficiency of the sequential procedures is of interest. Efficiency in this connection must be seen with respect to the stopping behaviour of the procedures. If two procedures V1 and V2 satisfy the same probability condition, then procedure V1 is called more efficient than procedure V2, if the means of their stoppin variables N1 and N2, respectively, satisfy the inequality: $E(N1) < E(N2)$.

The sequential selection procedures for one-factor-models can be generalized in such a way that they are also applicable in two-factor-designs.
As already seen, in two-factor-designs apart from the interesting treatment a second factor influences the observable variables. To be able to select the best treatment, the unknown and non-interesting effect of the second factor must be

eliminated. Because of the linear form of the observable random vectors, this elimination can be reached by constructing appropriate vectors of differences. The distributions of the components of such difference-vectors are always symmetric. Therefore, the generalized procedure can also be applied if the distributions of the original random variables are not symmetric.

Contents of Part 2

This part is also divided into three chapters, where in the first two chapters the nonparametric selection procedures are considered and in chapter 3 two methods are described that allow the construction of so-called adaptive nonparametric selection procedures in continuous-response-models.

Chapter 1 deals with subset-selection procedures based on rank-order statistics. After the definition of linear rank-order statistics and the derivation of the asymptotic distribution of two special types of such statistics in case of non-identically distributed populations, depending on the experimental design, three groups of subset-selection procedures are considered. For each of the considered classes of subset-selection procedures extended Monte-Carlo-studies are carried out in order to investigate the influence of the used scorefunction on the efficiency of the procedures. One section is reserved to the analytic investigation of the optimality of subset-selection procedures.

Chapter 2 deals with sequential selection procedures based on the indifference-zone model. Since the procedures which are only usable in population-models may be considered as special case of the procedures also applicable in two-factor-designs, these generalized procedures take the greatest part of chapter 2. Though these procedures are only asymptotic distribution-free and terminate with probability one only, they are nevertheless very useful for solving practical selection problems. This result is founded on extended Monte-Carlo-studies, which also show that the efficiency of the sequential selection procedures is influ enced by the choice of the special scorefunction.

Adaptive nonparametric selection procedures are procedures which allow a preliminary selection of the "best" scorefunction by using the same data, which are then used to select the best alternative or a subset, which contains the best alternative. Methods which can be used to select the optimal scorefunction are considered in chapter 3.

CHAPTER 1

Subset-Selection Procedures Based on Linear Rank-Order Statistics

1 Linear Rank-Order Statistics and their Asymptotic Distributions

1.1 The general linear rank-order statistic

Let $R=(R_1,R_2,\ldots,R_m)$ be a random vector as described in the introduction, associated with the random vector $(X_1,X_2,\ldots,X_m)$. The general linear rank-order-statistic based on R is defined by

$$S:=\sum_{i=1}^{m} c_i a(R_i) , \qquad (1.1)$$

where the constants c_i, $i\varepsilon\{1,\ldots,m\}$, are called regression constants. The function $a:\{1,2,\ldots,m\}\rightarrow \mathbb{R}$ yields the so-called scores $a(i)$. We sometimes write a_i instead of $a(i)$.

remark: $a(R_i)$ and $a(i)$ are different in general.

This form of statistics is widely used in statistical inference, especially in testing statistical hypotheses.
Let $X_1,X_2,\ldots,X_m$ be independent random variables, where X_i describes the outcome of the i-th performance of some random experiment. We cannot generally expect the experimental conditions to remain unchanged at each performance of the random experiment. Properly speaking, it seems to be obvious that no experiment may be carried out twice under exactly the same conditions, at least a small variation is more plausible than the belief in the ability to reproduce the same experimental conditions at each performance of the random experiment.

Let X_i denote the outcome of the i-th performance of the experiment. If the experimental conditions remain unchanged, then X_i and X_j, $i \neq j$, have the same distribution. On the other hand, the experimental conditions may be altered by the experimenter himself, that is the case if one is interested in the question whether a variation of the experimental conditions causes a variation of the distributions of the X_i's. In rolling a die, the time of day has no influence on the outcome of that experiment, whereas the material of the table, on which the die is rolled, may have a strong influence on the possible outcome. Recalling what we have considered throughout part 1, further fundamental examples of testing whether the variation of the experimental conditions causes a variation of the distributions of the X_i's are immediately available. In part 1, a variation of the experimental conditions means to change the treatment on which a single trial is carried out, and a variation of the distributions of the X_i's means that the success parameter of the binomial chance variables X_i (index n=1) has changed. In part 2, where the response X_i is assumed to be continuous, a variation of the distributions of the X_i means for example that the median of the distributions is no longer the same.

Our goal is, therefore, to develop a procedure that allows us to detect existing connections between variations of the experimental conditions on the one hand and variations of the distributions of the random variables X_i on the other hand. Such a procedure can be based on the general linear rank-order statistic S defined above. The quality of the procedure is highly connected with the choice of the scores, and that choice again depends mainly on the underlying distribution of the X_i and on which parameter is assumed to be influenced by the experimental conditions. Most frequently that parameter will be some location- and/or some scale-parameter.

Prior to some remarks upon the distribution of S, we derive mean and variance of the general linear rank-order statistic (r.-o.s.) in case that R is uniformly distributed over $\mathcal{T}_m$.

For the mean of S, we can write

$$E(S)=E(\sum_{i=1}^{m} c_i a(R_i))=\sum_{i=1}^{m} c_i \; E(a(R_i)).$$

With A4/2 it follows

$$E(a(R_i))=\sum_{j=1}^{m} a_j \; P(R_i=j)=\sum_{j=1}^{m} \frac{a_j}{m} =:\bar{a}$$

and therefore

$$E(S)=\sum_{i=1}^{m} c_i\bar{a}. \tag{1.2}$$

In addition

$$Var(S)=\sum_{i=1}^{m} c_i^2\ Var(a(R_i))+\sum_{i=1}^{m}\sum_{\substack{j=1\\ j\neq i}}^{m} c_ic_j\ Cov(a(R_i),a(R_j)).$$

Again with A4/2, we obtain:

$$Var(a(R_i)) = E((a(R_i)-\bar{a})^2)=\sum_{j=1}^{m}\frac{(a_j-\bar{a})^2}{m}$$

and

$$Cov(a(R_i),\ a(R_j))=E((a(R_i)-\bar{a})(a(R_j)-\bar{a}))=\sum_{\ell=1}^{m}\sum_{\substack{h=1\\ h\neq\ell}}^{m}\frac{(a_\ell-\bar{a})(a_h-\bar{a})}{m(m-1)}$$

$$=\frac{1}{m(m-1)}\left(\sum_{\ell=1}^{m}(a_\ell-\bar{a})\right)^2-\frac{1}{m(m-1)}\sum_{\ell=1}^{m}(a_\ell-\bar{a})^2$$

$$=-\frac{1}{m(m-1)}\sum_{\ell=1}^{m}(a_\ell-\bar{a})^2.$$

Thus the variance of S is given by

$$Var(S)=\frac{1}{m(m-1)}\sum_{\ell=1}^{m}(a_\ell-\bar{a})^2\left((m-1)\sum_{i=1}^{m}c_i^2-\sum_{i=1}^{m}\sum_{\substack{j=1\\ j\neq i}}^{m}c_ic_j\right)$$

$$=\frac{1}{m-1}\sum_{i=1}^{m}(c_i-\bar{c})^2\sum_{i=1}^{m}(a_i-\bar{a})^2. \tag{1.3}$$

In case that R is uniformly distributed over $\mathcal{T}_m$, the distribution of S can be given exactly for some special scores. The exact determination of the distribution of S is such laborious for moderate m that even the enormous capacities of modern computers are not sufficient.
A rather good approximation of the probability $P(S\leq s)$ is given by the normal approximation.
Let $J(u)$, $0<u<1$, be a real-valued function, which is either nondecreasing for $u\varepsilon(0,1)$, or nonincreasing for $u\varepsilon(0,x]$ and nondecreasing for $u\varepsilon(x,1)$, for some $x\varepsilon(0,1)$. Moreover J satisfies:

$$0<\int_0^1(J(u)-\bar{J})^2du<\infty,\ \text{with}\ \bar{J}:=\int_0^1 J(u)du.$$

In the following we consider scores[1] given by

$$a(i)=J\left(\frac{i}{m+1}\right),\quad 1\le i\le m,$$

and we say the scores are generated by the function J. J is called scorefunction or sometimes score-generating function.

According to S, we define the sequence of rank-order statistics

$$S_m := \sum_{i=1}^{m} c_{mi} J\left(\frac{R_{mi}}{m+1}\right).$$

Then with A4/5 it follows: If

$$\lim_{m\to\infty} \frac{\max\limits_{1\le i\le m} \{(c_{mi}-\bar{c}_m)\}}{\sum\limits_{i=1}^{m} (c_{mi}-\bar{c}_m)^2} = 0, \tag{1.4}$$

then

$$\lim_{m\to\infty} \sup_{-\infty<s<\infty} \left|P(S_m<s) - \Phi\left(\frac{s-E(S_m)}{\sqrt{Var(S_m)}}\right)\right| = 0. \tag{1.5}$$

If (1.5) holds, then S_m is said to be asymptotically normal distributed with parameters $E(S_m)$ and $\sqrt{Var(S_m)}$. In the following, we usually ignore the subscript m, and thus the random variable $(S-E(S))/\sqrt{Var(S)}$ is said to be asymptotically N(0,1) distributed, i.e. $(S-E(S))/\sqrt{Var(S)}$ has asymptotically the standard normal distribution.

There have been made many investigations concerning the accuracy of the normal approximation. Most of them are empirical and heuristic considerations about the size of m necessary to guarantee an acceptable small difference between the exact distribution and the asymptotic one. The results obtained so far are rules of thumb that provide lower bounds for m, beyond which the approximation is accurate enough. The bounds given in literature[2] are such that we can easily make use of the normal approximation for solving the problems to be considered in the following.

1.2 Some special linear rank-order statistics

In this section we consider the asymptotic distribution of the linear r.-o.s.

---

1 *In Chapter 2 we consider scores which are somewhat different.*

2 *See for example GIBBONS [69].*

S in different situations. Making the assumption that a population-model is appropriate, we specify the distribution of S for some special scorefunctions J in case that the cdf's of the k populations $\Pi_1,\ldots,\Pi_k$ are identical. For two of these functions J, we also give the asymptotic distribution of S in case that the cdf's of the k populations are different.

Let $F_1,F_2,\ldots,F_k$ be the cdf's of the k populations $\Pi_1,\Pi_2,\ldots,\Pi_k$, respectively, and let $\vec{X}_1,\vec{X}_2,\ldots,\vec{X}_k$ be the k random samples of size n, drawn from the k populations. Then the random vector $\vec{X}$ (total sample) is defined by

$$\vec{X}:=(X_1,X_2,\ldots,X_m):=(X_{11},X_{12},\ldots,X_{1n},X_{21},\ldots,X_{2n},\ldots,X_{k1},\ldots,X_{kn}), \quad (1.6)$$

and the total sample size is m=nk.

As outlined in the introduction, one possible way of selecting a random subset of populations can be based on the comparison of each population $\Pi_i, i\varepsilon\{1,\ldots,k\}$, with a hypothetical population resulting from the combination of the remaining k-1 populations Π_j, $j=1,\ldots,k$, $j\neq i$. Because of the infinity of the populations $\Pi_1,\ldots,\Pi_k$, an element ω drawn from the hypothetical population may be drawn from each population Π_j, $1\leq j\neq i\leq k$, with probability 1/(k-1). Hence, the cdf of the hypothetical population is given by

$$G(x):=\frac{1}{k-1}\sum_{\substack{j=1\\ j\neq i}}^{k} F_j(x). \quad (1.7)$$

Without loss of generality, in the following we consider the comparison of population Π_1 with the hypothetical population consisting of the populations $\Pi_2,\ldots,\Pi_k$.

For reason of simplification, the cdf of population Π_1 will be denoted by F in the remaining section. Thus "$F(x)=G(x)$, $\forall x\varepsilon\mathbb{R}$" is equivalent to "$F_1,F_2,\ldots,F_k$ are identical".

We are interested in the asymptotic distribution of the r.-o.s. S based on the random vector $\vec{X}$, given in (1.6), and on the following regression constants:

$$c_{11}=1,\ldots,c_{1n}=1,c_{21}=0,\ldots,c_{2n}=0,\ldots,c_{k1}=0,\ldots,c_{kn}=0\,. \quad (1.8)$$

With (1.8) the r.-o.s. S can be written as

$$S=\sum_{i=1}^{n} a(R_i), \quad (1.9)$$

where $R=(R_1,\ldots,R_m)$ is the rank vector associated with the random vector $\vec{X}$ defined by (1.6).

If $F(x)=G(x)$, $\forall x \in \mathbb{R}$, then by recalling A4/4, R is known to be uniformly distributed over $\mathcal{T}_m$, and thus we have the following

Theorem 1.1:

Let S be given by (1.9), and let the scores a_i, $1 \leq i \leq m$, be generated by a scorefunction J. Then for each $\varepsilon > 0$ there exists a $M:=M(\varepsilon,J)$ such that $nk/(k-1) > M$ implies

$$\sup_{-\infty<S<\infty} \left|P(S \leq s) - \Phi\left(\frac{s-E(S)}{\sqrt{Var(S)}}\right)\right| < \varepsilon .$$

Proof:

Let c_{ij} be the regression constants given by (1.8), and define $c_1:=c_{11},\ldots,c_n:=c_{1n},c_{n+1}:=c_{21},\ldots,c_m:=c_{kn}$, then

$$\bar{c}:=\frac{1}{m}\sum_{i=1}^{m} c_i=\frac{1}{k},\quad \sum_{i=1}^{m}(c_i-\bar{c})^2=\frac{n(k-1)}{k} \quad \text{and} \quad \max_{1\leq i\leq m}\{(c_i-\bar{c})^2\}=\left(\frac{k-1}{k}\right)^2 .$$

From this,we get

$$\frac{\max\{(c_i-\bar{c})^2\}}{\sum_{i=1}^{m}(c_i-\bar{c})^2}=\frac{k-1}{nk},$$

and consequently with $M=\varepsilon^{-1}$ the proof of the theorem follows from A4/5. ■

According to 1.1 S is said to be asymptotically normal distributed, if theorem 1.1 is valid.

We will make intense use of this general result, in considering the asymptotic distribution of S (given by (1.9)) for some special scorefunctions J.

Definition 1.2: (Wilcoxon-statistic)

Let the regression constants be given by (1.8), and let the scorefunction J be given by $J(u)=u$, $0<u<1$. The resulting r.-o.s.

$$SW:=\sum_{i=1}^{n}\frac{R_i}{m+1} \tag{1.10}$$

is usually referred to as Wilcoxon-statistic. ■

Let $F(x)=G(x)$, $\forall x \in \mathbb{R}$. Because of $0<\int_0^1 (u-\bar{J})^2 du = \frac{1}{12} < \infty$, $\bar{J}=\frac{1}{2}$, and by using theorem 1.1, (1.2) and (1.3), SW is easily seen to be asymptotically normal distributed with mean $E(SW)=\frac{1}{2}n$, and variance $Var(SW)=\frac{n^2(k-1)}{12(m+1)}$.

Instead of the statistic SW the so-called Mann-Whitney-statistic

$$S_{XY} := \frac{1}{m+1} \sum_{i=1}^{n} \sum_{j=n+1}^{nk} U_{ij}, \quad U_{ij} = \begin{cases} 1, & \text{if } X_i \geq X_j \\ 0, & \text{if } X_i < X_j \end{cases} \tag{1.11}$$

is often used in literature. It is easily seen that

$$SW = S_{XY} + \frac{n(n+1)}{2(m+1)} .$$

Thus the results concerning SW are also valid for S_{XY}, when substituting E(SW) by $E(S_{XY}) = \frac{n^2(k-1)}{2(m+1)}$.

The statistic S_{XY} is chosen to investigate the case of non-identical cdf's $F_1,\ldots,F_k$, i.e. when $F(x)\neq G(x)$ for at least one $x\in\mathbb{R}$.

Theorem 1.3:

Let S_{XY} be given by (1.11). With $p:=P(X_i>X_j,\ 1\leq i\leq n,\ n+1\leq j\leq m)$, we have:

(a) $E(S_{XY}) = p\,\frac{n^2(k-1)}{m+1}$,

$$Var(S_{XY}) = \frac{n^2(k-1)}{(m+1)^2}\,(p(1-p)+(n(k-1)-1)\beta+(n-1)\gamma), \text{ where}$$

$$\beta := Cov(U_{ik},U_{i\ell}) = \int_{-\infty}^{+\infty} (G(x))^2 dF(x) - p^2, \quad 1\leq i\leq n,\ n+1\leq k\neq \ell\leq m$$

$$\gamma := Cov(U_{ik},U_{jk}) = \int_{-\infty}^{+\infty} (F(x))^2 dG(x) - (1-p)^2, \quad 1\leq i\neq j\leq n,\ n+1\leq k\leq m.$$

(b) $\sqrt{n}(S_{XY}-E(S_{XY}))$ is asymptotically normal distributed with mean 0 and variance $nVar(S_{XY})$.

Proof:

(a) With $E(U_{ij})=p$, we get $E(S_{XY}) = p\,\frac{n^2(k-1)}{m+1}$.

The variance of S_{XY} can be written as

$$Var(S_{XY}) = \frac{1}{(m+1)^2}\,(n^2(k-1)\ Var(U_{ik}) + n^2(k-1)(n(k-1)-1)Cov(U_{ik},U_{i\ell}) +$$

$+ n^2(k-1)(n-1)\mathrm{Cov}(U_{ik},U_{jk}))$, $1\leq i\neq j\leq n$, $n+1\leq k\neq \ell\leq m$.

In using

$$\mathrm{Cov}(U_{ik},U_{i\ell})=E(U_{ik}U_{i\ell})-(E(U_{ik}))^2 = \int_{-\infty}^{+\infty} (G(x))^2 dF(x)-p^2,$$

$$\mathrm{Cov}(U_{ik},U_{jk})=E(U_{ik}U_{jk})-(E(U_{ik}))^2 = \int_{-\infty}^{+\infty} (F(x))^2 dG(x)-(1-p)^2,$$

the proof of part (a) is obvious.

(b) With $t(X_i,X_j):=U_{ij}$, $1\leq i\leq n$, $n+1\leq j\leq m$, the proof follows directly from A1/6.■

The statistic S_{XY} was used by Zaremba [201], in testing the general hypothesis, whether the probability $p:=P(X<Y)$ is equal to $\frac{1}{2}$, where X and Y are random variables with cdf's F and G respectively. This is equivalent to the question, whether the two populations have the same median, and in case of symmetric distributions this is even equivalent to the question, whether the two populations have the same mean.

We see from theorem 1.3 that for $F\neq G$ the variance of S_{XY} depends on the underlying cdf's F and G. Some lower and upper bounds for the variance of S_{XY}, in dependence upon p, are given by Birnbaum and Klose [41].

Another special form of the r.-o.s. S, given in (1.9), is formulated in

<u>Definition 1.4</u>: (Median-statistic)

Let the regression constants be given by (1.8), and let $J(u)=c(u-\frac{1}{2})$, $0<u<1$, where

$$c(x) = \begin{cases} 1, & \text{if } x\geq 0 \\ 0, & \text{if } x<0, \end{cases}$$

then

$$SM:=\sum_{i=1}^{n} c\left(\frac{R_i}{m+1}-\frac{1}{2}\right)$$

is called <u>Median-statistic</u>. ■

We consider the case that $F(x)=G(x)$, $\forall x\in\mathbb{R}$. In using $0<\int_0^1 (u-\bar{J})^2 du = \frac{1}{4}<\infty, \bar{J}=\frac{1}{2}$, theorem 1.1, (1.2) and (1.3), we find that SM is asymptotically normal distributed with mean

$$E(SM) = \begin{cases} \frac{1}{2}n & \text{, if } m \text{ even} \\ \frac{(m-1)n}{2m}, & \text{if } m \text{ odd} \end{cases} \text{ ,and variance } Var(SM) = \begin{cases} \frac{n^2(k-1)}{4(m-1)} & \text{,if } m \text{ even} \\ \frac{n^2(k-1)(m+1)}{4m^2}, & \text{if } m \text{ odd.} \end{cases}$$

Analogous to the Mann-Whitney-Statistic, we consider the asymptotic distribution of SM in case that $F(x) \neq G(x)$ for at least one $x \in \mathbb{R}$.

For this purpose recall that F is the continuous cdf of the first n random variables $X_1,..,X_n$ in (1.6) and that G is the continuous cdf of the remaining $(k-1)n$ random variables $X_{n+1},...,X_m$. Let Z denote the median of $X_1,...,X_m$, and let A denote the number of X_i's, $1 \le i \le n$, smaller than Z.
Then, to derive the asymptotic distribution of SM for $F \neq G$, we first consider the joint distribution of the random vector (A,Z). We find

Theorem 1.5:

With the preceding assumptions and random variables V and W given by

$$A = nF(c) + \sqrt{n}V, \quad Z = c + W/\sqrt{n}, \quad \text{where } nF(c) + n(k-1)G(c) = \frac{m}{2},$$

we have

$$\lim_{n \to \infty} P(A=a, Z=z) = f_{VW}(\sqrt{n}F(c) - a/\sqrt{n}, \sqrt{n}(z-c)),$$

where

$$f_{VW}(v,w) = \left(\frac{1}{4\pi^2}\left(\frac{f^2}{(k-1)F(1-F)G(1-G)} + \frac{(k-1)g^2}{F(1-F)G(1-G)} + \frac{2fg}{F(1-F)G(1-G)}\right)\right)^{\frac{1}{2}} \cdot$$

$$\cdot \exp\left(-\frac{1}{2}\left(v^2\left(\frac{1}{F(1-F)} + \frac{1}{(k-1)G(1-G)}\right) - 2vw\left(\frac{f}{F(1-F)} - \frac{1}{G(1-G)}\right) + \right.\right.$$

$$\left.\left. + w^2\left(\frac{f^2}{F(1-F)} + \frac{(k-1)g^2}{G(1-G)}\right)\right)\right).$$

For simplification, we have written f,g,F and G instead of f(c),g(c),F(c) and G(c), respectively.

Proof:

We make the unessential restriction that $m=kn=2r+1$, $r \in \mathbb{N}$, i.e. m is odd, and use the notations $n1:=n$ and $n2:=n(k-1)$. With the assumptions made above we get:

$$P(A=a,Z=z) = P(A=a,Z=z,Z=X_i, 1 \le i \le n1) + P(A=a,Z=z,Z=X_j, n1+1 \le j \le n1+n2)$$

$$= n1 \binom{n1-1}{a}\binom{n2}{r-a} F(z)^a(1-F(z))^{n1-a-1} G(z)^{r-a}(1-G(z))^{n2-r+a} f(z) +$$

$$+ n2 \binom{n1}{a}\binom{n2-1}{r-a} F(z)^a(1-F(z))^{n1-a}G(z)^{r-a}(1-G(z))^{n2-r+a-1} g(z).$$

Using Stirling's formula (cf. A2/14), and taking logarithms, we get:

$$P(A=a,Z=z) = \left(\frac{n1n2}{4\pi^2 a(n1-a)(r-a)(n2-r+a)}\right)^{1/2} \Delta(n1,n2) \cdot$$
$$\cdot \exp(-n1H(p_1^*,F)-n2H(p_2^*,G)) \cdot \left(\frac{(n1-a)f(z)}{1-F(z)} + \frac{(n2-r+a)g(z)}{1-G(z)}\right), \tag{1.12}$$

where

$$\Delta(n1,n2):=\exp(\theta(n2)+\theta(n1)-\theta(a)-\theta(r-a)-\theta(n1-a)-\theta(n2-r+a)),$$

$$p_1^*:=\frac{a}{n1},\ p_2^*:=\frac{r-a}{n2},\ H(x,y):=x\ell n\frac{x}{y}+(1-x)\ell n\frac{1-x}{1-y}.$$

With p_1^* and p_2^* close to F(z) and G(z) respectively, the right side of (1.12) can be written in another form.
The function H(x,y) is analytic in (0,1) x (0,1). Thus in using first and second partial derivatives of H(x,y) with respect to x, and noting that H(y,y)= =H'(y,y)=0, we get:

$$H(p_1^*,F(z)) = \frac{1}{2F(z)(1-F(z))}(p_1^*-F(z))^2 + O(|p_1^*-F(z)|^3), \text{ and}$$
$$H(p_2^*,G(z)) = \frac{1}{2G(z)(1-G(z))}(p_2^*-G(z))^2 + O(|p_2^*-G(z)|^3). \tag{1.13}$$

With the definition of Z and in using Taylors formula again, we have:

$$F(z)=F(c)+\frac{w}{\sqrt{n1}}f(c)+o\left(\frac{1}{\sqrt{n1}}\right),\ G(z)=G(c)+\frac{w}{\sqrt{n1}}g(c)+o\left(\frac{1}{\sqrt{n1}}\right). \tag{1.14}$$

Using $p_1^* = \frac{n1F(c)+\sqrt{n1}v}{n1}$, $p_2^* = \frac{n2G(c)-\sqrt{n1}v-1/2}{n2}$, (1.12), (1.13), and (1.14), we get:

$$\lim_{n1,n2\to\infty} P(A=a,Z=z) = \left(\frac{1}{4\pi^2 B1}\right)^{1/2} \exp(-\tfrac{1}{2}B2), \quad \text{where} \tag{1.15}$$

$$B1:=\frac{n1^2f^2+n2^2g^2+2n1n2fg}{n1n2F(1-F)G(1-G)}, \text{ and}$$

$$B2:=v^2\left(\frac{1}{F(1-F)}+\frac{n1}{n2(G(1-G))}\right)-2vw\left(\frac{f}{F(1-F)}-\frac{g}{G(1-G)}\right)+w^2\left(\frac{f^2}{F(1-F)}+\frac{n2g^2}{n1G(1-G)}\right).$$

This proves the theorem. ■

The following theorem gives the asymptotic distribution of SM in case that $F \neq G$.

Theorem 1.6:

With the assumptions made in theorem 1.5, the random variable $SM^* = \sqrt{n}(SM-E(SM))$, $E(SM)=E(n-A)$, is asymptotically normal distributed with mean 0 and variance

$$Var(SM^*) = n^2 Var(V) = n^2(k-1)(G(1-G)+(k-1)(\tfrac{g}{f})^2 F(1-F))/(1+\tfrac{g(k-1)}{f})^2 .$$

Proof:

With $\sqrt{n}(SM-E(SM))=\sqrt{n}((n-A)-E(n-A))= -nV$ the first part of the theorem follows from theorem 1.5.

The general quadratic form of a two-dimensional normal density is

$$\alpha(v-\mu_1)^2 - 2\beta(v-\mu_1)(w-\mu_2) + \gamma(w-\mu_2)^2, \qquad (1.16)$$

where μ_1,μ_2 are the means of V,W respectively,

$$\alpha := \frac{1}{(1-\rho^2)\sigma_1^2}, \quad \beta := \frac{\rho}{(1-\rho^2)\sigma_1^2\sigma_2^2}, \quad \gamma := \frac{1}{(1-\rho^2)\sigma_2^2},$$

σ_1^2,σ_2^2 are the variances of V,W respectively, and ρ is the correlation coefficient between V and W.

Comparing (1.16) with the expression B2 in (1.15), the second part of the theorem follows, that is

$$E(V)=0,\ Var(V)=(k-1)(G(1-G)+(k-1)(\tfrac{g}{f})^2 F(1-F))/(1+\tfrac{(k-1)g}{f})^2 .$$

■

Remark:

In case of $F(x)=G(x)$, $\forall x \in \mathbb{R}$, the asymptotic variance of SM is immediately found by using the preceding theorems, i.e.

$$Var(SM) = \frac{n(k-1)}{4k} .$$

A rather well-known form of the linear r.-o.s. is given in

Definition 1.7: (Van-der-Waerden-statistic)

Let the regression constants be given by (1.8). In using the scorefunction $J(u)=\Phi^{-1}(u)$, $0<u<1$, the resulting r.-o.s.

$$SN:=\sum_{i=1}^{n} \Phi^{-1}\left(\frac{R_i}{m+1}\right)$$

is called Van-der Waerden statistic. ■

For this statistic, we give the asymptotic distribution only in case that $F(x)=G(x)$, $\forall x\in \mathbb{R}$.
It is easily seen that $E(SN)=0$ and $Var(SN)=\frac{n^2(k-1)}{m(m-1)}\sum_{i=1}^{m}\left(\Phi^{-1}\left(\frac{i}{m+1}\right)\right)^2$. Noting in addition that $0<\int_0^1 (\Phi^{-1}(u))^2 du = 1<\infty$ and making use of theorem 1.1, $SN/\sqrt{Var(SN)}$ is immediately seen to be asymptotically standard normal distributed.

Finally we define a special r.-o.s. first considered by Gastwirth [64].

Definition 1.8: (Percentile-modified statistic)

Let the regression constants be given again by (1.8). For preassigned values $p,q\in[0,\frac{1}{2}]$, let $P:=[mp]$ and $Q:=[mq]$.
Let $TP:=\sum_{i=1}^{n} c(R_i-m+P)(m-P-R_i)$, and $BQ:=\sum_{i=1}^{n} c(Q-R_i)(R_i-Q-1)$.
Then

$$ST:=TP-BQ$$

is called Percentile-modified statistic.

The associated score-generating function is given by

$$J(u):=\begin{cases} u-q & , \text{ if } 0<u\leq q \\ 0 & , \text{ if } q<u\leq 1-p \\ u-(1-p) & , \text{ if } 1-p<u<1 \end{cases},$$

and therefore the statistic ST can be rewritten as:

$$ST=(m+1)\sum_{i=1}^{n} J\left(\frac{R_i}{m+1}\right).$$

■

The special effect of this scorefunction is that only random variables with ranks not greater than Q or with ranks greater than N-P get scores greater than zero.

In case that $F(x)=G(x)$, $\forall x \in \mathbb{R}$, the asymptotic normality of ST is easily proved by using theorem 1.1, (1.2) and (1.3). Mean and variance of ST are as follows:

$$E(ST)= \frac{1}{k} \left(\frac{Q(Q+1)-P(P+1)}{2}\right);$$

$$Var(ST)= \frac{(k-1)}{k^2(m-1)} \left(\frac{P(P+1)}{12}(-3P^2-3P+4mP+2m) + \frac{Q(Q+1)}{12}(-3Q^2-3Q+4mQ+2m)+ \frac{PQ}{2}(P+1)(Q+1)\right).$$

One of the main advantages of this statistic is that a judicious choice (with respect to the possible shape of F and G) of p and q may lead to more effective selection rules. A more detailed discussion of this question can be found in sections 3 and 4.

1.3 The joint asymptotic distribution of the vector of rank-order statistics $(S_1,S_2,\ldots,S_k)$ based on joint ranks

Up to now, we have been only interested in the distribution of exactly one r.-o.s., which is the r.-o.s. associated with population Π_1. This we have done since one selection rule can be based on the comparison of each single population with the combined remaining k-1 populations. Another selection rule can be derived from the comparison of each population with that one, which yields the maximum r.-o.s.. To derive the mentioned selection rule, we must evaluate the joint distribution of $(S_1,S_2,\ldots,S_k)$, the vector of r.-o.s.'s calculated from the total sample given in (1.6), where the r.-o.s.'s S_j, $1 \le j \le k$, are based on the regression constants

$$\begin{aligned} &c_{(j-1)n+1}=1,\ c_{(j-1)n+2}=1,\ldots,c_{jn}=1, \text{ and} \\ &c_i=0, \text{ for } i \notin \{(j-1)n+1, (j-1)n+2,\ldots,jn\}, \text{ respectively.} \end{aligned} \tag{1.17}$$

S_j is hence given by:

$$S_j := \sum_{\ell=(j-1)n+1}^{jn} J\left(\frac{R_\ell}{m+1}\right). \tag{1.18}$$

Recalling a result of 1.1, we know that

$$Cov(S_h,S_\ell) = -\frac{n^2}{m(m-1)} \sum_{i=1}^{m} (a_i-\bar{a})^2, \text{ where } a_i=J\left(\frac{i}{m+1}\right) \text{ in this special case.}$$

The joint asymptotic distribution of $(S_1,\ldots,S_k)$ is now given by

Theorem 1.9:

Let $S_1,S_2,\ldots,S_k$ be given by (1.18), where the scorefunction J satisfies the conditions given in 1.1 and A4/7. If the cdf's of the populations $\Pi_1,\ldots,\Pi_k$ are all equal to some continuous cdf F, then the random vector $(V_1,V_2,\ldots,V_k)$, where $V_j=(S_j-E(S_j))/\sqrt{\frac{k}{k-1}\mathrm{Var}(S_j)}$, $1\leq j\leq k$, has asymptotically a k-variate normal distribution with mean vector $\vec{0}$ and covariance-matrix $\Gamma = ((\gamma_{ij}))$, where

$$\gamma_{ij}=\begin{cases}\frac{k-1}{k}, \text{ if } i=j\\ -\frac{1}{k}, \text{ if } i\neq j .\end{cases}$$

The proof is a direct consequence of A4/7. ■

The results obtained in this section will be used in section 4, to calculate the asymptotic critical values required by the selection rule to be considered there.

1.4 The treatment of ties

In the preceding two sections, we have always assumed that the random variables $X_1,X_2,\ldots,X_m$ have continuous cdf's, i.e.

$$P(X_i=X_j, \text{ for at least one pair } (i,j),\ 1\leq i\neq j\leq m)=0. \quad (1.19)$$

But even if (1.19) is valid, the occurrence of the event $\{X_i=X_j\}$, $1\leq i\neq j\leq m$, is not impossible. In case that such events occur, we speak of ties or of tied observations of X_i and X_j.

In practice the accuracy of measurement is limited, and thus (1.19) is not longer valid.

In the presence of ties among the random vector $(X_1,\ldots,X_m)$ the assignment of the ranks $R_1,\ldots,R_m$ is not clearly defined. For solving this difficulty, two different methods are discussed in the following. For this purpose, we assume that the components of the random vector $(X_1,\ldots,X_m)$ can be divided into G groups in such a way that all variables to be included in the same group have the same value. The random variables $T_1,T_2,\ldots,T_G$, respectively, denote the number of random variables in the "groups of tied variables" 1,2,...,G, where group 1 contains the smallest X_i's, group 2 the second smallest, and so on.

If $g, \tau_1, \ldots, \tau_g$ are the realizations of the random variables $G, T_1, \ldots, T_G$, then we always have $1 \leq g \leq m$, $\tau_1, \ldots, \tau_g \varepsilon \{1, \ldots, m\}$ and $\sum_{i=1}^{g} \tau_i = m$.

One of the possibilities to deal with ties is the method of randomization. Randomization we call the random assignment of ranks to the random variables included in some group of tied variables. For example let the random variables X_1, X_2, X_3, X_4 be such that $X_1 < X_2 = X_3 < X_4$. Then we have $G=3$, $T_1=1$, $T_2=2$, $T_3=1$, and the ranks in the second group of tied variables are not clearly defined.
If we want to break the tie between X_2 and X_3 by randomization, we have to determine the ranks R_2 and R_3 by a random experiment. One possibility would be the tossing of a coin, where R_2 takes the value 2 if the coin shows head, and 3 if tail appears. Thereby the value of R_3 is also determined. It takes the values 3 or 2 respectively.

Another possibility of randomization in this example consists in generating two uniformly distributed random numbers U_1, U_2, where we put $R_2=2$, $R_3=3$ if $U_1 < U_2$, and $R_2=3$, $R_3=2$ if $U_1 > U_2$.
The second kind of random experiment can be generalized to more than two random variables contained in some group of tied variables, by generating as much random numbers as variables are included in the group. There may be more than one group of tied variables consisting of more than one variable, therefore we fix the following general method of randomization.
Let $U_1, \ldots, U_m$ be random numbers, uniformly distributed over (0.1), and different from each other. Then the ranks $R_1^*, \ldots, R_m^*$ of the random variables $X_1^*, \ldots, X_m^*$, given by $X_i^* = X_i + U_i$, $1 \leq i \leq m$, are clearly defined. Thus R_i^* is the rank of X_i.
It is easily seen that $R^* = (R_1^*, \ldots, R_m^*)$ is an element of $\mathcal{T}_m$ with probability 1, and is moreover independent of $(X_1, \ldots, X_m)$, i.e. independent of $G, T_1, \ldots, T_G$.
According to (1.18), we define the r.-o.s.'s

$$S_j^* = \sum_{i=(j-1)n+1}^{jn} a(R_i^*), \quad 1 \leq j \leq k. \tag{1.20}$$

If the underlying cdf's are identical, then with A4/2:

$$E(S_j^*) = E(S_j), Var(S_j^*) = Var(S_j), Cov(S_h^*, S_\ell^*) = Cov(S_h, S_\ell), \quad 1 \leq j \leq k, \quad 1 \leq h \neq \ell \leq k.$$

The asymptotic distributions of the single r.-o.s. S_j^* and of the random vector $(S_1^*, \ldots, S_k^*)$ will be given later.

We first consider a further method of dealing with ties. The method described in the following is based on averaged scores.

Let g and $\tau = (\tau_1,\ldots,\tau_g)$ be the realizations of G and $T=(T_1,\ldots,T_G)$ respectively. We use the r.-o.s.'s

$$\bar{S}_j := \sum_{i=(j-1)n+1}^{jn} a(R_i,\tau),\ 1\le j\le k, \text{ where the} \tag{1.21}$$

averaged scores $a(R_i,\tau)$ are defined as follows:

$$a(R_i,\tau) := \frac{1}{\tau_\ell} \sum_{j=h_u}^{h_o} a(j), \text{ for } X_i \text{ element of group } \ell \text{ of size } \tau_\ell \text{ ,where}$$

$$h_u := \sum_{\alpha=2}^{\ell} \tau_{\alpha-1}+1,\ h_o := \sum_{\alpha=1}^{\ell} \tau_\alpha . \tag{1.22}$$

We get:

Theorem 1.10:

If the underlying cdf's are all identical for arbitrary τ, the conditional mean, variance and covariance of $\bar{S}_j$, τ fixed, $1\le j\le k$, are given by:

$$E(\bar{S}_j|\tau)=E(S_j),\ 1\le j\le k,$$

$$Var(\bar{S}_j|\tau)= \frac{n^2(k-1)}{m(m-1)} \sum_{\ell-1}^{m} (a_\ell-\bar{a})^2 - \sum_{i=1}^{m} (a(i,\tau)-a_i)^2,\ 1\le j\le k,$$

$$Cov(\bar{S}_h,\bar{S}_\ell|\tau)= \frac{n^2}{m(m-1)} \sum_{j=1}^{m} (a_j-\bar{a})^2 - \sum_{i=1}^{m} (a(i,\tau)-a_i)^2,\ 1\le h\ne\ell\le k.$$

Proof:

Recalling the definitions of the rank vectors R^*,R, and of the averaged scores, $\bar{S}_j$ can be rewritten as

$$\bar{S}_j= \sum_{i=(j-1)n+1}^{jn} a(R_i^*,\tau),\ 1\le j\le k.$$

R^* and T are independent. Thus with A2/8 we get:

$$E(\bar{S}_j|\tau)= \sum_{i=1}^{n} E(a(R_i^*,\tau)).$$

With (1.21) we have $E(a(R_i^*,\tau))=\bar{a}$, and hence the first assertion of the theorem is proved.

Using A2/8 and the following identity

$$\sum_{i=1}^{m} (a(i,\tau)-\bar{a})^2 = \sum_{i=1}^{m} (a_i-\bar{a})^2 - \sum_{i=1}^{m} (a(i,\tau)-a_i)^2$$

the proof for $Var(\bar{S}_j|\tau)$ and $Cov(\bar{S}_h,\bar{S}_\ell|\tau)$ is immediately obvious.

For the special scorefunctions considered in section 1.2 we give the explicit expression of $\sum_{i=1}^{m}(a(i,\tau)-\bar{a})^2$:

Wilcoxon-scores

With $J(u) = u$, we have $\sum_{i=1}^{m}(a(i,\tau)-\bar{a})^2 = \frac{1}{12}\left(\frac{m(m-1)}{m+1} - \frac{1}{(m+1)^2}\sum_{\ell=1}^{g}(\tau_\ell^2-1)\tau_\ell\right)$.

Median-scores

Let $X^{(1)},\dots,X^{(m)}$ be the order-statistics associated with the random vector $(X_1,\dots,X_m)$. If $X^{(1/2m)} \neq X^{(1/2m+1)}$ for m even or $X^{(1/2m+1/2)} \neq X^{1/2m+3/2}$ for m odd, we have $\sum_{i=1}^{m}(a(i,\tau)-\bar{a})^2 = \sum_{i=1}^{m}(a_i-\bar{a})^2$.

If on the contrary

$$X^{(\frac{1}{2}m-\lambda)} < X^{(\frac{1}{2}m-\lambda+1)} = \dots = X^{(\frac{1}{2}m+K)} < X^{(\frac{1}{2}m+K+1)}, \text{ if m even}$$

or

$$X^{(\frac{1}{2}m+\frac{1}{2}-\lambda)} < X^{(\frac{1}{2}m+\frac{1}{2}-\lambda+1)} = \dots = X^{(\frac{1}{2}m+\frac{1}{2}+K)} < X^{(\frac{1}{2}m+\frac{3}{2}+K)}, \text{ if m odd,}$$

we have

$$\sum_{i=1}^{m}(a(i,\tau)-\bar{a})^2 = \begin{cases} \frac{m}{4} - \frac{\lambda K}{\lambda+K} & \text{, if m even} \\ \frac{(m+1)^2}{4m} - \frac{\lambda K}{\lambda+K} & \text{, if m odd.} \end{cases}$$

Van-der-Waerden-scores

With $J(u) = \Phi^{-1}(u)$, we have $\sum_{i=1}^{m}(a(i,\tau)-\bar{a})^2 = \sum_{\ell=1}^{g}\tau_\ell(a(\ell,\tau))^2$.

Percentile-modified-statistic

The statistics BQ and TQ are considered separately. To obtain a clear definition in the presence of ties, the statistics BQ and TP are substituted by $B\bar{Q}$ and $T\bar{P}$ respectively, where

$$\bar{P} = \sum_{\ell=\pi}^{g}\tau_\ell, \text{ with } \pi \text{ such that } \bar{P} \leq P \text{ and } \bar{P}+\tau_{\pi-1} > P, \text{ and}$$

$$\bar{Q} = \sum_{\ell=1}^{\rho}\tau_\ell, \text{ with } \rho \text{ such that } \bar{Q} \leq Q \text{ and } \bar{Q}+\tau_{\rho+1} > Q.$$

Then we have

$$E(T\bar{P}|\tau)=E(T\bar{P}),\ E(B\bar{Q}|\tau)=E(B\bar{Q}),$$

$$Var(T\bar{P}|\tau)=Var(T\bar{P}) - \frac{n^2(k-1)}{m(m-1)} \sum_{\ell=\pi}^{g} \frac{1}{12}(\tau_\ell^2-1)\tau_\ell,$$

$$Var(B\bar{Q}|\tau)=Var(B\bar{Q}) - \frac{n^2(k-1)}{m(m-1)} \sum_{\ell=1}^{\rho} \frac{1}{12}(\tau_\ell^2-1)\tau_\ell,$$

$$Cov(T\bar{P},B\bar{Q}|\tau)=Cov(T\bar{P},B\bar{Q}),$$

where $E(T\bar{P})$, $E(B\bar{Q})$, $Var(T\bar{P})$, $Var(B\bar{Q})$ and $Cov(T\bar{P},B\bar{Q})$ are the moments of $T\bar{P}$ and $B\bar{Q}$ in case of no ties.

The asymptotic distribution of a single r.-o.s. S_j and that one of the vector $(S_1,\ldots,S_k)$ in the presence of ties have been considered by Vorlickowa [197] and Conover [50]. Using their results, we get the following:
If the underlying cdf's are all identical, and if the assumptions made above hold, then

(a) the statistics S_j^* and $\bar{S}_j$ are asymptotically normal distributed with mean and variance given in (1.20) and theorem 1.10, respectively,

(b) the vectors $(S_1^*,\ldots,S_k^*)$ and $(\bar{S}_1,\ldots,\bar{S}_k)$ have asymptotically a k-variate normal distribution with mean vectors $(E(S_1^*),\ldots,E(S_k^*))$, $(E(\bar{S}_1|\tau),\ldots,E(\bar{S}_k|\tau))$ and covariance matrices $\Sigma^* = ((\sigma_{ij}^*))$ and $\bar{\Sigma} = ((\bar{\sigma}_{ij}))$, respectively, where

$$\sigma_{ij}^* = \begin{cases} Var(S_i^*) & , i=j \\ Cov(S_i^*,S_j^*), & i\neq j \end{cases} \quad \text{and} \quad \bar{\sigma}_{ij} = \begin{cases} Var(\bar{S}_i|\tau) & , i=j \\ Cov(\bar{S}_i,\bar{S}_j|\tau), & i\neq j. \end{cases}$$

In the following sections, we will give only a short remark concerning the question, which one of the two methods of dealing with ties is appropriate in connection with the considered selection rule.

2 Two Subset-Selection Procedures in One-Factor Designs Including the General Behrens-Fisher-Problem

In this section, we consider two special procedures, based on a rule for selecting a subset from k populations, which we have shortly explained in the introduction. As already outlined, the procedures looked for should be proper to select a random subset from k populations, which contains the "best" population with a probability of atleast P*, where the population denoted as best is that one, which is associated with the greatest median.

2.1 The selection rule R1

Let the independent random samples $\vec{X}_i$, $1 \leq i \leq k$, from populations $\Pi_1,\dots,\Pi_k$ be combined in the total sample $\vec{X}$ given by (1.6). Furthermore the cdf's F_j, $1 \leq j \leq k$, of the populations $\Pi_1,\dots,\Pi_k$ are assumed to be continuous and identical except of possible differences between their location-parameters $\Delta_1,\Delta_2,\dots,\Delta_k$ and/or their scale-parameters $\sigma_1,\sigma_2,\dots,\sigma_k$, (p.301). In addition, we restrict our considerations to such populations for which all median configurations $\theta=(\theta_1,\dots,\theta_k)$ can be received by a variation of the location-parameters only. Let the r.-a.s.'s $S_1,\dots,S_k$, based on the total sample $\vec{X}$, be defined according to (1.18), and let the scorefunction J(u), 0<u<1, be nondecreasing in u. Then the general selection rule is defined as follows:

$$\text{R1: Select population } \Pi_j \Longleftrightarrow S_j \geq D_j,\ 1 \leq j \leq k,\ D_j \in \mathbb{R}\,.$$

To be able to derive the critical values D_j, $1 \leq j \leq k$, we first consider the probability P(CS|R1), (p.305), i.e. the probability that the subset selected according to rule R1 includes the population with the greatest median. The values D_j, $1 \leq j \leq k$, have to be chosen such that P(CS|R1) does not fall below the preassigned probability P*, i.e.:

$$\inf_{\Omega} P(CS|R1) \geq P^*, \text{ with } \Omega=\{\theta=(\theta_1,\dots,\theta_k):\theta_j \in I,\ j=1,\dots,k,\ I \subseteq \mathbb{R}\} \tag{2.1}$$

Prior to the further discussion of P(CS|R1), we should make some notes concerning the choice of the median as a selection criterion.
In the introduction, we have already seen that the median is a suitable criterion for ranking the populations, if their scale-parameters coincide. The situation to be considered in this section is not just as clear as before, because

of the differences between medians and scale-parameters. To answer the question whether the median is a proper criterion in this case, too, we have to distinguish between symmetrically and non-symmetrically distributed populations.

In the first case, the median is determined only by the location-parameter, and thus the median is a proper selection criterion as long as some subjective limit for the difference between the scale-parameters is not exceeded (cf. figure 1 and figure 2).

fig. 1

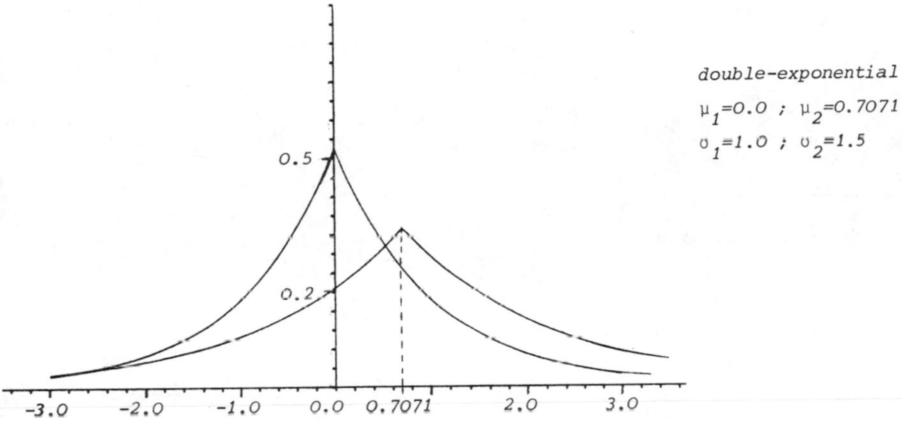

double-exponential
$\mu_1=0.0$; $\mu_2=0.7071$
$\sigma_1=1.0$; $\sigma_2=1.5$

fig. 2

normal
$\mu_1=0.0$; $\mu_2=0.5$
$\sigma_1=1.0$; $\sigma_2=1.5$

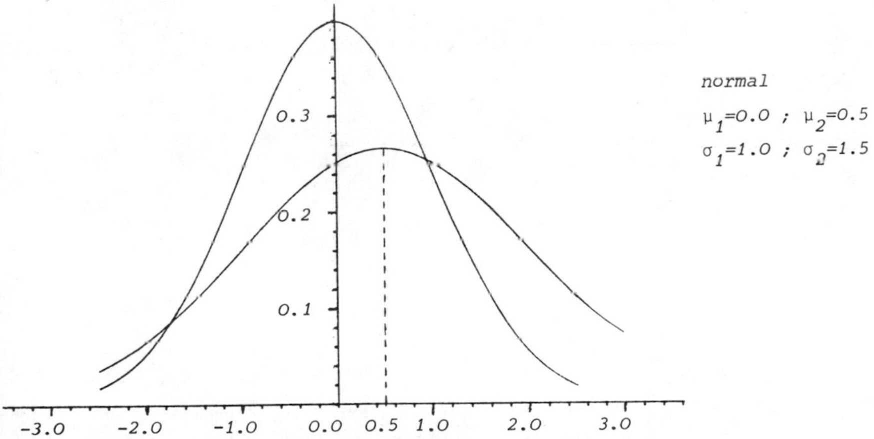

If on the contrary the difference between the scale-parameters is "too large" (cf. figure 3 and figure 4), the selection based on the size of the medians only, does not seem to be meaningful. For this reason, some investigations concerning the scale-parameters of the considered populations should be carried out prior to application of the selection procedure.

fig. 3

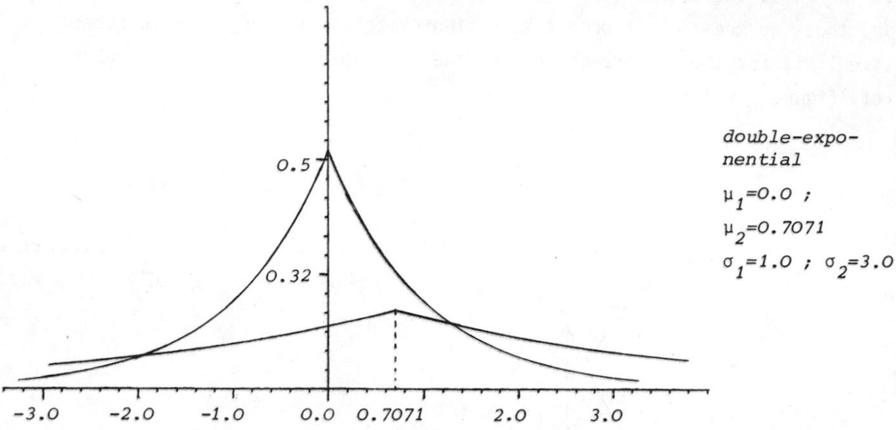

double-expo-
nential

μ_1=0.0 ;

μ_2=0.7071

σ_1=1.0 ; σ_2=3.0

fig. 4

normal

μ_1=0.0 ; μ_2=0.5

σ_1=1.0 ; σ_2=3.0

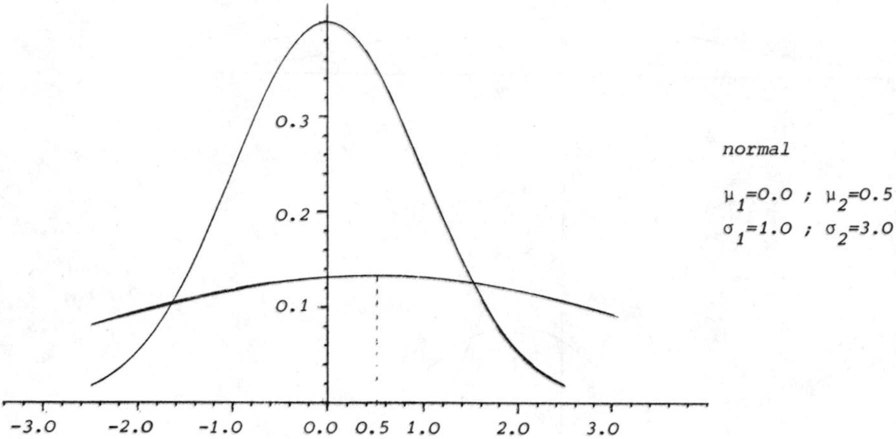

If the underlying distributions are not symmetrical, the size of the medians depends on both, location- and scale-parameters. Thus the median of a population can be kept constant in case of a simultaneous increase (decrease) of the location-parameter _and_ decrease (increase) of the scale-parameter. For instance, the populations with densities given in figure 5 possess the same median. Figures 6 and 7 give rise to the assumption that also in case of non-symmetrically distributed populations, the selection based on the size of the median is convenient only if the differences between the scale-parameters are not "too large".

fig. 5

Gamma α=2

$\mu_1=0.0$; $\mu_2=0.8176$
$\sigma_1=2.0$; $\sigma_2=1.0$

fig. 6

Gamma α=2

$\mu_1=0.0$; $\mu_2=1.3176$
$\sigma_1=2.0$; $\sigma_2=1.0$

fig. 7

Gamma α=2

$\mu_1=0.0$; $\mu_2=1.0$
$\sigma_1=3.0$; $\sigma_2=1.0$

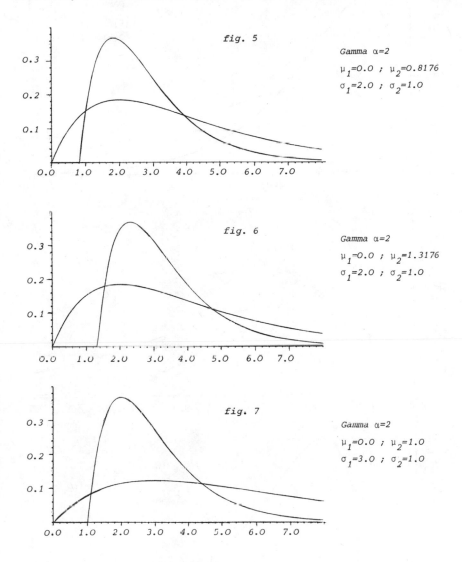

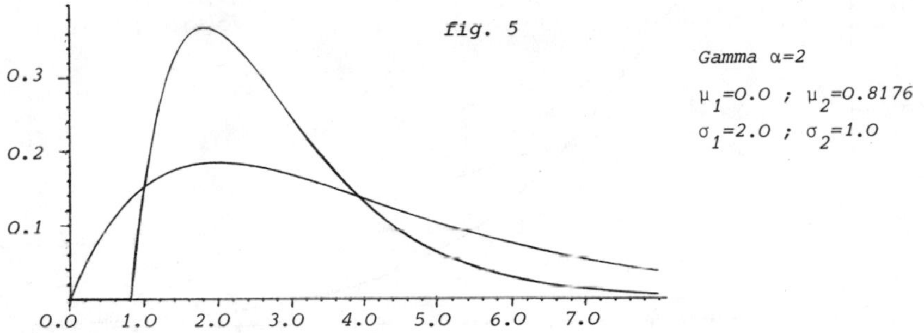

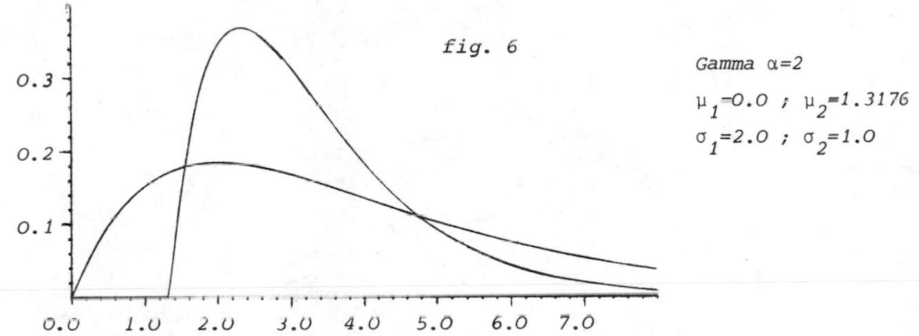

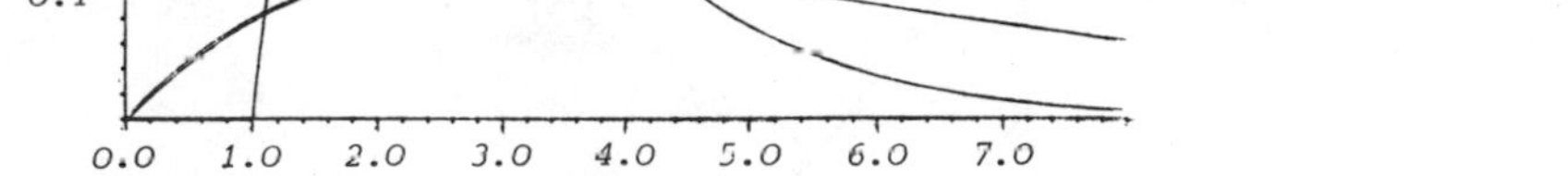

On the contrary the criterion "median" is quite convenient, if the reason of a larger median is a larger location- and/or scale-parameter. Three configurations of this type in case of the gamma-distribution with shape-parameter α=2.0 are illustrated in figures 8, 9 and 10.

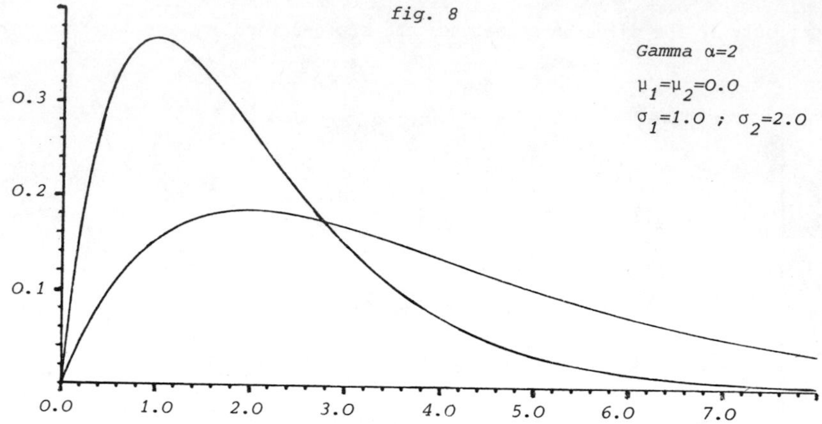

fig. 8

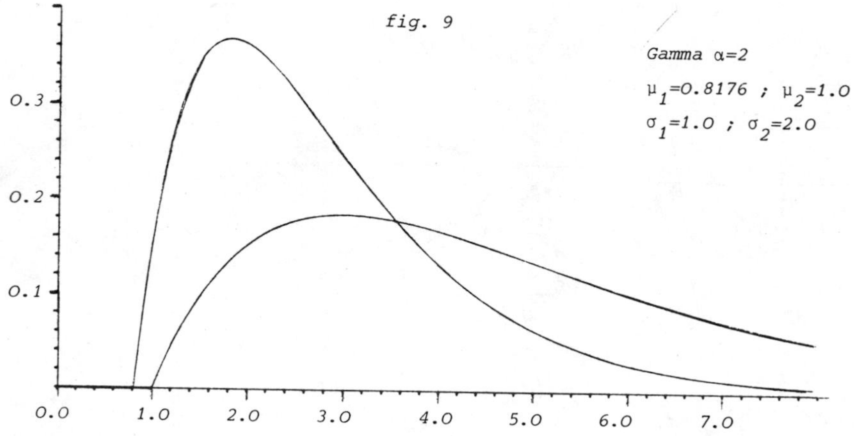

fig. 9

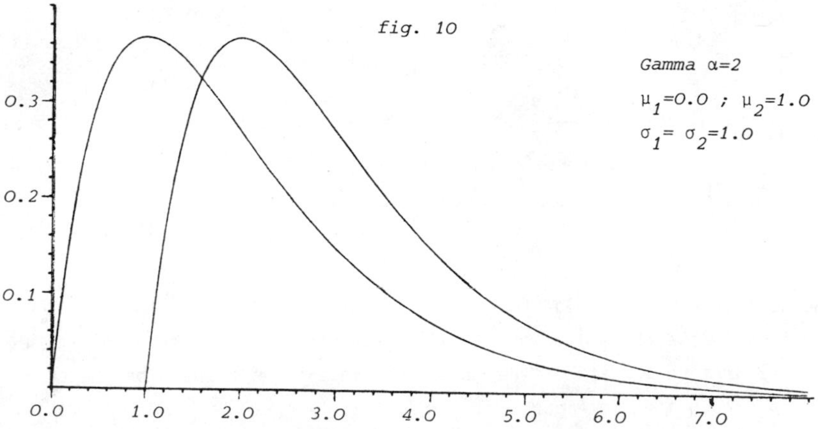

fig. 10

An extra position take distributions skewed to the left, with some fixed maximal value. For such populations the median increases if the scale-parameter decreases. For this type of distributions the criterion "median" is convenient, too.

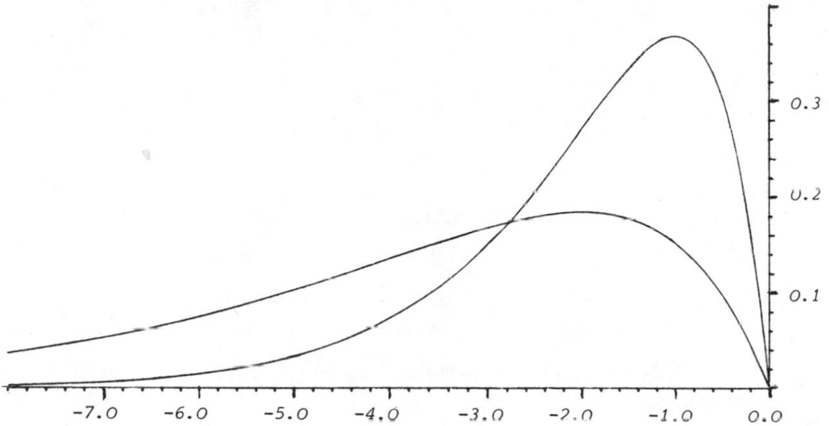

We now assume that preliminary investigations have shown that the criterion "median" can be used to select a subset, and in the following we consider the probability $P(CS|R1)$.

2.2 The infimum of the probability P(CS|R1)

Let $\theta_{(1)} \leq \theta_{(2)} \leq \ldots \leq \theta_{(k)}$ be the ordered medians of the k populations, and let $\Pi_{(i)}$ denote the population associated with median $\theta_{(i)}$. Then the event CS is realized if population $\Pi_{(k)}$ is included in the selected subset. Let $S_{(k)}$ be the r.-o.s. associated with population $\Pi_{(k)}$, then we have

$$P(CS|R1)=P(S_{(k)} \geq D_{(k)}), \text{ where } D_{(k)} \in \{D_1,\ldots,D_k\}, \qquad (2.2)$$

and with the assumptions, made in 2.1, we obtain

Theorem 2.1:

With the assumptions made above for any but fixed scale-parameters $\sigma_1,\ldots,\sigma_k$ it follows:

$$\inf_{\Omega} P(CS|R1)=\inf_{\Omega_0} P(CS|R1), \text{ where } \Omega_0=\{\theta=(\theta_1,\ldots,\theta_k):\theta_{(1)}=\theta_{(2)}=\ldots=\theta_{(k)}=\theta_0\}.$$

Proof:

The m components of the random vector $\vec{X}=(X_{11},\ldots,X_{1n},\ldots,X_{k1},\ldots,X_{kn})$ are independent, and $X_{j.}$, $j=1,\ldots,k$, has the continuous cdf $F_{j\Delta_j}(x)$, where the indices j and Δ_j indicate the dependence of F upon the scale-parameter σ_j and the loca-

tion parameter Δ_j. Furthermore we have: $F_{j\Delta_j}(x) \geq F_{j\Delta_j'}(x)$, $\forall x \in \mathbb{R}$, if $\Delta_j < \Delta_j'$.
Let

$$H(\vec{X}) := \begin{cases} 1, \text{ if } S_{(\ell)} \geq D_{(\ell)} \\ 0, \text{ otherwise} \end{cases},$$

and let $R_{(j)i}$ denote the rank of $X_{(j)i}$, (the i-th component of the sample, associated with $\Pi_{(j)}$), i=1...,n. In case, if $x_{(\ell)i}$, the realization of $X_{(\ell)i}$, increases and the other realizations remain fixed, there are three possibilities:

(a) $x_{(\ell)i}$ exceeds first a $x_{(j)h}$, $\ell \neq j$, $1 \leq i,h \leq n$, and consequently $R_{(\ell)i}$ increases by 1, and $R_{(j)h}$ decreases by 1;

(b) $x_{(\ell)i}$ exceeds first a $x_{(\ell)h}$, $1 \leq i \neq h \leq n$, and consequently $R_{(\ell)i}$ increases by 1, and $R_{(\ell)h}$ decreases by 1;

(c) $x_{(\ell)i}$ exceeds no other realization, and consequently all ranks remain unchanged.

In all three cases, the statistic $S_{(\ell)}$ is nondecreasing in $x_{(\ell).}$, and nonincreasing in $x_{(j).}$, $1 \leq j \neq \ell \leq k$. Thus with A4/8 it follows:

$$E_{(\Delta_{(\ell)})}(H(\vec{X})) \text{ is nondecreasing in } \Delta_{(\ell)}, \ \ell=1,\dots,k, \text{ and}$$
$$\text{nonincreasing in } \Delta_{(j)}, \ 1 \leq j \neq \ell \leq k,$$

and the proof therefore follows by using (2.2). ■

Theorem 2.1 means, that the probability P(CS|R1) is minimal if with any but fixed scale-parameters the location-parameters $\Delta_1,\dots,\Delta_k$ are such that the medians of the k populations are all equal to some value θ_0. In this case, we speak of a least favorable configuration (LFC), (cf. introduction).The r.-o.s.'s are invariant to variations of θ_0. Hence $\theta_0=0$ can be assumed without loss of generality in the following.

2.3 The asymptotic distributions of two special rank-order statistics in case of consistent estimation of the unknown parameters

The rule R1 is based on the comparison of each population Π_i, $i \in \{1,\dots,k\}$ with the hypothetical population resulting from the combination of the remaining k-1 populations Π_j, j=1,...,k, j≠i, (cf. 1.2). In comparing population Π_j with the hypothetical one, we denote the cdf of Π_j with F_j and that one of the hypothetical population with G_j, where G_j is defined by (1.7).

With the assumption that the medians of $\Pi_1,\dots,\Pi_k$ are all equal to zero, we have

$$p_j := P(X_{j\cdot} > X_{i\cdot},\ 1 \leq i \neq j \leq k) = \int_0^1 G_j(x)dF_j(x) = \frac{1}{2},\ 1 \leq j \leq k. \tag{2.3}$$

For the derivation of P(CS| R1) in case of equal medians, we need the distribution of S_j on the premises given above. As we have seen in 1.2, the asymptotic distribution of S_j can be specified in case of two special scorefunctions.

Mann-Whitney-statistic

Let the total sample $\vec{X}$ be given by (1.6). With the assumptions made above and by using theorem 1.3, we find that

$$S_j := S_{XY_j} := \frac{1}{m+1} \sum_{\ell=(j-1)n+1}^{jn} \sum_{h \in I_j} U^j_{\ell h},\ j=1,\dots,k,$$

where

$$U^j_{\ell h} := \begin{cases} 1, \text{ if } X_\ell \geq X_h \\ 0, \text{ otherwise} \end{cases},\quad I_j = \{i : i \in \{1,\dots,m\} \setminus \{(j-1)n+1,\dots,jn\}\}, \tag{2.4}$$

is asymptotically normal distributed with mean

$$E(S_j) = \frac{n^2(k+1)}{2(m+1)} \tag{2.5}$$

and variance

$$Var(S_j) = \frac{n^2(k+1)}{(m+1)^2}\left(\frac{1}{4} + ((k-1)n-1)\beta_j + (n-1)\gamma_j\right),$$

where

$$\beta_j := E(U^j_{\ell h} U^j_{\ell g}) - (E(U^j_{\ell h}))^2,\quad \gamma_j := E(U^j_{\ell h} U^j_{ph}) - (E(U^j_{\ell h}))^2.$$

In case of different-scale parameters, the cdf's F_j and G_j are different and hence the distribution of S_j depends on F_j and G_j, i.e. on the scale parameters.

In using the upper bound for the variance $Var(S_j)$ formulated by Birnbaum and Klose [41], we get a conservative selection rule, and the efficiency of the selection procedure will be reduced. A solution of the problem to get an explicit expression for the distribution of S_j independent of F_j and G_j lies in the estimation of the variance $Var(S_j)$, $1 \leq j \leq k$, based on the vector $\vec{X}$.

Considering the expression for $Var(S_j)$, we see that the estimation problem reduces to the estimation of β_j and γ_j. Before giving proper estimators, two useful theorems will be proved.

Theorem 2.2:

With the notations given above,

$$W_n^j := \frac{2}{n^2(k-1)(n(k-1)-1)} \sum_{\ell=(j-1)n+1}^{jn} \sum_{h\in I_j} \sum_{g\in I_j\setminus\{1,\dots,h\}} U_{\ell h}^j U_{\ell g}^j,$$

$$V_n^j := \frac{2}{n(k-1)(n-1)n} \sum_{\ell=(j-1)n+1}^{jn-1} \sum_{p=\ell+1}^{jn} \sum_{h\in I_j} U_{\ell h}^j U_{ph}^j,$$

$$Q_n^j := \frac{S_j(m+1)}{n^2(k-1)} = \frac{1}{n^2(k-1)} \sum_{\ell=(j-1)n+1}^{jn} \sum_{h\in I_j} U_{\ell h}^j ,$$

are unbiased estimators of $E(U_{\ell h}^j U_{\ell g}^j)$, $E(U_{\ell h}^j U_{ph}^j)$ and $p_j=E(U_{\ell h}^j)$, $j=1,\dots,k$, respectively.

The proof follows directly by using the additivity of means. ■

Theorem 2.3:

The sequences of estimators $(W_n^j)_{n\in\mathbb{N}}$, $(V_n^j)_{n\in\mathbb{N}}$ and $(Q_n^j)_{n\in\mathbb{N}}$ are consistent for $E(U_{\ell h}^j U_{\ell g}^j)$, $E(U_{\ell h}^j U_{ph}^j)$ and p_j, $1\leq j\leq k$, respectively.

Proof:

The variance of W_n^j can be written as

$$\mathrm{Var}(W_n^j) = \sum_{\ell} \sum_{p} \sum_{q} \sum_{h} \sum_{g} \sum_{d} \mathrm{Cov}((U_{\ell p}^j U_{\ell q}^j)(U_{hg}^j U_{hd}^j)).$$

Since any two pairs $(U_{\ell p}^j U_{\ell q}^j)$ and $(U_{hg}^j U_{hd}^j)$ are independent, unless one of the equations $\ell=h$, $p=g$, $q=d$, $p=d$, $q=g$ is satisfied, and since the covariances are all finite, it follows:

$$\mathrm{Var}(W_n^j) = o(\tfrac{1}{n}).$$

Hence the sequence $(W_n^j)_{n\in\mathbb{N}}$ is consistent for $E(U_{\ell p}^j U_{\ell q}^j)$. The same argument provides the consistency of $(V_n^j)_{n\in\mathbb{N}}$ for $E(U_{\ell p}^j U_{hp}^j)$. The consistency of $(Q_n^j)_{n\in\mathbb{N}}$ for $E(U_{\ell p}^j)$ follows by using the results concerning the statistic S_{XY} given in 1.2. ■

With the two theorems 2.2 and 2.3, we know that

$$B_n^j := W_n^j - (Q_n^j)^2, \quad C_n^j := V_n^j - (Q_n^j)^2 \tag{2.6}$$

are unbiased and consistent estimators of β_j and γ_j, $1\leq j\leq k$, respectively.

Theorem 2.4:

Let S_j, $j=1,\dots,k$, be defined by (2.4). Then with (2.6) the estimator

$$\hat{\sigma}_j^2 := \frac{n^2(k-1)}{(m+1)^2}\left(\frac{1}{4} + (n(k-1)-1)B_n^j + (n-1)C_n^j\right) \tag{2.7}$$

is consistent for $\sigma_j^2 := \mathrm{Var}(S_j)$, $j=1,\dots,k$, and thus it follows:

$$P((S_j - E(S_j))/\sqrt{\hat{\sigma}_j^2} < x) \xrightarrow[n\to\infty]{d} \Phi(x),\ 1\leq j\leq k.$$

Proof:

Since $(\hat{\sigma}_j^2)_{n\in\mathbb{N}}$ is a consistent sequence of estimators for σ_j^2, we have by using A2/9

$$P((S_j - E(S_j))/\sqrt{\hat{\sigma}_j^2} < x) \xrightarrow[n\to\infty]{d} P((S_j - E(S_j))/\sqrt{\sigma_j^2} < x)$$

and hence with theorem 1.3, the assertion of the theorem follows. ■

Another special subset-selection procedure based on the rule R1 results from the use of the

Median-statistic

Again let the total sample $\vec{X}$ be given by (1.6). With the assumptions made above and by using theorem 1.6 it can easily be shown that the random variables

$$S_j := \sqrt{n}(SM_j - E(SM_j)),\ j=1,\dots,k,$$

where (2.8)

$$SM_j := \sum_{\ell=(j-1)n+1}^{jn} c\left(\frac{R_\ell}{m+1} - \frac{1}{2}\right),\ j=1,\dots,k,$$

are asymptotically normal distributed with mean and variance given by

$$E(S_j)=0,\ \mathrm{Var}(S_j) = n^2(k-1)\left(\frac{1}{4} + \frac{k-1}{4}\left(\frac{g_j(0)}{f_j(0)}\right)^2\right)\Big/\left(1+(k-1)\frac{g_j(0)}{f_j(0)}\right)^2. \tag{2.9}$$

Considering the expression for $\mathrm{Var}(S_j)$, we see that it depends on the densities g_j and f_j.

As for the Mann-Whitney-statistic there exist consistent estimators for the variances $\mathrm{Var}(S_j)$, $1\leq j\leq k$, based on the given total sample $\vec{X}$. The only unknown expression in (2.9) is the quotient $\frac{g_j(0)}{f_j(0)}$, and thus the estimation problem reduces to the estimation of this quotient.

Theorem 2.5:

Let $X^{(1)} < X^{(2)} < \ldots < X^{(n)}$ be the order-statistics associated with the original sample $(X_1,\ldots,X_n)$, with cdf F and density f, and suppose that S has a binomial distribution with parameters n and $p=\frac{1}{2}$, and let $\alpha\varepsilon(0,\frac{1}{2})$. If $\Phi(t_{\alpha/2})=1-\frac{\alpha}{2}$, $F(\overline{\theta}_X)=\frac{1}{2}$ and $P(S<c)=P(S>n-c)=\frac{\alpha}{2}$, (i.e. $(X^{(c)}, X^{(n-c+1)})$ is a $(1-\alpha)$%-confidence interval for the median $\overline{\theta}_X$), then

$$\frac{1}{2}\sqrt{n}(X^{(n-c+1)}-X^{(c)})/t_{\alpha/2} \xrightarrow[n\to\infty]{p} (2f(\overline{\theta}_X))^{-1}.$$

Proof:

Let $B(F):=\int_{-\infty}^{+\infty}\frac{d}{dx}J(F(x))dF(x)$ and $A^2:=\int_0^1 J(F(x))^2dF(x)$, where the scorefunction $J(u)$, $0<u<1$, is continuous and the first derivative exists for $u\varepsilon(0,1)$. A result of Sen [165] says, that

$$\sqrt{n}(X^{(n-c+1)}-X^{(c)})/2t_{\alpha/2} \xrightarrow[n\to\infty]{p} A(B(F))^{-1}. \tag{2.10}$$

The Median-statistic is based on the scorefunction $J(u)=c(u-\frac{1}{2})$, $u\varepsilon(0,1)$, and hence $A^2=\frac{1}{4}$. For proving the theorem, it remains to show that for this special scorefunction $B(F)=f(\overline{\theta}_X)$. Defining $G(x):=\frac{d}{dx}J(F(x))$, where $J(F(x))=c(F(x)-\frac{1}{2})$, $G(x)$ is a generalized function and can be represented by the sequence of regular functions $(f_n(x))_{n\varepsilon\mathbb{N}}$, where $f_n(x):=\delta(x):=\sqrt{\frac{n}{2\pi}}\,e^{-nx^2/2}$ and $\delta(x)$ denotes Dirac's delta function. Using this representation we get:

$$B(F)=\int_{-\infty}^{+\infty}G(x)f(x)dx=\lim_{n\to\infty}\int_{-\infty}^{+\infty}\delta(x)f(x)dx=f(\overline{\theta}_X),$$

and with (2.10) the proof follows. ■

Let $X_j^{(1)} < X_j^{(2)} < \ldots < X_j^{(n)}$ and $Y_j^{(1)} < Y_j^{(2)} < \ldots < Y_j^{(n(k-1))}$ denote the order-statistics associated with the samples drawn from population Π_j and the k-1 remaining populations Π_i, $1\le i\ne j\le k$, respectively, and let (cf. 1.2) F_j, G_j, f_j and g_j be the cdf's and densities associated with them. Then, in using theorem 2.5 with $\theta_0=0$ (cf. 2.2),

$$\sqrt{n}(X_j^{(n-c+1)}-X_j^{(c)})/t_{\alpha/2} \text{ and } \sqrt{n(k-1)}(Y_j^{(n(k-1)-c+1)}-Y_j^{(c)})/t_{\alpha/2}$$

are seen to be consistent estimators of $f_j(0)^{-1}$ and $g_j(0)^{-1}$, j=1,...,k, respectively.

To estimate the unknown quotient $\frac{g_j(0)}{f_j(0)}$, we use the estimator

$$QU_j := \frac{\sqrt{n}(X_j^{(n-c+1)} - X_j^{(c)})}{\sqrt{n(k-1)}(Y_j^{(n(k-1)-c+1)} - Y_j^{(c)})} . \tag{2.11}$$

Applying A2/10 when $f_j(0) > 0$, we get:

$$QU_j \xrightarrow[n\to\infty]{p} \frac{g_j(0)}{f_j(0)} . \tag{2.12}$$

Theorem 2.6:

Let S_j be defined by (2.8). With the assumptions made above and with the estimator

$$\hat{\sigma}_j^2 := \frac{n^2(k-1)}{4} (1+(k-1)QU_j^2)/(1+(k-1)QU_j)^2 \tag{2.13}$$

for $\sigma_j^2 := Var(S_j)$, $1 \le j \le k$, it follows:

$$P(S_j / \sqrt{\hat{\sigma}_j^2} < x) \xrightarrow[n\to\infty]{p} \Phi(x).$$

Proof:

Because of (2.11) the sequence of estimators $(\hat{\sigma}_j^2)_{n \in \mathbb{N}}$ is consistent for σ_j^2. Thus the assertion of the theorem follows by using A2/9 and theorem 1.6. ∎

The main result of the preceding derivations is that the standardized random variables

$$U_j := (S_j - E(S_j)) / \sqrt{\hat{\sigma}_j^2} , \; j=1,\dots,k \tag{2.14}$$

where S_j, $E(S_j)$ and $\hat{\sigma}_j^2$ are given by (2.4), (2.5), (2.7) or by (2.8), (2.9), (2.13), respectively, have asymptotically the standard normal distribution.

2.4 The probability P(CS|R1) in the LFC

As already seen in (2.2),

$$P(CS|R1) = P(S_{(k)} \ge D_{(k)}) = 1 - P(S_{(k)} < D_{(k)}).$$

Let $E(S_{(k)})$ be the mean, and let $\hat{\sigma}_{(k)}^2$ denote the variance-estimator of the r.-o.s. $S_{(k)}$. Then by using the results in 2.3, we obtain:

$$P((S_{(k)}-E(S_{(k)}))/\sqrt{\hat{\sigma}^2_{(k)}} < (D_{(k)}-E(S_{(k)}))/\sqrt{\hat{\sigma}^2_{(k)}}) \xrightarrow[n\to\infty]{d} \Phi((D_{(k)}-E(S_{(k)}))/\sqrt{\hat{\sigma}^2_{(k)}}).$$

Condition (2.1) is also satisfied, if in the LFC, $P(S_{(k)} \geq D_{(k)}) = 1-P^*$, i.e. (2.1) is satisfied for

$$D_{(k)} = \psi(1-P^*)\sqrt{\hat{\sigma}^2_{(k)}} + E(S_{(k)}),$$

where ψ denotes the inverse of Φ.

In the LFC, each r.-o.s. S_j, $j=1,\ldots,k$, can be the r.-o.s. associated with the "best" population $\Pi_{(k)}$, and all standardized random variables given by (2.14) have the same asymptotic distribution.
Thus the probability condition (2.1) is satisfied by critical values

$$D_j = \psi(1-P^*)\sqrt{\hat{\sigma}^2_j} + E(S_j),\ j = 1,\ldots,k,$$

where P^* is the preassigned minimum probability of CS.

The two selection procedures considered in this section are said to be asymptotically distribution-free, because of the independence of the asymptotic distribution of the statistics S_j, $j=1,\ldots,k$, of distributions of the populations in research.

In the presence of ties, the two proposed selection procedures can be applied after breaking the ties by randomization.

2.5 A numerical example

Let Π_1,Π_2 and Π_3 denote different varieties of wheat. There are a lot of homogeneous fields to investigate the three varieties. Thus the assumption of a population model (cf. introduction) seems to be appropriate. The yield corresponding to field α, $1\leq\alpha\leq n$, and variety Π_i, $1\leq i\leq 3$, is described by the random variable $X_{i\alpha} := \mu+e_{i\alpha}$, where μ is the overall mean, i.e. the mean of the three populations associated with Π_1,Π_2 and Π_3, and the residual $e_{i\alpha}$ is a random variable, which has a continuous distribution with mean μ_i and variance σ_i^2.
In this example, we consider random samples of size n=12, i.e. m=3n=36. The following table gives the realizations of twelve residuals for each Π_i, $1\leq i\leq 3$.
The values are obtained by using a method for generating random numbers, described in appendix 5. The random numbers used for Π_1,Π_2,Π_3 are realizations of normal distributed populations with $(\mu_1=0,\sigma_1^2=1)$, $(\mu_2=0,\sigma_2^2=1)$ and $(\mu_3=0.5,\sigma_3^2=1)$, respectively.

	ordered realizations of the residuals											
Π_1	-1.862	-0.955	-0.723	-0.517	-0.495	0.111	0.148	0.290	0.515	0.707	1.045	1.523
Π_2	-0.400	-0.330	-0.128	0.123	0.386	0.576	0.594	1.088	1.651	1.685	1.777	2.048
Π_3	-0.732	-0.480	-0.478	-0.036	0.198	0.440	0.731	0.919	0.972	1.155	2.503	3.060

We assume that we do not have any knowledge about the distributions of the populations associated with Π_1,Π_2 and Π_3, and therefore the procedures discussed in this section are used, where $P^* = 0.90$.

For simplification of the following calculations, we first combine the three ordered samples into one ordered total sample. Then, in writing below each component of the ordered total sample the number of its original sample, we get the vector $K=(K_1,K_2,\ldots,K_m)$, which we call rank-configuration.
In the present example we get:

rank	ordered total sample	rank-configuration	rank	ordered total sample	rank-configuration
1	-1.862	1	19	0.440	3
2	-0.955	1	20	0.515	1
3	-0.732	3	21	0.576	2
4	-0.723	1	22	0.594	2
5	-0.517	1	23	0.707	1
6	-0.495	1	24	0.731	3
7	-0.480	3	25	0.919	3
8	-0.478	3	26	0.972	3
9	-0.400	2	27	1.045	1
10	-0.330	2	28	1.088	2
11	-0.128	2	29	1.155	3
12	-0.036	3	30	1.523	1
13	0.111	1	31	1.651	2
14	0.123	2	32	1.685	2
15	0.148	1	33	1.777	2
16	0.198	3	34	2.048	2
17	0.290	1	35	2.503	3
18	0.386	2	36	3.060	3

First we carry out the calculations for the selection procedure based on the <u>Mann-Whitney-statistic</u>.

Using the rank-configuration K, the estimates W_{12}^{i}, V_{12}^{i} and Q_{12}^{i}, i=1,2,3 can easily be calculated. We get:

$$W_{12}^{1} = 0.14825 \quad W_{12}^{2} = 0.46226 \quad W_{12}^{3} = 0.39644$$

$$V_{12}^{1} = 0.13699 \quad V_{12}^{2} = 0.49747 \quad V_{12}^{3} = 0.38005$$

$$Q_{12}^{1} = 0.29514 \quad Q_{12}^{2} = 0.64236 \quad Q_{12}^{3} = 0.56250$$

From this,we get the following estimates for the variances:

$$\hat{\sigma}_1^2 = 0.46388 \quad \hat{\sigma}_2^2 = 0.48907 \quad \hat{\sigma}_3^2 = 0.58700.$$

With $\Phi(-1.28125) = 0.1 = 1-P^*$ and $E(S_j) = 3.89189$, the critical values are given by:

$$D_1 = 3.01925 \quad D_2 = 2.99587 \quad D_3 = 2.91025,$$

and with

$$S_1 = 2.29729 \quad S_2 = 5.00000 \quad S_3 = 4.37838,$$

populations Π_2 and Π_3 form the selected subset.
Using the selection procedure based on the Median-statistic, we again estimate first the variances of the r.-o.s.'s S_1, S_2 and S_3. With $\alpha = 0.1$, we get the following confidence intervals for the medians $\overline{\theta}_X$ and $\overline{\theta}_Y$ (cf. 2.3):

$$i=1: \; (X_1^{(2)}, X_1^{(11)}) = (-0.955, 1.045), \; (Y_1^{(7)}, Y_1^{(18)}) = (-1.128, 1.088)$$

$$i=2: \; (X_2^{(2)}, X_2^{(11)}) = (-0.330, 1.777), \; (Y_2^{(7)}, Y_2^{(18)}) = (-0.478, 0.972)$$

$$i=3: \; (X_3^{(2)}, X_3^{(11)}) = (-0.480, 2.503), \; (Y_3^{(7)}, Y_3^{(18)}) = (-0.400, 0.707).$$

The estimates of $\frac{g_i(0)}{f_i(0)}$, i=1,2,3, are

$$QU_1 = 1.16282 \quad QU_2 = 1.02679 \quad QU_3 = 1.90279,$$

and consequently

$$\hat{\sigma}_1^2 = 0.16746 \quad \hat{\sigma}_2^2 = 0.16669 \quad \hat{\sigma}_3^2 = 0.17843.$$

Hence for $P^* = 0.90$, the critical values are

$$D_1 = -0.52432 \quad D_2 = -0.52311 \quad D_3 = -0.54121.$$

The r.-o.s.'s have the values

$S_1 = -0.57735 \quad S_2 = 0.28867 \quad S_3 = 0.28867$,

and thus we see again, that Π_2 and Π_3 form the selected subset.
In this example both procedures yield the same subset, i.e. with both procedures,we obtain a subset of size 2, which includes the best variety of wheat Π_3.

2.6 Some Monte-Carlo studies

In the following, we write R1u, if we speak of selection procedures based on rule R1 in case that the scale-parameters of the considered populations are allowed to be different.
The analytic investigation of R1u seems to be very difficult. But there exists a lot of questions concerning use and efficiency of R1u. The fact that selection procedures R1u are only asymptotically distribution-free yields the first question:

(a) Is the probability condition (2.1) satisfied,if the sample sizes are finite and how good is the approximation of the probability P* in the LFC?

Further questions are related to be efficiency of R1u, i.e. the average number T of populations included in the selected subset.

(b) How large is the number T of populations included in the selected subset?

(c) What kind of influence on the number T has the preassigned probability P*?

(d) Are there any differences between the efficiencies of procedures based on different scorefunctions?

(e) Has the variation of the scale-parameter-configuration any influence on the selection procedures?

To answer this questions we have carried out some Monte-Carlo studies, based on the method for generating random numbers, described in appendix 5.

The following types of distributions have been included in our studies: uniform distribution (UNI), normal distribution (NOR), double-exponential- or Laplace distribution (DE), Cauchy distribution (CAU), gamma distribution with shape parameter $\alpha = 2$ (GAM).

For each of these types of distributions,we have generated realizations of size n for k populations with parameter-configuration $(\sigma_1,\ldots,\sigma_k)/(a\Delta_1,\ldots,a\Delta_k)$

in such a way that the realization of population Π_j, $1\leq j\leq k$, is represented by n random numbers, which are distributed according to the distribution of Π_j. The factor a is chosen so that the results for the various distributions are comparable.[1] We have chosen: a=1 for the normal distribution, $a=\sqrt{2}$ for DE, a=2 for CAU, a=1 for GAM and a=0.28 for UNI. The two procedures denoted by R1u have been applied to the k realizations. For each simulated parameter-configuration there have been done 100 repetitions.
The following tables contain values of $\hat{P}^*$, the "relative frequency of CS" and of $\hat{ET}$, the "average number of populations in the selected subset".
Tables 1,2 give values of $\hat{P}^*$ and $\hat{ET}$, obtained by simulation of four different configurations, where n=12, 20 and 40, $P^*=0.90$, and $\alpha=0.1$ have been chosen.
Choosing n=20, we made some studies, the results of which are useful in comparing R1u with procedures R1g, which are only usable for populations with equal scale-parameters. Table 3 contains values $\hat{P}^*$ and $\hat{ET}$ for two possible LFC's.
To investigate the behaviour of R1u in case that the scale-parameters are equal, we have carried out some studies whose results are given in table 4.
Table 5 has been established, to show the influence of P^* on R1u. In table 6 some results for k=6 populations are given.
The special procedures based on Mann-Whitney-statistics and Median-statistics are denoted by WILC and MED, respectively.
Considering the results of the Monte-Carlo studies with regard to questions (a)-(e), we may conclude the following.

(a) For all parameter-configurations with $\theta_{(k)}>\theta_{(i)}$, $i\neq k$, $\hat{P}^*$ is substantially greater than P^*. Only in few cases $\hat{P}^*$ is smaller than P^* in the LFC. The small variation of $\hat{P}^*$ about P^* in the LFC indicates that the asymptotically distribution-free procedure yields correct results in case of finite sample-sizes, too.

(b) The average number of populations in the selected subset is rather large for k=4 and samples of size $n\leq 20$. For n=40 the values of $\hat{ET}$ are smaller than the corresponding values for $n\leq 20$, and $\hat{ET}/k$ seems to decrease as k becomes large.

(c) The influence of P^* on $\hat{P}^*$ is very small, if we have no LFC and as long as $P^*\geq 0.70$. On the contrary, with decreasing P^*, $\hat{ET}$ becomes substantially smaller.

---

[1] *If X is N(0,1)-distributed and Y possesses a double-exponential distribution with parameter $\mu=0$ and $\sigma=1$, then we have Var(X)=1, but Var(Y)=2·1. To make location shifts of X and Y comparable we have to shift X by Δ and Y by $\sqrt{2}\Delta$.*

(d) Considering the results of all our Monte-Carlo-studies in this section, we generally get the smallest $\widehat{ET}$, if the selection is based on Mann-Whitney-statistics. In case of cauchy distributed populations, the smallest $\widehat{ET}$ is generally obtained, if the selection is based on Median-statistics.

(e) The results of our studies give no rise to the assumption that the scale-parameter-configuration has any influence on R1u. The desirable independence of the procedures of the given configuration seems to be true.

Table 1

n=	12				20				40			
P* = 0.90	parameter-configuration (1,0,1.2,1.6,2.0)/(0.0,0.3a,0.6a,0.9a)											
	WILC		MED		WILC		MED		WILC		MED	
	$\widehat{P^*}$	$\widehat{ET}$	$\widehat{P^*}$	$\widehat{ET}$	$\widehat{P^*}$	$\widehat{ET}$	$\widehat{P^*}$	$\widehat{ET}$	$\widehat{P^*}$	$\widehat{ET}$	$\widehat{P^*}$	$\widehat{ET}$
NOR	0.99	3.31	0.98	3.37	1.00	3.27	1.00	3.20	1.00	2.87	1.00	2.94
DE	1.00	3.22	0.98	3.21	1.00	3.09	1.00	3.16	1.00	2.72	1.00	2.57
CAU	0.99	3.46	0.99	3.53	0.96	3.39	1.00	3.42	1.00	2.60	1.00	2.45
GAM	1.00	3.26	0.99	3.32	1.00	3.25	1.00	3.40	1.00	3.00	1.00	2.95
P* = 0.90	parameter-configuration (1.0,1.2,1.6,2.0)/(0.0,0.3a,0.6a,0.9a)											
NOR	0.99	3.23	0.98	3.28	0.99	3.01	0.99	3.20	1.00	2.77	1.00	2.94
DE	0.99	3.08	0.99	3.12	0.99	2.97	1.00	3.06	1.00	2.77	1.00	2.83
CAU	0.97	3.32	0.98	3.36	0.97	3.28	1.00	3.28	1.00	2.79	1.00	2.80
GAM	0.98	3.33	0.98	3.26	1.00	3.13	1.00	3.32	1.00	3.20	1.00	3.26

Table 2

n=	12				20				40			
$P^* = 0.90$	parameter-configuration (2.0,1.6,1.2,1.0)/(0.0,0.0,0.0,0.5a)											
	WILC		MED		WILC		MED		WILC		MED	
	$\widehat{P^*}$	$\widehat{ET}$	$\widehat{P^*}$	$\widehat{ET}$	$\widehat{P^*}$	$\widehat{ET}$	$\widehat{P^*}$	$\widehat{ET}$	$\widehat{P^*}$	$\widehat{ET}$	$\widehat{P^*}$	$\widehat{ET}$
NOR	0.99	3.43	0.99	3.40	1.00	3.24	1.00	3.45	1.00	3.12	1.00	3.16
DE	1.00	3.34	0.98	3.30	1.00	3.26	1.00	3.41	1.00	2.93	1.00	2.88
CAU	0.99	3.35	1.00	3.44	1.00	3.31	1.00	3.25	1.00	2.95	1.00	2.72
GAM	1.00	3.24	1.00	3.26	1.00	2.99	0.99	3.09	1.00	2.73	1.00	2.66
$P^* = 0.90$	parameter-configuration (1.0,1.2,1.6,2.0)/(0.0,0.0,0.0,1.0a)											
NOR	1.00	3.21	1.00	3.24	1.00	3.01	1.00	3.31	1.00	2.56	1.00	3.04
DE	1.00	3.17	1.00	3.21	1.00	3.02	1.00	3.16	1.00	2.53	1.00	2.93
CAU	0.99	3.42	0.99	3.43	1.00	3.33	1.00	3.25	1.00	3.10	1.00	3.11
GAM	0.97	3.25	0.97	3.27	1.00	3.29	1.00	3.40	1.00	3.13	1.00	3.20

Table 3

P* = 0.90	n = 20							
parameter-configuration	(2.0,1.6,1.2,1.0)/ (0.0,0.0,0.0,0.0)				(1.0,1.2,1.6,2.0)/ (0.0,0.0,0.0,0.0)			
	WILC		MED		WILC		MED	
	$\widehat{P^*}$	$\widehat{ET}$	$\widehat{P^*}$	$\widehat{ET}$	$\widehat{P^*}$	$\widehat{ET}$	$\widehat{P^*}$	$\widehat{ET}$
UNI	0.90	3.59	0.92	3.62	0.85	3.57	0.88	3.64
NOR	0.92	3.52	0.93	3.60	0.92	3.59	0.91	3.65
DE	0.92	3.55	0.95	3.52	0.95	3.64	0.89	3.58
CAU	0.89	3.65	0.91	3.67	0.91	3.57	0.90	3.59
GAM	0.92	3.51	0.92	3.57	0.91	3.64	0.90	3.65

Table 4

P* = 0.90		n = 20							
parameter-configu-ration		(0.0,0.0, 0.0,0.0)		(0.0,0.0, 0.0,0.1a)		(0.0,0.0, 0.0,0.3a)		(0.0,0.0, 0.0,0.5a)	
		$\widehat{P^*}$	$\widehat{ET}$	$\widehat{P^*}$	$\widehat{ET}$	$\widehat{P^*}$	$\widehat{ET}$	$\widehat{P^*}$	$\widehat{ET}$
UNI	WILC	0.92	3.59	0.94	3.58	0.99	3.42	1.00	3.36
	MED	0.91	3.64	0.95	3.66	0.96	3.63	1.00	3.53
NOR	WILC	0.91	3.58	0.93	3.55	1.00	3.39	1.00	3.18
	MED	0.88	3.62	0.95	3.61	1.00	3.50	1.00	3.34
DE	WILC	0.88	3.66	0.93	3.50	1.00	3.31	1.00	3.21
	MED	0.88	3.63	0.94	3.52	1.00	3.40	1.00	3.43
CAU	WILC	0.89	3.56	0.95	3.53	1.00	3.18	1.00	3.13
	MED	0.93	3.66	0.98	3.59	1.00	3.26	1.00	3.05
GAM	WILC	0.90	3.57	0.93	3.59	1.00	3.45	1.00	3.35
	MED	0.90	3.67	0.93	3.54	0.98	3.53	1.00	3.43

Table 5

parameter-configuration (0.0,0.0,0.0,0.5a) n=20								
P* =	0.80				0.70			
	WILC		MED		WILC		MED	
	$\widehat{P^*}$	$\widehat{ET}$	$\widehat{P^*}$	$\widehat{ET}$	$\widehat{P^*}$	$\widehat{ET}$	$\widehat{P^*}$	$\widehat{ET}$
UNI	1.00	2.60	1.00	2.87	1.00	2.43	1.00	2.90
NOR	1.00	2.90	0.93	2.96	1.00	2.27	1.00	2.80
DE	1.00	2.73	1.00	2.70	1.00	2.20	1.00	2.66
CAU	1.00	2.50	1.00	2.46	1.00	2.70	1.00	2.53
GAM	1.00	3.03	1.00	3.00	0.97	2.40	0.96	2.96

Table 6

P* = 0.90 n = 20								
param.-conf.	(2.0,2.0,2.0,1.6,1.2,1.0)/ (0.0,0.0,0.0,0.3a,0.6a,0.9a)				(1.0,1.0,1.0,1.2,1.6,2.0)/ (0.0,0.0,0.0,0.3a,0.6a,0.9a)			
	WILC		MED		WILC		MED	
	$\widehat{P^*}$	$\widehat{ET}$	$\widehat{P^*}$	$\widehat{ET}$	$\widehat{P^*}$	$\widehat{ET}$	$\widehat{P^*}$	$\widehat{ET}$
NOR	1.00	4.70	1.00	4.71	1.00	4.70	1.00	4.90
DE	1.00	4.50	1.00	4.56	1.00	4.53	1.00	4.59
CAU	1.00	4.25	1.00	4.10	1.00	4.35	1.00	4.12

In all tables we have used:

Distribution	UNI	NOR	DE	CAU	GAM
a	0.28	1.0	$\sqrt{2}$	2.0	1.0

3 The Selection Rule R1 in the Case of Equal Scale-Parameters

In this section, we make direct use of the results of section 2, to derive a class of subset-selection procedures based on rule R1, which is usable, if we may assume that the parameters of the populations in research differ only in such kind that an ordering of the populations is possible.

3.1 The probability P(CS|R1) in the LFC

Let $F_1,\dots,F_k$ be the continuous cdf's of the populations $\Pi_1,\dots,\Pi_k$, respectively, and let the F_j be identical apart from some parameter δ, which allows the ordering of the cdf's of $\Pi_1,\dots,\Pi_k$ in the following way:

$$F_\delta(x) \leq F_{\delta'}(x),\ \forall x\in\mathbb{R},\ \text{if } \delta > \delta'.^{1} \tag{3.1}$$

In the following, we choose the medians as selection criterion, because all parameters δ, which satisfy (3.1), have a direct influence on the size of the median and, therefore, the ordering of the populations with respect to some parameter δ is the same as that one with respect to the median. Hence with the same notations as in section 2, it is immediately seen that

$$\inf_{\Omega} P(CS|R1)=\inf_{\Omega_0} P(CS|R1),\quad \Omega_0=\{\theta=(\theta_1,\dots,\theta_k):\theta_{(1)}=\dots=\theta_{(k)}=\theta_0\}, \tag{3.2}$$

i.e. the minimal probability of the event "CS" is reached, if the medians of the k populations are identical. Without loss of generality we assume $\theta_0=0$. Thus in case of a LFC the continuous cdf's $F_1,F_2,\dots,F_k$ are identical, and therefore, as outlined in section 1, the r.-o.s.'s

$$S_j := \sum_{i=(j-1)n+1}^{jn} J\left(\frac{R_i}{m+1}\right),\quad j=1,\dots,k \tag{3.3}$$

are asymptotically normal distributed with mean and variance:

$$E(S_j)=n\bar{a},\quad Var(S_j)=\frac{n^2(k-1)}{m(m-1)}\sum_{i=1}^{m}(a_i-\bar{a})^2, \tag{3.4}$$

where $a_i=J(\frac{i}{m+1})$, and $\bar{a}=\bar{J}$.

In a LFC the r.-o.s.'s S_j, $j=1,\dots,k$, are interchangeable and the critical values D_j, $1\leq j\leq k$, are all equal to D. Thus to guarantee the validity of the pro-

---

[1] *In case of gamma-distributed populations this condition is satisfied by location- and scale-parameters.*

bability condition

$$\inf_{\Omega} P(CS|R1) \geq P^* , \tag{3.5}$$

we have to determine only that value D, which satisfies

$$P(S_j \geq D) = P^* . \tag{3.6}$$

Using the asymptotic distribution of S_j, we get the critical value

$$D = x\sqrt{Var(S_j)} + E(S_j), \text{ where } x = \psi(1-P^*). \tag{3.7}$$

Both methods of breaking ties, given in section 1, can be used. The preceding derivations remain valid, if the parameters $E(S_j)$ and $Var(S_j)$ are substituted by those which are associated with the statistics S_j^* and $\bar{S}_j$, respectively (cf. 1.4).

Up to now we have discussed the selection rule based on the premises that statistics of the form (3.3) are used. An interesting extension of the class of procedures based on rule R1 can be obtained by use of a statistic, which is no r.-o.s., but which possess an asymptotic distribution that is also independent of the underlying cdf's.

3.2 The Haga-statistic

A special procedure based on the selection R1, usable in situations described in 3.1, results from the use of a statistic, proposed by Haga [84] for two-sample location tests.

Let $(x_{i1},\dots,x_{in})$, $1\leq i\leq k$, be the realization of the sample drawn from population Π_i, $1\leq i\leq k$. As we have already seen in 2.3, for each index j, $1\leq j\leq k$, the k samples can be divided in such a way that $(x_{j1},\dots,x_{jn})$ is the realization of a first sample $(X_1,\dots,X_n)$, and the combined remaining k-1 realizations are the realization of a second sample $(Y_1,\dots,Y_{(k-1)n})$. The Haga-statistic associated with this two samples is defined as follows:

$SK_j := A_j - A_j^c - B_j + B_j^c$, $j=1,\dots,k$, where

A_j:=number of X_i's, $1\leq i\leq n$, greater than $\max\{Y_1,\dots,Y_{(k-1)n}\}$,

A_j^c:=number of Y_i's, $1\leq i\leq (k-1)n$, greater than $\max\{X_1,\dots,X_n\}$,

B_j:=number of X_i's, $1\leq i\leq n$, smaller than $\min\{Y_1,\dots,Y_{(k-1)n}\}$,

B_j^c:=number of Y_i's, $1\leq i\leq (k-1)n$, smaller than $\min\{X_1,\dots,X_n\}$.

The statistic SK_j is a nondecreasing function of X_{ji}, $1 \le i \le n$, if all other components of the k samples are held fixed. Thus with A4/8, (3.1) can easily be seen to be satisfied by the selection procedure based on the Haga-statistic.

To compute the critical value D in the LFC, we use the limit distribution of the statistic SK_j, j=1,...,k.
In order to derive the limit distribution of SK_j, we consider the problem of assigning n X-values and (k-1)n Y-values to m places 1,2,...,m. Let p:=1/k and q:=(k-1)/k denote the probabilities of placing an X-value or an Y-value, respectively, to any of the places 1,...,m. Then, we have as $n \to \infty$:

$$P(A_j=i)=p^i q,\ P(A_j^c=i)=pq^i,\ P(B_j=i)=p^i q,\ P(B_j^c=i)=pq^i.$$

First we determine the probability function of SK_j for $k>2$, $s \in \mathbb{Z}$.
We get

$$P(SK_j=s)=\begin{cases} p^2q^2\left(\frac{1}{1-p^2}+\frac{1}{1-q^2}\right) & \text{, if } s=0 \\ \frac{pq(q^2+p)}{(q-p)(1+q)}\,q^s - \frac{pq(p^2+q)}{(q-p)(1+p)}\,p^s, & \text{if } s\neq 0. \end{cases}$$

Using this, one can easily show that

$$P(SK_j\geq s)=\begin{cases} \frac{q^2(q^2-p^2+p+pq)}{(q-p)(1+q)} - \frac{p^2(p^2-q^2+q+pq)}{(q-p)(1+p)}, & \text{if } s=0 \\ \frac{(q^2+p)q}{(q-p)(1+q)}\,q^s - \frac{(p^2+q)p}{(q-p)(1+p)}\,p^s & \text{, if } s\geq 1 \\ 1-P(SK_j\geq -s+1) & \text{, if } s\leq -1. \end{cases}$$

Using statistics SK_j, j=1,...,k, for a selection procedure based on rule R1, the condition (3.5) is satisfied, if the critical value D is the solution of

$$P(SK_j\geq D) = P^*. \qquad (3.8)$$

Since the SK_j, j=1,...,k, are discrete random variables, there exists no solution of (3.8) in general. Therefore, we choose D such that the inequalities $P(SK_j\geq D)\geq P^*$ and $P(SK_j\geq D+1)<P^*$ are satisfied. Table I in appendix 6 gives values of $P(SK_j\geq D)$ for k=3,...,10.

In case of ties, the procedure based on Haga-statistics remains correct, if the ties are broken by randomization.

3.3 A numerical example

To show the practical use of the class of selection procedures based on rule R1, we take up the example given in 2.5, and assume additionally, that preliminary investigations justify the assumption that the three scale parameters are equal. Again with $P^*=0.90$, we want to select a subset of the three varieties of wheat.

We consider the application of the 4 different scorefunctions given in section 1 and the Haga-statistic.

Again the implementation of the five special procedures is simplified, by calculating first the rank-configuration $K=(K_1,K_2,\ldots,K_m)$. In our example, we have

rank vector	1	2	3	4	5	6	7	8	9	10	11	12	13	14	15	16	17	18
rank-configuration	1	1	3	1	1	1	3	3	2	2	2	3	1	2	1	3	1	2
rank vector	19	20	21	22	23	24	25	26	27	28	29	30	31	32	33	34	35	36
rank-configuration	3	1	2	2	1	3	3	3	1	2	3	1	2	2	2	2	3	3

This table contains all information,we must have for computing the five different triples of statistics. To compute D, we need the value x, with $x=\psi(1-P^*)$. For $P^*=0.90$, we get $x=-1.28125$.

The following table shows the results of the five special procedures in short form, where WILC, WAE, MED, PMS and HAGA are used as abbreviations for the procedures based on Wilcoxon-, Van-der-Waerden-, Median-, Percentile-modified $(p=0.0,q=0.5)$- and Haga-statistics, respectively.

		WILC	WAE	MED	PMS	HAGA
S_1		4.405	-5.675	4	-89	-8
S_2		7.108	3.644	7	-33	6
S_3		6.486	2.031	7	-49	4
$E(S_i)$	$P(SK_i)$	6.0	0.0	6	-57	$P(SK_i\geq -5)=0.92$
$Var(S_i)$		0.648	6.872	2.057	296	$P(SK_i\geq -4)<0.90$
D		4.968	-3.359	4.162	-79.06	-5
subset		Π_2,Π_3	Π_2,Π_3	Π_2,Π_3	Π_2,Π_3	Π_2,Π_3

The last line in the preceding table shows that the 5 procedures yield all the same subset, containing Π_2 and Π_3. This may give rise to the assumption that there are no differences between the efficiencies of the considered procedures. But this conjecture is false.

To illustrate the differences, we determine for each procedure the probability P_i^{crit}, i=1,...,5, of just excluding population Π_1, when the special statistic i has been used. For the r.-o.s.'s, we get these probabilities by using the following formula:

$$P(S_j \geq s_1) = P\left(\frac{S_j - E(S_j)}{\sqrt{Var(S_j)}} \geq \frac{s_1 - E(S_j)}{\sqrt{Var(S_j)}}\right) = 1-\Phi(x) = P_i^{crit},$$

where $x=(s_1-E(S_j))/\sqrt{Var(S_j)}$ and s_1 is the realization of S_1.
In detail we get:

i	procedure	x/P(SK)	P_i^{crit}
1	WILC	-1.9799	0.9761
2	WAE	-2.1649	0.9848
3	MED	-1.3945	0.9184
4	PMS	-1.8587	0.9684
5	HAGA	$P(SK \geq -9)=0.976$	0.976

Comparing the critical probabilities P_i^{crit}, WAE seems to be the best and MED seems to be the worst procedure in this case. As for the differences between the efficiencies of the special procedures, we refer the reader to the following part of the present section. An analytic consideration of the interdependence of special statistics and the underlying distribution will be given in section 5.

3.4 Some Monte-Carlo studies

The procedures, considered in this section, are denoted by R1g in the following.
Just as for R1u, the analytic investigation of R1g seems to be very difficult. Hence, to answer the questions (a)-(d), given in 2.6, we have carried out some Monte-Carlo studies for R1g.

The simulation-model, given in 2.6, has been used to simulate, besides the distributions considered there, the exponential distribution (EXP).
The five special procedures considered in 3.3, too, have been applied to the k realizations obtained in the known way.
For all parameter-configurations $\omega=(a\Delta_1, a\Delta_2, \ldots, a\Delta_k)$, $\sigma_i=\sigma$, i=1,...,k and $(\sigma_1, \sigma_2, \ldots, \sigma_k)/(a\Delta_1, a\Delta_2, \ldots, a\Delta_k)$, we have carried out 100 repetitions. The factor a is defined as in 2.6, and the chosen sample size was always n=20.

Table 1 contains values of $\widehat{P^*}$ and $\widehat{ET}$ (cf. 2.6), in case of parameter-configurations with scale-parameters all equal to 1.0.
For the purpose of comparison, we have simulated two configurations with different scale-parameters but equal location-parameters (in case of GAM- and EXP-distributed populations the parameters are chosen in such a way that the medians are equal). The corresponding values of $\widehat{P^*}$ and $\widehat{ET}$ are given in table 2.
To prove the influence of P* on R1g, we have computed table 3. Finally table 4 contains some results for k=6 populations.
After considering the results of our studies, the following remarks, concerning questions (a)-(d), may be given:

(a) Except of those, given in table 2, the results show that $\widehat{P^*}$ is substantially greater than P*, as long as we have no LFC. In case that the medians are all equal, the values of $\widehat{P^*}$ vary in the interval [P*-0.03,P*+0.07]. Thus we may conclude, that the approximation of the minimal probability of CS yields rather good results in case of finite sample sizes.

(b) The number of selected populations depends on the given parameter-configuration. For k=4, only one population is eliminated on an average. In case of configuration (0.0,0.3a,0.6a,0.9a), however, at least one population is eliminated. $\widehat{ET}/k$ decreases as k becomes large.

(c) As for R1u,a decrease of P* has scarcely influence on $\widehat{P^*}$. On the contrary, $\widehat{ET}$ becomes smaller, as P* decreases.

(d) The results of our studies show clearly that the choice of the special statistic has influence on the efficiency of the selection procedure, i.e. on the size of $\widehat{ET}$. The efficiency of a special procedure seems to depend on the underlying distribution. Only in case of configurations with small differences $(\Delta_{(k)}-\Delta_{(i)})$, $i\neq k$, the differences between efficiencies are blurred.

On an average, the procedure based on Wilcoxon-statistics seems to be the most efficient. Its use can be called advantageous, especially in case of NOR- and DE-distributed populations.
For UNI- and NOR-distributed populations, the procedure based on Van-der-Waerden-statistics yields the smallest values of $\widehat{ET}$. The procedure based on Haga-statistics seems to be of equivalent efficiency, if the populations are UNI-distributed.
The Median-form of the selection procedure can be called most efficient in case of CAU-distributed populations.
The use of Percentile-modified-statistics (p=0.0,q=0.5) yields the best

procedure, if the distributions of the populations are skewed to the right.

The results in table 2 show that, depending on the parameter-configuration, the probability condition is not always satisfied, if, in case of equal medians, the scale-parameters are different. This gives rise to the assumption that in case of different scale-parameters and only small differences between the medians, the procedures R1g do not always satisfy the probability condition.

Table 1

$P^* = 0.90$ $n = 20$													
$\omega =$		ω_1		ω_2		ω_3		ω_4		ω_5		ω_6	
		$\widehat{P^*}$	$\widehat{ET}$	$\widehat{P^*}$	$\widehat{ET}$	$\widehat{P^*}$	$\widehat{ET}$	$\widehat{P^*}$	$\widehat{ET}$	$\widehat{P^*}$	$\widehat{ET}$	$\widehat{P^*}$	$\widehat{ET}$
	HAGA	0.92	3.76	0.98	3.69	1.00	3.62	1.00	3.38	0.97	3.46	1.00	2.90
U	WAE	0.91	3.57	0.98	3.64	0.99	3.36	1.00	3.18	0.98	3.42	1.00	2.88
N	WILC	0.92	3.59	0.96	3.65	0.99	3.40	0.99	3.27	0.97	3.48	1.00	2.95
I	MED	0.89	3.56	0.93	3.62	0.97	3.54	0.99	3.40	0.93	3.50	0.99	3.21
	PMS	0.90	3.54	0.94	3.69	0.99	3.45	0.99	3.31	0.98	3.52	1.00	3.04
	HAGA	0.95	3.77	0.97	3.75	0.99	3.60	0.99	3.61	0.97	3.66	1.00	3.29
N	WAE	0.90	3.66	0.96	3.54	1.00	3.43	1.00	3.29	0.98	3.55	1.00	2.82
O	WILC	0.90	3.64	0.96	3.56	1.00	3.40	1.00	3.24	0.97	3.53	1.00	2.85
R	MED	0.91	3.63	0.93	3.54	1.00	3.42	1.00	3.35	0.96	3.56	1.00	3.02
	PMS	0.90	3.63	0.93	3.55	0.99	3.45	1.00	3.32	0.96	3.59	1.00	2.97
	HAGA	0.97	3.77	0.95	3.74	0.98	3.71	1.00	3.63	0.96	3.66	1.00	3.38
D	WAE	0.92	3.59	0.96	3.64	0.99	3.48	1.00	3.18	0.99	3.39	1.00	2.81
E	WILC	0.92	3.55	0.96	3.65	0.98	3.48	1.00	3.17	0.99	3.42	1.00	2.78
	MED	0.91	3.57	0.96	3.63	0.99	3.58	0.99	3.25	0.99	3.38	1.00	2.63
	PMS	0.89	3.57	0.96	3.63	0.98	3.49	0.99	3.27	1.00	3.48	1.00	2.85
	HAGA	0.95	3.79	0.94	3.76	0.96	3.76	0.98	3.71	0.98	3.78	0.98	3.69
C	WAE	0.91	3.60	0.94	3.60	0.99	3.35	1.00	3.24	0.99	3.53	1.00	2.87
A	WILC	0.87	3.58	0.95	3.59	1.00	3.31	1.00	3.15	0.99	3.44	1.00	2.73
U	MED	0.89	3.62	0.97	3.57	1.00	3.30	1.00	3.09	0.99	3.38	1.00	2.63
	PMS	0.87	3.56	0.95	3.54	0.99	3.38	0.99	3.31	0.98	3.48	1.00	2.85
	HAGA	0.95	3.74	0.99	3.75	1.00	3.65	1.00	3.69	1.00	3.68	1.00	3.26
E	WAE	0.92	3.62	0.98	3.39	1.00	3.26	1.00	2.95	1.00	3.21	1.00	2.75
X	WILC	0.91	3.60	0.98	3.38	1.00	3.27	1.00	2.90	1.00	3.22	1.00	2.71
P	MED	0.87	3.57	0.99	3.58	0.99	3.51	1.00	3.16	0.97	3.44	1.00	2.78
	PMS	0.91	3.51	0.98	3.42	1.00	3.17	1.00	2.69	1.00	3.17	1.00	2.69
	HAGA	0.94	3.77	0.98	3.80	0.98	3.72	0.99	3.68	0.98	3.74	1.00	3.55
G	WAE	0.89	3.62	0.97	3.60	0.99	3.47	0.99	3.29	0.96	3.45	1.00	2.92
A	WILC	0.89	3.62	0.96	3.59	0.99	3.46	0.99	3.29	0.96	3.44	1.00	2.92
M	MED	0.89	3.65	0.95	3.66	0.98	3.51	0.99	3.35	0.94	3.45	1.00	3.13
	PMS	0.88	3.59	0.96	3.53	0.99	3.45	1.00	3.22	0.98	3.46	1.00	2.84

$\omega_1 = (0.0,0.0,0.0,0.0)$; $\omega_3 = (0.0,0.0,0.0,0.3a)$; $\omega_5 = (0.0,0.1a,0.2a,0.3a)$

$\omega_2 = (0.0,0.0,0.0,0.1a)$; $\omega_4 = (0.0,0.0,0.0,0.5a)$; $\omega_6 = (0.0,0.3a,0.6a,0.9a)$

Table 2

$P^* = 0.90$ $n = 20$					
parameter-configuration		(2.0,1.6,1.2,1.0)/ (0.0,0.0,0.0,0.0)		(1.0,1.2,1.6,2.0)/ (0.0,0.0,0.0,0.0)	
		$\widehat{P^*}$	$\widehat{ET}$	$\widehat{P^*}$	$\widehat{ET}$
	HAGA	0.88	3.82	0.98	3.83
U	WAE	0.95	3.62	0.86	3.69
N	WILC	0.94	3.60	0.89	3.64
I	MED	0.87	3.57	0.92	3.65
	PMS	0.99	3.26	0.52	3.27
	HAGA	0.85	3.70	0.97	3.69
N	WAE	0.93	3.68	0.86	3.58
O	WILC	0.92	3.68	0.85	3.54
R	MED	0.89	3.63	0.88	3.56
	PMS	0.99	3.44	0.63	3.37
	HAGA	0.83	3.66	0.98	3.65
D	WAE	0.92	3.63	0.87	3.58
E	WILC	0.90	3.64	0.87	3.57
	MED	0.85	3.57	0.89	3.57
	PMS	0.99	3.49	0.70	3.47
	HAGA	0.89	3.72	0.96	3.69
C	WAE	0.90	3.61	0.88	3.54
A	WILC	0.89	3.57	0.87	3.52
U	MED	0.90	3.55	0.92	3.59
	PMS	0.93	3.54	0.78	3.48
	HAGA	0.96	3.88	0.95	3.83
G	WAE	0.97	3.55	0.79	3.59
A	WILC	0.95	3.56	0.82	3.58
M	MED	0.90	3.56	0.89	3.57
	PMS	0.99	3.29	0.61	3.39

Table 3

parameter-configuration		(0.0,0.0,0.0,0.5a) n = 20			
		P*=0.80 in case of HAGA P*=0.84		P*=0.70 in case of HAGA P*=0.71	
		$\widehat{P^*}$	$\widehat{ET}$	$\widehat{P^*}$	$\widehat{ET}$
	HAGA	1.00	2.90	1.00	2.37
U	WAE	1.00	2.67	0.99	2.30
N	WILC	0.99	2.81	0.98	2.44
I	MED	0.96	2.99	0.96	2.97
	PMS	0.99	2.86	0.99	2.52
	HAGA	0.99	3.29	0.98	2.97
N	WAE	1.00	2.72	0.99	2.48
O	WILC	1.00	2.78	0.98	2.52
R	MED	0.99	2.86	1.00	2.86
	PMS	0.99	2.84	0.98	2.62
	HAGA	1.00	3.34	0.96	2.95
D	WAE	1.00	2.78	1.00	2.28
E	WILC	1.00	2.77	1.00	2.28
	MED	1.00	2.77	1.00	2.75
	PMS	1.00	2.92	0.99	2.45
	HAGA	0.94	3.44	0.83	3.11
C	WAE	1.00	2.82	0.99	2.32
A	WILC	1.00	2.67	0.99	2.26
U	MED	1.00	2.49	1.00	2.28
	PMS	0.99	2.90	0.99	2.44
	HAGA	0.97	3.47	0.99	3.31
G	WAE	0.99	2.93	1.00	2.61
A	WILC	0.98	2.89	1.00	2.62
M	MED	0.98	2.92	1.00	2.89
	PMS	0.99	2.79	1.00	2.61

Table 4

P* = 0.90	n = 20			in case of HAGA P* = 0.91						
parameter-configuration: (0.0,0.0,0.0,0.25a,0.375a,0.5a)										
	HAGA		WAE		WILC		MED		PMS	
	$\hat{P^*}$	$\hat{ET}$	$\hat{P^*}$	$\hat{ET}$	$\hat{P^*}$	$\hat{ET}$	$\hat{P^*}$	$\hat{ET}$	$\hat{P^*}$	$\hat{ET}$
UNI	1.00	4.91	1.00	4.87	1.00	5.04	0.97	5.09	0.99	5.05
NOR	0.97	5.08	0.99	4.39	0.99	4.46	0.99	4.61	0.99	4.67
DE	0.98	5.49	1.00	4.91	1.00	4.91	1.00	4.85	1.00	4.97
CAU	0.94	5.44	0.99	4.97	1.00	4.77	1.00	4.40	1.00	4.81
EXP	1.00	5.45	1.00	0.42	1.00	4.44	0.98	4.69	1.00	4.13
GAM	0.98	5.49	1.00	5.05	1.00	5.08	0.99	5.13	1.00	4.92
parameter-configuration: (0.0,0.0,0.0,0.5a,0.75a,1.0a)										
UNI	1.00	3.63	1.00	3.95	1.00	4.20	1.00	4.75	1.00	4.43
NOR	1.00	4.68	1.00	3.92	1.00	3.92	1.00	4.19	1.00	4.31
DE	1.00	4.95	1.00	3.85	1.00	3.73	1.00	3.84	1.00	4.08
CAU	1.00	5.49	1.00	3.96	1.00	3.72	1.00	3.49	1.00	4.07
EXP	1.00	5.17	1.00	3.48	1.00	3.45	1.00	3.71	1.00	3.31
GAM	1.00	5.22	1.00	3.88	1.00	3.86	1.00	3.97	1.00	3.87
Differences between medians caused by scale parameters (1.0,1.0,1.0,1.2,1.6,2.0), equal location parameters										
EXP	1.00	5.14	1.00	4.89	1.00	4.87	1.00	4.86	1.00	5.04
GAM	1.00	5.49	1.00	5.23	1.00	5.12	1.00	5.07	0.95	5.31

4 A Further Class of Subset-Selection Procedures in One-Factor Designs

The rule R1, discussed in sections 2 and 3 can be considered as the reduction of a k-sample problem to a two-sample problem. In this section we will propose a class of procedures based on a selection rule, which consists in the comparison of the r.-o.s. of each population with the maximal r.-o.s. of the remaining k-1 populations.

4.1 The selection rule R2

The derivations, we give in the following are valid only in case of the special problems, already considered in section 3. Hence, we assume that all assumptions made in 3.1 are satisfied. Then we define the selection rule

$$\text{R2: Select population } \Pi_i \iff S_i \geq \max_{1\leq j\leq k} S_j - d, \quad 1\leq i\leq k, \ d\geq 0,$$

where S_i is the r.-o.s. associated with Π_i, $1\leq i\leq k$, and the constant d is chosen as small as possible, satisfying

$$\inf_{\Omega} P(CS|R2)\geq P^*, \quad \Omega:=\{\theta=(\theta_1,\ldots,\theta_k):\theta_j\varepsilon I,\ 1\leq j\leq k,\ I\subseteq\mathbb{R}\}. \qquad (4.1)$$

In other words, in using rule R2, a population will be eliminated if the r.-o.s. associated with it takes a value smaller than $\max_{1\leq j\leq k} S_j - d$.

To derive the value d, we first must know, what $\theta\varepsilon\Omega$ minimizes the probability $P(CS|R2)$.

4.2 The infimum of the probability P(CS|R2)

The probability condition (4.1) means that $P(CS|R2)\geq P^*$ must be satisfied for all possible parameter-configurations $\theta\varepsilon\Omega$. Hence, we have to determine the LFC. In doing this, we get

Theorem 4.1:

Let $p_s(R2)$ denote the probability that the population $\Pi_{(s)}$ is included in the selected subset, if rule R2 is used. $p_s(R2)$ is nondecreasing in $\theta_{(s)}$, i.e. we have

$$\inf_{\Omega} p_s(R2) = \inf_{\Omega_s} p_s(R2),\ s=1,\ldots,k,$$

where $\Omega_s:=\{\theta\varepsilon\Omega:\theta_{(s)}=\theta_{(s-1)}\}$ and $\theta_{(o)}$ denotes the least admissible value of the median.

Proof:

Using a similar argument as in the proof of theorem 2.1, we immediately see that the function

$$H_s(\vec{X}):=\begin{cases} 1, \text{ if } S_{(s)}\geq \max\limits_{1\leq j\leq k} S_j-d \\ 0, \text{ otherwise}\end{cases}$$

is nondecreasing in $x_{(s)i}$, $1\leq i\leq n$. With A4/8 it follows:

$$E(H_s(\vec{X})) = p_s(R2) \text{ is nondecreasing in } \theta_{(s)}. \qquad \blacksquare$$

Using theorem 4.1, we have for s=k: $\inf\limits_{\Omega} P(CS|R2)=\inf\limits_{\Omega_k} P(CS|R2)$. This result is not as strong as that one for rule R1. A general expansion of theorem 2.1 to rule R2 is not possible[1], because of the dependence of the r.-o.s.'s $S_j, 1\leq j\leq k$.

To be able to implement selection procedures based on rule R2, we must evaluate the distribution of $\max\limits_{1\leq j\leq k} S_j-S_i$, $1\leq i\leq k$, in the LFC. But this seems to be very difficult. Therefore in the following we assume that a LFC is realized in case of equal medians, i.e. we assume that in the LFC all distributions of the k populations are identical. An heuristic argument for the validity of this assumption will be given in 4.5.

4.3 Exact and asymptotic distribution of $\max S_j-S_1$ for identically distributed populations

Assuming that the k populations are identically distributed, the statistics S_i are interchangeable. Hence without loss of generality, we choose i=1.
First we consider the exact distribution of $\max\limits_{1\leq j\leq k} S_j-S_1$.

---

1 *Rizvi and Woodworth [154] have chosen a special distribution, which has two finite disjoint intervals as support and does not possess the MLR-property. They have shown that in case of populations following this special distribution, the minimal probability P(CS|R2) is not obtained if all medians are equal.*

Let K be one of the $\underbrace{\binom{m}{n,n,\ldots,n}}_{k\text{-times}}$ equally likely rank-configurations (cf. 2.5).

To compute the probability $P(S_1 \geq \max_{1 \leq j \leq k} S_j - d)$ it is enough to evaluate the number of rank-configurations satisfying:

$$S_2-S_1 \leq d,\ S_3-S_1 \leq d, \ldots, S_k-S_1 \leq d.$$

This number of rank-configurations is obtained in using the following recursion formula I:

We define

$N(n_1,n_2,\ldots,n_k|d_2,d_3,\ldots,d_k)$:= number of rank-configurations with n_1 components equal to 1, n_2 components equal to 2,...,n_k components equal to k, satisfying $S_j-S_1 \leq d_j$, $2 \leq j \leq k$.

The following symmetry holds:

$$N(n_1,\ldots,n_i,n_{i+1},\ldots,n_k|d_2,\ldots,d_i,d_{i+1},\ldots,d_k) =$$
$$= N(n_1,\ldots,n_{i+1},n_i,\ldots,n_k|d_2,\ldots,d_{i+1},d_i,\ldots,d_k).$$

With respect to the number of population, indicated by the last component K_S, $S := \sum_{i=1}^{k} n_i$, of the present rank-configuration K, we get:

$$N(n_1,\ldots,n_k|d_2,\ldots,d_k) = N(n_1-1,n_2,\ldots,n_k|d_2+a_S,\ldots,d_k+a_S) +$$
$$+ \sum_{i=2}^{k} N(n_1,\ldots,n_i-1,\ldots,n_k|d_2,\ldots,d_i-a_S,\ldots,d_k),$$

where $a_S = J(\frac{S}{m+1})$ denotes the score associated with S.

Thereby, with $Q_i := \sum_{j-1}^{n_i} a_j$, $1 \leq i \leq k$, the following boundary conditions must be verified:

(a) If for any $i \geq 2$, $d_i \leq Q_i - \sum_{j=S-n_1+1}^{S} a_j$, then $N(n_1,\ldots,n_k|d_2,\ldots,d_k) = 0$.

(b) If for each $i \geq 2$, $d_i \geq \sum_{j=S-n_i+1}^{S} a_j - Q_1$, then $N(n_1,\ldots,n_k|d_2,\ldots,d_k) = \binom{S}{n_1,n_2,\ldots,n_k}$

(c) If an expression of the form $N(0,n_2,\ldots,n_k|d_2,\ldots,d_k)$ is reached, which counts all rank-configurations, satisfying $S_i \le d_i$, $2 \le i \le k$, then we make use of

<u>recursion formula II:</u>

We replace

$$N(0,n_2,\ldots,n_k|d_2,\ldots,d_k) \text{ by } N(n_2,\ldots,n_k|d_2,\ldots,d_k),$$

and we have

$$N(n_2,\ldots,n_k|d_2,\ldots,d_k)=\sum_{i=2}^{k} N(n_2,\ldots,n_i-1,\ldots,n_k|d_2,\ldots,d_i-a_S,\ldots,d_k),$$

with boundary conditions:

(c1) If for any $i \ge 2$, $d_i < Q_i$, then $N(n_2,\ldots,n_k|d_2,\ldots,d_k)=0$.

(c2) If for each $i \ge 2$, $d_i \ge Q_i$, then

$$N(n_2,\ldots,n_{i-1},0,n_{i+1},\ldots,n_k|d_2,\ldots,d_{i-1},d_i,d_{i+1},\ldots,d_k) =$$
$$= N(n_2,\ldots,n_{i-1},n_{i+1},\ldots,n_k|d_2,\ldots,d_{i-1},d_{i+1},\ldots,d_k).$$

(c3) For $d_i \ge Q_i$, we have $N(n_i|d_i)=1$, $2 \le i \le k$.

(d) If an expression $N(n_1,\ldots,n_{i-1},0,n_{i+1},\ldots,n_k|d_2,\ldots,d_k)$ is reached, including all rank-configurations, satisfying

$S_2-S_1 \le d_2,\ldots,S_{i-1}-S_1 \le d_{i-1}, S_i \ge -d_i, S_{i+1}-S_1 \le d_{i+1},\ldots,S_k-S_1 \le d_k$, then the recursion formula I is used again.

(e) If an expression $N(n_1,0,\ldots,0,n_i,0,\ldots,0|d_2,\ldots,d_k)$, including all rank-configurations, satisfying the inequalities: $S_1 \ge -d_j$, $2 \le j \ne i \le k$, and $S_i - S_1 \le d_i$, i.e.

$$S_1 \ge M := \max\{\max_{j \ne i}(-d_j), \sum_{j=1}^{S} \frac{a_j - d_i}{2}\}$$, then the following cases are possible:

(e1) If $M > \sum_{j=n_i+1}^{n_i+n_1} a_j$, then $N(n_1,0,\ldots,0,n_i,0,\ldots,0|d_2,\ldots,d_k)=0$.

(e2) If $M < Q_1$, then $N(n_1,0,\ldots,0,n_i,0,\ldots,0|d_2,\ldots,d_k) = \binom{S}{n_1}$.

(e3) If (e1) and (e2) fail to hold, then $N(n_1,0,\ldots,0,n_i,0,\ldots,0|d_2,\ldots,d_k)$ is the number of rank-configurations, satisfying $S_i \le \sum_{j=1}^{n_1+n_i} a_j - M$. This number is obtained in evaluating for two samples of sizes n_1 and n_i, the number of rank-configurations, satisfying $S_1 \ge M$.

The exact distribution of $\max_{1\leq j\leq k} S_j - S_1$ is given by

$$P(S_1 \geq \max_{1\leq j\leq k} S_j - d) = \frac{N(\overbrace{n,\ldots,n}^{k\text{-times}}|\overbrace{d,\ldots,d}^{(k-1)\text{-times}})}{\binom{m}{\underbrace{n,n,\ldots,n}_{k\text{-times}}}} . \quad (4.2)$$

A short example may illustrate the use of the recursion formulae:
The number $N(2,2,2|\frac{1}{7}, \frac{1}{7})$ is evaluated for the special scorefunction $J(u)=u$ (Wilcoxon-scores), i.e. for scores $a_i=i/(m+1)$, with $k=3$, $n=2$ and $d=1/(m+1)$.

For simplification of writing, a_i and d are replaced by $i=(m+1)a_i$ and $(m+1)d=1$, respectively. $N(\ldots|\ldots)$ is shortly written as AN.
Adding up the values AN, given at the endpoints of the tree, given in figure 1, we get $N(2,2,2|\frac{1}{7},\frac{1}{7})=44$. Noting that $\binom{6}{2,2,2} = 90$ rank-configurations are possible, we get the probability

$$P(S_1 \geq \max_{j=2,3} S_j - \frac{1}{7}) = \frac{44}{90} = 0.4\bar{8}.$$

The evaluation of the exact distribution is very extensive even if k and n are small. A practicable way to determine critical values d is offered by the asymptotic distribution of $\max_j S_j - S_1$, formulated in

Theorem 4.2:

Let S_j, $1\leq j\leq k$, be the r.-o.s.'s defined in section 1. If the cdf's of the populations $\pi_1, \ldots, \pi_k$ are identical, then

$$\lim_{n\to\infty} P(S_1 \geq \max_{2\leq j\leq k} S_j - d) = \int_{-\infty}^{+\infty} \Phi(x+\frac{d}{z})^{k-1} \, d\Phi(x),$$

where

$$z^2 = \frac{k}{k-1}\sigma^2 \text{ and } \sigma^2 = \mathrm{Var}(S_j),\ 1\leq j\leq k.$$

Proof:

Let Y_{ij}, $1\leq i\leq k$, $1\leq j\leq n$, be $N(0,1)$-distributed random variables. It is easy to verify that the random variables

$$Z_i = \sqrt{n}(Y_{i.} - Y_{..}),\text{ where } Y_{i.} = \frac{1}{n}\sum_{j=1}^{n} Y_{ij},\ Y_{..} = \frac{1}{m}\sum_{i=1}^{k}\sum_{j=1}^{n} Y_{ij},\ i=1,\ldots,k,$$

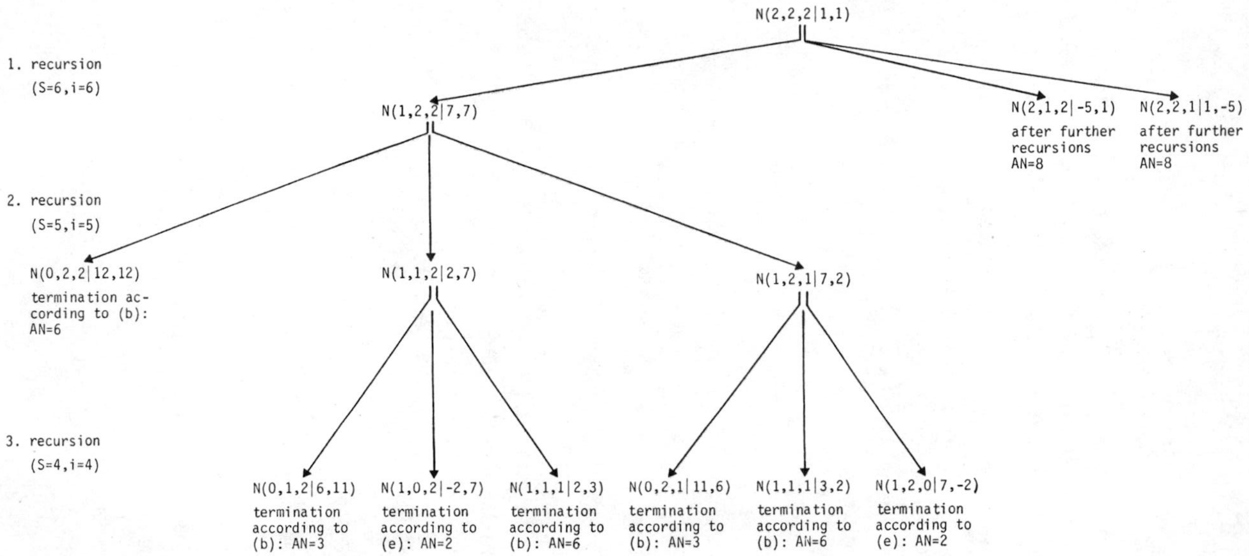
N(2,2,2|1,1)
1. recursion
(S=6,i=6)
N(1,2,2|7,7)
N(2,1,2|-5,1)
after further recursions AN=8
N(2,2,1|1,-5)
after further recursions AN=8
2. recursion
(S=5,i=5)
N(0,2,2|12,12)
termination according to (b): AN=6
N(1,1,2|2,7)
N(1,2,1|7,2)
3. recursion
(S=4,i=4)
N(0,1,2|6,11)
termination according to (b): AN=3
N(1,0,2|-2,7)
termination according to (e): AN=2
N(1,1,1|2,3)
termination according to (b): AN=6
N(0,2,1|11,6)
termination according to (b): AN=3
N(1,1,1|3,2)
termination according to (b): AN=6
N(1,2,0|7,-2)
termination according to (e): AN=2

have a k-variate normal distribution, with

$$E(Z_i)=0,\ Var(Z_i)=\frac{k-1}{k},\ 1\leq i\leq k, \text{ and } Cov(Z_i,Z_\ell)=-\frac{1}{k},\ 1\leq i\neq \ell\leq k.$$

Recalling theorem 1.9, we know that the random variables

$$V_j=(S_j-E(S_j))/\sqrt{\frac{k}{k-1}Var(S_j)},\ j=1,\dots,k,$$

have asymptotically a k-variate normal distribution, with

$$E(V_j)=0,\ Var(V_j)=\frac{k-1}{k},\ 1\leq j\leq k, \text{ and } Cov(V_i,V_\ell)=-\frac{1}{k},\ 1\leq i\neq \ell\leq k.$$

Hence the asymptotic distribution of $(V_1,\dots,V_k)$ and that one of $(Z_1,\dots,Z_k)$ coincide, and we have

$$\lim_{n\to\infty} P(\max_{2\leq j\leq k} S_j-S_1\leq d) = \lim_{n\to\infty} P(\max_{2\leq j\leq k} V_j-V_1 \leq \frac{d}{\sqrt{\frac{k}{k-1}\sigma^2}}) =$$

$$= P(\max Z_j-Z_1\leq \frac{d}{z}) = \int_{-\infty}^{+\infty} \Phi(x+\frac{d}{z})^{k-1}\, d\Phi(x),$$

where $z:=\sqrt{\frac{k}{k-1}\sigma^2}$.

This proves the assertion of the theorem. ■

To preassigned probability P^*, the critical value d is given by $d=h\sqrt{\frac{k}{k-1}\sigma^2}$, where h is the solution of $\int_{-\infty}^{+\infty} \Phi(x+h)^{k-1}d\Phi(x)=P^*$.

For various values of k and P^* some h-values are given in table II, appendix 6.

Tables III, IV and V in appendix 6 contain critical values d, associated with various values k, n and some special scorefunctions. There are values, computed by using the exact- and the asymptotic distribution of $\max_{2\leq j\leq k} S_j-S_1$. Additionally in column N(..n..|..d..) the number of "favorable" rank-configurations, associated with d, is given. From the comparison of the asymptotic and exact values, we see that the approximation of the probability yields rather good results, even for very small sample sizes.

Considering the tables, we note that for some fixed critical value d, the exact probability is always smaller than the asymptotic one. Thus the use of critical values, asymptotically determined, leads to somewhat conservative procedures. To get an idea of the effect of using asymptotic critical values, the reader is referred to 4.5.

In the presence of ties both methods of breaking ties, given in section 1, may be used.

4.4 A numerical example

Again we consider the example given in 2.5, and assume additionally that the scale-parameters of the three populations Π_1,Π_2 and Π_3 are equal.

In using rule R2, we want to select a subset from the three varieties of wheat. The same four types of r.-o.s.'s, also used in 3.3, are considered, and $P^* = 0.90$ is chosen.

Except of the determination of critical values d, the calculations for rule R2 are the same as in 3.3.

Using table II, given in appendix 6, we see that h=2.2302 solves the equation $\int_{-\infty}^{+\infty} \Phi(x+h)^{k-1} d\Phi(x)=0.90$. Hence, the critical values d are given by $d=2.2302\cdot\sqrt{1.5\sigma^2}$, where σ^2 denotes the variance of the specified r.-o.s..

The following table shows in short form the results of the four special procedures, where we use the same abbreviations as in section 3.

	WILC	WAE	MED	PMS
S_1	4.405	-5.675	4	-89
S_2	7.108	3.644	7	-33
S_3	6.486	2.031	7	-49
$\max_{1\le j\le 3} S_j$	7.108	3.644	7	-33
σ^2	0.643	6.872	2.057	296
d	2.199	7.160	3.917	47.025
subset	Π_2,Π_3	Π_2,Π_3	Π_2,Π_3	Π_2,Π_3

It can be seen from the table that each of the four procedures selects the same subset, that is the subset consisting of Π_2 and Π_3. But in using rule R2 the efficiency of the selection procedure is influenced by the specified score-function, too. To support this assertion, we have carried out some Monte-Carlo-studies.

4.5 Some Monte-Carlo studies

Just as for rule R1, it is very difficult to give any analytic statements concerning the efficiency of rule R2. To answer the questions (a)-(d), formulated in section 2, with respect to rule R2, we have carried out some Monte-Carlo simulations.

In addition question (a) is supplemented by the question, whether the evaluation of critical values, based on equal medians can be justified.

To get comparable results, the simulation of R2 has been carried out, in using the same simulation-model and the same parameter-configurations, which have been chosen for rule R1 in section 3. The four special procedures based on rule R2, considered in 4.4, have been applied to the k realizations, generated by use of the method described in appendix 5. For each configuration, the process is repeated 100-times. The PMS-statistic is used in two cases, that is p=0, q=0.5 and p=q=0.25. Therefore the actual values of p and q are additionally given at the end of the tables.

Corresponding to table 1 in section 3, we have evaluated the first table of this section. To answer the supplementary question to (a), we have simulated two real (cf. theorem 4.1) LFC's, i.e. two configurations for which the two greatest medians are equal but slightly larger than the remaining k-2 medians. The results are given in table 2. Table 3 can be used to answer question (c). For k=6, we have considered two different configurations. The corresponding values for $\widehat{P^*}$ and $\widehat{ET}$ are combined in table 4.

Concerning questions (a)-(d), the following may be concluded:

(a) Apart from LFC's (we speak of LFC's if all medians are equal), all simulated parameter-configurations are associated with values of $\widehat{P^*}$, which are greater than P*. Also in cases, where only the two largest medians are equal, the $\widehat{P^*}$-values are almost without exception substantially greater than P*. In a LFC, the values of $\widehat{P^*}$ differ only slightly from P*. Thus we see:

The minimal probability of CS seems to be realized, if all medians are equal, therefore it seems to be justified to speak of a LFC in this special case. The critical values, evaluated by using the asymptotic distribution of $\max S_j - S_1$, yield correct selection procedures with finite sample sizes, too.

(b) The magnitude of $\widehat{ET}$ depends very strongly on the actual parameter-configuration. For k=4 and configurations not far from a LFC, no population is excluded in nearly fifty per cent. If the differences between the medians become larger, $\widehat{ET}$ takes smaller values. In case of configuration (0.0,0.3a, 0.6a,0.9a), nearly 2 populations are excluded on an average. $\widehat{ET}/k$ decreases as k becomes larger.

(c) The results, given in table 3, show that the influence of P* over $\widehat{P^*}$ is very small, if the parameter-configuration is different from a LFC. On the contrary, the values of $\widehat{ET}$ become substantially smaller, if P* decreases.

(d) The selection procedures based on different types of r.-o.s.'s differ also in their efficiencies. Sometimes this differences are very distinct. For example, the procedure based on PMS-statistics with p=q=0.25 is clearly the "best" procedure, if the populations are uniformly distributed.

The application of Van-der-Waerden-scores yields the smallest values of $\widehat{ET}$, if the populations are uniformly - or normally distributed. In general, the best procedure based on rule R2 is obtained when the Wilcoxon-statistic is used. Its application can be recommended especially, if the underlying distributions are normal or double-exponential.

In case of distributions, which are skewed to the right, $\widehat{ET}$ takes its smallest values, if the PMS-statistic with p=0,q=0.5 is used.

The preceding statements are valid only if the differences between the medians are caused by differences of the location-parameters. If the underlying distributions are skewed to the right, and the differences between the medians result from different scale-parameters, the most efficient form of rule R2 is obtained by using Van-der-Waerden- or Wilcoxon-scores (cf. table 4).

Table 1

P^* = 0.90 n = 20													
ω =		ω_1		ω_2		ω_3		ω_4		ω_5		ω_6	
		$\widehat{P^*}$	$\widehat{ET}$	$\widehat{P^*}$	$\widehat{ET}$	$\widehat{P^*}$	$\widehat{ET}$	$\widehat{P^*}$	$\widehat{ET}$	$\widehat{P^*}$	$\widehat{ET}$	$\widehat{P^*}$	$\widehat{ET}$
UNI	WAE	0.89	3.58	0.95	3.44	1.00	3.22	1.00	2.51	0.95	3.35	1.00	2.37
	WILC	0.89	3.58	0.95	3.49	1.00	3.39	0.98	2.80	0.92	3.39	1.00	2.51
	MED	0.90	3.56	0.95	3.51	0.98	3.52	0.98	3.17	0.93	3.61	0.99	3.03
	PMS_1	0.88	3.49	0.99	3.44	1.00	2.80	1.00	2.08	0.95	3.14	0.99	2.02
NOR	WAE	0.95	3.63	0.96	3.57	1.00	3.33	1.00	2.62	0.98	3.36	1.00	2.31
	WILC	0.93	3.57	0.96	3.55	1.00	3.30	1.00	2.65	0.97	3.37	1.00	2.31
	MED	0.93	3.61	0.95	3.57	0.98	3.28	1.00	2.97	0.91	3.29	0.98	2.48
	PMS_1	0.91	3.50	0.91	3.36	1.00	3.25	1.00	2.60	0.94	3.28	1.00	2.26
DE	WAE	0.92	3.60	0.97	3.59	1.00	3.17	1.00	2.49	0.97	3.46	1.00	2.34
	WILC	0.93	3.60	0.98	3.58	0.99	3.12	1.00	2.42	0.99	3.43	1.00	2.37
	MED	0.91	3.63	0.95	3.59	0.98	3.14	0.99	2.68	0.98	3.47	1.00	2.46
	PMS_1	0.90	3.48	0.92	3.60	0.98	3.13	1.00	2.55	0.93	3.27	0.99	2.28
CAU	WAE	0.87	3.56	0.93	3.54	0.95	3.23	0.99	3.35	0.97	3.52	0.99	2.45
	WILC	0.87	3.60	0.94	3.47	0.97	3.13	1.00	3.22	0.96	3.42	1.00	2.22
	MED	0.86	3.60	0.93	3.40	0.98	2.98	0.99	2.93	0.97	3.33	0.97	2.16
	PMS_1	0.84	3.42	0.91	3.50	0.96	3.24	0.99	3.41	0.96	3.55	0.97	2.86
EXP	WAE	0.93	3.61	0.98	3.46	1.00	2.74	1.00	2.00	0.99	3.17	1.00	2.18
	WILC	0.93	3.63	0.96	3.45	1.00	2.73	1.00	1.89	0.99	3.28	1.00	2.07
	MED	0.91	3.61	0.92	3.49	0.97	2.95	1.00	2.35	0.99	3.50	1.00	2.13
	PMS_2	0.94	3.69	1.00	3.54	1.00	2.32	1.00	1.38	1.00	3.15	1.00	1.98
GAM	WAE	0.95	3.64	0.89	3.45	1.00	3.20	1.00	2.84	0.98	3.37	1.00	2.61
	WILC	0.94	3.61	0.88	3.45	1.00	3.21	1.00	2.91	0.98	3.33	0.99	2.58
	MED	0.89	3.62	0.91	3.56	0.99	3.20	0.99	3.03	0.95	3.36	0.98	2.80
	PMS_2	0.94	3.63	0.94	3.55	0.99	3.29	1.00	2.84	0.98	3.26	1.00	2.55

[1] p=q=0.25 [2] p=0.0 q=0.5

ω_1=(0.0,0.0,0.0,0.0), ω_3=(0.0,0.0,0.0,0.3a), ω_5=(0.0,0.1a,0.2a,0.3a)

ω_2=(0.0,0.0,0.0,0.1a), ω_4=(0.0,0.0,0.0,0.5a), ω_6=(0.0,0.3a,0.6a,0.9a)

Table 2

P* = 0.90 n = 20					
parameter-configuration		(0.0,0.0,0.1a,0.1a)		(0.0,0.0,0.5a,0.5a)	
		$\widehat{P^*}$	$\widehat{ET}$	$\widehat{P^*}$	$\widehat{ET}$
UNI	WAE	0.94	3.63	0.96	2.51
	WILC	0.93	3.59	0.97	2.73
	MED	0.94	3.60	0.96	3.17
	PMS p=q=0.25	0.91	3.39	0.93	2.29
NOR	WAE	0.94	3.56	0.99	2.99
	WILC	0.94	3.55	0.99	3.01
	MED	0.94	3.57	0.95	3.10
	PMS p=q=0.25	0.89	3.40	0.97	2.81
DE	WAE	0.96	3.59	0.98	2.59
	WILC	0.96	3.60	0.97	2.60
	MED	0.93	3.60	0.95	2.68
	PMS p=q=0.25	0.95	3.53	0.93	2.54
CAU	WAE	0.92	3.41	0.95	2.90
	WILC	0.93	3.41	0.95	2.76
	MED	0.95	3.43	0.94	2.59
	PMS p=q=0.25	0.90	3.40	0.94	3.20
EXP	WAE	0.98	3.44	0.94	2.29
	WILC	0.98	3.53	0.95	2.34
	MED	0.94	3.56	0.91	2.70
	PMS q=0.5 p=0.0	0.98	3.38	1.00	2.17
GAM	WAE	0.92	3.55	0.99	3.06
	WILC	0.92	3.55	0.99	3.04
	MED	0.90	3.59	0.96	3.27
	PMS q=0.5 p=0.0	0.88	3.51	0.99	2.85

Table 3

parameter-configuration		(0.0,0.0,0.0,0.5a)		n = 20	
P* =		0.80		0.86	
		$\widehat{P^*}$	$\widehat{ET}$	$\widehat{P^*}$	$\widehat{ET}$
UNI	WAE	1.00	1.88	1.00	2.22
	WILC	1.00	2.19	0.99	2.47
	MED	0.98	2.73	0.99	2.88
	PMS p=q=0.25	0.99	1.59	1.00	1.93
NOR	WAE	1.00	2.03	1.00	2.36
	WILC	1.00	2.02	1.00	2.31
	MED	1.00	2.52	1.00	2.52
	PMS p=q=0.25	0.99	1.88	1.00	2.34
DE	WAE	1.00	2.14	1.00	2.18
	WILC	1.00	2.12	1.00	2.15
	MED	1.00	2.55	1.00	2.11
	PMS p=q=0.25	0.98	2.19	1.00	2.34
CAU	WAE	0.99	2.59	0.99	2.66
	WILC	0.99	2.31	0.99	2.48
	MED	1.00	2.13	1.00	2.23
	PMS p=q=0.25	0.96	2.84	0.98	3.02
EXP	WAE	1.00	1.60	0.99	1.75
	WILC	0.99	1.56	0.99	1.71
	MED	1.00	2.14	0.98	1.76
	PMS q=0.5 p=0.0	1.00	1.34	1.00	1.36
GAM	WAE	1.00	2.39	1.00	2.50
	WILC	0.99	2.47	1.00	2.54
	MED	0.99	2.67	0.99	2.69
	PMS q=0.5 p=0.0	0.99	2.32	1.00	2.41

Table 4

$P^* = 0.90$	n = 20							
parameter-configuration: (0.0,0.0,0.0,0.25a,0.375a,0.5a)								
	WAE		WILC		MED		PMS p=0 q=0.5	
	$\widehat{P^*}$	$\widehat{ET}$	$\widehat{P^*}$	$\widehat{ET}$	$\widehat{P^*}$	$\widehat{ET}$	$\widehat{P^*}$	$\widehat{ET}$
UNI	0.98	4.17	0.98	4.49	0.97	5.22	0.98	4.88
NOR	1.00	4.18	1.00	4.23	0.99	4.79	1.00	4.63
DE	1.00	4.16	1.00	4.13	1.00	4.53	1.00	4.53
CAU	1.00	4.42	1.00	4.07	0.99	4.04	0.98	4.60
EXP	1.00	3.55	0.99	3.51	0.99	4.30	1.00	3.21
GAM	0.99	4.47	0.99	4.64	0.99	5.04	1.00	4.43
parameter-configuration: (0.0,0.0,0.0,0.5a,0.75a,1.0a)								
UNI	1.00	2.82	1.00	3.03	1.00	4.23	1.00	3.78
NOR	1.00	2.89	1.00	2.95	1.00	3.55	1.00	3.56
DE	1.00	2.54	1.00	2.47	1.00	3.05	1.00	3.25
CAU	1.00	3.22	1.00	2.87	1.00	2.89	1.00	3.52
EXP	1.00	2.66	1.00	5.27	1.00	2.52	1.00	2.86
GAM	1.00	2.19	1.00	2.00	1.00	2.25	1.00	2.51
different medians by different scale parameters (1.0,1.0,1.0,1.2,1.6,2.0), equal location parameters								
EXP	1.00	3.58	1.00	3.63	0.98	4.73	1.00	4.91
GAM	0.97	4.74	0.97	4.75	0.99	5.29	0.92	5.38

5 Some Properties of Optimality and a Brief Comparison of the Procedures from No.2 - No.4

5.1 Local optimality of selection rule R1

Before constructing an optimal rank selection rule δ, the following assumptions concerning the distributions of the populations must hold.

Let the distributions of the k populations $\Pi_1,\ldots,\Pi_k$, satisfy the assumptions made in section 2. Let in addition $f_j(x,\Delta_j)$, the density associated with Π_j, $1\le j\le k$, where $\Delta_j \varepsilon I$, $I\subseteq\mathbb{R}$, possess the properties:

(a) $f_j(x,\Delta_j) = \frac{1}{\sigma_j} f\left(\frac{x-\Delta_j}{\sigma_j}\right)$, $1\le j\le k$, where f is some density

(b) $f_j(x,\Delta_j)$ is absolutely continuous in Δ_j for almost all $x\varepsilon\mathbb{R}$;

(c) The limit $\dot{f}_j(x,0):=\lim\limits_{\Delta_j\to 0} \frac{1}{\Delta_j} (f_j(x,\Delta_j)-f_j(x,0))$ exists for almost all $x\varepsilon\mathbb{R}$;

(d) If $\dot{f}_j(x,\Delta_j)$ denotes the partial derivative of $f_j(x,\Delta_j)$ with respect to Δ_j, then :[1]

$$\lim_{\Delta_j\to 0} \int_{-\infty}^{+\infty} |\dot{f}_j(x,\Delta_j)|dx = \int_{-\infty}^{+\infty} |\dot{f}_j(x,0)|dx<\infty.$$

Let the scale-parameters $\sigma_1,\ldots,\sigma_k$ be arbitrary but fixed and the rank-configuration K, associated with the total sample $\vec{X}$, defined as in 2.5. It can easily be seen that there are $\binom{m}{n,n,\ldots,n}$ different rank-configurations.

We now turn to the construction of the desired rule δ. A rank selection rule can be generally defined as a measurable mapping of the form:

$$\delta:\quad \mathcal{K}\to \{(p_1(K),\ldots,p_k(K)):K\varepsilon\mathcal{K}\}$$

$$\delta:\mathcal{K}\ni K \to \delta(K):=(p_1(K),\ldots,p_k(K)),\ 0\le p_i(K)\le 1,$$

---

[1] *Recall a convergence theorem in Hájek and Sidák [86], page 64.*

where $p_i(K)$ denotes the probability of selecting a population Π_i, in case that the rank-configuration K is realized. In the following, the index ω indicates the dependence of $P_\omega(CS|\delta)$ on $\omega = (\Delta_1,\ldots,\Delta_k)$, the vector of location-parameters. The aim of the following derivations is the construction of a rank selection rule δ, which satisfies:

$$P_{\omega_0}(CS|\delta) \geq P^* \quad , \omega_o = (0,0,\ldots,0) \tag{5.1}$$

and

$$\frac{1}{k!} \sum_{g\varepsilon \mathcal{T}_k} \left. \frac{\partial P_{\alpha(g\omega)}(CS|\delta)}{\partial\alpha} \right|_{\alpha=0} , \text{ maximal for } \omega=(\Delta_1,\ldots,\Delta_k), \tag{5.2}$$

where $\mathcal{T}_k$ is the permutation-group of order k; α a scalar; $g\omega = (\Delta_{h1},\ldots,\Delta_{hk})$, $h=g^{-1}$, $g\varepsilon\mathcal{T}_k$.

Condition (5.2) is required, to derive a rule δ, for which $P_\omega(CS|\delta)$ has maximal slope in ω_0, if the parameter vector is replaced by $\omega=(\Delta_1,\Delta_2,\ldots,\Delta_k)$, and all permutations $g\omega$ have the same probability.

The probability $P_{\alpha\omega}(CS|\delta)$ can be written as

$$P_{\alpha\omega}(CS|\delta) = \sum_K p_k(K) P_{\alpha\omega}(K), \tag{5.3}$$

where

$$P_{\alpha\omega}(K) = \int_{-\infty}^{+\infty}\int_{-\infty}^{x_m} \ldots \int_{-\infty}^{x_2} \prod_{i=1}^{m} f_{K_i}(x_i,\alpha\Delta_{K_i})\, dx_1\ldots dx_{m-1}dx_m \tag{5.4}$$

is the probability of rank-configuration K, if $\alpha\omega$ is the actual parameter vector.

Because of conditions (b), (c) and (d), the densities f_j can be differentiated with respect to α under the integral sign. We get:

$$\dot{P}_{\alpha\omega}(K) := \int_{-\infty}^{+\infty}\int_{-\infty}^{x_m} \ldots \int_{-\infty}^{x_2} \sum_{j=1}^{m} \dot{f}_{K_j}(x_j,\alpha\Delta_{K_j})\Delta_{K_j} \prod_{\substack{i=1\\ i\neq j}}^{m} f_{K_i}(x_i,\alpha\Delta_{K_i})dx_1\ldots dx_{m-1}dx_m \tag{5.5}$$

and for $\alpha = 0$,

$$\dot{P}_0(K) := \sum_{j=1}^{m} \Delta_{K_j} A_j = \sum_{j=1}^{k} \Delta_j \sum_{\substack{\ell \\ K_\ell=j}} A_\ell \text{ , where} \tag{5.6}$$

$$A_j := \int_{-\infty}^{+\infty} \int_{-\infty}^{x_m} \dots \int_{-\infty}^{x_2} f_{K_j}(x_j,0) \prod_{\substack{i=1 \\ i \neq j}}^{m} f_{K_i}(x_i,0)\,dx_1 \dots dx_{m-1}\,dx_m, \quad j=1,\dots,m. \tag{5.7}$$

Defining $B_j(K) := \sum_{\substack{\ell \\ K_\ell = j}} A_\ell$, (5.6) can be rewritten as

$$\dot{P}_0(K) = \sum_{j=1}^{k} \Delta_j B_j(K). \tag{5.8}$$

Let now Π_k be the "best" population, that is $\Delta_{(k)} = \Delta_k$, then condition (5.2) can be rewritten as follows:

$$\frac{1}{(k-1)!} \sum_{g \in \mathfrak{F}_{k,k}} \sum_K p_k(K) \sum_{j=1}^{k} \Delta_{gj} B_j(K), \text{ maximal for } \omega = (\Delta_1,\dots,\Delta_k), \tag{5.9}$$

where $\mathfrak{F}_{k,k} := \{(g1,\dots,gk) : gk = k\}$.

If the index $j \in \{1,\dots,k-1\}$ is held fixed, there are $(k-2)!$ permutations g, with $gj = \nu$. This can be repeated for $\nu = 1,\dots,k-1$, where all $(k-1)(k-2)!$ permutations satisfy $gk = k$. Hence, (5.9) can be replaced by

$$\frac{1}{(k-1)!} \sum_K p_k(K)\Big((k-2)! \sum_{\nu=1}^{k-1} \Delta_\nu \sum_{i=1}^{k-1} B_i(K) + (k-1)!\, \Delta_k B_k(K)\Big), \text{ maximal for } \omega. \tag{5.10}$$

Defining $U := \sum_{i=1}^{k} \Delta_i$ and $V := \sum_{i=1}^{k} B_i(K) = n \sum_{i=1}^{k} A_i$, condition (4.2) is equivalent to

$$\sum_K p_k(K)\Big(\frac{1}{k-1}(U-\Delta_k)V + (k\Delta_k - U)B_k(K)\Big), \text{ maximal for } \omega. \tag{5.11}$$

Recalling $\Delta_k \geq \Delta_i$, $1 \leq i \leq k-1$, we see that $U - \Delta_k \geq 0$ and $k\Delta_k - U \geq 0$, i.e. taking into account, that

$$\sum_K p_k(K) P_{\omega_0}(K) = P^*, \tag{5.12}$$

condition (5.11) is satisfied, if we choose

$$p_k(K) = \begin{cases} 1, & \text{if } B_k(K) > c(K) \\ \rho, & \text{if } B_k(K) = c(K) \\ 0, & \text{if } B_k(K) < c(K), \end{cases}$$

where c(K)[1] and ρ are chosen such that

---

1 *In case of equal scale-parameters, all rank-configurations have the same probability to be realized, and then we can replace c(K) by c.*

$$\sum_{\{K:B_k(K)>c(K)\}} P_{\omega_0}(K) + \rho \sum_{\{K:B_k(K)=c(K)\}} P_{\omega_0}(K) = P^*$$

is satisfied (cf. (5.12)).

Using condition (a), (satisfied by the densities f_i, i=1,...,k) and substituting $F(x_j')$ by u, the expressions A_j can be rewritten as:

$$A_j = \frac{1}{\sigma_{K_j}(m-1)!}\binom{m-1}{j-1}\int_{-\infty}^{+\infty} u^{j-1}(1-u)^{m-j}\left(-\frac{\dot{f}(F^{-1}(u))}{f(F^{-1}(u))}\right) du.$$

It is gathered from this expression that the optimal selection rule is based on scores, which depend on the underlying distribution.

The important function $\varphi(u,f) := -\frac{\dot{f}(F^{-1}(u))}{f(F^{-1}(u))}$ is specified for some special distributions in the following.

Normal-distribution:

Let $f_1(x) = \frac{1}{\sqrt{2\pi}} \exp(-x^2/2)$ be the standard density of the populations in research, then we get $\varphi(u,f_1) = \psi(u)$, where ψ is the inverse of Φ, the cdf associated with f_1.

Logistic-distribution:

Let $f_2(x)=\exp(-x)/(1+\exp(-x))^2$, $F_2(x)=1/(1+\exp(-x))$ be the standard density, cdf of the populations in research, respectively. Because of $F_2^{-1}(u)=\ell n(u/(1-u))$, we get $\varphi(u,f_2)=2u-1$.

Double-exponential-distribution:

For $f_3(x) = \frac{1}{2}\exp(-|x|)$ and $F_3(x)=\begin{cases} \frac{1}{2}\exp(-x), & \text{if } x \leq 0 \\ 1-\frac{1}{2}\exp(-x), & \text{if } x > 0, \end{cases}$

i.e., $F_3^{-1}=\begin{cases} -\ell n(2u), & \text{if } 0 < u < \frac{1}{2} \\ -\ell n(2(1-u)), & \text{if } \frac{1}{2} \leq u < 1 \end{cases}$, we get $\varphi(u,f_3)=c(2u-1)$, 0<u<1, where c(x) is given by definition 1.4.

From these three examples, we see that the scorefunctions, given for the rule R1, are highly connected with the functions $\varphi(u,f)$. Thus, if we use the appropriate r.-o.s. (cf. 3.4), the procedures based on rule R1 may be called "locally optimal".

5.2 Influence of the scorefunction on the efficiency of procedures based on rules R1 and R2

At first we consider the interdependence between the shape of the density f and the rapidity of the growth of the function $\varphi(u,f)$, if u becomes larger.

Let f and g be two densities which are symmetric around zero, associated with the cdf's F and G, respectively, and let F^{-1}, G^{-1} denote the inverse functions, respectively. In the following, we say that f has shorter tails than g, if

$$F^{-1}(u) = b(u)G^{-1}(u),\ u > 0,$$

where the function b(u) is nonincreasing and nonconstant. In this sense we speak of short-tailed, medium-tailed and long-tailed distributions.

Figure 1 shows the tails of five distributions. According to the standard normal density, the densities of an uniformly-, logistic-, double-exponential- and a cauchy-distributed random variable are standardized in such a way that the medians and the 0.9-quantiles of the associated distributions are all equal to those of the standard normal distribution. Thereby the tails of the five densities become comparable.

From figure 1, we see that the density of the uniform distribution can be cal-

fig. 1

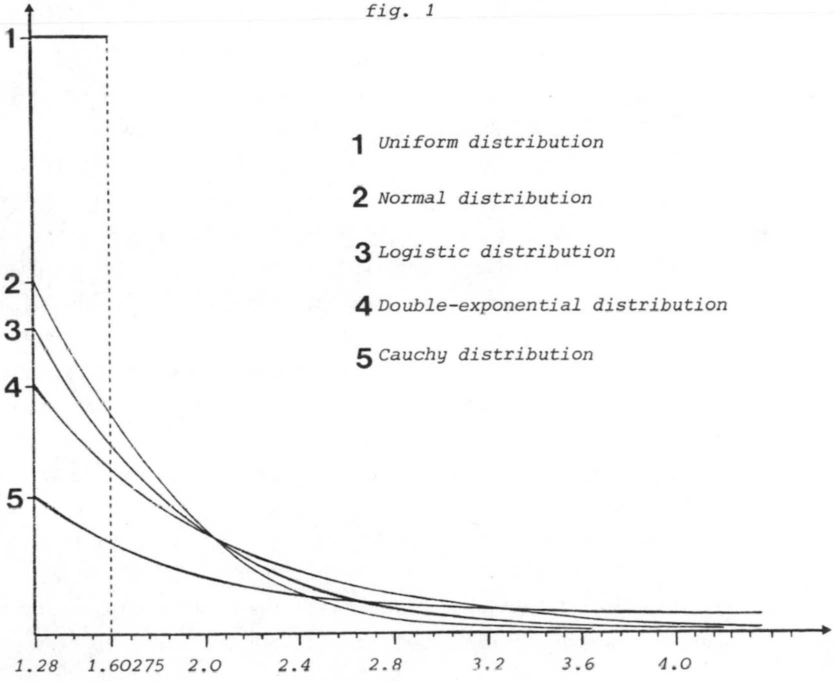

led "short-tailed", that of the logistic distribution "medium-tailed", and that of the cauchy-distribution "long-tailed".

Let $\varphi(u,f)$ and $\psi(u,g)$ be two skew symmetric nondecreasing functions, defined for $u\varepsilon(0,1)$, then we say that $\varphi(u,f)$ increases more rapidly than $\psi(u,g)$, if

$$\varphi(u,f) = d(u)\psi(u,g), \ \frac{1}{2} < u < 1,$$

where the function $d(u)$ is nondecreasing and nonconstant for $u\varepsilon(\frac{1}{2},1)$. It can be shown (cf. Hâjek [85]) that for strongly unimodal densities f,g and

$$\varphi(u,f) := -\frac{\dot{f}(F^{-1}(u))}{f(F^{-1}(u))}, \quad \psi(u,g) := -\frac{\dot{g}(G^{-1}(u))}{g(G^{-1}(u))},$$

the following statement is valid:

If $\varphi(u,f)$ increases more rapidly than $\psi(u,g)$, then f has shorter tails than g. Comparing this result with those of the examples given in 5.1, we can note that slowly increasing scorefunctions yield better selection procedures than more rapidly increasing scorefunctions, in case of medium- or long-tailed distributions. And on the contrary, rapidly increasing scorefunctions yield more efficient selection procedures than slowly increasing score-functions, if the underlying distributions are short-tailed.

In order to underpin the preceding statements, the following may be remarked:

By slowly increasing scorefunctions outliers are not (or only slightly) more weighted than the more "probable" values of the realization. Therefore, the result of the applied selection procedures is not falsified by outliers. On the other hand, if the larger values of some realization are caused by a shift of location, these values are not heavier weighted, if a slowly increasing scorefunction is used. This in mind, the property of optimality of the different scorefunctions can be explained by the fact that the small and large values in some realization are more affected by a shift in location, if the underlying distribution is medium- or long-tailed, than, if the underlying distribution is short-tailed. And in addition, the probability for outliers in some realization becomes the larger the longer the tails of the underlying density are.
The results of our Monte-Carlo studies, given in sections 2 and 3, support the preceding remarks for rule R1.
The validity of our statements for rule R2, we can gather from the Monte-Carlo results, given in section 4.
In a certain sense an outstanding position takes the Haga-statistic. Its application can be recommended, if the underlying distribution seems to be short-tailed.

If some prior information about the tails of the distribution of the populations in research is available, we ought to use this information for choosing an appropriate scorefunction.
A possibility to get such information is considered in chapter 3 of this part.

5.3 Comparison of the procedures given in sections 2,3, and 4

We start with the comparison of the two classes of subset selection procedures R1u and R1g, both based on rule R1. According to the results, given in table 2 of section 3, the application of procedures R1u must be recommended, if the hypothesis of equal scale-parameters must be rejected. The procedures R1g may yield incorrect results, if the scale-parameters are not equal. In case of equal scale-parameters, the procedures R1g and R1u approximately have the same efficiencies, but the implementation of procedures R1g is essentially easier than that of procedures R1u. This reason and the possibility to choose from a larger number of scorefunctions are arguments for the application of procedures R1g, if the assumption of equal scale-parameters seems to be justified.
The comparison of procedures R1g and those based on rule R2, gives rise to the following conclusions: If the actual parameter-configuration is close to some LFC, the two classes R1g and R2 have almost the same efficiency. Because of the easier implementation, the procedures R1g should be preferred.
If the differences $(\Delta_{(k)}-\Delta_{(i)})$, $i\neq k$, surpass the value $0.3a\sigma$, the procedures based on rule R2 are substantially more efficient than those based on rule R1.

6 The Selection Rules R1 and R2 in Case of Randomized-Block-Designs

In this section we modify the selection rules R1 and R2 given in sections 2 and 4, respectively, in such a way that they can be applied in randomized-block-designs. Because of this more general experimental design, in this section we speak of k treatments instead of k populations. The modification of the rules consists in another appointment of ranks to the realizations of the k given samples of size n.

6.1 Modified definition of ranks and the distribution of the resulting rank-order-statistics

Let $\Pi_1,\Pi_2,\ldots,\Pi_k$ be the k treatments. In order to select a subset, which contains the best treatment, for each treatment $\Pi_i, 1\leq i\leq k$, we have observed the realization $\vec{x}_i=(x_{i1},x_{i2},\ldots,x_{in})$ of the random vector $\vec{X}_i=(X_{i1},X_{i2},\ldots,X_{in})$, where $X_{ij}, 1\leq i\leq k, 1\leq j\leq n$, are independent random variables with the continuous distribution functions $F_{ij}, 1\leq i\leq k, 1\leq j\leq n$, satisfying the following assumptions:

(a) $F_{ij}(x) = F_j(x,\delta_i)$, $1\leq i\leq k$, $1\leq j\leq n$, where δ_i denotes a parameter associated with treatment Π_i, which allows the ordering of the cdf's $F_{ij}(x)$, and therefore the ordering of the treatments in the following way:
$F_j(x,\delta_i) \leq F_j(x,\delta_\ell)$, $\forall x\in\mathbb{R}$, if $\delta_i > \delta_\ell$, $1\leq i\neq\ell\leq k$ $\Longrightarrow$ Π_i is better than Π_ℓ.

(b) The distribution functions $F_j, 1\leq j\leq n$, are dependent on j.

These two assumptions mean that on the one hand the components of all random vectors $\vec{X}_i, 1\leq i\leq k$, with the same subscript j, $1\leq j\leq n$, possess distribution functions that differ only in one parameter, which allows the ordering of the k distributions in the way given above, and that on the other hand the distributions of components, with different subscripts, differ likely in their functional form as well as in their parameters.
This experimental design is a more general model than the linear two-factor-design to be considered in chapter 2.

To construct r.-o. s.'s allowing the use of the selection rules R1 and R2 in this general model, we first define the rank-vectors $R_j=(R_{j1},\ldots,R_{jk})$, $1\leq j\leq n$, as follows: Let R_{ji}, $1\leq i\leq k$, the rank of X_{ji}, denote the number of random variables $X_{j\ell}$, $1\leq\ell\leq k$, which are less than or equal to X_{ji}. Thus, R_{ji} can take the values 1,2,...,k. Recalling the continuity of the cdf's F_{ij}, $1\leq i\leq k$, $1\leq j\leq n$, and A4/4 we know that the possible realizations of R_j are the permutations of (1,2,...,k).

The r.-o. s. $S_i, 1\leq i\leq k$, associated with treatment Π_i, and basing on the rank-vectors R_j, $1\leq j\leq n$, is defined by

$$S_i := \sum_{j=1}^{n} J\left(\frac{R_{ji}}{k+1}\right), \quad 1\leq i\leq k, \tag{6.1}$$

where J denotes some scorefunction (cf. sec. 1).

If the cdf's $F_{j1},\ldots,F_{jk}$, are identical for each j, $1\leq j\leq n$, then the statistics S_i, $1\leq i\leq k$, are asymptotically normal distributed with mean and variance:

$$E(S_i) = n\bar{a},\ \mathrm{Var}(S_i) = \frac{n}{k}\sum_{\ell=1}^{k}(a_\ell-\bar{a})^2, \tag{6.2}$$

where $a_\ell := J\left(\frac{\ell}{k+1}\right)$ and $\bar{a}=\bar{J}$ (cf. sec. 1).

The asymptotic normality follows from the central limit theorem (cf. A2/1), and the mean and variance can be easily derived by using the more general form of the r.-o. s.'s

$$S_i = \sum_{j=1}^{n}\sum_{\ell=1}^{k} c^i_{j\ell} J\left(\frac{R_{j\ell}}{k+1}\right), \tag{6.3}$$

where $c^i_{j\ell}$, $1\leq j\leq n$, $1\leq \ell\leq k$, are the regression constants with

$$c^i_{j\ell} := \begin{cases} 1, & \text{if } \ell=i \\ 0, & \text{if } \ell\neq i \end{cases},\quad 1\leq j\leq n.$$

6.2 The rules R1 and R2

The k treatments $\Pi_1,\Pi_2,\ldots,\Pi_k$ are associated with the effects $\delta_1,\delta_2,\ldots,\delta_k$. If

$$\delta_{(1)} \leq \delta_{(2)} \leq \cdots \leq \delta_{(k)} \tag{6.4}$$

are the ordered treatment-effects, and if $\Pi_{(i)}$ denotes the treatment associated with $\delta_{(i)}$, then treatment $\Pi_{(k)}$ is the best.
To implement the selection rules, we must know the configurations $(\delta_1,\delta_2,\ldots,\delta_k)$ for which the probability of CS is minimal.

Rule R1:

In using the rule

R1: Select treatment $\Pi_i \iff S_i \geq B$, $1\leq i\leq k$, $B\in\mathbb{R}$, where S_i denotes the r.-o. s. given in 6.1, and B is chosen as large as possible, satisfying

$$\inf_{\Delta} P(CS|R1) \geq P^*,\ \Delta := \{\delta=(\delta_1,\ldots,\delta_k) : \delta_\ell\in I,\ 1\leq\ell\leq k,\ I\subseteq\mathbb{R}\}, \tag{6.5}$$

the infimum of $P(CS|R1)$ is given if the parameters δ_i, $1\leq i\leq k$, are identical (LFC).
Analogous to section 2, the proof can be based on the lemma A4/8 and the arguments used in the proof of theorem 2.1.

Since the statistics S_i, $1 \le i \le k$, are asymptotically normal distributed in the LFC, and since the treatments can be changed in this case without any loss of generality, we have

$$\inf P(CS|R1) = P(S_k \ge B),$$

and for preassigned value of P*, the critical value B can be derived as follows:

$$\inf P(CS|R1) = P^* \iff P(S_k \ge B) = P^* .$$

Using the asymptotic distribution of S_k we get:

$$B = x \cdot \sqrt{Var(S_k)} + E(S_k), \text{ where } x = \Psi(1-P^*).$$

Also in randomized-block-designs, the preceding derivations remain valid, if one of the two methods of breaking ties,given in section 1,is used.

Rule R2:

As in section 4, in case of the randomized-block-design, a similar result concerning the LFC cannot be given for the selection rule

$$\text{R2: Select treatment } \Pi_i \iff S_i \ge \max_{1 \le \ell \le k} S_\ell - b, \quad 1 \le i \le k, \; b \ge 0,$$

where S_i is given again by 6.1 and b is chosen as small as possible satisfying (6.5), with R1 substituted by R2.
Using the same arguments as in section 4,we get:

$$\inf_{\Delta} P(CS|R2) = \inf_{\Delta_k} P(CS|R2), \text{ where } \Delta_k := \{\delta : \delta_{(k-1)} = \delta_{(k)}\}.$$

To be able to implement the subset-selection procedures based on the rule R2 we assume that the LFC is realized in case of equal parameters δ_i, $1 \le i \le k$. An heuristic argument for the validity of this assumption (at least for some well-known distributions) is based on Monte-Carlo-studies whose results are given in 6.5.

6.3 The asymptotic and the exact distribution of max $S_\ell - S_1$ for identical parameters

When assuming that the parameters δ_i, $1 \le i \le k$, are all equal to some value δ_o, that means, the distributions F_{ji}, $1 \le i \le k$, are identical for each j, $1 \le j \le n$, then the statistics S_i are interchangeable.
Hence without loss of generality, we choose i=1.

Theorem 6.1:

Let S_i, $1\le i\le k$, be the r.-o. s.'s defined in 6.1. If the preceding assumptions are satisfied then

$$\lim_{n\to\infty} P(S_1 \ge \max_{1\le \ell \le k} S_\ell - b) = \int_{-\infty}^{+\infty} \Phi(x+\frac{b}{z})^{k-1} d\Phi(x),$$

where

$$z^2 = \frac{k}{k-1}\sigma^2 \text{ and } \sigma^2 = Var(S_i),\ 1\le i\le k.$$

Proof:

Recalling a generalized central limit theorem in Anderson [4], page 74, we know that the vector $(S_1,S_2,\dots,S_k)$ possesses asymptotically a k-variate normal distribution. Using a similar argument as in the proof of theorem 4.2, the assertion of the theorem follows. ■

To preassigned probability P*, the critical value b, needed for the implementation of rule R2, is given by $b=h\sqrt{\frac{k}{k-1}\sigma^2}$, where h is the solution of $\int_{-\infty}^{+\infty} \Phi(x+h)^{k-1} d\Phi(x)=P^*$, and $\sigma^2=Var(S_i)$, $1\le i\le k$.

For various values of k and P* some h-values are given in table II, appendix 6.

To get an idea of the accuracy of the approximation of the exact distribution of max $S_\ell - S_1$ by the preceding asymptotic distribution, we derive also the exact distribution and compare critical values computed by both, the asymptotic and the exact method.

To compute the exact distribution of max $S_\ell - S_1$, we consider the sequence of the n rank-vectors $R_1,R_2,\dots,R_n$, with length nk. Because of the assumptions made above, we know that each vector R_j, $1\le j\le n$, with equal probability takes the k! permutations of $(1,2,\dots,k)$. Thus the sequence $R_1,R_2,\dots,R_n$ takes with equal probability $(k!)^n$ different realizations, where each realization consists in the sequence of n permutations of $(1,2,\dots,k)$.
Hence to get the probability $P(S_1 \ge \max_{1\le \ell\le k} S_\ell - b)$ it is sufficient to evaluate the number of realizations of the sequence $R_1,R_2,\dots,R_n$, satisfying:

$$S_2-S_1 \le b,\ S_3-S_1 \le b,\dots,S_k-S_1 \le b.$$

This number of realizations can be obtained by using the following recursion formula:

We define

$$N(s|b_2,b_3,\dots,b_k) := \text{number of realizations of } R_1,R_2,\dots,R_s, \text{ satisfying } S_i-S_1 \le b_i,\ i=2,3,\dots,k.$$

The following symmetry holds:

$$N(s|b_2,b_3,\dots,b_k) = N(s|m_{j2},m_{j3},\dots,m_{jk}),$$

where (j2,j3,...,jk) denotes some permutation of (2,3,...,k). Moreover let $T:=\{(a_{r1},a_{r2},\dots,a_{rk})\}$ denote the set of the k! permutations of the scores $(a_1,a_2,\dots,a_k)$, where r is some permutation of (1,2,....,k).
Considering the last k places of the sequence $R_1,R_2,\dots,R_s$, we get:

$$N(s|b_2,\dots,b_k) = \sum_T N(s-1|b_2+a_{r1}-a_{r2},\dots,b_k+a_{r1}-a_{rk}).$$

If k, the scorefunction J, and the actual value of s is given, then for each difference $S_\ell-S_1$, $2\le\ell\le k$, a maximal value Max(k,s,J) and a minimal value Min(k,s,J) can be given in such a way that the number $N(s|b_2,\dots,b_k)$ is known whenever one of the following boundary conditions is satisfied:

(a) If for some $i\ge 2$, $b_i<\text{Min}(k,s,J)$, then $N(s|b_2,\dots,b_k) = 0$;

(b) If for each $i\ge 2$, $b_i\ge\text{Max}(k,s,J)$, then $N(s|b_2,\dots,b_k) = (k!)^s$;

(c) If an expression of the form $N(1|b_2,\dots,b_k)$ is reached, then this expression denotes the number of elements of T satisfying the conditions $a_{ri}-a_{r1}\le b_i$, i=2,...,k.

The extreme values Max(k,s,J) and Min(k,s,J) are given by:

$$\text{Max}(k,s,J) := \sum_{j=1}^{s} a_k - \sum_{j=1}^{s} a_1 = s(a_k-a_1)$$

$$\text{Min}(k,s,J) := \sum_{j=1}^{s} a_1 - \sum_{j=1}^{s} a_k = s(a_1-a_k).$$

In the following table 1 the extreme values are given for the three kinds of scores for which we have considered the exact distribution of $\max_{1\le\ell\le k} S_\ell-S_1$.

Table 1:

Scores / extreme values	WILC	WAE	MED
Min(k,s,J)	$-s\,\frac{k-1}{k+1}$	$-2s\Phi^{-1}\left(\frac{k}{k+1}\right)$	$-s$
Max(k,s,J)	$s\,\frac{k-1}{k+1}$	$2s\Phi^{-1}\left(\frac{k}{k+1}\right)$	s

The evaluation of the exact distribution is very extensive even if k and n are small. Tables VI, VII and VIII in appendix 6 contain critical values b associated with various values k, n and the three scorefunctions WILC, WAE and MED, respectively. There are also given some critical values computed by using the asymptotic distribution of $\max_{1\leq \ell \leq k} S_\ell - S_1$.

From the comparison of the asymptotic and critical values we see that the approximation yields rather good results if the sample size n is not too small.

For the implementation of rule R2 only the asymptotic distribution ought to be used, that is, the critical value b can be computed as mentioned in the proof of theorem 6.1.
In the presence of ties both methods of breaking ties given in section 1 may be used.

6.4 A numerical example

There are four varieties of wheat Π_1, Π_2, Π_3 and Π_4 (treatments). The aim is to select a subset that includes that variety which yields the greatest harvest on an average. The experimental conditions can be described as follows: There are 5 fields available which differ with respect to some very important properties as for example their size, quality of soil, etc. To be able to compare the four varieties of wheat, each field is divided into four homogeneous parts, and the varieties Π_1, Π_2, Π_3 and Π_4 are randomly sown to these four parts. Thereby the 5 fields are the blocks and the vector of the 5 harvests from the 5 parts on which variety Π_i, $1\leq i\leq 4$, has been sown, denotes the realization of the vector $\vec{X}_i$, $1\leq i\leq 4$.

Based on the four realizations $\vec{x}_i$, $1\leq i\leq 4$, a subset ought to be selected by using the rules R1 and R2, alternatively, where for each rule the three scorefunctions WILC, WAE and MED are used, and the minimal probability $P^*=0.90$ is chosen. The following realizations have been observed (in 100 lb):

Variety \ Field	1	2	3	4	5
Π_1	10.5	4.3	8.7	20.8	1.9
Π_2	11.7	5.9	7.1	30.1	1.7
Π_3	9.8	5.1	8.9	35.7	2.4
Π_4	12.1	6.2	8.1	35.9	2.5

The realizations of the rank-vectors R_j, $1\leq j\leq 5$ are hence as follows:

Variety \ R_j	1	2	3	4	5
Π_1	2	1	2	1	2
Π_2	3	3	1	2	1
Π_3	1	2	4	3	3
Π_4	4	4	3	4	4

Using the asymptotic distributions of the r.-o. s. S_k, and $\max\limits_{1\leq \ell\leq k} S_\ell - S_1$, the critical values for each special scorefunction are given by:

$$B = \Phi^{-1}(0.1)\cdot\sqrt{Var(S_1)} + E(S_1)$$

$$= -1.281\cdot\sqrt{Var(S_1)} + E(S_1)$$

$$b = h(4,0.9)\sqrt{\tfrac{4}{5}Var(S_1)} = 2.4516\sqrt{\tfrac{4}{5}Var(S_1)}.$$

The following table shows in short form the results for both rules R1 and R2, each applied in connection with the three special scorefunctions WILC, WAE, MED.

scores	S_1	S_2	S_3	S_4	$\max_{1\leq\ell\leq 4} S_\ell$	$E(S_1)$	$Var(S_1)$	B	b	selected subset R_1	R_2
WILC	1.6	2.0	2.6	3.8	3.8	2.5	0.25	1.86	1.42	Π_2 Π_3 Π_4	Π_3 Π_4
WAE	-2.44	-1.43	0.25	3.62	3.62	0	1.93	-1.78	3.93	Π_2 Π_3 Π_4	Π_3 Π_4
MED	0	2	3	5	5	2.5	3.165	1.07	3.17	Π_2 Π_3 Π_4	Π_3 Π_4

In order to investigate the influence of the chosen rule and the special scorefunction upon the efficiency of the resulting subset-selection procedure we have carried out some Monte-Carlo-studies whose results are given below.

6.5 Some Monte-Carlo studies

Just as in sections 3 and 4 it is very difficult to give analytic statements concerning the efficiency of the rules R1 and R2 in case of randomized-block-designs. But as in the preceding sections it is very interesting to investigate the questions (a)-(d), formulated in section 2. This investigation is based on the results of some Monte-Carlo-studies.
Supplementary to question (a) we ask again whether the evaluation of critical values for rule R2, based on equal parameters $\delta_1,...,\delta_k$, can be justified.

The simulations in this section are carried out by using the same simulation-model which has been also used in the preceding sections. Both rules R1 and R2 in connection with the scorefunctions WILC, WAE and MED are applied to the realizations generated by use of the method described in appendix 5.

For each of the six distributions: UNI, NOR, DE, CAU, EXP, and GAM different configurations $\delta=(\delta_1,...,\delta_k)$ have been simulated, and for each configuration the simulation process has been repeated 100-times. For k=4,5,7 and 10 samples of sizes n=20 and n=40 have been chosen.
The results in table 1 can be used to answer the question (a). The answer of the supplementary question to (a) can be based on the results given in table 3. The influence of the preassigned value P* on the magnitudes $\widehat{ET}$ and $\widehat{P^*}$ can be recognized by considering table 2. Finally, in table 4 results are given for distributions EXP and GAM, where the parameters δ_i, $1\leq i\leq k$, are scale-para-

meters.

Concerning questions (a)-(d) the following may be concluded.

(a) Apart from LFC's (LFC denotes that configuration where all parameters δ_i, $1 \leq i \leq k$, are equal), all simulated configurations are associated with values of $\widehat{P^*}$, which are substantially larger than P^*. In most cases $\widehat{P^*}$ is equal to one. In a LFC the values of $\widehat{P^*}$ differ slightly from P^*. Only in few cases the absolute difference between $\widehat{P^*}$ and P^* is greater than 0.5. Hence we may conclude: The minimal probability of CS seems to be realized for both rules (also rule R2), if all parameters are equal. That is, also in randomized-block-designs, the critical values for rule R2, evaluated by using the asymptotic distribution of $\max_{\ell} S_{\ell}-S_1$ in case of equal parameters δ_i, $1 \leq i \leq k$, yield correct selection procedures with finite sample sizes, too.

(b) The magnitude $\widehat{ET}$ depends very strongly upon the chosen sample size and the selection rule. In almost all cases the rule R2 yields much smaller values of $\widehat{ET}$ than rule R1. If the two greatest parameters $\delta_{(k-1)}$ and $\delta_{(k)}$ are equal for both rules, then the values of $\widehat{ET}$ are much smaller than in case of configurations, where $\delta_{(k)} > \delta_{(k-1)}$.

(c) The results, given in table 2, show that the influence of P^* over $\widehat{P^*}$ is very small if the configuration is different from a LFC. On the contrary, the values of $\widehat{ET}$ become substantially smaller in case of both rules R1 and R2, if P^* decreases.

(d) In randomized-block-designs, the influence of the special form of the applied r.-o. s.'s on the efficiency of the resulting subset-selection procedure seems to be smaller than in case of population models.
For both rules R1 and R2, the scorefunctions WILC and WAE yield values $\widehat{ET}$, which are only slightly different. In almost all cases the scorefunction WAE seems to be slightly better than the scorefunction WILC.
The application of Median-scores can be recommended only if a very long-tailed (Cauchy-) distribution must be supposed. Also in cases, where the differences among the treatments are caused by different scale-parameters (cf. table 4) the procedures based on Van-der-Waerden-scores seem to be more efficient than those based on Wilcoxon- or Median-scores.

Table 1

$P^* = 0.90$ $n = 20$ configuration $(0.0, 0.0, \ldots, 0.0, 0.5a\sigma)$

rule		R1						R2					
		WILC		WAE		MED		WILC		WAE		MED	
	k	$\widehat{P^*}$	$\widehat{ET}$	$\widehat{P^*}$	$\widehat{ET}$	$\widehat{P^*}$	$\widehat{ET}$	$\widehat{P^*}$	$\widehat{ET}$	$\widehat{P^*}$	$\widehat{ET}$	$\widehat{P^*}$	$\widehat{ET}$
UNI	4	1.00	3.42	1.00	3.42	1.00	3.33	1.00	3.16	1.00	3.08	1.00	3.31
	5	1.00	4.24	1.00	4.17	1.00	4.15	1.00	3.26	1.00	3.58	1.00	3.98
	7	1.00	5.96	1.00	5.94	1.00	6.38	1.00	5.42	1.00	5.15	1.00	5.75
	10	1.00	8.68	1.00	8.65	1.00	8.39	1.00	7.32	1.00	7.15	1.00	8.41
NOR	4	1.00	3.28	1.00	3.28	1.00	3.20	1.00	2.83	1.00	2.78	1.00	3.13
	5	1.00	4.28	1.00	4.20	1.00	4.09	1.00	3.78	1.00	3.74	1.00	3.92
	7	1.00	6.04	1.00	6.02	1.00	6.23	1.00	5.30	1.00	5.13	0.99	5.32
	10	1.00	8.76	1.00	8.72	1.00	8.39	1.00	7.22	1.00	7.22	1.00	8.32
DE	4	1.00	3.30	1.00	3.30	1.00	3.08	1.00	2.88	1.00	2.82	1.00	3.06
	5	1.00	4.20	1.00	4.12	1.00	3.92	1.00	3.26	1.00	3.20	1.00	3.43
	7	1.00	6.04	1.00	6.05	1.00	6.25	1.00	5.22	1.00	5.12	1.00	5.17
	10	1.00	8.80	1.00	8.81	1.00	8.37	1.00	7.36	1.00	7.39	1.00	8.37
CAU	4	0.99	3.27	0.99	3.27	0.99	3.11	0.99	2.92	0.99	2.92	0.99	3.01
	5	1.00	4.35	1.00	4.26	1.00	4.22	1.00	3.85	1.00	3.82	1.00	3.93
	7	1.00	6.05	1.00	6.07	1.00	6.07	1.00	5.26	1.00	5.20	1.00	4.90
	10	1.00	8.72	1.00	8.73	0.99	8.27	1.00	7.49	1.00	7.66	1.00	8.07
EXP	4	1.00	3.10	1.00	3.10	1.00	2.98	1.00	2.32	1.00	2.24	1.00	2.49
	5	1.00	4.08	1.00	3.98	0.99	3.83	1.00	2.72	1.00	2.68	1.00	3.19
	7	1.00	5.81	1.00	5.81	1.00	5.99	1.00	3.65	1.00	3.58	1.00	4.23
	10	1.00	8.63	1.00	8.61	1.00	8.34	1.00	5.86	1.00	5.94	1.00	7.07
GAM	4	0.99	3.35	0.99	3.35	0.99	3.33	0.99	3.09	0.99	3.06	0.99	3.19
	5	1.00	4.26	1.00	4.17	0.99	4.10	1.00	3.75	1.00	3.77	0.99	3.95
	7	1.00	6.02	1.00	6.02	1.00	6.20	1.00	5.46	1.00	5.42	0.99	5.50
	10	1.00	8.90	1.00	8.87	1.00	8.43	1.00	8.18	1.00	8.18	1.00	8.66

Table 2

$P^* = 0.90$ $n = 40$ configuration $(0.0,\ldots,0.0,0.5a_\sigma)$

rule			R1						R2					
	k	P^*	WILC		WAE		MED		WILC		WAE		MED	
			$\widehat{P^*}$	$\widehat{ET}$	$\widehat{P^*}$	$\widehat{ET}$	$\widehat{P^*}$	$\widehat{ET}$	$\widehat{P^*}$	$\widehat{ET}$	$\widehat{P^*}$	$\widehat{ET}$	$\widehat{P^*}$	$\widehat{ET}$
UNI	5	0.9	1.00	4.12	1.00	4.08	1.00	4.07	1.00	2.73	1.00	2.76	1.00	3.67
		0.8	1.00	3.45	1.00	3.46	1.00	3.53	1.00	2.20	1.00	2.16	1.00	2.98
	10	0.9	1.00	8.46	1.00	8.36	1.00	9.07	1.00	6.01	1.00	5.57	1.00	7.48
		0.8	1.00	7.61	1.00	7.49	1.00	7.50	1.00	4.95	1.00	4.43	1.00	7.03
NOR	5	0.9	1.00	3.91	1.00	3.89	1.00	3.84	1.00	2.80	1.00	2.82	1.00	3.35
		0.8	1.00	3.31	1.00	3.34	1.00	3.25	1.00	2.09	1.00	2.06	1.00	2.35
	10	0.9	1.00	8.47	1.00	8.41	1.00	8.75	1.00	5.42	1.00	5.37	1.00	6.20
		0.8	1.00	7.38	1.00	7.35	1.00	7.38	1.00	4.34	1.00	4.35	1.00	5.69
DE	5	0.9	1.00	3.91	1.00	3.91	1.00	3.78	1.00	2.35	1.00	2.42	1.00	2.83
		0.8	1.00	3.28	1.00	3.28	1.00	3.21	1.00	1.88	1.00	1.87	1.00	2.18
	10	0.9	1.00	8.41	1.00	8.37	1.00	9.07	1.00	4.66	1.00	4.57	1.00	7.48
		0.8	1.00	7.31	1.00	7.30	1.00	7.11	1.00	3.36	1.00	3.43	1.00	4.19
CAU	5	0.9	1.00	4.04	1.00	4.05	1.00	3.81	1.00	2.79	1.00	2.85	1.00	2.91
		0.8	1.00	3.42	1.00	3.46	1.00	3.39	1.00	2.42	1.00	2.43	1.00	2.63
	10	0.9	1.00	8.58	1.00	8.63	1.00	8.74	1.00	6.10	1.00	6.40	1.00	5.51
		0.8	1.00	7.35	1.00	7.29	1.00	7.08	1.00	4.61	1.00	4.76	1.00	4.83
EXP	5	0.9	1.00	3.66	1.00	3.68	1.00	3.53	1.00	1.55	1.00	1.61	1.00	2.48
		0.8	1.00	2.90	1.00	2.91	1.00	3.08	1.00	1.28	1.00	1.27	1.00	1.77
	10	0.9	1.00	8.18	1.00	8.12	1.00	8.46	1.00	2.61	1.00	2.82	1.00	2.81
		0.8	1.00	6.93	1.00	6.94	1.00	6.90	1.00	1.98	1.00	2.13	1.00	2.54
GAM	5	0.9	1.00	4.08	1.00	4.08	1.00	4.03	1.00	2.93	1.00	2.97	1.00	3.51
		0.8	1.00	3.39	1.00	3.43	1.00	3.38	1.00	2.38	1.00	2.34	1.00	2.98
	10	0.9	1.00	8.51	1.00	8.48	1.00	8.83	1.00	6.73	1.00	6.59	1.00	6.51
		0.8	1.00	7.62	1.00	7.59	1.00	7.45	1.00	5.41	1.00	5.37	1.00	6.35

Table 3

P* = 0.90 n = 40

rule			R1						R2					
	k	configurat.	WILC		WAE		MED		WILC		WAE		MED	
			$\widehat{P^*}$	$\widehat{ET}$	$\widehat{P^*}$	$\widehat{ET}$	$\widehat{P^*}$	$\widehat{ET}$	$\widehat{P^*}$	$\widehat{ET}$	$\widehat{P^*}$	$\widehat{ET}$	$\widehat{P^*}$	$\widehat{ET}$
UNI	4	a	0.93	3.71	0.91	3.65	0.91	3.72	0.88	3.64	0.88	3.59	0.90	3.54
		b	1.00	2.84	1.00	2.79	1.00	3.16	0.96	2.50	0.96	2.44	0.98	2.74
	7	a	0.91	6.39	0.89	6.35	0.89	6.19	0.95	6.47	0.95	6.46	0.91	6.41
		b	1.00	5.34	1.00	5.24	0.99	5.43	0.98	3.48	0.97	3.40	0.95	4.50
NOR	4	a	0.88	3.61	0.87	3.57	0.90	3.71	0.89	3.63	0.87	3.59	0.88	3.55
		b	1.00	2.86	1.00	2.80	0.99	3.03	0.97	2.53	0.97	2.49	0.96	2.57
	7	a	0.82	6.32	0.81	6.33	0.86	6.20	0.84	6.21	0.84	6.23	0.85	6.26
		b	1.00	5.25	1.00	5.21	1.00	5.20	0.99	3.57	1.00	3.49	0.99	3.96
DE	4	a	0.93	3.70	0.93	3.68	0.94	3.74	0.90	3.68	0.88	3.64	0.92	3.65
		b	1.00	2.78	1.00	2.75	1.00	2.91	0.98	2.49	0.97	2.46	0.98	2.41
	7	a	0.90	6.28	0.91	6.25	0.88	6.12	0.95	6.29	0.96	6.29	0.90	6.27
		b	1.00	5.11	1.00	5.06	1.00	5.11	0.99	3.32	0.99	3.33	0.98	3.60
CAU	4	a	0.94	3.74	0.93	3.67	0.94	3.80	0.94	3.73	0.92	3.68	0.90	3.66
		b	0.99	2.84	0.99	2.81	1.00	2.96	0.97	2.65	0.97	2.61	0.94	2.44
	7	a	0.86	6.24	0.85	6.20	0.84	6.21	0.89	6.42	0.89	6.39	0.92	6.48
		b	1.00	5.58	1.00	5.57	1.00	5.14	0.98	4.13	0.98	4.21	0.95	3.90
EXP	4	a	0.87	3.58	0.86	3.54	0.87	3.60	0.91	3.63	0.89	3.58	0.82	3.50
		b	1.00	2.50	1.00	2.49	1.00	2.62	0.99	2.14	0.99	2.14	0.94	2.13
	7	a	0.93	6.36	0.90	6.27	0.91	6.12	0.91	6.34	0.92	6.34	0.91	6.26
		b	1.00	4.62	1.00	4.54	1.00	4.71	0.99	2.33	0.99	2.36	0.97	2.87
GAM	4	a	0.95	3.72	0.94	3.64	0.97	3.78	0.95	3.70	0.95	3.70	0.90	3.58
		b	1.00	2.93	1.00	2.88	1.00	3.11	0.98	2.65	0.98	2.60	0.96	2.68
	7	a	0.88	6.33	0.88	6.21	0.90	6.12	0.94	6.39	0.92	6.35	0.90	6.35
		b	1.00	5.58	1.00	5.50	1.00	5.68	0.98	4.33	0.97	4.35	0.96	5.28

b := (0.0,...,0.0,0.5,0.5) a := (0.0,...,0.0,0.0,0.0)

Table 4

$P^* = 0.90 \quad k = 4 \qquad$ scale-parameters $(1.0\sigma, 1.0\sigma, 1.0\sigma, 2.0\sigma)$

rule	n =	20				40			
		EXP		GAM		EXP		GAM	
	scores	$\widehat{P^*}$	$\widehat{ET}$	$\widehat{P^*}$	$\widehat{ET}$	$\widehat{P^*}$	$\widehat{ET}$	$\widehat{P^*}$	$\widehat{ET}$
R1	WILC	1.00	3.21	1.00	3.46	1.00	2.91	1.00	3.33
	WAE	1.00	3.21	1.00	3.46	1.00	2.83	1.00	3.25
	MED	1.00	3.13	1.00	3.23	1.00	3.17	1.00	3.41
R2	WILC	1.00	2.66	1.00	3.12	1.00	2.02	1.00	2.72
	WAE	1.00	2.63	1.00	3.08	1.00	1.96	1.00	2.67
	MED	1.00	2.93	1.00	3.25	1.00	2.24	1.00	2.84

CHAPTER 2

Asymptotic Distribution-Free Sequential Selection Procedures Based on an Indifference-Zone Model

1 Introduction

As distinguished from chapter 1 of the present part, where we have considered subset-selection procedures, in this chapter we discuss the problem to select the best of k populations - because of reasons given below, in this chapter we speak of k treatments (cf. introduction) - where the probability of a correct selection shall not be smaller than some preassigned value P*.

Instead of populations, we speak of k treatments, because we consider the selection problem in a more general sampling design, the so-called linear two-factor design.

Let $\Pi_1,\Pi_2,\dots,\Pi_k$ be the k different treatments, and assume that from preliminary investigations we know that the observable variables $X_{i\alpha}$, $1\leq i\leq k$, $1\leq\alpha\leq n$, are of the form

$$X_{i\alpha} = \mu+\beta_\alpha+\tau_i+e_{i\alpha}, \quad \sum_{\alpha=1}^{n} \beta_\alpha = 0, \quad \sum_{i=1}^{k} \tau_i = 0, \qquad (1.1)$$

where the parameters have the following meanings:

μ - the mean effect,

β_α - the effect of block α,

τ_i - the effect of treatment Π_i,

$e_{i\alpha}$ - the residual.

If the block effects β_α, $1 \leq \alpha \leq n$, are all equal to zero, we have an one-factor model, i.e. a population-model.

We assume that the random variables $e_{i\alpha}$, $1 \leq i \leq k$, $1 \leq \alpha \leq n$, are independent and identically distributed with the cdf F(e). The function F, the treatment effects τ_i, $1 \leq i \leq k$, the block effects β_α, $1 \leq \alpha \leq n$, and the mean effect μ are assumed to be unknown. Let

$$\tau_{(1)} \leq \tau_{(2)} \leq \cdots \leq \tau_{(k)}$$

denote the ordered treatment effects τ_i, $1 \leq i \leq k$. As we have already outlined in the introduction of part two, the treatment effects are chosen as selection criterion, and hence $\Pi_{(k)}$ denotes the <u>best</u> treatment, if $\tau_{(k)} > \tau_{(k-1)}$, where $\Pi_{(i)}$ is the population associated with $\tau_{(i)}$.

The selection problem can be formulated as follows:
Let $\tau_{(k)} - \tau_{(k-1)} \geq \Delta^*$, where Δ^* is some preassigned constant (chosen with respect to some practical arguments, cf. introduction), let $\Omega := \{\omega = (\tau_1, \ldots, \tau_k): \tau_{(k)} - \tau_{(k-1)} \geq \Delta^*\}$, and let $\mathcal{F}_0(J^*)$ denote a class of cdf's, to be specified later on. Then we want to obtain a procedure, which can be used to select the best treatment $\Pi_{(k)}$, where the probability P(CS) of a correct selection does not fall below the preassigned value P*, as n becomes large. That is, the desired procedure must satisfy the condition:

$$P(CS) \geq P^*, \text{ for all } \omega \in \Omega \text{ , and } F \in \mathcal{F}_0(J^*). \tag{1.2}$$

In order to estimate the treatment effects from the k samples, corresponding to the k treatments, we must eliminate the noninteresting block-effects β_α, $1 \leq \alpha \leq n$. For this purpose, we consider the <u>differences</u>:

$$X_{ij\alpha} := X_{i\alpha} - X_{j\alpha} = (\tau_i - \tau_j) + (e_{i\alpha} - e_{j\alpha}) =: \Delta_{ij} + e_{ij\alpha}, \tag{1.3}$$

where the cdf of the new residuals $e_{ij\alpha}$ is denoted by G(e).

<u>Remark</u>:

The distribution of the residuals $e_{ij\alpha}$, $1 \leq i \neq j \leq k$, $1 \leq \alpha \leq n$, is symmetric around zero, because the $e_{i\alpha}$, $1 \leq i \leq k$, $1 \leq \alpha \leq n$, are independent and identically distributed.

Thus we can note that the random variables $X_{ij\alpha}$ possess the cdf $G(x - \Delta_{ij})$, $1 \leq i \neq j \leq k$. For reasons, which the reader will understand in the following section, we replace the selection criterion τ_i by the functions

$$f_i(\tau_1,\dots,\tau_k) := \tau_i - \frac{1}{k}\sum_{j=1}^{k}\tau_j,\ 1\leq j\leq k. \qquad (1.4)$$

This function f_i, $1\leq i\leq k$, takes its maximum value, if the differences $\tau_i-\tau_j$ are maximal. In other words, if the effects τ_j, $1\leq j\neq i\leq k$, are fixed, then the function f_i is strictly increasing in τ_i, $1\leq i\leq k$.
Let $f_{(i)}$, $1\leq i\leq k$, denote the function associated with $\tau_{(i)}$, then we have

$$f_{(k)}-f_{(k-1)} \geq \Delta^*, \text{ for all } \omega\varepsilon\Omega . \qquad (1.5)$$

The sequential selection procedures, to be evaluated in the following sections are based on a class of estimators of the functions f_i, $1\leq i\leq k$, which on their part are based on an extended class of one-sample-r.-o.s.'s.

2 A Class of Estimators of the Functions $f_i(\tau_1,\dots,\tau_k)$

In order to derive a class of estimators of the new selection criterion $f_i(\tau_1,\dots,\tau_k)$, $1\leq i\leq k$, we first define the one-sample-r.-o.s., based on the so-called <u>sample of differences</u>, which we get from the given k samples, associated with the treatments $\pi_1,\dots,\pi_k$. Using the class of one-sample-r.-o.s's, we define the general Hodges-Lehmann-estimator of the differences Δ_{ij}. The estimators of the functions $f_i(\tau_1,\dots,\tau_k)$, $1\leq i\leq k$, are obtained from the appropriate convex linear combination of the k Hodges-Lehmann-estimators of the differences Δ_{ij}, $1\leq i\leq k$, $j=1,\dots,k$.

2.1 General one-sample-rank-order statistics

Let $\vec{X}_{ij} := (X_{ij1},X_{ij2},\dots,X_{ijn})$, $1\leq i\neq j\leq k$, be a sample of differences, where the components $X_{ij\alpha}$, $1\leq\alpha\leq n$, are defined by (1.3). We know that the random variables $X_{ij\alpha}$ are symmetrically distributed with cdf $G(x-\Delta_{ij})$. The one-sample-r.-o.s., to be defined in the following, will be based on the <u>single ranks</u>, that is, the ranks of the random variables $|X_{ij\alpha}|$ in the absolutely ordered single sample of differences, obtained from $\vec{X}_{ij}$. Formally the <u>single ranks</u> $R_{n\alpha}(ij)$, $1\leq\alpha\leq n$, associated with $\vec{X}_{ij}$, the sample of differences of size n, are defined by

$$R_{n\alpha}(ij) := \sum_{\beta=1}^{n} c(|X_{ij\alpha}|-|X_{ij\beta}|),\ 1\leq\alpha\leq n,$$

where c(x) is the function given in definition 1.4, chapter 1 of the present part.

The general one-sample-r.-o.s. $T_n(\vec{X}_{ij})$, associated with $\vec{X}_{ij}$, is then defined as follows:

$$T_n(\vec{X}_{ij}) := \frac{1}{n}\sum_{\alpha=1}^{n} c(X_{ij\alpha})J_n^*(R_{n\alpha}(ij)/(n+1)),\ 1\leq i\neq j\leq k, \tag{2.1}$$

where

$J_n^*(u) := E(J^*(U_{n\alpha}))$, if $(\alpha-1)/n \leq u \leq \alpha/n$, $\alpha=1,\dots,n$, and $U_{n1} < U_{n2} < \dots < U_{nn}$, are n ordered $U(0,1)$-distributed random variables. $U(a,b)$ denotes the uniform distribution over the interval (a,b). $c(x)$ is given again by definition 1.4, in chapter 1.

Prior to some remarks concerning the scorefunction J^*, we will try to put out the connection between this one-sample-r.-o.s. and the r.-o.s. based on joint ranks, which we have considered in chapter 1.
According to the r.-o.s.'s, which we have used in chapter 1, the regression constants of the one-sample-r.-o.s. take on the values 0 and 1 only. According to the definition $c(x)$, they take on the value 1 for all components of $\vec{X}_{ij}$, which are not negative, and they take on the value 0 for all negative components of $\vec{X}_{ij}$. Instead of scores a_i, we now use the scores $J_n^*(i)$, $1\leq i\leq n$, generated by the scorefunction J^*, where

$J^*(u) := H^{-1}((1+u)/2)$, $0<u<1$,

and $H(x)$ is some continuous cdf, which satisfies the following assumptions.

Assumption I: $H(x)$ is symmetric around zero.

Assumption II: Defining $r(x) := \frac{dH(x)}{dx}$, then $r(x)/(1-H(x))$ is nondecreasing for all $x\geq x_0$, $x\in\mathbb{R}$, where x_0 is some real number.

Remark:

The functions J_n^* and J^* also satisfy the assumptions:

(IIa) The i-th derivative of J^* with respect to u satisfies: $J^{*(i)}(u) \leq$ $\leq K(u(1-u))^{\delta-i-1/2}$, $i=0,1,2$, $0<K<\infty$, $0<u<1$, $\delta>0$;

(IIb) $\frac{1}{n}\sum_{\alpha=1}^{n} c(u)(J_n^*(u)-J^*(u))=o_p(n^{-1/2})$, $(\alpha-1)/n<u<\alpha/n$, $1\leq\alpha\leq n$.

Let $\mathcal{G}_0$ be the class of all absolutely continuous cdf's G, with the corresponding density g, which is symmetric around zero and for which the first derivative g' exists. Both, g and g' are bounded.

Let in addition

$$\mathcal{G}_0(J^*) := \{G: G\varepsilon\mathcal{G}_0 \text{ and } \lim_{x\to\infty} g(x)J^{*\prime}(G(x)-G(-x)) \text{ is finite}\},$$

then $\mathcal{F}_0(J^*)$, (cf. section 1), is the class of all cdf's F, associated with the cdf G, which is included in $\mathcal{G}_0(J^*)$.
In the following we always assume that the cdf F of the random variables $X_{i\alpha}$, $1\leq i\leq k$, $1\leq\alpha\leq n$, is included in $\mathcal{F}_0(J^*)$.

Prior to further derivations, we give some definitions, which are needed in the following. Let

$$\alpha_{ij} := \int_0^{\infty} J^*(G_{ij}(x)-G_{ij}(-x-))dG_{ij}(x), \text{ where}$$
$$G_{ij}(x) := G(x-\Delta_{ij}),\ 1\leq i\neq j\leq k. \tag{2.2}$$

Let $J(u)=H^{-1}(u)$, and

$$A^2 := \int_0^1 J^2(u)du,\ B := \int_{-\infty}^{+\infty} \frac{d}{dx} J(G(x))dG(x),$$
$$\lambda_J(G) := \int_{-\infty}^{+\infty}\int_{-\infty}^{+\infty} J(G(x))J(G(y))dG^*(x,y), \tag{2.3}$$

where $G^*(x,y)$ denotes the joint cdf of the residuals $e_{ij\alpha}$ and $e_{i\ell\alpha}$, $j\neq\ell$, which have the marginal cdf's G(x) and G(y), respectively.

In using the preceding assumptions and definitions, Puri and Sen [143] have proved

Lemma 2.1:

Let $G(x-\Delta_{ij})$ denote the cdf of the random variable $X_{ij\alpha}$, $1\leq i<j\leq k$, where $\Delta_{ij}= -\sqrt{n}a_{ij}$, and a_{ij} some finite real number. Let the statistics $T_n(\vec{X}_{ij})$ be defined according to (2.1), and let H(u) be such that the assumptions I and II are satisfied. Then the random variables $(\sqrt{n}(T_n(\vec{X}_{ij})-\alpha_{ij}))$, $1\leq i<j\leq k$, where α_{ij} is defined by (2.2), have asymptotically a $\binom{k}{2}$-variate normal distribution with mean vector $\vec{0}$ and covariance-matrix $\Gamma = ((\gamma_{ij,rs}))$, where

$$\gamma_{ij,rs}= \begin{cases} \frac{1}{4}A^2 & \text{, if } i=r,\ j=s \\ \frac{1}{4}\lambda_J(G)\text{, if } i=r,\ j\neq s, \text{ or } i\neq r,\ j=s \\ -\frac{1}{4}\lambda_J(G)\text{, if } i=s,\ j\neq r, \text{ or } i\neq s,\ j=r \\ 0 & \text{, if } i,j,r,s \text{ distinct ,} \end{cases}$$

and A^2, $\lambda_J(G)$ are defined by (2.3). ■

This result will be used to derive the distribution of the general Hodges-Lehmann-estimators, to be defined in the following. Prior to this definition, we will discuss a special form of an one-sample-r.-o.s., which takes a somewhat particular position in the sense that it cannot be integrated into the general class of one-sample-r.-o.s.'s, considered above.

2.2 The one-sample-rank-order statistics based on Median-scores

Especially with respect to the implementation of the selection procedure, to be evaluated in this chapter, the extension of the class of one-sample-r.-o.s.'s given in 2.1, by statistics based on Median-scores,- i.e. the addition of the one-sample-r.-o.s., which is well-known from the sign-test - is rather interesting.

Let $\vec{X}_{ij}$, $1 \leq i < j \leq k$, be again one of the $k(k-1)/2$ samples of differences, associated with the treatments $\Pi_1,\ldots,\Pi_k$. Using Median scores, we get the special one-sample-r.-o.s.'s

$$T_n(\vec{X}_{ij}) := \frac{1}{n}\sum_{\alpha=1}^{n} c(X_{ij\alpha}),\ 1 \leq i < j \leq k. \tag{2.4}$$

Recalling that the random variables $X_{ij\alpha}$ possess the cdf $G(x-\Delta_{ij})$, $1 \leq i < j \leq k$, $1 \leq \alpha \leq n$, the following can be proved:

With $\alpha_{ij} := G(0-\Delta_{ij})$, $\Delta_{ij} := -\sqrt{n}a_{ij}$, where a_{ij} is some finite real number, the $\binom{k}{2}$ random variables $(\sqrt{n}(T_n(\vec{X}_{ij})-\alpha_{ij}))$, $1 \leq i < j \leq k$, have asymptotically a $\binom{k}{2}$ -variate normal distribution with mean vector $\vec{0}$ and covariance matrix $\Gamma = ((\gamma_{ij,rs}))$, where

$$\gamma_{ij,rs} = \begin{cases} \frac{1}{4} & \text{, if } i=r,\ j=s \\ G^*(0,0)-\frac{1}{4}, & \text{if } i=r,\ j\neq s, \text{ or } i\neq r,\ j=s \\ \frac{1}{4}-G^*(0,0), & \text{if } i=s,\ j\neq r, \text{ or } i\neq s,\ j=r \\ 0 & \text{, if } i,j,r,s \text{ distinct,} \end{cases}$$

and $G^*(0,0)$ denotes the value of the joint cdf of the residuals $e_{ij\alpha}$ and $e_{i\ell\alpha}$, $j\neq\ell$, at the point (0,0).

Speaking of the extended class of one-sample-r.-o.s's, in the following we mean the combination of the class of general one-sample-r.-o.s.'s, (defined in 2.1) and the special statistics based on Median-scores (considered above).

2.3 The general Hodges-Lehmann-estimator

Let $T_n(\vec{X}_{ij})$ be an element of the extended class of one-sample-r.-o.s.'s. For simplification of writing, we denote the class of general one-sample-r.-o.s.'s by CA, and the one-sample-r.-o.-s.'s based on Median-scores, is denoted by CM.

At first for $1 \leq i < j \leq k$, we define the magnitudes

$$\begin{aligned} \Delta^*_{ij}(n) &:= \sup\{a: T_n(\vec{X}_{ij} - a\vec{1}) > \mu/2\}, \\ \Delta^{**}_{ij}(n) &:= \inf\{a: T_n(\vec{X}_{ij} - a\vec{1}) < \mu/2\}, \end{aligned} \tag{2.5}$$

where

$$\mu := \begin{cases} E_H(|X_{ij1}|) = 2\int_0^\infty J^*(G(x)-G(-x-))dG(x), & \text{if } T_n(\vec{X}_{ij}) \in CA \\ 1 & , \text{if } T_n(\vec{X}_{ij}) \in CM. \end{cases}$$

Then the <u>general Hodges-Lehmann-estimator of the differences Δ_{ij}</u>, $1 \leq i < j \leq k$, is given by

$$Y_{ij(n)} := \frac{1}{2}(\Delta^*_{ij}(n) + \Delta^{**}_{ij}(n)). \tag{2.6}$$

Remarks:

(a) $Y_{ij(n)}$ is translation-invariant, i.e. $Y_{ij(n)}(\vec{X}_{ij} + b\vec{1}) = Y_{ij(n)}(\vec{X}_{ij}) + b$.

(b) On the premises, made above, the distribution of $Y_{ij(n)}$ is (absolutely) continuous (cf. Hodges and Lehmann [89]).

(c) Let $T_n(\vec{X}_{ij})$ be an element of CA, and let H denote the cdf of an U(-1,1)-distributed random variable, then the estimator $Y_{ij(n)}$ is equal to the median of the $n(n+1)/2$ averages $D_{ij} := (X_{ij\alpha} + X_{ij\beta})/2$, $1 \leq \alpha \leq \beta \leq n$, if $n(n+1)/2$ is odd, and $Y_{ij(n)}$ is equal to the $(n(n+1)/4)$-largest average D_{ij}, if $n(n+1)/2$ is even.
Choosing another function H, for example the cdf of a N(0,1)-distributed random variable, the estimator $Y_{ij(n)}$ cannot be given in an explicit form. For each sample of differences and each new sample size n, the values a (cf.(2.5)) have to be evaluated by trial and error. In a later section, we will return to this problem.
If $T_n(\vec{X}_{ij})$ is an one-sample-r.-o.s. based on Median-scores, we have:

$$Y_{ij(n)} = \begin{cases} X_{ij}^{((n+1)/2)} & , \text{ if } n \text{ odd} \\ X_{ij}^{(n/2+1)} & , \text{ if } n \text{ even.} \end{cases}$$

In the following, we give two theorems concerning the asymptotic distribution of the estimators $Y_{ij(n)}$.

Theorem 2.2:

Let $T_n(\vec{X}_{ij})$ be contained in the class CA, and let $\Delta_{ij} := -\sqrt{n}\, a_{ij}$, $1 \le i < j \le k$, where a_{ij} is some finite real number, then the $\binom{k}{2}$ random variables $(\sqrt{n}(Y_{ij(n)} - \Delta_{ij}))$, $1 \le i < j \le k$, have asymptotically a $\binom{k}{2}$-variate normal distribution with mean vector $\vec{0}$ and covariance matrix $\Sigma = ((\sigma_{ij,rs}))$, where

$$\sigma_{ij,rs} = \begin{cases} \dfrac{A^2}{B^2} & , \text{ if } i=r,\ j=s \\[2ex] \dfrac{\lambda_J(G)}{B^2} & , \text{ if } i=r,\ j\neq s, \text{ or } i\neq r,\ j=s \\[2ex] -\dfrac{\lambda_J(G)}{B^2} & , \text{ if } i=s,\ j\neq r, \text{ or } i\neq s,\ j=s \\[2ex] 0 & , \text{ if } i,j,r,s \text{ distinct,} \end{cases}$$

and A^2, B^2 and $\lambda_J(G)$ are defined by (2.3).

Proof:

Using the definition of $Y_{ij(n)}$, we get:

$$\begin{aligned} &\lim_{n\to\infty} P(\sqrt{n}(Y_{ij(n)} - \Delta_{ij}) \le a_{ij},\ \forall 1 \le i < j \le k) = \\ &\quad = \lim_{n\to\infty} P_n(\sqrt{n}(T_n(\vec{X}_{ij}) - \mu/2) \le 0,\ \forall\ 1 \le i < j \le k), \end{aligned} \tag{2.7}$$

where $\mu = E_H(|X_{ij1}|)$, and P_n denotes the probability, evaluated for the sequence of shifts in location, $\{\Delta_{ij}\}_{n\in\mathbb{N}} = \{-\sqrt{n}\, a_{ij}\}_{n\in\mathbb{N}}$, $1 \le i < j \le k$. For α_{ij}, defined by (2.2) and μ, we get:

$$\lim_{n\to\infty} \sqrt{n}(\mu/2 - \alpha_{ij}) = a_{ij} \int_0^\infty \frac{d}{dx} J(G(x))dG(x) = \frac{1}{2} Ba_{ij},\ 1 \le i < j \le k.$$

Hence, using (2.7), it follows:

$$\begin{aligned} &\lim_{n\to\infty} P(\sqrt{n}(Y_{ij(n)} - \Delta_{ij}) \le a_{ij},\ \forall\ 1 \le i < j \le k) = \\ &\quad = \lim_{n\to\infty} P_n(\sqrt{n}(T_n(\vec{X}_{ij}) - \alpha_{ij}) \le \frac{1}{2} Ba_{ij},\ \forall\ 1 \le i < j \le k). \end{aligned} \tag{2.8}$$

Recalling lemma 2.1, we see that the right side of (2.8) is equal to $Q(\frac{1}{2} Ba_{12}, \ldots, \frac{1}{2} Ba_{k-1k})$, where $Q(\vec{x})$ denotes the cdf of the $\binom{k}{2}$-variate normal distribution with mean vector $\vec{0}$ and covariance-matrix Γ, given in lemma 2.1.

From this the assertion of the theorem follows. ■

Theorem 2.3:

Let $T_n(\vec{X}_{ij})$ be contained in CM, and let $\Delta_{ij} := -\sqrt{n}\, a_{ij}$, $1\le i<j\le k$, where a_{ij} is some finite real number, then the random variables $(\sqrt{n}(Y_{ij(n)}-\Delta_{ij}))$, $1\le i<j\le k$, have asymptotically a $\binom{k}{2}$-variate normal distribution with mean vector $\vec{0}$ and covariance matrix $\sum = ((\sigma_{ij,rs}))$, where

$$\sigma_{ij,rs} = \begin{cases} \dfrac{1}{4g(0)^2}, & \text{if } i=r,\ j=s \\ \dfrac{G^*(0,0)-1/4}{g(0)^2}, & \text{if } i=r,\ j\neq s, \text{ or } i\neq r,\ j=s \\ \dfrac{1/4-G^*(0,0)}{g(0)^2}, & \text{if } i=s,\ j\neq r, \text{ or } i\neq s,\ j=r \\ 0, & \text{if } i,j,r,s \text{ distinct.} \end{cases}$$

In using the results in 2.2 and $\lim_{n\to\infty} \sqrt{n}(\mu/2-\alpha_{ij}) = g(0)a_{ij}$, where $\alpha_{ij}=G(0-\Delta_{ij})$, the proof of the preceding theorem is similar to that of theorem 2.2. ■

2.4 A class of compatible estimators of the functions $f_i(\tau_1,\dots,\tau_k)$

To construct compatible estimators of the functions $f_i(\tau_1,\dots,\tau_k)$, based on the general Hodges-Lehmann-estimators of the differences Δ_{ij}, $1\le i\neq j\le k$, we use the least squares method.
By minimizing the k functions

$$Q(i) := \sum_{\substack{j=1\\ j\neq i}}^{k} (Y_{ij(n)}-(\tau_i-\tau_j))^2,\quad i=1,\dots,k,$$

with respect to τ_i, $1\le i\le k$, we get the solutions:

$$\Delta_{ij} = \tau_i-\tau_j = Y_{i.(n)}-Y_{j.(n)} =: Z_{ij(n)},\quad 1\le i\neq j\le k,$$

where $Y_{\ell\ell(n)}:=0$, and

$$Y_{\ell.(n)} := \frac{1}{k}\sum_{p=1}^{k} Y_{\ell p(n)},\quad 1\le \ell\le k. \tag{2.9}$$

Hence, as compatible estimators of the differences Δ_{ij}, we use

$$Z_{ij(n)} = Y_{i.(n)}-Y_{j.(n)},\quad 1\le i\neq j\le k. \tag{2.10}$$

The estimators $Z_{ij(n)}$ are linear functions of general Hodges-Lehmann-estimators of the differences Δ_{ij}, defined in 2.3. It can easily be seen that the estimators $Z_{ij(n)}$ possess the same linear properties as the differences Δ_{ij}, i.e. we have

$$Z_{ij(n)} = -Z_{ji(n)},\ Z_{ij(n)}+Z_{jk(n)} = Z_{ik(n)}.$$

The estimators $Z_{ij(n)}$ are called compatible, because by its use, any contrast of the form

$$\theta := \sum_{i=1}^{k} \ell_i\tau_i, \text{ where } \sum_{i=1}^{k} \ell_i=0,$$

can be evaluated. In particular the estimators of the contrasts

$$f_i(\tau_1,\ldots,\tau_k) = \tau_i - \frac{1}{k}\sum_{j=1}^{k} \tau_j,\ 1\leq i\leq k,$$

can be got instantly by the $Z_{ij(n)}$. Because of $\tau_i-\tau_j = f_i-f_j$, the variables $Y_{i.(n)}$ are compatible estimators of the functions $f_i(\tau_1,\ldots,\tau_k)$, $i=1,\ldots,k$. At this point the reader may realize the reasons why the functions $f_i(\tau_1,\ldots,\tau_k)$ have been used as selection criterion.
In the following, we consider some asymptotic properties of the random variables $Z_{ik(n)}$, $1\leq i\leq k-1$.

Theorem 2.4:

The random variables $(\sqrt{n}(Z_{ik(n)}-\Delta_{ik}))$, $i=1,\ldots,k-1$, have asymptotically a $(k-1)$-variate normal distribution with mean vector $\vec{0}$ and covariance matrix $T = ((\tau_{ij}))$, $1\leq i,j\leq k-1$, where

$$\tau_{ii} = 2\sigma_0^2\,(1-\rho),\ \tau_{ij} = \sigma_0^2(1-\rho),\ \rho = -1/(k-1),$$

and

$$\sigma_0^2 := \begin{cases} \frac{k-1}{k^2}\,(A^2+(k-2)\lambda_J(G))/(B(G))^2 & , \text{ if } T_n(\vec{X}_{ij})\varepsilon CA \\ \frac{k-1}{k^2}\,(1+(k-2)(4G^*(0,0)-1))/4g(0)^2, & \text{ if } T_n(\vec{X}_{ij})\varepsilon CM, \end{cases}$$

where A^2, $B(G)$, $\lambda_J(G)$ and $G^*(0,0)$ are defined in (2.3) and 2.2.

Proof:

The proof is essentially based on the theorems 2.2 and 2.3. As already outlined, the estimators $Z_{ik(n)}$ are linear functions of the asymptotically normal distributed random variables $Y_{ij(n)}$, $1\leq i\neq j\leq k$.

In case that the estimators $Y_{ij(n)}$, $1 \leq i \neq j \leq k$, are based on one-sample-r.-o.s.'s, which are contained in CA, we get:

$$\mathrm{Var}(\sqrt{n}(Y_{i.(n)}-f_i)) = \frac{k-1}{k^2} (A^2+(k-2)\lambda_J(G))/B(G)^2 =: \sigma_0^2 ,$$

$$\mathrm{Cov}(\sqrt{n}(Y_{i.(n)}-f_i),\sqrt{n}(Y_{j.(n)}-f_j)) = -\frac{1}{k^2} (A^2+(k-2)\lambda_J(G))/B(G)^2 =: \rho\sigma_0^2 .$$

If the estimators $Y_{ij(n)}$ are based on r.-o.s's, which are elements of CM, we get the expression σ_0^2, by substituting A^2,B^2 and $\lambda_J(G)$ by $\frac{1}{4}$, $g(0)^2$ and $G^*(0,0)-\frac{1}{4}$, respectively.In using these results and the asymptotic normality of the random variables $Y_{ij(n)}$, $1 \leq i \neq j \leq k$, the proof of the theorem is straightforward.■

This result will be used in section 4, where the class of sequential selection procedures will be defined.

Another asymptotic property of the estimators $Z_{ik(n)}$, also needed in section 4, is given by

Theorem 2.5:

The sequence of random variables $\{Z_{ik(n)}\}_{n\in\mathbb{N}}$ is uniformly continuous in probability with respect to the sequence $\{n^{-1/2}\}_{n\in\mathbb{N}}$, i.e., for each ε and $\eta > 0$, there exists a $\delta > 0$, such that

$$P(\sup_{n':|n'-n|<\delta n} |\sqrt{n}(Z_{ik(n')}-Z_{ik(n)})| > \varepsilon) < \eta ,$$

as $n\to\infty$.

Proof:

Since the $Z_{ik(n)}$ are linear functions of the random variables $Y_{ij(n)}$ and $Y_{jk(n)}$, $1 \leq i$, $j \leq k$, the assertion follows directly from A4/10, which means that the sequences of random variables $\{Y_{\ell p(n)}\}_{n\in\mathbb{N}}$, $1 \leq \ell < p \leq k$, are uniformly continuous in probability with respect to $\{n^{-1/2}\}_{n\in\mathbb{N}}$. ■

3 Several Strongly Consistent Estimators

In 2.4,we have given the asymptotic distribution of the compatible estimators $Z_{ik(n)}$, $1\le i\le k-1$. But from theorem 2.4,we see that the elements of the covariance matrix T depend on σ_0^2, which on its part depends on the parameters $B^2(G)$, $\lambda_J(G)$, if the used one-sample-r.-o.s's are contained in CA, and on the parameters $g(0)$, $G^*(0,0)$, in case that the used one-sample-r.-o.s.'s are elements of CM.

In this section,we derive strongly consistent estimators of the unknown parameters, which we use in section 4 to evaluate the actual estimates for σ_0^2.

3.1 An estimator of $(B^2(G))^{-1}$

Let $\vec{X}_{ij}$ be some sample of differences, $1\le i<j\le k$, then according to (2.5), we define

$$\begin{aligned} \Delta_{ijL}(n) &:= \inf\{a:T_n(\vec{X}_{ij}-a\vec{1}) < T_{n\alpha}^{(2)}\}, \\ \Delta_{ijU}(n) &:= \sup\{a:T_n(\vec{X}_{ij}-a\vec{1}) > T_{n\alpha}^{(1)}\}, \end{aligned} \tag{3.1}$$

where

$$T_{n\alpha}^{(2)} = \frac{\mu}{2} - \frac{At_{\alpha/2}}{2\sqrt{n}}, \quad T_{n\alpha}^{(1)} = \frac{\mu}{2} + \frac{At_{\alpha/2}}{2\sqrt{n}}$$

are determined such that

$$\lim_{n\to\infty} P(T_{n\alpha}^{(2)} \le T_n(\vec{X}_{ij}) \le T_{n\alpha}^{(1)}) = 1-\alpha, \text{ if } \Delta_{ij}=0 \text{ , where}$$

$$\Phi(t_{\alpha/2}) = 1-\alpha/2, \ \alpha\varepsilon(0,\tfrac{1}{2}).$$

Choosing

$$B_n^2 := \frac{n}{(2At_{\alpha/2})^2\binom{k}{2}} \sum_{i=1}^{k-1}\sum_{j=i+1}^{k}(\Delta_{ijU}(n)-\Delta_{ijL}(n))^2, \tag{3.2}$$

as estimator of $(B^2(G))^{-1}$, we have

Theorem 3.1:

Let B_n^2 be the estimator, given by (3.2), then it follows

$$B_n^2 \xrightarrow[n\to\infty]{a.s.} (B^2(G))^{-1}.$$

Proof:

Recalling A4/11, we see that for $s>0$, there exist positive constants c_{s1}, c_{s2} and $n_s \in \mathbb{N}$ such that for $n \geq n_s$ and $B_{ni}^2 := \frac{n}{(2At_{\alpha/2})^2} (\Delta_{ijU}(n)-\Delta_{ijL}(n))^2$, the following inequality holds:

$$P(B_{ni}^2-(B(G)^2)^{-1} > c_{s1}n^{-1/4}(\ell n n)^3) \leq c_{s2}n^{-s}.$$

This together with the Borel-Cantelli-lemma (A2/6) yields:

$$P(\limsup_{n\geq n_s}(B_{ni}^2-(B(G)^2)^{-1}) > c_{s1}n^{-1/4}(\ell n n)^3) = 0.$$

The assertion of the theorem follows, because B_n^2 is the arithmetic mean of $\binom{k}{2}$ terms B_{ni}^2. ∎

Hence $(B_n^2)^{-1}$ is a strongly consistent estimator of the unknown parameter $(B(G)^2)^{-1}$.

3.2 An estimator of $(g(0)^2)^{-1}$

Let $\vec{X}_{ij}$ be again a sample of differences, $1\leq i<j\leq k$, and let $X_{ij}^{(1)}<X_{ij}^{(2)}<\ldots<X_{ij}^{(n)}$ denote the ordered elements of $\vec{X}_{ij}$.
Choosing

$$C_n^2 := \frac{2}{k(k-1)} \sum_{i=1}^{k-1} \sum_{j=i+1}^{k} C_{ni}^2 , \tag{3.3}$$

as estimator of $(g(0))^{-1}$, where

$$b(n) := \max\{1,\frac{n}{2} - t_{\alpha/2} \frac{\sqrt{n}}{2}\} , \quad a(n) := n-b(n)+1, \text{ and}$$
$$C_{ni} := \sqrt{n}(X_{ij}^{(a(n))} - X_{ij}^{(b(n))})/t_{\alpha/2} \tag{3.4}$$

then the following theorem holds:

Theorem 3.2:

Let C_n^2 be given by (3.3), and assume that $g=G'$ and $g'=G''$ exist in a neighbourhood of zero, and that g' is bounded in a neighbourhood of zero, then it follows

$$C_n^2 \xrightarrow[n\to\infty]{a.s.} (g(0)^2)^{-1}.$$

Proof:

Using the preceding assumptions and a result of Bahadur [25], we get

$$X_{ij}^{(c(n))} = \left(\frac{c(n)}{\sqrt{n}} - G_n(\xi)\right)/g(\xi) + O(n^{-3/4}\ell n n) \text{ a.s.}, \tag{3.5}$$

where $\{c(n)\}_{n\in\mathbb{N}}$ is some sequence of positive integers, and $c(n)=np+o(\sqrt{n}\,\ell n n)$ $0\leq p\leq 1$, $G(\xi)=p$, and G_n denotes the empirical cdf of the random variables $X_{ij\alpha}$, $1\leq\alpha\leq n$. From (3.5) it follows:

$$\sqrt{n}(X_{ij}^{(a(n))}-X_{ij}^{(b(n))}) = \frac{a(n)-b(n)}{\sqrt{n}\,g(0)} + O(n^{-1/4}\ell n n) \text{ a.s.}. \tag{3.6}$$

Using $a(n)\approx\frac{n}{2}+t_{\alpha/2}\frac{\sqrt{n}}{2}$ and $b(n)\approx\frac{n}{2}-t_{\alpha/2}\frac{\sqrt{n}}{2}$, we get:

$$\frac{a(n)-b(n)}{\sqrt{n}\,g(0)} + O(n^{-1/4}\ell n n) \xrightarrow[n\to\infty]{} \frac{t_{\alpha/2}}{g(0)}.$$

This together with (3.6) yields the assertion of the theorem. ■

Hence we see that C_n^2 is a strongly consistent estimator of $(g(0)^2)^{-1}$.

3.3 Two estimators of $\lambda_J(G)$ and $G^*(0,0)$

Let $J_n(u) = E(J(U_{n\alpha}))$, if $(\alpha-1)/n < u < \alpha/n$, where $U_{n\alpha}$, $1\leq\alpha\leq n$, denote the ordered $U(0,1)$-distributed random variables. Based on the two samples of differences $\vec{X}_{ij}$ and $\vec{X}_{i\ell}$, $1\leq i\leq k$, $\ell,j\neq i$, $1\leq j\neq\ell\leq k$, we define the $k(k-1)(k-2)/2$ estimators

$$L_{ij\ell,n} := \frac{1}{n}\sum_{\alpha=1}^{n} J_n\left(\frac{R_{ij\alpha}}{n+1}\right)J_n\left(\frac{S_{i\ell\alpha}}{n+1}\right), \tag{3.7}$$

where $R_{ij\alpha}$ denotes the rank of the difference $X_{ij\alpha}$ in the sample of differences $\vec{X}_{ij}$, and $S_{i\ell\alpha}$ denotes the rank of $X_{i\ell\alpha}$ in $\vec{X}_{i\ell}$.

Using the preceding definitions, we get

Theorem 3.3:

Let the estimators $L_{ij\ell,n}$ be defined by (3.7), then it follows:

$$L_{n1} := \frac{2}{k(k-1)(k-2)}\sum_{i=1}^{k}\sum_{\substack{j=1\\ j\neq i}}^{k}\sum_{\substack{\ell=j+1\\ \ell\neq i}}^{k} L_{ij\ell,n} \xrightarrow[n\to\infty]{\text{a.s.}} \lambda_J(G),$$

where $\lambda_J(G)$ is given by (2.3).

Proof:

In proving

$$L_{ij\ell,n} \xrightarrow[n\to\infty]{a.s} \lambda_J(G), \tag{3.8}$$

the assertion of the theorem follows, because L_{n1} is the arithmetic mean of $k(k-1)(k-2)/2$ estimators $L_{ij\ell,n}$.

The ranks $R_{ij\alpha}$ and $S_{i\ell\alpha}$ remain invariant under shifts of location of the underlying distributions. Defining $Z_{ij\alpha}:=X_{ij\alpha}-\Delta_{ij}$, $1\leq\alpha\leq n$, the rank of $Z_{ij\alpha}$ with respect to the new sample $\vec{Z}_{ij}$ is given by $R_{ij\alpha}$, too. Let

$$F_n(x) := \frac{1}{n}\,(\text{number of } Z_{ij\alpha}\leq x),\quad G_n(y) := \frac{1}{n}\,(\text{number of } Z_{i\ell\alpha}\leq y)$$

and

$$H_n(x,y) := \frac{1}{n}\,(\text{number of } (Z_{ij\alpha},Z_{i\ell\alpha}) \text{ with } Z_{ij\alpha}\leq x \text{ \underline{and} } Z_{i\ell\alpha}\leq y)$$

denote the empirical cdf's of $Z_{ij\alpha}$, $Z_{i\ell\alpha}$ and $(Z_{ij\alpha},Z_{i\ell\alpha})$, respectively, then the estimator $L_{ij\ell,n}$ can be rewritten as:

$$\int_{-\infty}^{+\infty}\int_{-\infty}^{+\infty} J_n(nF_n(x))J_n(nG_n(y))\,dH_n(x,y) =: n^{-1}T_n.$$

Furthermore, we define:

$$n^{-1}T_n^* := \int_{-\infty}^{+\infty}\int_{-\infty}^{+\infty} J(nF_n(x)/(n+1))J(nG_n(y)/(n+1))\,dH_n(x,y).$$

Then, in proving

$$n^{-1}(T_n-T_n^*) \xrightarrow[n\to\infty]{a.s} 0 \text{ and } n^{-1}T_n^* \xrightarrow[n\to\infty]{a.s.} \lambda_J(G), \tag{3.9}$$

the assertion (3.8) follows.

Using the definition of F_n, G_n and H_n, we get

$$n^{-1}(T_n-T_n^*) = I_{n1}+I_{n2}, \text{ where}$$

$$I_{n1} := n^{-1}\sum_{i=1}^{n} J_n(i)(J_n(i)-J(i/(n+1))),$$

$$I_{n2} := n^{-1}\sum_{i=1}^{n} J(i/(n+1))(J_n(i)-J(i/(n+1))).$$

Recalling the Schwarz inequality, the following inequalities are immediately seen to hold:

$$I_{n1}^2 \leq (n^{-1}\sum_{i=1}^{n}(J_n(i))^2)(n^{-1}\sum_{i=1}^{n}(J_n(i)-J(i/(n+1)))^2), \tag{3.10}$$

$$I_{n2}^2 \leq (n^{-1}\sum_{i=1}^{n}(J(i/(n+1)))^2)(n^{-1}\sum_{i=1}^{n}(J_n(i)-J(i/(n+1)))^2). \tag{3.11}$$

Because of

$$n^{-1}\sum_{i=1}^{n}(J_n(i))^2 \leq n^{-1}\sum_{i=1}^{n}(J(i/(n+1)))^2 \xrightarrow[n\to\infty]{} \int_0^1 J(u)du=:c<\infty, \tag{3.12}$$

and

$$\lim_{n\to\infty} n^{-1}\sum_{i=1}^{n}(J_n(i)-J(i/(n+1)))^2 = 0, \text{ (cf. Hoeffding [91])}.$$

we have:

$$\lim_{n\to\infty} I_{n1} = \lim_{n\to\infty} I_{n2} = 0.$$

From Khintchine's law of large numbers and A2/11 we get:

$$\int_{-\infty}^{+\infty}\int_{-\infty}^{+\infty} J(G(x))J(G(y))dH_n(x,y) = n^{-1}\sum_{\alpha=1}^{n} J(G(Z_{ij\alpha}))J(G(Z_{i\ell\alpha})) \xrightarrow[n\to\infty]{a.s.} \lambda_J(G).$$

To prove the second part of (3.9), we only have to show that

$$I_n := \int_{-\infty}^{+\infty}\int_{-\infty}^{+\infty}(J(nF_n(x)/(n+1))J(nG_n(y)/(n+1))-J(G(x))J(G(y)))dH_n(x,y) \xrightarrow[n\to\infty]{a.s.} 0.$$

At first, we put:

$I_n := I_{n3}+I_{n4}$, where

$$I_{n3} := \int_{-\infty}^{+\infty}\int_{-\infty}^{+\infty} J(nF_n(x)/(n+1))(J(nG_n(y)/(n+1))-J(G(y)))dH_n(x,y),$$

$$I_{n4} := \int_{-\infty}^{+\infty}\int_{-\infty}^{+\infty} J(G(y))(J(nF_n(x)/(n+1))-J(G(x)))dH_n(x,y).$$

Using the Schwarz inequality and (3.12), we get:

$$I_{n3}^2 \leq c\int_{-\infty}^{+\infty}(J(nG_n(y)/(n+1))-J(G(y)))^2 dG_n(y).$$

Because of (3.12), for each $\varepsilon>0$, there exists a $\delta_1(0<\delta_1<1/2)$, such that

$$\int_0^{\delta_1}(J(u))^2du + \int_{1-\delta_1}^{1}(J(u))^2du < \frac{1}{8}\varepsilon.$$

Let a,b be chosen, such that $G(a)=1-G(b)=\frac{1}{2}\delta_1$, then

$$\int_a^b (J(nG_n(y)/(n+1))-J(G(y)))^2 dG_n(y) = \tag{3.13}$$
$$= \int_{\frac{\delta_1}{2}}^{1-\frac{\delta_1}{2}} (J(\tfrac{n}{n+1} G_n(G^{-1}(u)))-J(G(G^{-1}(u))))^2 dG_n(G^{-1}(u)).$$

Since J(u) is continuous in the interval (0,1), it is uniformly continuous in all intervals $[\eta, 1-\eta]$, $0<\eta<\frac{1}{2}$. Using in addition the theorem of Glivenko, we have

$$\sup_{u\in[0,1]} \left|\frac{n}{n+1} G_n(G^{-1}(u))-u\right| \xrightarrow[n\to\infty]{a.s.} 0,$$

and hence, the right side of (3.13) converges to zero almost sure, as $n\to\infty$. Furthermore, we have

$$\int_b^\infty (J(nG_n(y)/(n+1))-J(G(y)))^2 dG_n(y) \leq \tag{3.14}$$
$$\leq 2(\int_b^\infty (J(nG_n(y)/(n+1)))^2 dG_n(y) + \int_b^\infty (J(G(y)))^2 dG_n(y)).$$

Because of $nG_n(b)/(n+1) \xrightarrow[n\to\infty]{a.s.} G(b)=1-\delta_1/2$, the inequality $G_n(b) \geq 1-\delta_1$ holds almost sure as $n\to\infty$.
In writing $n^* = [nG_n(b)] + 1$, we get:

$$\int_b^\infty (J(nG_n(y)/(n+1)))^2 dG_n(y) = n^{-1} \sum_{i=n^*}^{n} (J(i/(n+1)))^2 \overset{a.s.}{\leq}$$
$$\overset{a.s.}{\leq} n^{-1} \sum_{i=[n(1-\delta_1)]+1}^{n} (J(i/(n+1)))^2 \xrightarrow[n\to\infty]{} \int_{1-\delta_1}^{1} (J(u))^2 du < \frac{1}{8}\varepsilon.$$

Recalling the strong law of large numbers, we see that

$$\int_b^\infty (J(G(y)))^2 dG_n(y) \xrightarrow[n\to\infty]{a.s.} \int_{1-\frac{\delta_1}{2}}^{1} (J(u))^2 du < \frac{1}{8}\varepsilon.$$

Thus, in choosing an appropriate b, the right side of (3.14) can be made smaller than $\varepsilon/2$ almost sure as n becomes large. Similar arguments yield:

$$\int_{-\infty}^{a} (J(nG_n(y)/(n+1))-J(G(y)))^2 dG_n(y) \underset{n\to\infty}{\overset{a.s}{\leq}} \frac{1}{2}\varepsilon .$$

Hence, it follows

$$I_{n3} \xrightarrow[n\to\infty]{a.s.} 0.$$

Because of

$$I_{n4}^2 \leq c \int_{-\infty}^{+\infty} (J(nF_n(x)/(n+1))-J(F(x)))^2 dF_n(x),$$

we also have:

$$I_{n4} \xrightarrow[n\to\infty]{a.s.} 0.$$

This completes the proof of the theorem. ■

In using median-scores, the parameter $G^*(0,0)$ must be estimated. Let $Z_{ij\alpha}$ and $H_n(x,y)$ be defined as above, and let $G_{ij\ell}(x,y)$ denote the cdf of the random vector $(X_{ij\alpha},X_{i\ell\alpha})$, $1\leq i\leq k$, $1\leq j<\ell\leq k$, $j,\ell\neq i$, $1\leq\alpha\leq n$, then we have:

$$H_n(\Delta_{ij},\Delta_{i\ell}) = \frac{1}{n}\sum_{\alpha=1}^{n} c(X_{ij\alpha})\cdot c(X_{i\ell\alpha}),$$

where $c(x)$ is given by definition 1.4, chap. 1.

Because of $G^*(0,0)=G_{ij\ell}(\Delta_{ij},\Delta_{i\ell})$, $1\leq i\leq k$, $1\leq j<\ell\leq k$, $j,\ell\neq i$, and by use of the strong law of large numbers, it follows:

$$H_n(\Delta_{ij},\Delta_{i\ell}) \xrightarrow[n\to\infty]{a.s.} G^*(0,0). \tag{3.15}$$

Defining

$$L_{n2}:=\frac{2}{k(k-1)(k-2)}\sum_{i=1}^{k}\sum_{\substack{j=1\\ j\neq i}}^{k}\sum_{\substack{\ell=j+1\\ \ell\neq i}}^{k} H_n(\Delta_{ij},\Delta_{i\ell}),$$

(3.15) yields:

$$L_{n2} \xrightarrow[n\to\infty]{a.s.} G^*(0,0).$$

The main result, to be used in the following section, is that L_{n1} and L_{n2} are strongly consistent estimators of the parameters $\lambda_J(G)$ and $G^*(0,0)$, respectively.

4 A Class of Sequential Selection Procedures

4.1 Definition of the selection procedures

In this section we consider a class of sequential selection procedure based on the indifference-zone model (cf. introduction) which can be applied in two-factor designs. Thereby, we make use of the results of the preceding sections 1, 2 and 3.

Let h be the solution of the equation

$$\int_{-\infty}^{+\infty} \Phi^{k-1}(y+h\sqrt{2})\, d\Phi(y) = P^*, \tag{4.1}$$

where P^* is the preassigned probability (cf. section 1). For various values of P^* and k, the values $d=h\sqrt{2}$ are given in table II, appendix 6.
The class of selection procedures, based on the class of compatible estimators of the functions $f_i(\tau_1,\ldots,\tau_k)$, $1\leq i\leq k$, is defined as follows:
Starting with some minimal[1] sample size n_0, at each stage of the sequential sampling, one observation of the random variables, defined by (1.1), is made for each treatment Π_i, $1\leq i\leq k$.
Thereafter, we examine, whether the reached sample size $n\geq n_0$ satisfies the inequality

$$n \geq h^2\hat{\sigma}^2/(\Delta^*)^2, \tag{4.2}$$

where

$$\hat{\sigma}^2 := \begin{cases} \frac{2}{k}\,(A^2+(k-2)L_{n1})B_n^2 & \text{, if } T_n(\vec{X}_{ij})\in CA \\ \frac{2}{k}\,(\frac{1}{4}+(k-2)(L_{n2}-\frac{1}{4}))C_n^2, & \text{if } T_n(\vec{X}_{ij})\in CM. \end{cases}$$

The <u>stopping rule</u> prescribes that sampling must be continued at the next stage, if the inequality (4.2) does not hold, and that sampling must be terminated, i.e. the <u>stopping variable N</u> takes the value n, in case that (4.2) holds. Then, based on the samples of differences $\vec{X}_{ij}$, (of size N), $1\leq i\neq j\leq k$, the estimates $Y_{i.(N)}$ for the functions $f_i(\tau_1,\ldots,\tau_k)$, can be evaluated, and that treatment Π_j is selected as <u>best</u> treatment, which is associated with $Y_{j.(N)}$, satis-

---

[1] *Some remarks, concerning the choice of n_o, we make in section 6.*

fying the inequalities

$$Y_{j.(N)} \geq Y_{i.(N)}, \text{ for each } i\varepsilon\{1,\ldots,k\}, \tag{4.3}$$

where $Y_{i.(N)}$ denotes the estimate of $f_i(\tau_1,\ldots,\tau_n)$, $1 \leq i \leq k$.
If more than one treatment is selected, "one best" is selected at random.

4.2 Some important properties of the sequential selection procedures

The essential properties of the sequential selection procedures, defined in 4.1, are given in

Theorem 4.1:

For all $\omega\varepsilon\Omega$, the following statements are valid for the class of selection procedures, defined in 4.1:

(i) $P(N<\infty) = 1$;

(ii) $N \xrightarrow[\Delta^*\to 0]{a.s.} \infty$; $E(N) \xrightarrow[\Delta^*\to 0]{} \infty$;

(iii) $\frac{N}{\nu} \xrightarrow[\Delta^*\to 0]{a.s.} 1$;

(iV) $\frac{E(N)}{\nu} \xrightarrow[\Delta^*\to 0]{} 1$;

(V) $\lim\limits_{\Delta^*\to 0} \inf\limits_{\omega\varepsilon\Omega} P(CS) \geq P^*$, where

$$\nu := \nu(\Delta^*,P^*) = h^2\sigma^2/(\Delta^*)^2, \text{ and}$$

$$\sigma^2 := \begin{cases} \frac{2}{k}\,(A^2+(k-2)\lambda_J(G))/(B(G))^2 & , \text{ if } T_n(\vec{X}_{ij})\varepsilon CA \\ \frac{2}{k}\,(\frac{1}{4}+(k-2)(G^*(0,0)-\frac{1}{4}))/(g(0))^2, & \text{if } T_n(\vec{X}_{ij})\varepsilon CM. \end{cases}$$

Proof:

(i) follows from the definition of N and the almost sure convergence of $\hat{\sigma}^2$ to σ^2 as n becomes large.

(ii) follows like (i)

(iii) follows from the almost sure convergence of $\hat{\sigma}^2$ to σ^2 and A4/12

(iV) Because of A2/13, we must only show that the sequence of random variables $\{N/\nu\}$ is uniformly integrable for $\Delta^* > 0$. Recalling A2/12, we see that the validity of

$$\sum_{n=1}^{\infty} \sup_{0<\Delta^*<\Delta_0^*} P(N/\nu>n) < \infty \text{ for some } \Delta_0^* > 0, \tag{4.4}$$

is sufficient for the uniformly integrability of $\{N/\nu\}$ for all $\Delta^* > 0$. Thus, only (4.4) must be proved.
According to the two classes of one-sample-r.-o.s.'s CA and CM, we have two different proofs of (4.4) :

(a) $\underline{T_n(\vec{X}_{ij}) \varepsilon CM,\ 1 \leq i \neq j \leq k}$:

Let $m:=m_n(\nu)=n\nu$, then we have:

$$P(N>n\nu) \leq P(N>m) \leq P(\hat{\sigma}^2 > \Delta^{*2} m/h^2) =$$

$$= P(\tfrac{2}{k}(\tfrac{1}{4}+(k-2)(L_{m2}-\tfrac{1}{4}))C_m^2 > \Delta^{*2} m/h^2) \leq$$

$$\leq P(\tfrac{9}{4} C_m^2 > \frac{\Delta^{*2} m}{h^2}) \leq \sum_{i=1}^{k-1} \sum_{j=i+1}^{k} P(X_{ij}^{(a(m))} - X_{ij}^{(b(m))} > \frac{2\Delta^* t_{\alpha/2}}{3h})$$

If we put $2\delta := 2\Delta^* t_{\alpha/2}/3h$, with $S_{ijm}(t) = \sum_{\alpha=1}^{m} c(X_{ij\alpha}-t)$, where $c(x)$ is given by definition 1.4, chap. 1, we get:

$$P(N>n\nu) \leq \sum_{i=1}^{k-1} \sum_{j=i+1}^{k} P(S_{ijm}(\delta) < a(m)) + P(S_{ijm}(-\delta) \geq b(m)) =$$

$$= \sum_{i=1}^{k-1} \sum_{j=i+1}^{k} 2P(B(m,G_{ij}(-\delta)) \geq b(m)),$$

where $B(m,p)$ is a binomial random variable with parameters m and p, and $G_{ij}(x)$ denotes the cdf of the random variables $X_{ij\alpha}$, $1 \leq \alpha \leq m$.
Recalling A2/7 it is easily seen that

$P(B(m,G_{ij}(-\delta)) \geq b(m)) \leq \exp(-2mt_m^2)$, where

$$t_m := (b(m)-mG_{ij}(-\delta))/m = \frac{1}{2} - \frac{t_{\alpha/2}}{2\sqrt{m}} - G_{ij}(-\delta) \geq \gamma,\ \gamma > 0,$$

$m \geq M_0$, $\Delta^* < \Delta_0^*$, and $\Delta_0^* > 0, M_0$ are chosen large enough.

For $\Delta^* < \Delta_0^*$, $m \geq M_0$ and a constant A we find:

$$P(N>n\nu) \leq \sum_{i=1}^{k-1} \sum_{j=i+1}^{k} A\exp(-2m\gamma^2),$$

and hence, the validity of (4.4) follows from $\sum_{m=M_0}^{\infty} A\exp(-2m\gamma^2) < \infty$.

(b) $T_n(\vec{X}_{ij}) \varepsilon CA$, $1\leq i\neq j\leq k$:

Using the same notations as in (a), we get:

$$P(N>n\nu) \leq P((A^2+(k-2)L_{m1})B_m^2 > \frac{k(\Delta^*)^2 m}{2h^2}) = \tag{4.5}$$

$$= P(B_m^2 + \frac{(k-2)L_{m1}B_m^2}{A^2} > \frac{k(\Delta^*)^2 m}{2h^2A^2}).$$

It is easily verified that $L_{m1} \leq A^2$, and this implies:

$$P(B_m^2 + \frac{(k-2)L_{m1}B_m^2}{A^2} > \frac{k(\Delta^*)^2 m}{2h^2A^2}) \leq P(B_m^2 > \frac{k(\Delta^*)^2 m}{2(k-1)A^2h^2})$$

$$= P(\sum_{i=1}^{k-1}\sum_{j=i+1}^{k}(\Delta_{ijU}(m)-\Delta_{ijL}(m))^2 > \frac{k^2(k-1)(\Delta^*)^2 4(t_{\alpha/2})^2}{2(k-1)h^2 2}) \tag{4.6}$$

$$\leq \sum_{i=1}^{k-1}\sum_{j=i+1}^{k} P((\Delta_{ijU}(m)-\Delta_{ijL}(m))^2 > \frac{2k(\Delta^*)^2(t_{\alpha/2})^2}{(k-1)h^2})$$

$$\leq \sum_{i=1}^{k-1}\sum_{j=i+1}^{k} P(m(\Delta_{ijU}(m)-\Delta_{ijL}(m)) > \frac{\sqrt{m}\Delta^* t_{\alpha/2}}{h}).$$

Recalling the definition of ν, for some constant $\mu>1$ and all $n\geq\mu^2$, it follows:

$$P(\sqrt{m}(\Delta_{ijU}(m)-\Delta_{ijL}(m)) > \frac{\sqrt{m}t_{\alpha/2}\Delta^*}{h}) \leq P(\sqrt{m}(\Delta_{ijU}(m)-\Delta_{ijL}(m)) > t_{\alpha/2}\mu\sigma),$$

where σ is given in theorem 4.1. Puri and Sen [143] have shown that $\lambda_J(G) \leq \frac{1}{2}A^2$. Using this result and the identity $\sigma = \eta\frac{A}{B(G)}$, where $\eta<1$, we get:

$$P(\sqrt{m}(\Delta_{ijU}(m)-\Delta_{ijL}(m)) > \frac{\sqrt{m}t_{\alpha/2}\Delta^*}{h}) \leq$$

$\leq P(\sqrt{m}(\Delta_{ijU}(m)-\Delta_{ijL}(m)) > t_{\alpha/2}\mu\eta\frac{A}{B(G)}) \leq c_s m^{-s}$, for all $m\geq m_s$ (independent of Δ^*), and $m\geq n\nu-1\geq\nu-1\geq m_s$, for $\Delta^*\leq\Delta_0^*$ and all n, where s is some constant >0. The last inequality follows by using A4/11.

Let $m>n\nu(\Delta_0^*,P^*)$, then for all $\Delta^*\leq\Delta_0^*$, we have:

$$P(\sqrt{m}(\Delta_{ijU}(m)-\Delta_{ijL}(m)) > \sqrt{m}t_{\alpha/2}\Delta^*/h) \leq Kn^{-s}, \tag{4.7}$$

where K denotes some constant, which is independent of Δ^*. Thus, the proof of (4.4) follows from (4.5), (4.6) and (4.7).

(V) The random variables $V_{i(n)} := \sqrt{n}(Z_{ik(n)}-\Delta_{ik})/\sigma$, $1\leq i\leq k-1$, have (cf. theorem 2.4), asymptotically a (k-1)-variate normal distribution with mean vector $\vec{0}$ and covariance matrix $\Gamma' = ((\gamma'_{ij}))$, where

$\gamma'_{ii} = 1$ and $\gamma'_{ij} = \frac{1}{2}$, $1 \leq i \neq j \leq k-1$.

In making the assumption that Π_k is the <u>best</u> treatment, we have:

$$P(CS)=P(Z_{ik(n)}<0,\ 1\leq i\leq k-1)=P(V_{i(n)}<-\Delta_{ik}\sqrt{n}/\sigma,\ 1\leq i\leq k-1). \tag{4.8}$$

Furthermore, for all $\omega\varepsilon\Omega$, the inequalities $\Delta^* \leq -\Delta_{ik}$, $1\leq i\leq k-1$, hold. Hence, it follows:

$$P(V_{i(n)}<-\Delta_{ik}\sqrt{n}/\sigma, 1\leq i\leq k-1) \geq P(V_{i(n)}<\Delta^*\sqrt{n}/\sigma, 1\leq i\leq k-1). \tag{4.9}$$

Part (iii) of the present theorem means that $\frac{N}{\nu} \xrightarrow[\Delta^*\to 0]{a.s.} 1$, and hence, $\frac{\Delta^*\sqrt{n}}{\sigma} \xrightarrow[\Delta^*\to 0]{a.s.} h$.

This together with theorem 2.5 and A4/13, yields:

$$\lim_{\Delta^*\to 0}\ \inf_{\omega\varepsilon\Omega} P(CS) \geq P(U_i<h,\ 1\leq i\leq k-1), \tag{4.10}$$

where $(U_1,\ldots,U_{k-1})$ has a (k-1)-variate normal distribution with mean vector $\vec{0}$ and covariance matrix Γ'. Let $T_1,T_2,\ldots,T_k$ be independent $N(0,\frac{1}{2})$-distributed random variables, then the random vector $(U_1,U_2,\ldots,U_{k-1})$ can be replaced by $((T_1-T_k),(T_2-T_k),\ldots,(T_{k-1}-T_k))$. Hence, we get:

$$\lim_{\Delta^*\to 0}\ \inf_{\omega\varepsilon\Omega} P(CS) \geq P(T_i<T_k+h,\ 1\leq i\leq k-1). \tag{4.11}$$

Using $S_i:=\sqrt{2}T_i$, $1\leq i\leq k$, it follows:

$$\lim_{\Delta^*\to 0}\ \inf_{\omega\varepsilon\Omega} P(CS) \geq P(S_i<S_k+h\sqrt{2},\ 1\leq i\leq k-1). \tag{4.12}$$

Recalling that the random variables S_i, $1\leq i\leq k$, are independent and N(0,1)-distributed, and in using (4.12) and

$$P(S_i<S_k+h\sqrt{2},\ 1\leq i\leq k-1) = \int_{-\infty}^{+\infty} P(S_i<x+h\sqrt{2},\ 1\leq i\leq k-1|S_k=x)d\Phi(x),$$

it is immediately seen that the following inequality holds:

$$\lim_{\Delta^*\to 0}\ \inf_{\omega\varepsilon\Omega} P(CS) \geq \int_{-\infty}^{+\infty} (\Phi(x+h\sqrt{2}))^{k-1}\, d\Phi(x). \tag{4.13}$$

Hence, if h satisfies equation (4.1), the assertion (V) of the theorem follows. ∎

The validity of statement (i) in theorem (4.1) is most important for the implementation of the sequential selection procedures. It means that the described <u>selection procedures terminate with probability one</u>, if all assumptions are satisfied.

But this property of the procedures is only of little avail, because it makes no statement about the sample size needed, to select the best treatment and does not exclude that the sampling does not terminate in case of unfavourable observations. Therefore we have carried out some Monte-Carlo-simulations to investigate the question, how large is the average sample size needed for a selection?

From these Monte-Carlo studies we have got also some information about the relative frequency of CS, and the minimal and maximal values of the stopping variable N.
Prior to the discussion of the results of our Monte-Carlo-studies, we give an example in order to illustrate the implementation of the procedures.

5 A Numerical Example and some Remarks Concerning the Practical Working with the Sequential Procedures

5.1 An example

An agricultural plant, specialized to the culture of sweet turnips, wants to select from three fertilizers, indifferent otherwise, that one, which yields the greatest increase of produce.
For the investigations, there are ten fields of equal square dimension available which can all be devided into 3 homogenous parts to be treated with the three varieties of fertilizer. Because of differences among the 10 fields and the possibility that the investigations may go over two or more years, the randomized-block-design, seems to be adequate. As minimal sample size, we choose $n_0=10$.
The units of treatment are of such size that an average produce of $\mu=100$ two-hundredweight (th) can be expected from each unit. This expected produce is already subtracted from the numbers given in the following table such that only the block effect (field, year), the treatment effect and a random residual is still included.
If the treatment effect associated with the best fertilizer is about $\Delta^*=0.75$ [t

greater than that of the second-best fertilizer, then with probability P*=0.90 the best fertilizer is to be selected.
For comparison, we apply the procedures based on the Wilcoxon- and the Median-scorefunction.

We have the following realizations (dimension: th):

stage of sampling	fertilizer 1	fertilizer 2	fertilizer 3
1	0.172	1.727	-0.818
2	0.303	1.315	-1.441
3	0.064	-1.019	0.453
4	-1.389	1.796	0.069
5	0.889	-0.562	1.663
6	0.687	-0.425	2.243
7	1.664	-0.349	1.348
8	1.728	0.302	0.161
9	0.846	-0.427	0.088
10	0.511	-0.392	1.154

With $t_{\alpha/2}$=1.645, (α=0.1), and $h=2.2302/\sqrt{2}$, we can evaluate the estimators for σ^2 and the stopping-inequalities at the stage $n=n_0=10$.

	Wilcoxon-scores	Median-scores
estimates	$B_{10}^2 = 13.00$	$C_{10}^2 = 50.48$
	$L_{10,1} = 0.26$	$L_{10,2} = 0.33$
	$\hat{\sigma}^2 = 5.14$	$\hat{\sigma}^2 = 11.22$
stopping-inequality	$n \geq 27.46$	$n \geq 49.59$

We realize that with both special procedures the stopping-inequality is not satisfied, i.e. sampling must be continued at the next stage n=11.
Also for n=11 until n=18, we get stopping-inequalities, which are not satisfied.
The observed realizations up to n=19 are:

stage of sampling	fertilizer 1	fertilizer 2	fertilizer 3
11	-0.637	-0.518	0.973
12	0.849	0.294	0.541
13	-0.171	0.827	2.327
14	-0.435	-2.258	0.279
15	0.047	-1.525	3.342
16	0.751	0.209	1.309
17	-0.861	-1.060	0.771
18	-0.602	-1.065	1.450
19	-0.275	-0.696	1.217

As estimates for σ^2 and as stopping-inequalities at stage n=19, we get:

	Wilcoxon-scores	Median-scores
estimates	$B_{19}^2 = 7.65$	$C_{19}^2 = 23.44$
	$L_{19,1} = 0.28$	$L_{19,2} = 0.28$
	$\hat{\sigma}^2 = 3.12$	$\hat{\sigma}^2 = 4.37$
stopping-inequality	$n \geq 16.67$	$n \geq 23.34$

We see from this table that the stopping-inequality is satisfied when the procedure is based on Wilcoxon-scores. Hence, sampling is terminated, and we have to evaluate the estimates $y_{i.(19)}$, i=1,2,3. We get:

$$y_{1.(19)} = -0.036,\ y_{2.(19)} = -0.657,\ y_{3.(19)} = 0.679,$$

and therefore fertilizer 3 is selected as best.

The procedure based on Median-scores must be continued at stage n=20, because the stopping-inequality is not satisfied.
At the stages n=20 and n=21 the stopping-inequality is not satisfied, too. The observed realizations up to n=22 are:

stage of sampling	fertilizer 1	fertilizer 2	fertilizer 3
20	1.284	-2.446	-0.137
21	-0.314	-0.021	0.893
22	0.081	0.920	1.591

As estimate for σ^2 and as stopping-inequality at stage n=22, we get:

estimates	$C^2_{22} = 18.41$, $L_{22,2} = 0.33$, $\hat{\sigma}^2 = 4.05$
stopping-inequality	$n \geq 21.63$

We see from this results that sampling must be terminated, and the estimates $y_{i.(22)}$ must be evaluated. We get:

$$y_{1.(22)} = -0.065,\ y_{2.(22)} = -0.681,\ y_{3.(22)} = 0.746$$

and therefore, also in case of using the procedure based on Median-scores, fertilizer 3 is selected as best.

5.2 The implementation of the procedures

Considering the definition of the sequential selection procedures and the preceding example, it can be seen that the implementation of the procedures is associated with a lot of computations. The main reason is that at each stage of sampling the stopping-inequality, i.e. σ^2, must be estimated.
The effect of this cause is intensified, because the estimation of σ^2 in case of k treatments is based on k(k-1) samples of differences. For k=5 we have already 20 samples of differences, which must be ordered for each $n \geq n_0$.
If the Wilcoxon-scorefunction is applied, for k(k-1)/2 samples of differences $\vec{X}_{ij}$, $1 \leq i < j \leq k$, n(n+1)/2 averages $D_{ij,\alpha\beta} = (X_{ij\alpha} + X_{ij\beta})/2$, $1 \leq \alpha \leq \beta \leq n$, must be ordered for each n.
In case of other scorefunctions contained in CΛ, the expenditure of computation associated with the implementation of the procedure is substantially larger, because the estimates must be evaluated by trial and error. Some further remarks to these cases are given in the next section. In the following, we consider only the application of the Median- and the Wilcoxon-scorefunction.
Since the implementation of the procedures is only possible, by using a computer, in the following we assume that a programmable computer is available.

The main problem associated with the implementation of the special procedures consists in the organization of the data that is the ordering of the k(k-1) samples of differences and the ordering of the (n(n+1)/2)-variate vectors $D_{ij} := (D_{ij,\alpha\beta})$, $1 \leq i < j \leq k$, $1 \leq \alpha \leq \beta \leq n$. An additional problem consists in the ordering of the vectors D_{ij}, because their dimension increases very rapidly, as n becomes larger. For n=100 the vectors D_{ij} have already the dimension 5050.

A rather efficient ordering-procedure is given by the "ordering by means of binary trees". This method is to be recommended for the ordering of the samples of differences.

The ordering of the vectors D_{ij} can be done by the application of the following algorithm for the ordering of the vectors D_{ij}.

For simplification, we write $D:=(D_{\alpha\beta})$, $1\leq\alpha\leq\beta\leq n$, instead of $D_{ij}=(D_{ij,\alpha\beta})$. The vector D can be written in the following form:

$$
\begin{array}{rccccl}
D = (D_{11} & D_{12} & \cdots\cdots\cdots & & & D_{1n} \\
 & D_{22} & D_{23} & \cdots\cdots & & D_{2n} \\
 & & & & & \vdots \\
 & & & & D_{n-1n-1} & D_{n-1n} \\
 & & & & & D_{nn}).
\end{array}
$$

Let $DS:=(DS_1,DS_2,\ldots,DS_{n(n+1)/2})$ be the vector of the ordered averages $D_{\alpha\beta}$, $1\leq\alpha\leq\beta\leq n$.

The nodes $K_1:=(W1,J1)$, $K_2:=(W2,J2),\ldots,K_n:=(Wn,Jn)$ are ordered in a binary tree with threading in such way that the symmetric order of the n nodes gives the ordered values $W(1) \leq W(2) \leq \ldots \leq W(n)$. The variable Z indicates that node, which contains the value W(1). I denotes a counting number.

Algorithm:

1. For $\ell=1,2,\ldots,n$, $W\ell:=D_{\ell\ell}$, and $J\ell:=\ell$; I:=0, Z:=1 ;
2. I:=I+1; $DS_I:=WZ$;

   If JZ < n, then $WZ:=D_{Z,JZ+1}$, JZ:=JZ+1 ;

   If JZ=n, then $WZ:=D_{nn}+1$;

   If $I\geq n(n+1)/2$, then the ordering is finished ;

   If $I<n(n+1)/2$, then continue with 3. ;
3. The changed node K_Z must be added to the binary tree in such a way that the symmetric order again gives the ordered values $W(1) \leq W(2) \leq \ldots \leq W(n)$, and that Z is equal to the number of that node which contains the value W(1). Continue with 2..

A short example may illustrate the application of the algorithm:
Let n=3, and let the elements of the sample of differences be given by: 1.0 1.5 3.1. Hence, we have:

$$D = \begin{pmatrix} D_{11} & D_{12} & D_{13} \\ & D_{22} & D_{23} \\ & & D_{33} \end{pmatrix} = \begin{pmatrix} 1.00 & 1.25 & 2.05 \\ & 1.50 & 2.30 \\ & & 3.10 \end{pmatrix}.$$

After step 1 of the algorithm, we have:

I=0; DS=(0,0,0,0,0,0);

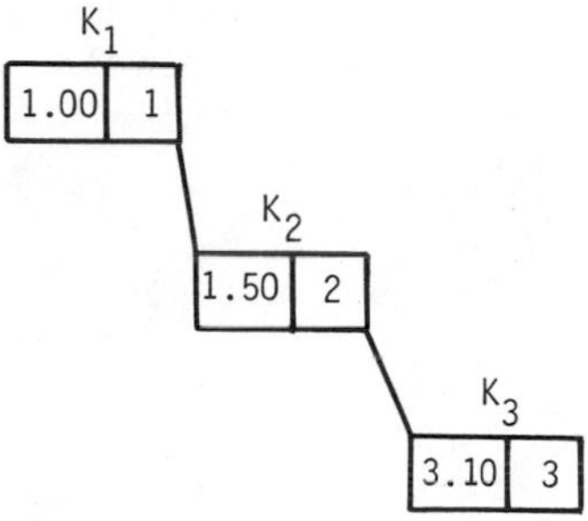

Z=1;

The results of the repeated performances of step 2 and step 3 are summarized on the following page.

Step 2:

I=1;

DS=(1.00,0,0,0,0,0);

J1=1<3;

W1=1.25;

J1=2;

I=1<6;

Step 3:

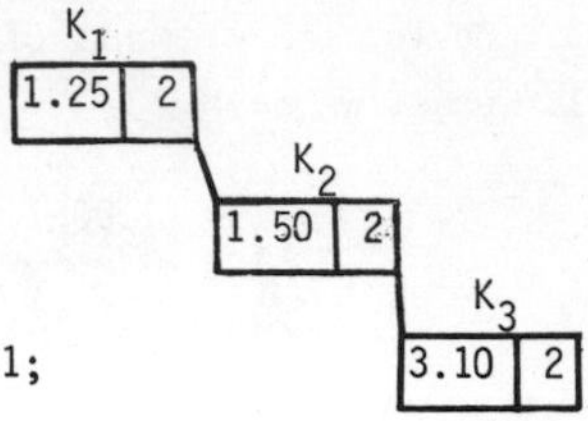

Z=1;

Step 2:

I=2;

DS=(1.00,1.25,0,0,0,0);

J1=2<3;

W1=2.05;

J1=3;

1=2<6;

Step 3:

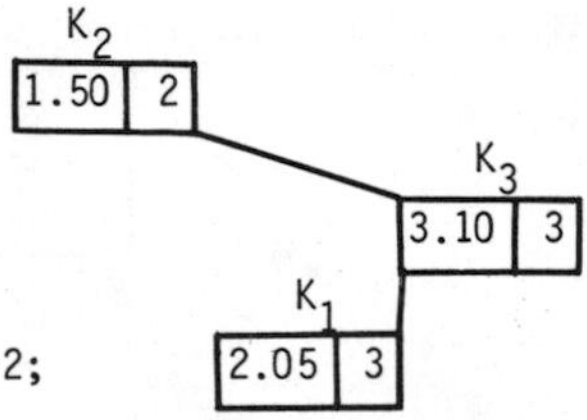

Z=2;

Step 2:

I=3;

DS=(1.00,1.25,1.50,0,0,0);

J2=2<3;

WS=2.30;

J2=3;

I=3<6;

Step 3:

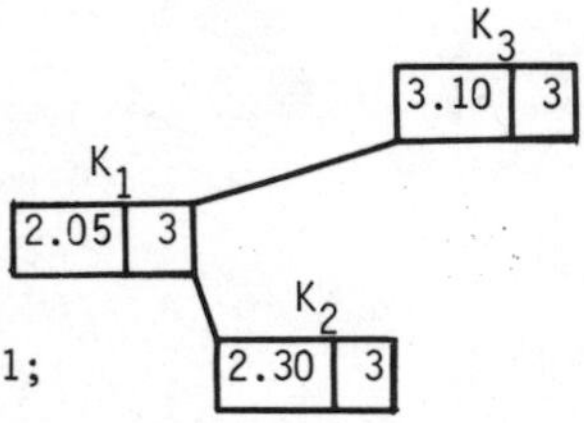

Z=1;

Step 2:

I=4;

DS=(1.00,1.25,1.50,2.05,0,0);

J1=3=3;

W1=3.10+1.00=4.10;

I=4<6;

Step 3:

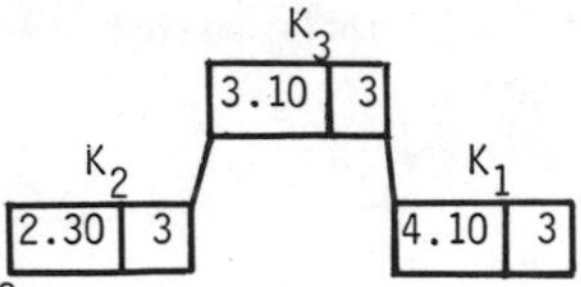

Z=2;

Step 2:

I=5

DS=(1.00,1.25,1.50,2.05,2.30,0);

J2=3=3;

W2=3.10+1.00=4.10;

I=5<6;

Step 3:

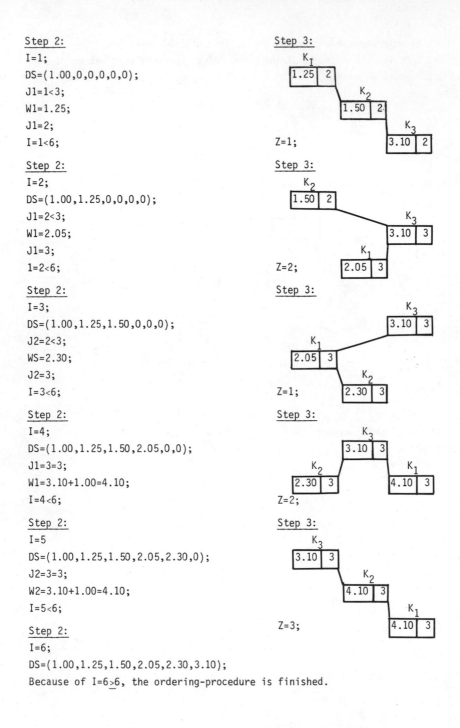

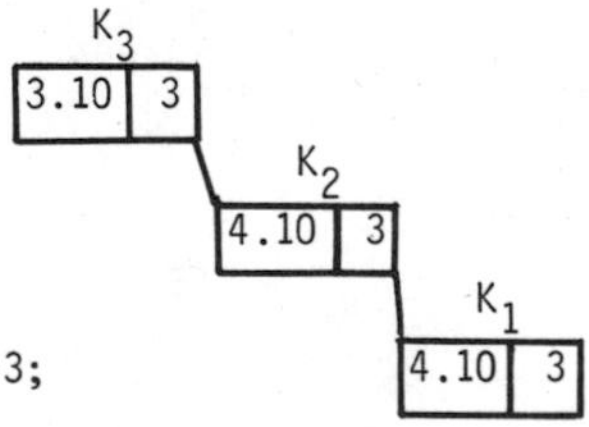

Z=3;

Step 2:

I=6;

DS=(1.00,1.25,1.50,2.05,2.30,3.10);

Because of I=6$\geq$6, the ordering-procedure is finished.

The whole implementation of the selection procedures based on the Median- or Wilcoxon-scorefunction respectively, is illustrated in the following two flow-diagrams (figure 1 and figure 2).

Figure 1: (Median-scorefunction)

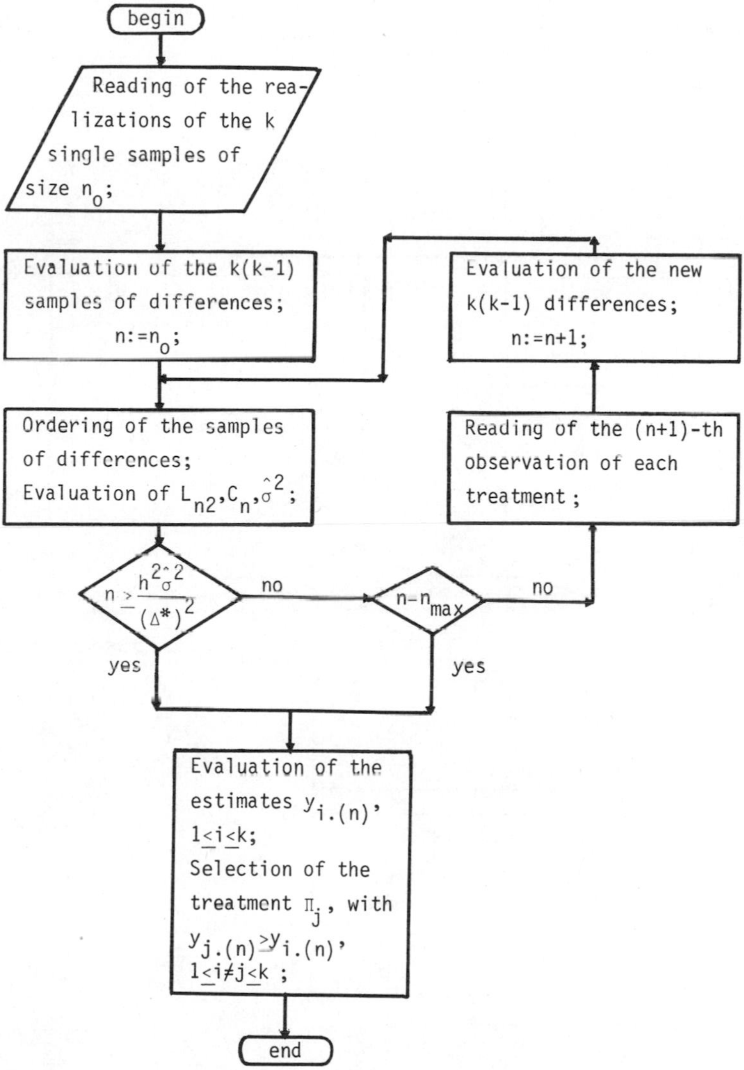

Figure 2: (Wilcoxon-scorefunction)

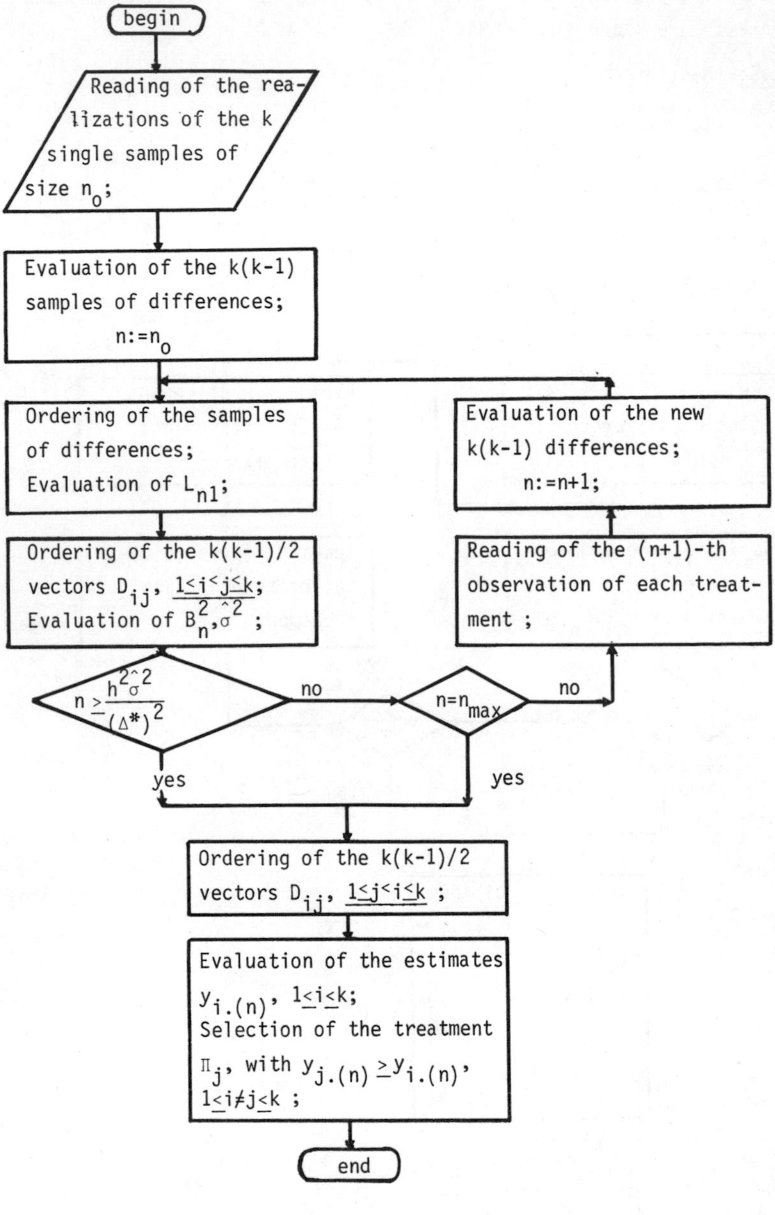

In spite of the termination of the sequential selection procedures with probability 1, we cannot exclude the case that the stopping- inequality does not hold for a practicable value of n. Therefore, some maximal sample size n_{max} should be determined, and sampling should be terminated, if this upper bound is reached. Some remarks concerning the choice of n_{max} are given in section 7. If sampling is terminated because of $n>n_{max}$, the selection based on samples of size n_{max} does not satisfy the probability condition, but the selection is usually correct.

6 Another Class of Sequential Selection Procedures

In this section, we give in short form the extended class of sequential selection procedures, proposed by Geertsema [72] , and Swanepoel and Geertsema [191] This class is only usable in case of one-factor-designs (we called it: population-model), i.e. the observable variables $X_{i\alpha}$, $1\leq i\leq k$, $1\leq\alpha\leq n$, are of the form

$$X_{i\alpha} := \mu+\tau_i+e_{i\alpha}, \quad \sum_{i=1}^{k} \tau_i = 0, \tag{6.1}$$

where μ,τ_i, and $e_{i\alpha}$ have the same meaning as in section 1. The random variables $e_{i\alpha}$ are independent and identically distributed with the absolutely continuous cdf F(e). Let the density f, associated with cdf F, be symmetric around zero, and let $F\varepsilon\,\mathcal{F}_0(J^*)$, where $\mathcal{F}_0(J^*)$ denotes the class of cdf's given in section 2, and J^* is the scorefunction also defined there.

Let ω,Ω be defined as in section 1, then the selection procedures, to be given in the following, satisfy the probability condition (1.2).

As selection criterion, we choose the treatment effects τ_i, $1\leq i\leq k$. The selection procedures are based on the extended class of one-sample-r.-o.s.'s, associated with the single samples $\vec{X}_i=(X_{i1},\dots,X_{in})$, $1\leq i\leq k$, instead of the samples of differences, which we have used in the preceding sections. As estimators of the treatment effects τ_i, $1\leq i\leq k$, we use the general Hodges-Lehmann-estimators $Y_{i(n)}$, which is defined in 2.3.
With respect to the properties of the estimators $Y_{i(n)}$, we can make essentially the same statement as in 2.4.

The independent random variables $\sqrt{n}(Y_{i(n)}-\tau_i)$, i=1,...,k, are asymptotically normal distributed with mean 0 and variance

$$\sigma^2 = \begin{cases} A^2/(4B(F)^2), & \text{if } T_n(\vec{X}_i)\varepsilon CA \\ 1/(4f(0)^2), & \text{if } T_n(\vec{X}_i)\varepsilon CM, \end{cases} \tag{6.2}$$

where A^2 and $B(F)^2$ are given by (2.3).

Let B_n^2 and C_n^2 be the estimators of $(B(F)^2)^{-1}$ and $(f(0)^2)^{-1}$, respectively, evaluated from the single samples $\vec{X}_i$, $1\leq i\leq k$, then the class of sequential selection procedures is defined as follows:

Let d be the solution of the equation:

$$\int_{-\infty}^{+\infty} \Phi^{k-1}(y+d)d\Phi(y) = P^*, \tag{6.3}$$

(some values of d for various values of k and P* are given in table II, appendix 6), then the stopping-inequality is given by:

$$n \geq d^2\hat{\sigma}^2/4(\Delta^*)^2$$
$$\hat{\sigma}^2 := \begin{cases} \frac{1}{4}A^2B_n^2, & \text{if } T_n(\vec{X}_i)\varepsilon CA \\ \frac{1}{4}C_n^2, & \text{if } T_n(\vec{X}_i)\varepsilon CM \end{cases} \tag{6.4}$$

instead of (4.2). The proceeding with this class of selection procedures is the same as that of the procedures defined in section 4.

With $\nu := d^2\sigma^2/4(\Delta^*)^2$, Swanepoel and Geertsema[191] and Geertsema[67] have proved a theorem, which gives the same properties for the present class of procedures, as we have given in theorem 4.1 for the procedures defined there.

To illustrate the application of this class of procedures, we take up the example given in section 5, where we additionally assume that all block-effects (field, year) are equal to zero, such that we can use the one-factor-design.

With $t_{\alpha/2}=1.645$, we only evaluate the estimates $\hat{\sigma}^2$ and the stopping-inequalities at stage $n=n_0=10$.

With $P^* = 0.90$, from table I, appendix 6 we get $d = 2.2302$. Using the estimators C_n^2 and B_n^2, given in section 4, we can evaluate the estimates of $(f(0)^2)^{-1}$ and $(B(F)^2)^{-1}$ and the stopping-inequalities associated with the given samples.

	Wilcoxon-scores	Median-scores
estimates	$B_{10}^2 = 5.0396$	$C_{10}^2 = 17,19$
	$\hat{\sigma}^2 - 0.4199$	$\hat{\sigma}^2 = 4.2975$
stopping-inequality	$n \geq 3.71$	$n \geq 9.50$

7 Asymptotic Efficiency and some Monte-Carlo Studies

7.1 The asymptotic efficiency of the procedures of section 4 with respect to the procedures of section 6

In this section we assume that the block-effects $\beta_\alpha, 1 \leq \alpha \leq n$, are all equal to zero. Then the class of selection procedures defined in section 4 (procedures B) can be compared with the class of procedures, explained in section 6 (procedures SG)

Let N1, N2 be the stopping-variables of the two procedures V1, V2, and let P1, P2 denote the corresponding probabilities of CS, respectively.
As asymptotic efficiency ($\Delta^* \to 0$) of procedure V1 with respect to V2, we define

$$e(V1,V2) := \lim_{\Delta^* \to 0} \frac{E(N2)}{E(N1)},$$

provided that the limit exists and the inequalities $\liminf_{\Delta^* \to 0} P1 > P^*$, $\liminf_{\Delta^* \to 0} P2 \geq P^*$ hold for all $\omega \in \Omega$ and $F, G \in \mathcal{G}_0(J^*)$ (cf.2.1).
Then the asymptotic efficiency of procedures B with respect to procedures SG is given by

$$e(B,SG) = \begin{cases} \dfrac{kA^2B(G)^2}{4B(F)^2(A^2+(k-2)\lambda_J(G))} & ,\text{if } T_n(.) \in CA \\ \dfrac{kg(0)^2}{f(0)^2(1+(k-2)(4G^*(0,0)-1))} & ,\text{if } T_n(.) \in CM \end{cases}$$

Because of $\lambda_J(G) \leq \frac{1}{2} A^2$ and $G^*(0,0) \leq \frac{1}{2}$, we have:

$$e(B,SG) \geq \begin{cases} \dfrac{B(G)^2}{2B(F)^2} & , \text{ if } T_n(.) \in CA \\ \dfrac{g(0)^2}{f(0)^2} & , \text{ if } T_n(.) \in CM. \end{cases}$$

For the Wilcoxon-scorefunction J(u)=u and the standard normal density, we get:

$$e(B,SG) \geq \begin{cases} \frac{1}{4} , & \text{if } J(u) = u \\ \frac{1}{2} , & \text{if } T_n(.) \in CM. \end{cases}$$

This result may give rise to the assumption that the procedures B are substantially less effective than the procedures SG. The following results of some Monte-Carlo-studies indicate that the asymptotic properties are only of restricted meaning or even misleading.

7.2 Some Monte-Carlo studies of the procedures given in section 4 and 6

In the preceding sections only asymptotic results, concerning the two classes of selection procedures, are given. To examine whether the used approximations are exactly enough, and whether the defined procedures stop at practicable sample sizes, we have carried out a lot of Monte-Carlo-studies.
To be able to compare the results for procedures B and procedures SG, only one-factor-designs (population-models) are simulated (this restriction is of no importance).
Using the method of generating random numbers described in appendix 5 , we have generated the realizations of k samples. Because of our interest in the influence of the used scorefunction over the efficiency of the selection procedures, we have simulated normal (NOR)-, double-exponential (DE)- and cauchy (CA)-distributed populations, and have applied the special procedures based on Median (MED)- and Wilcoxon (WILC)-scorefunctions to each type of the three distributions.

Since one simulation-run needs very much computer-time, only 120 or 60 repetitions are done, but in considering the results of the Monte-Carlo-studies, it seems that the results can be assumed to be representative.

For the probabilities $P^* = 0.90$, $P^* = 0.95$ and the Δ^*-values $0.4a\sigma$, $0.5a\sigma$, $0.6a\sigma$ and $0.75a\sigma$, we have applied the generalized procedures B to the following parameter-configurations:

(a) $\tau_1 = \dots = \tau_{k-1}$, $\tau_k - \tau_{k-1} = 0.4a\sigma$,

(b) $\tau_1 = \dots = \tau_{k-1}$, $\tau_k - \tau_{k-1} = 0.5a\sigma$,

(c) $\tau_1 = \dots = \tau_{k-1}$, $\tau_k - \tau_{k-1} = 0.75a\sigma$.

σ denotes some scale-parameter and a is the factor, which we have already used in chap. 1. In the following, we have always used $\sigma=1$.

As minimal sample sizes, we have used $n_0=5$, $n_0=10$ for k=3, k=4, respectively. The tables 1-4 contain the Monte-Carlo-results for procedures B.
$\widehat{P^*}$ denotes the relative frequency of CS.
$\widehat{EN}$ denotes the average of the realizations of N (rounded to the next larger integer).
Table 5 contains the observed minimal values of the stopping-variable N, in case of the LFC for various values of Δ^* and P^*.
For comparison, the special procedures SG based on MED and WILC have been applied to NOR- and DE-distributed populations, for the configurations (b) and (c), with $P^* = 0.90$, and $\Delta^* = \tau_k - \tau_{k-1}$. The results are given in table 6.
Considering the results of the Monte-Carlo-studies, the following may be concluded:

1. Outside the indifference-zone, $\widehat{P^*}$ is not less than the preassigned probability P^*. Inside the indifference-zone, $\widehat{P^*}$ is usually smaller than P^*. In case of the LFC, the values of $\widehat{P^*}$ are slightly greater than the preassigned P^*.

2. The average size of N depends very much on Δ^* and k. The influence of P^* over the size of N is of such an extent, that the nessecity of an increase of P^* (for example 0.95 instead of 0.90) should be examined very carefully. On the contrary the real parameter-configuration seems to have no influence over the size of $\widehat{EN}$.

3. For NOR- and DE-distributed populations, the procedure WILC yields generally better results than procedure MED. However, it can be observed that the superiority of WILC is smaller in case of DE-distributed populations than in

case of NOR-distributed populations. For k=4 the differences between the efficiencies of the two procedures are very small, if the populations are DE-distributed. For CA-distributed populations, the procedure MED yields substantially more efficient results than procedure WILC. This gives rise to the assumption that the efficiency of the procedures B depends on the choice of the scorefunction. As for the subset-selection procedures, in case of short-tailed and medium-tailed distributions, rapidly increasing scorefunctions yield the more efficient sequential selection procedures, and in case of long-tailed distributions slowly increasing scorefunctions yield the more efficient sequential selection procedures.

4. The minimal values, observed for the stopping-variable N, depend very much on the values of Δ^* and P^*. Since the implementation of procedures B is associated with relatively much computer-time, it is inefficient to choose a too small value of n_o, i.e. n_o should be chosen with respect to P^* and Δ^* (cf. table 5).

5. On an average the procedures SG stop at lower stages than the procedure B. But for the simulated parameter-configurations, the ratios EN_{SG}/EN_B are considerably smaller than one would expect by virtue of asymptotic results in 7.1. Therefore, the procedures B are a reasonable alternative to procedures SG in case of population models, too.

A remarkable advantage of procedures B, apart from the more general sampling-design, is that they can also be used in case of non-symmetric distributions.

The great expense for the implementation of the two classes of sequential selection procedures (especially for large k) can be reduced by the application of an appropriate subset-selection procedure based on samples of size n_o (minimal sample size). For example, let k=6 and assume that a subset of size 4 has been selected by a preliminary subset-selection procedure, then we have to consider 4·3=12 instead of 5·6=30 samples of differences. The reduction of labour in case of procedures SG is smaller (4 instead of 6 samples) but still interesting. Concerning the choice of n_{max} it is difficult to give a general rule. It depends very much on the preassigned values of Δ^* and P^*. Therefore, corresponding to Δ^*, n_{max} should be chosen with respect to economical aspects. In our Monte-Carlo-studies, we have chosen n_{max}=95,70,50 for Δ^*=0.4aσ,0.5aσ,0.75aσ, respectively. Only in case of CA-distributed populations (long-tailed), the upper bound n_{max} was reached in some cases, if we have applied the selection procedure WILC. In the tables 1-4 for this cases the values of $\widehat{P^*}$ and $\widehat{EN}$ are evaluated from the simulation-runs where n_{max} was not reached.

Since the efficiency of the sequential selection procedures depends on the applied scorefunction, each knowledge about the underlying distribution should be used for its choice. A method for the production of such knowledges will be given in chap. 3.

7.3 Some remarks concerning the application of general scorefunctions

We have already mentioned that in applying general scorefunctions the evaluation of the estimator B_n^2 is only possible by trial and error, i.e. at each stage of sampling, the magnitudes $\Delta_{ijU}(n)$ and $\Delta_{ijL}(n)$ must be evaluated by trial and error.

To examine whether the large expense of calculation is worth while, i.e. whether the choice of a more expensive form of the selection procedure causes an increase of the efficiency, we have written a simulation-program for procedures SG with general scorefunctions. The procedures SG are chosen, because they are connected with less expense of calculation (only k instead of k(k-1)/2 samples) than procedures B, and secondly they are comparable with respect to the effect of some special scorefunction.

The programming of the general procedures SG is very expensive. Our program needs 1200 cards, and for k=4, and only one simulation-run already 6 minutes of computer-time.

Applying the special scorefunction $J(u)=\Phi^{-1}(u)$, $0<u<1$, for NOR-distributed populations and parameter-configuration (b), with k=3 and k=4, $\Delta^*=\tau_k-\tau_{k-1}$ and $P^* = 0.90$, we have got:

	$\widehat{P^*}$	$\widehat{EN}$
k=3	0.90	25
k=4	0.91	38

A comparison with the results in table 6 gives rise to the assumption that the application of the scorefunction $J(u)=\Phi^{-1}(u)$ does not cause an important increase of efficiency. A general statement is not possible. Because of the very long computer-times, we have not done further investigations of general scorefunctions. But because of the rapid development in the field of high-speed-computers, the extended investigation of general procedures SG and procedures B may be possible and interesting in the near future.

Table 1

$k = 3,\ P^* = 0.90,\ n_o = 5$																	
configuration	Δ^*	$0.4a\sigma$				$0.5a\sigma$				$0.6a\sigma$				$0.75a\sigma$			
		WILC		MED		WILC		MED		WILC		MED		WILC		MED	
		$\widehat{P^*}$	$\widehat{EN}$	$\widehat{P^*}$	$\widehat{EN}$	$\widehat{P^*}$	$\widehat{EN}$	$\widehat{P^*}$	$\widehat{EN}$	$\widehat{P^*}$	$\widehat{EN}$	$\widehat{P^*}$	$\widehat{EN}$	$\widehat{P^*}$	$\widehat{EN}$	$\widehat{P^*}$	$\widehat{EN}$
(a)	NOR	0.91	44	0.89	57	0.91	29	0.80	38	0.78	21	0.75	28	0.69	14	0.63	20
	DE	0.93	38	0.93	47	0.90	25	0.85	33	0.78	18	0.79	25	0.78	13	0.71	17
	CA	1.00	59	1.00	45	1.00	45	0.98	35	1.00	37	0.95	29	1.00	31	0.88	22
(b)	NOR	0.99	45	0.99	52	0.90	30	0.96	39	0.85	22	0.93	28	0.83	15	0.90	20
	DE	0.98	38	0.98	50	0.92	26	0.93	35	0.92	20	0.91	26	0.81	15	0.83	18
	CA	1.00	63	0.97	46	1.00	44	0.93	32	1.00	34	0.93	26	1.00	26	0.85	22
(c)	NOR	1.00	43	0.99	56	0.96	29	0.98	39	0.95	21	0.97	29	0.91	14	0.95	20
	DE	1.00	38	1.00	45	1.00	28	1.00	31	1.00	19	0.96	23	0.98	14	0.95	16
	CA	1.00	67	1.00	45	1.00	47	1.00	33	1.00	36	1.00	27	1.00	29	1.00	21

Table 2

k = 3, $P^* = 0.95$, $n_0 = 5$																	
con-figu-ration	Δ^*	$0.4a\sigma$				$0.5a\sigma$				$0.6a\sigma$				$0.75a\sigma$			
		WILC		MED		WILC		MED		WILC		MED		WILC		MED	
		$\widehat{P^*}$	$\widehat{EN}$	$\widehat{P^*}$	$\widehat{EN}$	$\widehat{P^*}$	$\widehat{EN}$	$\widehat{P^*}$	$\widehat{EN}$	$\widehat{P^*}$	$\widehat{EN}$	$\widehat{P^*}$	$\widehat{EN}$	$\widehat{P^*}$	$\widehat{EN}$	$\widehat{P^*}$	$\widehat{EN}$
(a)	NOR	0.95	62	0.98	77	0.93	41	0.89	54	0.93	30	0.81	39	0.77	20	0.71	27
	DE	0.98	54	0.95	64	0.93	36	0.94	45	0.86	26	0.87	34	0.78	18	0.77	24
	CA	1.00	82	1.00	57	1.00	58	1.00	44	0.98	35	0.98	35	1.00	36	0.92	27
(b)	NOR	0.99	63	0.99	75	0.97	43	0.99	52	0.91	39	0.97	39	0.85	21	0.93	27
	DE	0.99	55	0.99	68	0.98	36	0.96	47	0.92	36	0.93	36	0.90	19	0.88	25
	CA	1.00	88	1.00	58	1.00	60	0.97	44	1.00	45	0.95	35	1.00	33	0.92	27
(c)	NOR	1.00	62	0.99	77	1.00	41	0.99	53	0.96	29	0.98	40	0.95	20	0.97	28
	DE	1.00	55	1.00	62	1.00	36	1.00	43	1.00	27	1.00	31	1.00	19	0.97	22
	CA	1.00	88	1.00	60	1.00	62	1.00	44	1.00	47	1.00	33	1.00	35	1.00	27

Table 3

$k = 4,\ P^* = 0.90,\ n_0 = 10$																	
con-figu-ration	Δ^*	$0.4a\sigma$				$0.5a\sigma$				$0.6a\sigma$				$0.75a\sigma$			
		WILC		MED		WILC		MED		WILC		MED		WILC		MED	
		$\widehat{P^*}$	$\widehat{EN}$	$\widehat{P^*}$	$\widehat{EN}$	$\widehat{P^*}$	$\widehat{EN}$	$\widehat{P^*}$	$\widehat{EN}$	$\widehat{P^*}$	$\widehat{EN}$	$\widehat{P^*}$	$\widehat{EN}$	$\widehat{P^*}$	$\widehat{EN}$	$\widehat{P^*}$	$\widehat{EN}$
(a)	NOR	0.95	56	0.91	65	0.90	37	0.84	44	0.80	27	0.76	32	0.70	21	0.71	23
	DE	1.00	51	0.95	52	0.98	35	0.89	35	0.94	26	0.77	27	0.81	18	0.70	18
	CA	1.00	80	0.98	48	1.00	55	0.92	36	1.00	43	0.86	29	1.00	33	0.76	22
(b)	NOR	1.00	59	0.99	64	0.99	38	0.95	44	0.77	27	0.90	32	0.75	20	0.77	21
	DE	1.00	51	0.98	52	0.99	35	0.94	36	0.99	25	0.89	30	0.92	18	0.70	18
	CA	1.00	87	1.00	52	1.00	65	1.00	38	1.00	50	0.96	30	1.00	39	0.90	24
(c)	NOR	1.00	60	1.00	64	1.00	40	0.99	43	1.00	29	0.98	31	1.00	20	0.94	21
	DE	1.00	51	1.00	53	1.00	35	0.99	36	1.00	26	0.96	26	1.00	18	0.93	18
	CA	1.00	87	1.00	46	1.00	59	1.00	48	1.00	27	1.00	37	1.00	37	0.98	21

Table 4

k = 4, $P^* = 0.95$, $n_0 = 10$																	
con-figu-ration	Δ^*	$0.4a\sigma$				$0.5a\sigma$				$0.6a\sigma$				$0.75a\sigma$			
		WILC		MED		WILC		MED		WILC		MED		WILC		MED	
		$\widehat{P^*}$	$\widehat{EN}$	$\widehat{P^*}$	$\widehat{EN}$	$\widehat{P^*}$	$\widehat{EN}$	$\widehat{P^*}$	$\widehat{EN}$	$\widehat{P^*}$	$\widehat{EN}$	$\widehat{P^*}$	$\widehat{EN}$	$\widehat{P^*}$	$\widehat{EN}$	$\widehat{P^*}$	$\widehat{EN}$
(a)	NOR	0.95	76	0.95	87	0.95	50	0.91	60	0.90	36	0.84	44	0.80	26	0.75	30
	DE	1.00	70	0.96	69	0.99	47	0.96	47	0.99	34	0.87	34	0.94	23	0.78	24
	CA	1.00	95	0.98	62	1.00	76	0.92	46	1.00	55	0.92	36	1.00	40	0.86	27
(b)	NOR	1.00	78	1.00	86	0.98	54	0.98	59	0.97	37	0.95	43	0.69	25	0.86	29
	DE	1.00	69	0.99	70	1.00	47	0.98	48	1.00	34	0.93	36	0.93	22	0.38	24
	CA	1.00	95	1.00	65	1.00	79	1.00	48	1.00	65	1.00	38	1.00	48	0.96	29
(c)	NOR	1.00	79	1.00	86	1.00	55	1.00	59	1.00	40	0.99	39	1.00	28	0.97	28
	DE	1.00	70	1.00	70	1.00	47	1.00	49	1.00	34	0.98	35	1.00	24	0.95	24
	CA	1.00	95	1.00	58	1.00	80	1.00	42	1.00	59	1.00	33	1.00	46	1.00	25

Table 5

Δ*			0.4aσ			0.5aσ			0.75aσ		
k	Scores	P*	NOR	DE	CA	NOR	DE	CA	NOR	DE	CA
3	MED	0.90	29	20	29	17	22	17	5	5	5
		0.95	43	41	29	25	20	29	9	5	5
	WILC	0.90	24	19	-	12	7	-	6	5	-
		0.95	39	33	-	16	10	-	8	5	-
4	MED	0.90	47	39	29	29	15	27	10	10	15
		0.95	50	43	41	43	17	34	10	10	15
	WILC	0.90	27	21	-	21	14	-	10	10	-
		0.95	44	36	-	28	19	-	10	10	-

Table 6

$P^* = 0.90;\ \Delta^* = \tau_k - \tau_{k-1};$							
configuration		(b)				(c)	
k		3		4		3	4
distribution		NOR	DE	NOR	DE	NOR	NOR
WILC	$\widehat{P^*}$	0.94	0.91	0.93	0.92	0.95	0.92
	$\widehat{EN}$	24	20	28	23	13	15
MED	$\widehat{P^*}$	0.95	0.98	0.92	0.97	0.96	0.97
	$\widehat{EN}$	43	32	50	37	23	27

CHAPTER 3

Methods for Selecting an Optimal Scorefunction

1 The Basic Idea

In the chapters 1 and 2 of the second part, we have stated that the efficiency of the various selection procedures considered there, is influenced by the special form of the applied statistic (estimator). For all considered nonparametric procedures, the choice of a special statistic is identical with the choice of a special scorefunction. That means, there is a high connection between the chosen scorefunction and the efficiency of the various nonparametric selection procedures.[1]
As we have already seen in the chapters 1 and 2, the optimality of some special scorefunction is dependent on the form of the distribution of the populations under investigation. Therefore, if we associate each special statistic which can be chosen, with that subset of possible distributions for which the special statistic is optimal (among the possible statistics), then the choice of an optimal statistic is identical with the choice of that subset of distributions, which includes the distribution actually given. Therefore the problem of selecting an optimal scorefunction means to get information about the distribution of the populations in research. Whereas in case of subset-selection procedures it is also of interest whether a symmetric or skewed distribution is given, in using sequential selection procedures it is only of interest whether a short-tailed, medium-tailed or long-tailed distribution is given.

Prior to the discussion of two methods for selecting an optimal scorefunction,

[1] *Hájek and Sidák [86] have proved that the same dependence consists between scorefunctions and the efficiency of two- and one-sample-ranktests.*

we consider two statistics,[1] which allow the characterization of distributions.

2 Two Statistics for Characterizing a Distribution

2.1 An estimator for the skewness of some distribution

Let $z_1 \le z_2 \le \dots \le z_m$ be the realizations of the order-statistics $Z_i, 1 \le i \le m$, of some sample of size m. F denotes the underlying distribution function with the corresponding density f. The scale-invariant statistic

$$SKK := \begin{cases} \dfrac{\bar{O}_{0.05} - \bar{M}_{0.5}}{\bar{M}_{0.5} - \bar{U}_{0.05}}, & \text{if } m > 20 \\[2ex] \dfrac{Z_1 - \bar{M}_{0.5}}{\bar{M}_{0.5} - Z_m}, & \text{if } m \le 20 \end{cases}$$

can be used as indicator for the skewness of the distribution, where $\bar{O}_{0.05}$, $\bar{M}_{0.5}$ and $\bar{U}_{0.05}$ are the averages of the o.o5m largest, the o.5m medium and the o.o5m smallest order-statistics $Z_1, Z_2, \dots, Z_m$, respectively.
Computing the averages also fractionary rational multiples must be considered. Let for instance m=48, i.e. $\bar{U}_{0.05}$ be the average of the 2.4 smallest order-statistics, then we have

$$\bar{u}_{0.05} = (z_1 + z_2 + 0.4 z_3)/2.4.$$

Large values of SKK indicate a right-skewed distribution. If SKK takes a value considerably smaller than 1, then we may assume that the distribution is skewed to the left. In case of symmetric distributions the value of SKK will be nearly 1.

Approximate values for the discrimination among the three subsets of distributions have been determined by Monte-Carlo-studies. Values, which yield a good partition of the set of all possible values of SKK, are given in section 3.

Considering the form of SKK it can be easily seen that 1/SKK takes the same values in case of distributions skewed to the left as the statistic SKK in case of that distribution, which we get by reflection of the left-skewed one.

---

[1] *Doris M.Fischer [58] has considered these statistics in her thesis.*

2.2 An indicator for the peakedness of some distribution

Again let $z_1 \leq z_2 \leq \ldots \leq z_m$ be the realizations of the order-statistics of some sample of size m, with the distribution function F and the corresponding density f. As indicator for the peakedness of some distribution, i.e. the length of the tails of f, one may use the location- and scale-invariant statistic

$$RGG := \begin{cases} \dfrac{\bar{O}_{0.05}-\bar{U}_{0.05}}{\bar{O}_{0.5}-\bar{U}_{0.5}}, & \text{if } m > 20 \\[2ex] \dfrac{Z_m - Z_1}{\bar{O}_{0.5}-\bar{U}_{0.5}}, & \text{if } m \leq 20, \end{cases}$$

where $\bar{O}_{0.05}$, ($\bar{O}_{0.5}$) and $\bar{U}_{0.05}$, ($\bar{U}_{0.5}$) are the averages of the 0.05m (0.5m) largest and smallest order-statistics $Z_1, Z_2, \ldots, Z_m$, respectively. Also in computing RGG, fractionary rational multiples of the realizations must be considered.

The larger the value of RGG the longer are the tails of the underlying distributions in probability.

Using the Monte-Carlo-studies, we have appointed the different classes of distributions to certain intervals of possible values for RGG. Corresponding to the distinction between short-tailed, medium-tailed and long-tailed distributions, we have established three intervals, which will be given in the following section.

3 Two Methods for Selecting a Scorefunction

3.1 Selection based on the joint sample

At first the realizations of the given k single samples of size n must be combined to one joint ordered realization $z_1 \leq z_2 \leq \ldots \leq z_m$ of size m=kn. Based on the values z_i, $1 \leq i \leq m$, we compute the values of SKK and RGG.

Corresponding to the value of SKK, we decide, whether a left-skewed (L), symmetric (S) or right-skewed (R) distribution is given. The value of RGG determines, whether a short-tailed (1), medium-tailed (2) or long-tailed (3) distribution must be assumed. Using the abbreviations in parenthesis, we can give the following selection-scheme, which can be used for the distinction among the different classes of distributions. The given partition of the set of all possible realizations of (RGG,SKK) results from some Monte-Carlo-studies.

Scheme 1:

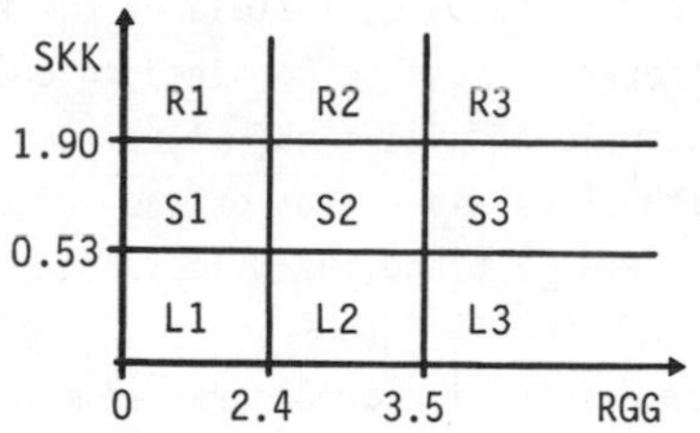

Depending on the selection procedure, the parts of the first quadrant given in scheme 1 must be appointed to the possible scorefunctions.
In the following we assume a population model. If a subset-selection procedure based on rule R1u (different scale parameters) or any of the sequential selection procedures given in chapter 2 ought to be applied, then it can be chosen only between two special scorefunctions. Taking into account the results of section 2 in chapter 1 and those of chapter 2, the Mann-Whitney- or Wilcoxon-scores should be used if RGG < 3.5 independently of the value of SKK. For values of RGG not less than 3.5, the Median-scores are more advantageous.
The value of SKK can be used to decide, whether the sequential procedure developed by Swanepoel and Geertsema (procedure SG) or the generalized procedure (procedure B) must be used. Assuming that the rule R1g or the rule R2 can be used, the special statistics given in sections 3 and 4 of chapter 1 can be chosen respectively. Thus the appointments given in table 1 can be recommended.

Table 1:

class of distribut. / rule	S1	S2	S3	R1,R2,R3	L1,L2,L3
R1g	HAGA or WAE	WILC	MED	WILC or PMS p=0.0,q=0.5	WILC or PMS p=0.5,q=0.0
R2	WAE or PMS p=q=0.25	WILC	MED	"	"

If a two-factor model is given, and a subset-selection procedure ought to be used, then the Wilcoxon-scores or the Van-der-Waerden-scores should be used in general, where the Van-der-Waerden-scores seem to be slightly preferable. The Median-scores yield the most efficient rule R2 if the value of RGG is greater than 3.5 or if SKK is greater than 1.90. In case of rule R1 the Median-scores

are the less efficient scores for all simulated distributions.
If the procedure B, given in chapter 2, ought to be used, instead of the k single samples the k(k-1)/2 samples of differences must be combined to one joint ordered sample. Based on this joint sample of size m=nk(k-1)/2 the values of SKK and RGG must be computed. Without consideration of the value of SKK, the Median-scores should be used if RGG ≥ 3.5 and the Wilcoxon-scores are more advantageous if RGG < 3.5.
In the following we prove that the considered selection procedures remain distribution-free or asymptotically distribution-free, in spite of the fact that a class of distributions, i.e. an optimal statistic is chosen.
Corresponding to the 9 subsets which may include the value of the vector (RGG,SKK), we divide the set of all possible distributions into nine disjoint classes of distributions $\Omega_1,\Omega_2,\dots,\Omega_9$. Further let $C_1,C_2,\dots,C_9$ be disjoint sets, which are a complete partition of the set that includes all possible realizations of the joint sample.
If the given realization of the joint sample is an element of $C_i, 1\leq i\leq 9$, then the class $\Omega_i, 1\leq i\leq 9$, will be chosen. Defining the regions of CS by $K_1,K_2,\dots,K_9$, corresponding to $\Omega_1,\Omega_2,\dots,\Omega_9$, then in case of some special selection rule we have

$$\inf P(K_i|\Omega_i) = P^*, \quad i=1,\dots,9.$$

Let Ω be the distribution actually given. If a LFC is given, then all considered selection procedures are (at least asymptotically) distribution-free, i.e. (at least for $n\to\infty$) independent of the order-statistics $Z_1,Z_2,\dots,Z_m$. Thus the pre-selection of a scorefunction based on the order-statistics is (at least for $n\to\infty$) independent of the selection procedure. Therefore, in case of the considered selection procedures, the minimal probability of CS is

$$\sum_{i=1}^{9} P(C_i\cap K_i|\Omega) = \sum_{i=1}^{9} P(C_i|\Omega) \inf P(K_i|\Omega).$$

Since the selection procedures are distribution-free, that is $\inf P(K_i|\Omega)=P^*$, $1\leq i\leq 9$, we have:

$$\sum_{i=1}^{9} P(C_i|\Omega) \inf P(K_i|\Omega) = \sum_{i=1}^{9} P(C_i|\Omega)\cdot P^* = P^*.$$

This proves that the selection procedures remain distribution-free in spite of the pre-selection of a special class of distributions.

Some falsification of the pre-selection arises from great differences among the location-parameters of the populations or the treatment-effects. The Monte-

Carlo-studies have shown, that the scheme 1 yields good results up to differences of 1.0σ.

3.2 Selection based on the single samples

In the following we assume that a population-model is adequate. Let SKK(i) and RGG(i) be the statistics defined in section 2, computed from the i-th single sample of size n, $1 \leq i \leq k$. For the selection of some special class of distributions, the averages

$$SK := \frac{1}{k}\sum_{i=1}^{k} SKK(i), \qquad RG := \frac{1}{k}\sum_{i=1}^{k} RGG(i)$$

can be used.
The selection-scheme 2 is the result of Monte-Carlo-studies. The notations are the same as in 3.1.

Scheme 2:

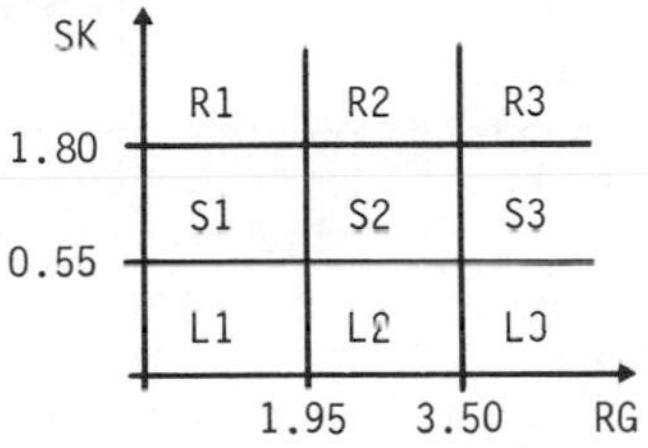

This method based on the single samples has the advantage that the magnitudes SK and RG are not influenced by differences among the location-parameters of the populations.
The main disadvantage of **this method** is that in case of single samples of sizes smaller than 40, the influence of very extreme realizations is very important. This is the main reason, that for instance in case of cauchy-distributed populations the value of SK is 2.95 on an average, in spite of the fact that the cauchy-distribution is symmetric.
A further disadvantage in comparison with the method given in 3.1 is, that the pre-selection based on the vector (RG,SK) influences the probability of CS, i.e. the selection procedures do not remain distribution-free. Because of this disadvantages, in case of subset-selection procedures, only the method considered in 3.1 should be used.

Randles and Hogg (1973) have stated, that the rank-order statistics used in the considered selection procedures and the random-variable RG are not correlated, if the underlying distribution is symmetric.
Monte-Carlo-studies have shown, that in spite of the asymptotic normality, in case of finite samples, the absence of correlation does not imply the independence, i.e. also the exclusive use of RG causes a variation of the probability of CS.
Since this influence is not very high, the following arguments may be given for the application of the statistic RG, if one of the sequential procedures ought to be used:

(a) The computation of the values of RGG(i), $1 \leq i \leq k$, is much less expensive than that of the value of RGG.
(b) Differences among location-parameters cause no falsification of the pre-selection.
(c) The exclusive use of RG is sufficient, because the sequential procedure B is based on samples of differences, which are always associated with symmetric distributions.
(d) If a two-factor-model is given, then the pre-selection by using the statistic RG, based on the k(k-1)/2 samples of differences, is possible. In using the method described in 3.1 the expense of computation is much smaller.

The selection-scheme 2 can be simplified as follows:

<u>Scheme 3:</u>

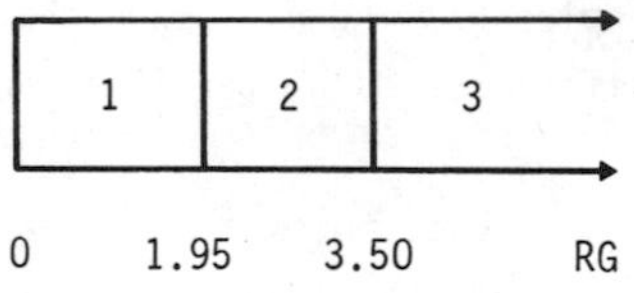

Assuming that only the two special procedures given in chapter 2 can be chosen the procedure based on Median-scores should be used if RG $\geq$ 3.5. In all other cases the Wilcoxon-scores yield the more efficient selection procedure.

The statistic SK can be used to decide whether the procedure SG can be applied.

APPENDIX 1

A1/1

$\Gamma(r) := \int_0^\infty x^{r-1}e^{-x}dx$; $r>0$; is called the (standardized) Gamma function. Because of $x^{r-1}e^{-x} \leq x^{r-1}$ for $x \in [0,1]$ and $x^{r-1}e^{-x} \leq e^{-x}$ for $x \in [1,\infty)$, the Gamma function is well defined for all positive numbers r, i.e. $\Gamma(r) \in \mathbb{R}$ for $r>0$.

■

A1/2

The random variable X is said to be <u>gamma distributed with parameter r</u> if X has the following density function:

$$f_X(x) := \begin{cases} \frac{1}{\Gamma(r)} x^{r-1}e^{-x}, & \text{if } x>0 \\ 0 & \text{, otherwise} \end{cases}$$

The distribution function of X is referred to as the <u>incomplete gamma function</u>.

■

A1/3

$\beta(r,s) := \int_0^1 x^{r-1}(1-x)^{s-1}dx; r,s>0$; is known as the (standardized) Beta function.

■

A1/4

The random variable X associated with the density function

$$f_X(x) := \begin{cases} \frac{1}{\beta(r,s)} x^{r-1}(1-x)^{s-1}, & \text{if } 0<x<1 \\ 0 & \text{, otherwise} \end{cases}$$

is said to be <u>beta distributed with parameters r,s.</u>

$J_p(r,s) := \frac{1}{\beta(r,s)} \int_0^p x^{r-1}(1-x)^{s-1}dx$ is referred to as the <u>incomplete beta function with parameters r,s</u>, $p \in (0,1)$; i.e., $J_p(r,s)$ conceived as a function of p is nothing but the distribution function of a beta distributed random variable in the interval (0,1).

■

A1/5

$\Gamma(r+s) \cdot \beta(r,s) = \Gamma(r) \cdot \Gamma(s)$; this equation is easily found to hold by noting that $\Gamma(r)\Gamma(s) = \int_0^\infty \int_0^\infty x^{r-1}y^{s-1}e^{-x-y}dxdy$ and by using the substitution $x=uv$; $y=(1-u)v$. ■

A1/6

X is called <u>negative binomial distributed with success parameter p and index r</u>, if X has the following probability function:

$$P(X=k) = \binom{r+k-1}{k} p^r q^k, \text{ if } k \in \mathbb{N}_0 \text{ ; } q := 1-p$$

Such a random variable is denoted by $X_{r,p}$.

We have: $E(X_{r,p}) = rq/p$; $Var(X_{r,p}) = rq/p^2$. ■

A1/7

$J_p(r,s) = 1-J_q(s,r)$, if $r,s>0$; for $s>0$, we define:

$J_p(0,s) := 1=1-J_q(s,0)$; $J_p(r,s) := 0$, if $r \geq 0$ <u>and</u> $s \leq 0$ or if $r<0$ <u>and</u> $s \geq 0$. ■

A1/8

$$q^r \sum_{j=0}^{s-1} \frac{\Gamma(r+j)}{\Gamma(r)j!} p^j = J_q(r,s) = p^s \sum_{j=r}^{\infty} \frac{\Gamma(s+j)}{\Gamma(s)j!} q^j ,$$

where the first equality holds for any real $r>0$ and the second for any real $s>0$. It is now obvious that $J_p(r,s)$ is a strictly increasing (decreasing) function of $s(r)$ holding r,s fixed respectively. ■

A1/9

$$J_p(r+1,s) = \sum_{\tau=0}^{s-1} \binom{r+s}{\tau} p^{r+s-\tau} q^\tau ; \; r,s \in \mathbb{N}$$

The truth of this is easily seen by partially integrating the incomplete beta function. ■

A1/10

$J_q(r,s) = J_q(r+1,s)(1+o(1))$;$J_q(r+1,s) = J_q(r,s)(1+o(1))$; $r \geq 1$, $s \in \mathbb{N}$

The above equations are seen to hold by using

$$J_q(r,s) = J_q(r+1,s) + p^s q^r \binom{r+s-1}{r}$$

∎

A1/11

$$\sum_{j=0}^{s-1} J_p(r,j+1) = (r+s)J_p(r,s) - \frac{r}{p} J_p(r+1,s);\ r,s>0$$

Proof:

$$\sum_{j=0}^{s-1} J_p(r,j+1) = \sum_{j=0}^{s-1} \frac{\Gamma(r+j+1)}{\Gamma(r)j!} \int_0^p x^{r-1}(1-x)^j dx =$$

$$= \int_0^p \frac{r}{x^2} x^{r+1} \sum_{j=0}^{s-1} \frac{\Gamma(r+j+1)}{\Gamma(r+1)j!} (1-x)^j dx = \int_0^p \frac{r}{x^2} J_x(r+1,s)dx = -\frac{r}{p} J_p(r+1,s) +$$

$$+ \underbrace{\lim_{x\to 0} \frac{rJ_x(r+1,s)}{x}}_{(1)} + \underbrace{\int_0^p \frac{r}{x} \frac{dJ_x(r+1,s)}{dx} dx}_{(2)}$$

(1)
$$0 \le \frac{rJ_x(r+1,s)}{x} = \underbrace{\frac{r\Gamma(r+s+1)}{\Gamma(r+1)\Gamma(s)}}_{:=\, k} \int_0^x \frac{1}{x} u^r(1-u)^{s-1}du \le \frac{k}{x} \int_0^x u^r du = \frac{k}{r+1} x^r$$

The last term tends to 0 as x becomes small.

(2)
$$\int_0^p \frac{r}{x} \frac{dJ_x(r+1,s)}{dx} dx = \int_0^p \left(\frac{r}{x} \frac{d}{dx} \left(\frac{\Gamma(r+1+s)}{\Gamma(r+1)\Gamma(s)} \int_0^x u^r(1-u)^{s-1}du\right)\right)dx =$$

$$= \int_0^p \frac{r}{x} \cdot \frac{\Gamma(r+1+s)}{\Gamma(r+1)\Gamma(s)} x^r(1-x)^{s-1}dx = (r+s)J_p(r,s)$$

∎

A1/12

$E(J_q(X_{r,p'},r)) = E(J_{p'}(r,X_{r,p}+1))$, if $r>0$ and $p,p'\varepsilon(0,1)$.

Proof:

Let $\alpha \ge 0$ and $s>0$, then we have:

$$E(J_q(X_{s,p'}+\alpha,r)) = \sum_{j=0}^{\infty} \left(\sum_{i=j+\alpha}^{\infty} \binom{i+r-1}{i} q^i p^r\right) \binom{j+s-1}{j} p'^s q'^j =$$

$$= \sum_{i=\alpha}^{\infty} \binom{i+r-1}{i} q^i p^r \sum_{j=0}^{i-\alpha} \binom{j+s-1}{j} p'^s q'^j = \sum_{i=\alpha}^{\infty} (p'^s \sum_{j=0}^{i-\alpha} \binom{j+s-1}{j} q'^j) \binom{i+r-1}{i} p^r q^i =$$

$$= \sum_{i=\alpha}^{\infty} J_{p'}(s,i-\alpha+1) \binom{j+r-1}{i} p^r q^i = E(J_{p'}(s,X_{r,p}-\alpha+1))$$

(note that $J_{p'}(s,i-\alpha+1) := 0$ if $i<\alpha$)

The above equality is obtained for $r=s$ and $\alpha=0$.

■

A1/13

$$\lim_{r\to\infty} rE(J_{p'}(r+\alpha,X_{r+\beta,p}+\gamma))=0, \quad \text{if } p'<p;\ \alpha,\beta\varepsilon \mathbb{N}_0;\ \gamma\varepsilon \mathbb{Z}$$

■

A1/14

$$\rho(X_{r,p_1}-X_{r,p_2},X_{r,p_1}-X_{r,p_3}) = \frac{q_1p_2p_3}{\sqrt{(q_1p_2^2+q_2p_1^2)(q_1p_3^2+q_3p_1^2)}}$$

This result is immediately seen by noting that

$$Cov(X_{r,p_1}-X_{r,p_2},X_{r,p_1}-X_{r,p_3}) = Var(X_{r,p_1})$$

■

APPENDIX 2

A2/1 (Central Limit Theorem)

Let $X_1,\ldots,X_n$ be independent identically distributed random variables with finite mean μ and finite variance $\sigma^2 > 0$, then:

$$F_n(x) := P(\frac{\bar{X}-\mu}{\sigma/\sqrt{n}} < x) \xrightarrow[n\to\infty]{} \Phi(x) := \frac{1}{\sqrt{2\pi}} \int_{-\infty}^{x} e^{-\frac{1}{2}t^2} dt,$$

uniformly in $x \in \mathbb{R}$.

■

A2/2 (Pólya's Lemma)

If the sequence of distribution functions $(F_n)_{n\in\mathbb{N}}$ converges to a continuous distribution function F for every $x\in\mathbb{R}$, then the convergence is uniform in $x\in\mathbb{R}$, that is

$$\lim_{n\to\infty} \sup_{x} |F_n(x)-F(x)| = 0$$

■

A2/3 (Helly-Bray)

If the sequence of distribution functions $(F_n)_{n\in\mathbb{N}}$ converges to the function F (we mean $F_n(x) \xrightarrow[n\to\infty]{} F(x)$ for all $x\in\mathbb{R}$ for which F is continuous), then

$$\lim_{n\to\infty} \int g(x)dF_n(x) = \int g(x)dF(x)$$

holds for every bounded and continuous function g.

■

A2/4

Let $V_{\lambda r}$ and V_λ be distribution functions or complements of distribution functions; $\lambda\in\{1,2,\ldots,k\}$; in addition, let V_λ be continuous, and $\lim\limits_{r\to\infty} V_{\lambda r}(x)=V_\lambda(x)$ is assumed to hold for all $x\in\mathbb{R}$, then

$$\lim_{r\to\infty} \int_{-\infty}^{+\infty} (\prod_{\lambda=2}^{k} V_{\lambda r}(x)) \, dV_{1r}(x) = \int_{-\infty}^{+\infty} (\prod_{\lambda=2}^{k} V_\lambda(x))dV_1(x)$$

Proof:

According to Pólya's lemma (cf. A2/2), $V_{\lambda r}$ converges to V_λ uniformly in $x \in \mathbb{R}$ for $\lambda \in \{1,\dots,k\}$. Using the inequality $\left| \prod_{\lambda=2}^{k} V_{\lambda r}(x) - \prod_{\lambda=2}^{k} V_\lambda(x) \right| \leq \sum_{\lambda=2}^{k} \left| V_{\lambda r}(x) - V_\lambda(x) \right|$ and the preceding statement, $W_r(x) := \prod_{\lambda=2}^{k} V_{\lambda r}(x)$ is immediately seen to converge to $W(x) := \prod_{\lambda=2}^{k} V_\lambda(x)$ uniformly in $x \in \mathbb{R}$. With this result, we obtain

$$\left| \int_{-\infty}^{+\infty} W_r(x) dV_{1r}(x) - \int_{-\infty}^{+\infty} W(x) dV_1(x) \right| \leq \left| \int_{-\infty}^{+\infty} (W_r(x) - W(x)) dV_{1r}(x) \right| +$$

$$+ \left| \int_{-\infty}^{+\infty} W(x) dV_{1r}(x) - \int_{-\infty}^{+\infty} W(x) dV_1(x) \right| < \varepsilon, \text{ provided } r \text{ is large enough.}$$

The first term of the right side tends to 0 because of the uniform convergence of W_r to W, and using the Lemma of Helly-Bray again, the second term is easily shown to tend to 0 as r becomes large.

■

A2/5

$$J_q(\alpha+1,r) = 1 - J_p(r,\alpha+1) = 1 - \sum_{j=0}^{\alpha} \binom{r+j-1}{j} p^r q^j = 1 - P(X_{r,p} < \alpha+1) =$$

$$= 1 - P\left(\frac{X_{r,p} - rq/p}{\sqrt{rq}/p} < \frac{\alpha+1-rq/p}{\sqrt{rq}/p} \right) \approx 1 - \Phi\left(\frac{p(\alpha+1) - rq}{\sqrt{rq}} \right)$$

A2/6 (Borel-Cantelli-Lemma)

Let $(A_n)_{n \in \mathbb{N}}$ be a sequence of events of the probability space $(\Omega, \mathcal{T}, P)$, then:

$$\sum_{n=1}^{\infty} P(A_n) < \infty \implies P(\limsup A_n) = 0$$

■

A2/7

Let $X_1, X_2, \dots, X_n$ be independent identically distributed random variables, and $0 \leq X_i \leq 1$ for $1 \leq i \leq n$, then for $0 < t < 1-\mu$:

$$P(\bar{X} - \mu \geq t) \leq e^{-2nt^2}, \text{ where } \mu := E(X_i),\ 1 \leq i \leq n.$$

(cf. Hoeffding [89])

■

A2/8

Let V and W be independent random vectors, assuming only a finite number of values with a probability >0, and let $T=t(V,W)$, then:

$$P(T=t_0 | W=w) = P(t(V,w) = t_0)$$

(cf. Hájek [85])

■

A2/9

Let $X_1, X_2, \ldots, X_n$ be a sequence of random variables with cdf's $F_1, F_2, \ldots, F_n$ respectively, and let $F_n(x) \xrightarrow[n\to\infty]{d} F(x)$. Let $Y_1, Y_2, \ldots, Y_n$ be another sequence of random variables, converging in probability to a constant $c > 0$. Define $Z_n := X_n / Y_n$, then the cdf of Z_n converges in distribution to $F(cx)$.
(cf. Cramér [52])

■

A2/10

Let $(X_{1n})_{n\in\mathbb{N}}, (X_{2n})_{n\in\mathbb{N}}, \ldots, (X_{rn})_{n\in\mathbb{N}}$ be sequences of random variables, converging in probability to the constants $x_1, x_2, \ldots, x_r$, respectively. Let $R(y_1, \ldots, y_r)$ be a rational function, then $R(X_{1n}, X_{2n}, \ldots, X_{rn})$ converges in probability to the constant $a := R(x_1, x_2, \ldots, x_r)$, provided a is finite.

■

A2/11

For independent random variables $X_1, X_2, \ldots, X_n$, "Convergence in probability" and "Almost sure convergence" are equivalent.
(cf. Loève [118])

■

A2/12

Let $(\Omega, \mathcal{T}, P)$ be a probability space, and let $(X_n)_{n\in\mathbb{N}}$ be a sequence of random variables defined on Ω with $X_n \xrightarrow{p} X$. Let $a_{mn} := P(|X_n| > m)$, and $\sup_n a_{mn} = a_m$, then the sequence $(X_n)_{n\in\mathbb{N}}$ is uniformly integrable with respect to P, if $\sum a_m < \infty$.

■

A2/13

Let $(X_n)_{n\in\mathbb{N}}$ be a sequence of random variables with finite first moments. If $X_n \xrightarrow{p} X$, and if $(X_n)_{n\in\mathbb{N}}$ is uniformly integrable, then $E(|X_n|) \xrightarrow[n\to\infty]{} E(|X|)$.

A2/14 (Stirlings formula)

$$n! = \sqrt{2\pi n} \cdot n^n e^{-n+\Theta(n)}, \text{ where } \frac{1}{12n+1} < \Theta(n) < \frac{1}{12n}$$

■

APPENDIX 3

A3/1 (Identity theorem for power series)

Suppose the series $A(x) := \sum_{n=0}^{\infty} a_n(x-x_0)^n$ and $B(x) := \sum_{n=0}^{\infty} b_n(x-x_0)^n$ converge in a circle K with center x_0 and radius ρ. Let $\mathcal{L}$ be the set of all $x \in K$ at which $A(x)=B(x)$. If $\mathcal{L}$ has a limit point in K, then $a_n=b_n$ for all $n \in \mathbb{N}_0$, i.e. $A(x) \equiv B(x)$.

■

A3/2

$$\int_0^1 \left(\frac{1}{x} - 1\right)^h x^{k-1} dx = \sum_{\nu=0}^{h} \binom{h}{\nu} (-1)^{k-\nu} \int_0^1 x^{k-1-\nu}\, dx = \sum_{\nu=0}^{h} \binom{h}{\nu} (-1)^{h-\nu} \frac{1}{k-\nu} =: a_h$$

for $0 \leq h \leq k-1$.

The following recursion formula for a_h is easily seen to hold

$a_h = \frac{h}{k} a_{h-1}$, and from this follows

$$a_h = \frac{h(h-1)\dots(h-(h-1))}{k(k-1)\dots(k-(h-1))} a_0, \text{ where } a_0 = \int_0^1 x^{k-h-1} dx = \frac{1}{k-h},$$

$$\text{i.e. } a_h = \frac{1}{\frac{k(k-1)\dots(k-(h-1))(k-h)}{h!}} = \frac{1}{k\binom{k-1}{h}}$$

■

A3/3

Let $A_{k-1}(\rho, D^*) := \int_{-\infty}^{+\infty} \left(\Phi\left(\frac{y\sqrt{\rho}+D^*}{\sqrt{1-\rho}}\right)\right)^{k-1} d\Phi(y)$ where

$$D^* := \frac{\Delta^* \sqrt{r}}{\sqrt{q_1 p_2^{*2} + q_2^* p_1^2}} \text{ and } \rho := \frac{q_1 p_2^{*2}}{q_1 p_2^{*2} + q_2^* p_1^2}\ ;\ k \geq 3;$$

then

$$\frac{d}{d\rho} A_{k-1}(\rho, D^*) = \frac{(k-1)(k-2)\varphi(D^*)\varphi\left(D^*\sqrt{\frac{1-\rho}{1+\rho}}\right)}{2\sqrt{1+\rho}} A_{k-3}\left(\frac{\rho}{1+2\rho}, D^*\sqrt{\frac{1-\rho}{(1+\rho)(1+2\rho)}}\right),$$

where φ denotes the standard normal density function. The above identity may be obtained by differentiating $A_{k-1}(\rho, D^*)$ under the integral sign and then using integration by parts. (cf. Sobel/Weiss [174])

■

A3/4

Let $a_n = 1+(a/(b+cn))$; $a\in\mathbb{R}$, $b>0$, $c>0$, and $n\geq 0$, and let $b_n \approx Bn^d$ with $B\neq 0$ and $d>a/c-1$, and let the sequence $(x_n)_{n\in\mathbb{N}_0}$ satisfy the recursion formula $x_{n+1}=a_n x_n+b_n$, then $x_n \approx (B/(d-(a/c)+1))n^{d+1}$, where $x_n \approx y_n$ means $x_n/y_n \to 1$ as $n\to\infty$.

The proof is given in [62].

■

APPENDIX 4

A4/1

Let f be a one-to-one transformation of $\mathcal{T}_m$, and let R be uniformly distributed over $\mathcal{T}_m$, where R and $\mathcal{T}_m$ are defined as in the introduction of part 2. Then D=f(R) has also the uniform distribution over $\mathcal{T}_m$.

■

A4/2

If R is uniformly distributed over $\mathcal{T}_m$, (cf. A4/1), then

(i) $P(R_i=g) = \frac{1}{m}$, $1\leq i$, $g\leq m$;

(ii) $P(R_i=g, R_j=h) = \frac{1}{m(m-1)}$, $1\leq i\neq j$, $g\neq h\leq m$;

(iii) $P(R_i=g, R_j=g) = 0$, $1\leq i\neq j$, $g\leq m$.

■

A4/3

If the random variables $X_1,X_2,\ldots,X_m$ are independent and possess continuous cdf's, then $P(X_i\neq X_j,\ 1\leq i\neq j\leq m)=1$.

■

A4/4

If the cdf's $F_1,F_2,\ldots,F_m$ of the random variables $X_1,X_2,\ldots,X_m$ are all equal to a continuous cdf F, then if R is the rank vector associated with $X_1,X_2,\ldots X_m$,

(i) R is a permutation of $(1,\ldots,m)$ with probability 1,

(ii) R is uniformly distributed over $\mathcal{T}_m$.

■

A4/5

Let S be a linear rank-order-statistic, with regression constants c_i and scores a_i, $1\leq i\leq m$, generated by the function $J(u)$, $0\leq u\leq 1$, possessing the properties given in section 1.1 of chap. 1, part 2. Then for each $\varepsilon>0$ there exists a $\delta := \delta(\varepsilon, J)$ such that

$$\max_{1\leq i\leq m} \{(c_i-\bar{c})^2\} < \delta \sum_{i=1}^{m} (c_i-\bar{c})^2$$

implies

$$\sup_{-\infty<S<\infty} |P(S\leq s) - \Phi((S-E(S))/\sqrt{\mathrm{Var} S})| < \varepsilon \quad .$$

(cf. Hâjek and Sidâk [86])

■

A4/6

Let $X_1, X_2, \ldots, X_m$; $Y_1, Y_2, \ldots, Y_n$ be independent random variables with the cdf's F and G respectively, and let $T(X_1,\ldots,X_r;\, Y_1,\ldots,Y_r) = T(X_{\Pi(1)},\ldots,X_{\Pi(r)};\, Y_{\Pi^*(1)},\ldots,Y_{\Pi^*(r)})$, where Π, Π^* are permutations of $\{1,\ldots,r\}$. Moreover let:

$$E(T(X_1,\ldots,X_r;\, Y_1,\ldots,Y_r)) = \theta(F,G) =: \theta,$$

$$E(T^2(X_1,\ldots,X_r;\, Y_1,\ldots,Y_r)) =: M < \infty.$$

Let $\frac{m}{n} =: c$ and n be sufficiently large so that $r\leq m$, n.
Define

$$U_n := \binom{m}{r}^{-1} \binom{n}{r}^{-1} \sum T(X_{\alpha_1},\ldots,X_{\alpha_r};\, Y_{\beta_1},\ldots,Y_{\beta_r}),$$

where the summation is extended over all subscripts $1\leq\alpha_1<\ldots<\alpha_r\leq m$, $1\leq\beta_1<\ldots<\beta_r\leq n$, then the random variable $\sqrt{n}(U_n-\theta)$ is asymptotically normal distributed.
(cf. Lehmann [115])

■

A4/7

Let $S_1, S_2, \ldots, S_k$ be linear rank-order statistics, where the scores

a) are generated by a function J(u), $0<u<1$, which is not constant and satisfies

$$\left|\frac{d^i J(u)}{du^i}\right| \leq K(u(1-u))^{-i-\frac{1}{2}+\delta} \quad ; \qquad i=0,1, \text{ for some } \delta > 0;$$

b) are the so-called median-scores generated by $J(u)=c(u-\frac{1}{2})$, $0<u<1$.

Then the random vector $\sqrt{m}(S_1-\mu,\ldots,S_k-\mu)$, where $\mu=E(S_i)$, $1\leq i\leq k$, has asymptotically a k-variate normal distribution with mean vector $\vec{0}$ and covariance matrix $\sum = ((\sigma_{ij}))$, where

$$\sigma_{ij}=\begin{cases}\dfrac{n^2(k-1)}{m-1}\sum\limits_{i=1}^{m}(a_i-\bar{a})^2 & \text{if } i=j\\[2ex] -\dfrac{n^2}{m-1}\sum\limits_{i=1}^{m}(a_i-\bar{a})^2 & \text{if } i\neq j\end{cases}$$

with $a_i:=J(\frac{i}{m+1})$ and $\bar{a}:=\bar{J}$.

■

A4/8

Let $\vec{X}=(X_{11},\dots,X_{1n},\dots,X_{k1},\dots,X_{kn})$ be a random vector with kn independent components X_{ij}, $1\leq i\leq k$, $1\leq j\leq n$, the cdf's of which are given by $F_{\delta_i}(x)$ and let $\delta:=(\delta_1,\dots,\delta_k)$.
Let $F_{\delta_i}(x)$, $1\leq i\leq k$, be a stochastically increasing family of distributions, i.e. if $\delta_i>\delta_i'$ then $F_{\delta_i}(x)$ and $F_{\delta_i'}(x)$ are distinct and $F_{\delta_i}(x)\leq F_{\delta_i'}(x)$ for all x.
Let H be a realvalued function of the realization $\vec{x}=(x_{11},\dots,x_{1n},\dots,x_{k1},\dots$ $\dots,x_{kn})$ of the random vector $\vec{X}$. If H is a nondecreasing function of x_{ij}, $1\leq j\leq n$, when all other components of $\vec{x}$ are held fixed, then $E_\delta(H(X))$ is a non-decreasing function of δ_i.
(cf. [123])

■

A4/9

Let f and g be two strongly unimodal symmetric densities with cdf's F and G, and quantile functions F^{-1} and G^{-1}. Define

$$\varphi(u) := -\frac{f'(F^{-1}(u))}{f(F^{-1}(u))}, \quad \psi(u) := -\frac{g'(G^{-1}(u))}{g(G^{-1}(u))},$$

and assume that $\varphi(u)$ increases more rapidly than $\psi(u)$, then f has shorter tails then g.
(cf. Hâjek [85])

■

A4/10

Let $Y_{ij(n)}$ be a general Hodges-Lehmann-estimator. Then the sequence $\{Y_{ij(n)}\}$ is uniformly continuous in probability with respect to $n^{-1/2}$ i.e., for each $\varepsilon>0$ and $\eta>0$, there exists a $\delta>0$, such that

$$P\Big(\sup_{\substack{n'\\|n-n'|<\delta\cdot n}} |n^{1/2}(Y_{ij(n')}-Y_{ij(n)}|>\varepsilon\Big) < \eta$$

as $n\to\infty$.

■

A4/11

Let $\Delta_{ijL}(n), \Delta_{ijU}(n)$ be the two estimators defined in (3.1), chapter 2, part 2. Then for each $s>0$ there exist positive constants c_{s1}, c_{s2} and n_s, such that for each $n \geq n_s$

$$P(|B(G)\sqrt{n}(\Delta_{ijU}(n)-\Delta_{ijL}(n))/At\alpha_{/2}-1| > c_{s1}n^{-1/4}(\ell n n)^3) \leq c_{s2}n^{-s}.$$

■

A4/12

Let $(Y_n)_{n\in\mathbb{N}}$ be any sequence of random variables such that $Y_n > 0$ a.s. and $\lim_{n\to\infty} Y_n = 1$ a.s.. Furthermore let $(f(u))_{n\in\mathbb{N}}$ be a sequence of constants with $f(n) > 0$, $\lim_{n\to\infty} f(n) = \infty$, and $\lim_{n\to\infty} \frac{f(n)}{f(n+1)} = 1$. For each $t > 0$, let $N=N(t)$ be the smallest positive integer $m\geq 1$, such that $Y_m \leq f(m)/t$, then

$$\lim_{t\to\infty} \frac{f(N)}{t} = 1 \quad \text{a.s.}$$

■

A4/13

Let $(X_n)_{n\in\mathbb{N}}$ be a infinite sequence of random variables, and let Y_n and Z_n be functions of $X_1, X_2, \ldots, X_n$, where Y_n is an arbitrary estimator of the parameter θ. Furthermore let $(Y_n)_{n\in\mathbb{N}}$ be a sequence of random variables, and let the associated sequence of cdf's $(F_n)_{n\in\mathbb{N}}$ converge to a continuous cdf F, which is uniformly continuous in probability with respect to the sequence $\{n^{-1/2}\}$ (cf. A4/11). The nonincreasing sequence of positive integers $(a_r)_{r\in\mathbb{N}}$ may converge to zero. The sequence of random variables $(N_r)_{r\in\mathbb{N}}$ is defined as follows: N_r is equal to the smallest positive integer, which satisfies $Z_n \leq a_r$. Let $(n_r)_{r\in\mathbb{N}}$ be a sequence of positive integers, where n_r is the smallest number n, such that $n^{-1/2} \leq a_r$. Then, if N_r is the random variable defined above and $\frac{N_r}{n_r} \xrightarrow[r\to\infty]{} 1$, it follows:

$$P(Y_{N_r} - \theta \leq x \cdot a_r) \xrightarrow[r\to\infty]{d} F(x).$$

■

APPENDIX 5

A Method for Generating Random Numbers

The method explained in the following can be used to generate random numbers for symmetric and skewed continuous distribution.
The algorithm is essentially based on the generalization of Tukey's lambda distribution (GLD:=Generalized Lambda Distribution), which possess the inverse

$$x := R(p) := \lambda_1+(p^{\lambda_3}-(1-p)^{\lambda_4})/\lambda_2, \quad 0\leq p\leq 1. \tag{5.1}$$

If p is the realization of an U(0,1)-distributed random variable, then the random variable X with the realization given by (5.1) follows the GLD. Skewness and peakedness of the GLD, derived by Ramberg and Schmeiser [148], are determined by the parameters λ_3 and λ_4. If their values are fixed, then the variance of X is specified by λ_2, and any mean of X can be gotten by the appropriate choice of λ_1. If $\lambda_3=\lambda_4$, then the GLD is symmetric and λ_1 is the mean of X.
In two articles [147],[148], Ramberg and Schmeiser have developed formulae and tables that allow to derive the values of $\lambda_1,\lambda_2,\lambda_3$ and λ_4, if the first four central moments of the distribution to be simulated are known in such a way that with the realization p of an U(0,1)-distributed random-variable P the value x, computed according to (5.1), can be considered as the realization of a random variable, possessing the distribution that should be simulated.
This method has been used for generating the random numbers needed for the various simulation-studies.
A random-number generator, implemented at the Computing Centre (Univac 1108) of the University of Karlsruhe, has been used to generate U(0,1)-distributed random numbers.
The λ-values and the corresponding central moments of the six simulated distributions are combined in the following table.

Distribution	λ_1	λ_2	λ_3	λ_4	E(X)	$E((X-E(X))^2)$	$E((X-E(X))^3)$	$E((X-E(X))^4)$
$U(\mu-h,\mu+h)$	μ	$\frac{1}{h}$	1.0	1.0	μ	$\frac{1}{3}h^2$	0	1.8
$N(\mu,\sigma)$	μ	$\frac{0.1975}{\sigma}$	0.135	0.135	μ	σ^2	0	3.0
D.-Exp(μ,σ)	μ	$\frac{-0.1192}{\sigma}$	-0.0802	-0.0802	μ	$2\sigma^2$	0	6.0
Cauchy(μ,σ)	μ	$\frac{-3.0674}{\sigma}$	-1.0	-1.0	(no moments; approximation by comparison of quantiles)[1]			
$\text{Exp}(\mu,\lambda)\lambda=\frac{1}{\sigma}$	$\mu+0.0004$	$\frac{0.0004}{\sigma}$	0.0	0.0004	$\mu+\alpha\sigma$	$\alpha\sigma^2$	$2\alpha^{-\frac{1}{2}}$	$3(1+\frac{2}{\alpha})$
Gamma(α,μ,σ)	$\mu+0.8624$	$\frac{0.029272}{\sigma}$	0.005675	0.04052				

α := shape-parameter, μ := location-parameter, σ := scale-parameter

1 *The Cauchy distribution possesses no finite moments. For such cases Ramberg and Schmeiser [147] have developed a method, which is based on the comparison of quantiles.*

APPENDIX 6

Table I:

For some values of D and k, the table contains the probability $P(SK_{\bullet} \geq D)$, where $SK_{\bullet}$ denotes some Haga-statistic (cf. section 3, part 2).

D \ k	3	4	5	6	7	8	9	10
1	0.428	0.441	0.451	0.459	0.464	0.469	0.472	0.475
2	0.350	0.372	0.389	0.402	0.414	0.422	0.429	0.435
3	0.255	0.289	0.317	0.340	0.357	0.371	0.383	0.393
4	0.177	0.219	0.255	0.283	0.306	0.325	0.340	0.354
5	0.121	0.165	0.204	0.236	0.262	0.284	0.303	0.318
6	0.081	0.124	0.163	0.197	0.225	0.249	0.269	0.286
7	0.054	0.093	0.130	0.164	0.193	0.218	0.239	0.258
8	0.036	0.070	0.104	0.137	0.165	0.190	0.213	0.232
9	0.024	0.052	0.084	0.114	0.142	0.167	0.189	0.209
10	0.016	0.039	0.067	0.095	0.121	0.146	0.168	0.188
15	0.002	0.009	0.022	0.038	0.056	0.075	0.093	0.111
20	0.000	0.002	0.007	0.015	0.026	0.038	0.052	0.065
30	0.000	0.000	0.001	0.002	0.006	0.010	0.016	0.023

Table II:

For given values of k and P^*, the table contains values of d, that solve the equation $\int_{-\infty}^{+\infty} \Phi^{k-1}(x+d)d\Phi(x) = P^*$.

k \ P^*	0.75	0.80	0.90	0.95	0.99
2	0.9539	1.1902	1.8124	2.3262	3.2900
3	1.4338	1.6524	2.2302	2.7101	3.6173
4	1.6822	1.8932	2.4516	2.9162	3.7970
5	1.8463	2.0528	2.5997	3.0552	3.9196
6	1.9674	2.1709	2.7100	3.1591	4.0121
7	2.0626	2.2639	2.7972	3.2417	4.0861
8	2.1407	2.3403	2.8691	3.3099	4.4175
9	2.2067	2.4049	2.9301	3.3679	4.1999
10	2.2637	2.4608	2.9829	3.4182	4.2456

Table III:

For given values of k and n (m=kn), the scorefunction J(u)=u and some values of d, the table contains the exactly (P_{exact}) and the asymptotically (P_{asympt}) computed probabilities for $\max_{2\leq j\leq k} S_j - S_1 \leq d$. N(..n..|..d..) denotes the number of favorable sequences.

d(m+1)	N(..n..|..d..)	P_{exact}	$d(m+1)_{asympt}$	P_{asympt}
k=3; n=3;	possible sequences: 1680			
10	1490	0.88690	-	-
11	1544	0.91905	11	0.90
12	1584	0.94286	-	-
13	1618	0.96310	13	0.95
14	1644	0.97857	-	-
15	1664	0.99048	15	0.97
16	1674	0.99643	-	-
17	1678	0.99881	-	-
18	1680	1.00000	18	0.99
k=3; n=3;	possible sequences: 34650			
15	30616	0.88358	-	-
16	31336	0.90436	-	-
17	31952	0.92214	17	0.90
18	32496	0.93784	-	-
19	32936	0.95053	-	-
20	33338	0.96214	20	0.95
21	33644	0.97097	-	-
22	33912	0.97870	22	0.97
23	34126	0.98488	-	-
24	34296	0.98978	-	-
25	34424	0.99348	-	-
26	34518	0.99619	-	-
27	34574	0.99781	27	0.99
28	34614	0.99896	-	-
29	34634	0.99954	-	-
30	34644	0.99983	-	-

Table III:(continued)

d(m+1)	N(..n..|..d..)	P_{exact}	$d(m+1)_{asympt}$	P_{asympt}
k=3; n=5;	possible sequences: 756756			
22	680548	0.89930	-	-
23	690952	0.91304	23	0.90
26	715862	0.94596	-	-
27	722454	0.95467	-	-
28	728098	0.96213	28	0.95
29	733120	0.96817	-	-
30	737408	0.97443	-	-
31	741054	0.97925	31	0.97
34	749008	0.98976	-	-
35	750826	0.99216	-	-
36	752296	0.99411	-	-
37	753464	0.99565	37	0.99
48	756750	0.99999	-	-
49	756756	1.00000	-	-
k=4; n=3;	possible sequences: 369600			
14	322872	0.87357	-	-
15	332796	0.90042	-	-
16	341094	0.92287	16	0.90
17	348396	0.94263	-	-
18	354228	0.95841	-	
19	359172	0.97179	19	0.95
20	362832	0.98169	-	-
21	365664	0.98935	21	0.97
22	367440	0.99416	-	-
23	368520	0.99708	-	-
24	369120	0.99870	24	0.99
25	369420	0.99951	-	-
26	369540	0.99984	-	-
27	369600	1.00000	-	-

Table IV:

Analogous to table III, for $J(u)=\Phi^{-1}(u)$ and some values of k and n the probabilities P_{exact} and P_{asympt} are given.

d	N(..n..|..d..)	P_{exact}	d_{asympt}	P_{asympt}
k=3; n=3;	possible sequences: 1680			
1.0	953	0.56726	-	-
2.0	1306	0.77738	2.1	0.75
3.0	1587	0.94464	3.2	0.90
4.0	1676	0.99762	-	-
5.0	1680	1.00000	5.2	0.99
k=3; n=4;	possible sequences: 34650			
2.0	24596	0.70984	-	-
3.0	29858	0.86170	2.5	0.75
4.0	33239	0.95928	-	-
5.0	34387	0.99530	4.6	0.95
6.0	34648	0.99994	6.2	0.99
k=3; n=5;	possible sequences: 756756			
2.0	501352	0.66250	-	-
3.0	609479	0.80538	2.8	0.75
4.0	687786	0.90886	4.4	0.90
5.0	734569	0.97068	-	-
6.0	752653	0.99458	5.9	0.97
7.0	756460	0.99961	7.1	0.99
8.0	756756	1.00000	-	-
k=4; n=3;	possible sequences: 369600			
2.0	248185	0.67150	-	-
3.0	316758	0.85703	2.5	0.75
4.0	357690	0.96778	4.3	0.95
5.0	369096	0.99864	-	-
6.0	369600	1.00000	5.6	0.99

Table V:

For the scorefunction J(u) = c(u-1/2) (Median-scores) the same magnitudes are given as in tables III and IV.

d	N(..n..|..d..)	P_{exact}	d_{asympt}	P_{asympt}
k=3; n=3;	possible sequences: 1680			
1	1360	0.80952	-	-
2	1600	0.95238	2	0.75
3	1680	1.00000	3	0.95
k=3; n=4;	possible sequences: 34650			
1	24750	0.71429	-	-
2	32400	0.93506	-	-
3	34200	0.98701	3	0.90
4	34650	1.00000	4	0.97
k=3; n=5;	possible sequences: 756756			
1	549192	0.72572	-	-
2	669144	0.88423	2	0.75
3	739704	0.97747	-	-
4	754404	0.99689	4	0.97
5	756756	1.00000	5	0.99
k=4; n=3;	possible sequences: 369600			
1	282000	0.76299	-	-
2	346800	0.93831	2	0.75
3	369600	1.00000	3	0.90
k=4; n=4;	possible sequences: 63063000			
1	42644700	0.67622	-	-
2	57344700	0.90932	-	-
3	62048700	0.98392	3	0.90
4	63063000	1.00000	4	0.97

Table VI:

The table below contains the exactly (P_{exact}) and the asymptotically (P_{asympt}) computed probabilities $P(\max_{2\leq \ell \leq k} S_\ell - S_1 \leq b)$ for given values of k and n, the scorefunction J(u)=u and some values of b.

b(k+1)	P_{exact}	$b(k+1)_{asympt.}$	P_{asympt}
k=3; n=5;			
2	0.66641	-	-
3	0.77186	3.20	0.75
4	0.85931	-	-
5	0.93030	4.98	0.90
6	0.96631	6.05	0.95
7	0.98688	-	-
k=3; n=7;			
3	0.71727	-	-
4	0.80164	3.79	0.75
5	0.87766	-	-
6	0.92717	5.90	0.90
7	0.96023	7.17	0.95
8	0.98169	-	-
9	0.99219	9.57	0.99
k=4; n=4;			
4	0.75635	3.76	0.75
5	0.84344	-	-
6	0.90824	5.48	0.90
7	0.95193	6.52	0.95
8	0.97951	-	-
9	0.99354	8.49	0.99

Table VII:

Analogous to table VI, for $J(u)=\Phi^{-1}(u)$ and some values of k and n, the probabilities P_{exact} and P_{asympt} are given in the following table

b(k+1)	P_{exact}	$b(k+1)_{asympt}$	P_{asympt}
k=3; n=5;			
2	0.66641	-	-
3	0.85931	2.16	0.75
4	0.93030	3.36	0.90
5	0.98688	4.55	0.97
6	0.99717	5.45	0.99
k=3; n=6;			
2	0.63490	-	-
3	0.82879	2.37	0.75
4	0.90081	3.68	0.90
5	0.97488	4.99	0.97
6	0.98993	5.97	0.99
7	0.99944	-	-
k=4; n=3;			
2	0.71050	2.09	0.75
3	0.89540	-	-
4	0.98785	3.99	0.97
5	0.99826	4.72	0.99
6	1.00000	-	-

Table VIII:

The same magnitudes as in tables VI and VII are given for the scorefunction $J(u)=c(u-\frac{1}{2})$ (Median-scores).

b	P_{exact}	b_{asympt}	P_{asympt}
k=3; n=5;			
1	0.66255	-	-
2	0.82716	1.851	0.75
3	0.95062	3.499	0.95
4	0.99177	4.670	0.99
5	1.00000	-	-
k=3; n=7;			
1	0.62140	-	-
2	0.78144	2.190	0.75
3	0.89666	-	-
4	0.96708	4.140	0.95
5	0.99268	5.526	0.99
6	0.99909	-	-
k=4; n=4;			
1	0.63194	-	-
2	0.86343	1.942	0.75
3	0.96528	3.367	0.95
4	1.00000	-	-

Abbreviations

CS	correct selection
FS	false selection
NS	none selection
P^*	significance level
Δ^*	discrimination level
NT=A	next trial is carried out on treatment A
$[x]$	greatest integer $\leq x$
$]x[$	smallest integer $\geq x$
LFC	least favorable configuration
$\bar{p}$	$:= \frac{1}{2}(p_A+p_B)$
N_A	number of patients treated with A
N_B	number of patients treated with B
$N=N_A+N_B$	patient horizon, number of patients involved in the experiment
$\approx$	approximately
$p_A \leftrightarrows p_B$	p_A and p_B as well as q_A and q_B are interchanged
PW	Play-the-winner sampling
PL	Play-the-loser sampling
VT	Vector-at-a-time sampling
FL	Follow-the-leader sampling
RPW	Randomized play the winner sampling
PW-mF	Play-the-winner sampling with change of the treatments after m failures
$X_{r,p}$	negative binomial chance variable with index r and success parameter p
$\underset{\sim}{J}_q(\cdot,\cdot)$	incomplete beta-function
$X_{r,p}$	binomial chance variable with index r and success parameter p
$E^{r-1}(Y)$	expectation of the discrete random variable Y truncated at r-1
S_A	current number of A-successes
S_B	current number of B-successes
F_A	current number of A-failures
F_B	current number of B-failures
$\tilde{F}_A$	number of A-failures preceding the r-th A-success

we sometimes use the abbreviation $F_A(r)$

$\tilde{F}_B$	number of B-failures preceding the r-th A-success
F_A^*	number of A-failures preceding the r-th B-success

F_B^*	number of B-failures preceding the r-th B-success

we sometimes use the abbreviation $F_B(r)$

$S_A(c)$	number of A-successes preceding the c-th A-failure
$F_A(r)$	number of A-failures preceding the r-th A-success

The random variables $S_i, S_i(c), F_i, F_i(r)$ are defined correspondingly

$S_{A,n}$	number of A-successes in n trials
$F_{A,n}$	number of A-failures in n trials
Φ	standard normal distribution function
φ	standard normal density function
R1,R2	selection rules
P(CS|R1)	probability for CS when rule R1 is used
$\mathfrak{S}_n$	symmetric group of order n
$c(u) := \begin{cases} 1, & \text{if } u \leq 0 \\ 0, & \text{if } u > 0 \end{cases}$	
■	end of a proof, end of a definition
r.-o.s.	rank-order statistic
cdf	cumulative distribution function
U(a,b)	uniform distribution over the interval [a,b]
$N(\mu,\sigma)$	normal distribution with mean μ and standard deviation $\sigma=1$
$\xrightarrow[n\to\infty]{}$	convergence
$\xrightarrow[n\to\infty]{a.s.}$	almost sure convergence
$\xrightarrow[n\to\infty]{p}$	convergence in probability
$\xrightarrow[n\to\infty]{d}$	convergence in distribution

References

Abbreviations:

ACM : Communications of the ACM (Association for Computing Machinery)
AMS : The Annals of Mathematical Statistics
AS : The Annals of Statistics
JASA: Journal of the American Statistical Association
JRSS: Journal of the Royal Statistical Society

[1] Alam, K. (1971)
On selecting the most probable category. Technometrics 13, 843-850.

[2] Alam, K./Ramey, J.T. (1979)
A sequential procedure for selecting the most probable multinomial event. Biometrika, 66, 1, 171-173.

[3] Alam, K./Thompson, J. (1972)
On selecting the least probable multinomial event. AMS, Vol. 43, No. 6, 1981-1990.

[4] Anderson, T.W. (1966)
An Introduction to multivariate statistical analysis. John Wiley, New York.

[5] Anscombe, F.J. (1952)
Large sample theory of sequential estimation. Proc. Cambridge Philos. Soc. 48, 600-607.

[6] Anscombe, F.J. (1963)
Sequential medical trials. JASA, Vol. 58, 365-383.

[7] Armitage, P. (1957)
Restricted sequential procedures, Biometrika 44, 9-26.

[8] Armitage, P. (1975)
Sequential medical trials. 2nd ed., Blackwell scientific publications.

[9] Asano, C./Goto, M./Sugimura, M. (1969)
Optimum sequential designs based on Markov Chains for selecting one of two clinical treatments. Mem.Fac.Engrg., Kumamoto University, Japan, XV, No. 3, 1-16.

[10] Asano, C./Goto, M./Sugimura, M. (1971)
Some sequential designs based on Markov Chains for selecting one of two medical treatments. J.Kobe University of Mercantile Marine, 19, 159-172.

[11] Asano, C./Goto, M./Sugimura, M. (1971)
Numerical tables of optimum sequential designs based on Markov Chains for selecting one of two medical treatments. Bull.Math.Stat., 14, No. 3~4, 27-56.

[12] Asano, C./Goto, M./Sugimura, M. (1972)
Some sequential designs based on Markov Chains for selecting one of two medical treatments II. J.Kobe University of Mercantile Marine, 20, 213-221.

[13] Asano, C./Jojima, K. (1975)
An optimum sequential selection plan based on vector-at-a-time sampling and inverse stopping rule. Res.Rep., Res.Inst.of Fund.Inf.Science, Fukuoka, Japan, No. 66.

[14] Asano, C./Jojima, K. (1975)
An optimum sequential selection plan based on play-the-winner sampling and inverse stopping rule in a finite population. Res.Rep., Res.Inst. of Fund.Inf.Science, Fukuoka, Japan, No. 52.

[15] Asano, C./Jojima, K. (1976)
An optimum sequential selection plan based on play-the-winner sampling and successive success stopping rule in a finite population. Res.Rep., Res.Inst.of Fund.Inf.Science, Fukuoka, Japan, No. 67.

[16] Asano, C./Jojima, K./Sugimura, M. (1976)
Extended versions of sequential optimum selection plan with play-the-winner sampling and the stopping rules in a finite population. Proceedings in computational statistics, 2nd Symposium held in Berlin (West) 1976, Physika-Verlag, Wien 1976, 277-284.

[17] Asano, C./Jojima, K./Osata, E./Ogawa, N. (1977)
A new sequential selection plan with play-the-winner sampling rule for m medical treatments in a finite population. Proceedings of the Second World Conference of Medical Informatics, MEDINFO 77, 713-717.

[18] Asano, C./Suesada, S./Sugimura, M. (1975)
A certain truncated sequential design based on Markov Chains for selecting one of two medical treatments. Rep.Fac.of Engrg. Oita University, No. 1, 43-53.

[19] Asano, C./Suesada, S./Sugimura, M. (1976)
Truncated and untruncated sequential designs based on Markov Chains for selecting one of two treatments. Rep.Fac.of Engrg. Oita University, No. 2, 37-56.

[20] Asano, C./Sugimura, M. (1967)
Optimum designs for selecting one of two treatments, fixed sample size plan 6 and sequential plan 3. Kumamoto J.Sci.Ser.A, 8. No. 1, 21-51.

[21] Asano, C./Sugimura, M. (1967)
Optimum designs for selecting one of two medical treatments, fixed sample size plan 3. Kumamoto J.Sci.Ser.A, 7. No. 4, 95-102.

[22] Asano, C./Sugimura, M. (1967)
Optimum designs for selecting one of two medical treatments, sequential plan 1. Bull.Math.Stat., 12, No. 3-4, 1-9.

[23] Atkinson, A.C./Pearce, M.C. (1976)
The computer generation of Beta, Gamma and normal random variables. JRSS, Series A, No. 139, pt. 4, 431-461.

[24] Atkinson, A.C./Whittaker, J. (1976)
A switching algorithm for the generation of Beta random variables with at least one parameter less than 1. JRSS, Series A, No. 139, pt. 4, 462-467.

[25] Bahadur, R.R. (1966)
A note on quantiles in large samples, AMS 37, 577-580.

[26] Bartlett, N.S./Govindarajulu, Z. (1965)
Some distribution-free statistics and their application to the selection problem. Abstract AMS, 36, 1597-1598.

[27] Bartlett, N.S. (1970)
Selecting a subset containing the best hypergeometric population. Sankhyā, Series B, Vol. 32, pt. 1 & 2, 341-352.

[28] Bechhofer, R.E. (1954)
A single-sample multiple decision procedure for ranking means of normal populations with known variances. AMS 25, 16-39.

[29] Bechhofer, R.E. (1970)
On ranking the players in a 3-playertournament. In: Nonparametric Techniques in Statistical Inference, pp. 545-559, Cambridge Univ.Press, London and New York.

[30] Bechhofer, R./Elmaghraby, S./Morse, N. (1959)
A single-sample multiple-decision procedure for selecting the multinomial event which has the highest probability. AMS, 30, 102-119.

[31] Bechhofer, R.E./Kiefer, J./Sobel, M. (1968)
Sequential Identification and Ranking Procedures. Chicago, University of Chicago Press.

[32] Begg, C.B./Mehta, C.R. (1979)
Sequential analysis of comparative clinical trials. Biometrika, 66, 1, 97-103.

[33] Benjamin, B. (1977)
Progress in Medical Statistics. JRSS, Ser. A, 140, Part 3, 366-376.

[34] Berry, D.A./Young, D.H. (1977)
A note on inverse sampling procedures for selecting the best binomial population. AS, Vol. 5, No. 1, 235-236.

[35] Berry, D.A./Sobel, M. (1973)
An improved procedure for selecting the better of two Bernoulli populations. JASA, Vol. 68, 979-984.

[36] Bhattacharjee, G.P./Majumder, K.L. (1973)
The incomplete Beta integral. JRSS, Series C, Vol.22, 409-414.

[37] Bickel, P.J./Yahav, J.A. (1968)
Asymptotically optimal Bayes and Minimax procedures in sequential estimation. AMS, 39, 972-994.

[38] Bickel, P.J./Yahav, J.A. (1977)
On selecting a set of good populations. In: Statistical Decision Theory and Related Topics II, ed. by S.S.Gupta and D.S.Moore, 37-55.

[39] Binns, M.R. (1974)
Approximating the negative binomial via the positive binomial. Technometrics, Vol. 16, No. 2, 323-324.

[40] Binns, M.R. (1975)
Sequential estimation of the mean of a negative binomial distribution. Biometrika, 62, 2, 433-440.

[41] Birnbaum, Z.W./Klose, O.M. (1957)
Bounds for the variance of the Mann-Whitney statistic. AMS 28, 933-945.

[42] Blackwell, D./Hodges, J.R.,jr. (1963)
Design for the control of selection bias. AMS 28, 449-460.

[43] Bross, J. (1952)
Sequential medical plans. Biometrics, 8, 188-205.

[44] Brown, G.W./Mood, A.M. (1951)
On median tests for linear hypotheses. Proc. Second Berkeley Symp. Math. Statist. Prob. 1, 159-166.

[45] Büringer, H. (1978)
Nonparametric sequential procedures for selecting the best of k treatments in two-way layouts. Inst. für Statistik und Math. Wirtschaftsth. Karlsruhe, Discussion paper 32/78.

[46] Canner, P.L. (1970)
Selecting one of two treatments when the responses are dichotomous. JASA, Vol. 65, 293-306.

[47] Choi, J. (1968)
Truncated sequential design for clinical trials based on Markov Chains. Biometrics 24, 159-166.

[48] Chow, Y.S./Robbins, H. (1965)
On the asymptotic theory of fixed-width sequential confidence intervals for the mean. AMS 36, 457-462.

[49] Colton, T. (1963)
A model for selecting one of two medical treatments. JASA, Vol. 58, 388-400.

[50] Conover, W.J. (1973)
Rank tests for one sample, two samples, and k samples without the assumption of a continuous distribution function. AS 1, 1105-1125.

[51] Cornfield, J./Greenhouse, S.W./Halperin, M. (1969)
An adaptive procedure for sequential clinical trial. JASA, Vol. 64, 759-770.

[52] Cramér, H. (1946)
Mathematical Methods in Statistics. Princeton Univ. Press.

[53] Cran, G.W./Martin, K.J./Thomas, G.E. (1977)
remark ASR 19 and algorithm AS 109. JRSS, Series C, Vol.26,No.1,111-114.

[54] Dunnett, C.W. (1960)
On selecting the largest of k normal population means. JRSS, Series B, 22, 1-40.

[55] Durham, S./Wei, L.J. (1978)
The randomized play-the-winner rule in medical trials. JASA, Vol. 73, 364, 840-843.

[56] Eckardt, W.L., jr. (1976)
A sequential method for selecting the best of three binomial populations. JASA, Vol. 71, No. 354, 473-474.

[57] Exton, H. (1976)
Multiple hypergeometric functions and applications. Ellis Horwood limited, Chichester, England.

[58] Fisher, D.M. (1972)
Classification, selection and testing procedures for asymmetric distributions. Ph.D., Department of Statistics, University of Iowa.

[59] Flehinger, B.J./Louis, T.A./Miller, J.M. (1969)
Clinical trials to test for difference in mean survival time - data dependent assignment of patients. IBM research report, december 16.

[60] Flehinger, B.J./Louis, T.A. (1971)
Sequential treatment allocation in clinical trials. Biometrika 58, 419-426.

[61] Flehinger, B.J./Louis, T.A. (1972)
Sequential medical trials with data dependent treatment allocation. Proceedings of the 6th Berkeley Symposium on Mathematical Statistics and Probability, Bd. 4, 43-52.

[62] Freedman, D.A. (1965)
Bernard Friedman's Urn. AMS, 36, 956-970.

[63] Fushimi, M. (1973)
An improved version of a Sobel-Weiss Play-the-winner procedure for selecting the better of two binomial populations. Biometrika, 60, 3, 517-523.

[64] Gastwirth, J.L. (1965)
Percentile modifications of two-sample rank tests. JASA 60, 1127-1141.

[65] Gebhardt, F. (1971)
Incomplete Beta-Integral $B(x;2/3,2/3)$ and $[p(1-p)]^{-1/6}$ for use with Borges' approximation of the binomial distribution. JASA,Vol.66,333,189-191.

[66] Geertsema, J.C. (1970)
Sequential confidence intervals based on rank tests. AMS 41, 1016-1026.

[67] Geertsema, J.C. (1972)
Nonparametric sequential procedures for selecting the best of k populations. JASA 67, 614-616.

[68] Ghosh, M. (1973)
Nonparametric selection procedures for symmetric location parameter populations. AS 1, 773-779.

[69] Gibbons, J.D. (1971)
Nonparametric Statistical Inference. McGraw-Hill, New York.

[70] Gibbons, J.D./Olkin, I./Sobel, M. (1977)
Selecting and Ordering Populations: A new statistical methodology. John Wiley, New York.

[71] Glazebrook, K.D. (1978)
On the optimal allocation of two or more treatments in a controlled clinical trial. Biometrika, 65, 2, 335-340.

[72] Goel, P.K./Rubin, H. (1976)
On selecting a subset containing the best population - a Bayesian approach. AS, Vol.5, No. 5, 969-983.

[73] Govindarajulu, Z./Le Cam, L./Raghavachari, M. (1966)
Generalizations of theorems of Chernoff and Savage on the asymptotic normality of test statistics. Proc. Fifth Berkeley Symp. on Math. Stat. and Prob. 1, 609-638.

[74] Gray, H.L./Lewis, T.O. (1971)
Approximation of tail probabilities by means of the B_n-transformation. JASA, Vol. 66, No. 336, 897-899.

[75] Grundy, P.M./Healy, M.J.R./Rees, D.H. (1954)
Decision between two alternatives - how many experiments? Biometrics, 10, 317-323.

[76] Grundy, P.M./Healy, M.J.R./Rees, D.H. (1956)
Economic choice of the amount of experimentation. JRSS, Series B, 18, 32-49.

[77] Gupta, S.S. (1963)
Probability integrals of multivariate normal and multivariate t. AMS 34, 792-828.

[78] Gupta, S.S. (1965)
On some multiple decision rules. Technometrics 7, 225-245.

[79] Gupta, S.S./Huang, D.-Y. (1975)
On some parametric and nonparametric sequential subset selection procedures. In: Statistical Inference and Related Topics, ed. M.L.Puri, Academic Press, New York.

[80] Gupta, S.S./McDonald, G. (1975)
On some classes of selection procedures based on ranks. In: Nonparametric techniques in statistical inference, ed. M.L.Puri, Cambridge University Press, London, 491-514.

[81] Gupta, S.S./Moore, D.S. (1977)
Statistical decision theory and related topics II. Proceedings of a symposium held at Purdue University, may 17-19, 1976, Academic Press, New York, San Francisco, London.

[82] Gupta, S.S./Sobel, M. (1960)
Selecting a subset containing the best of several binomial populations. Contributions to probability and statistics, ed. J. Olkin et al., Stanford studies in mathematics and statistics, 2.

[83] Gupta, S.S./Yackel, J. (1971)
Statistical decision theory and related topics I. Proceedings of a symposium held at Purdue University, november 23-25, 1970, Academic Press, New York and London.

[84] Haga, T. (1959/60)
A two-sample rank test on location. Ann.Inst.Stat.Math.11, 211-219.

[85] Hājek, J. (1969)
A course in Nonparametric Statistics. Holden-Day, San Francisco.

[86] Hājek, J./Sidāk, Z. (1967)
Theory of rank tests. Academia, Prag.

[87] Hettmansperger, T.P./Malin, J.S. (1974)
A modified Mood's test for location with no shape assumptions of the underlying distributions. JASA 69, 527-529.

[88] Hodges, J.L.,jr./Lehmann, E.L. (1963)
Estimates of location based on rank tests. AMS 34, 598-611.

[89] Hoeffding, W. (1963)
Probability inequalities for sums of bounded random variables. JASA 58, 13-30.

[90] Hoeffding, W. (1973)
On the centering of a simple linear rank statistic . AS 1, 54-66.

[91] Hoel, D.G. (1971)
A method for the construction of sequential selection procedures. AMS, Vol. 42, 630-642.

[92] Hoel, D.G. (1972)
An inverse stopping rule for play-the-winner sampling. JASA, Vol. 67, 148-151.

[93] Hoel, D.G./Simon, R./Weiss, G.H. (1975)
Sequential analysis of binomial clinical trials. Biometrika 62, 1, 195-200.

[94] Hoel, D.G./Simon, R./Weiss, G.H. (1976)
Sequential tests for composite hypotheses with two binomial populations. JRSS, Series B, Vol. 38, No. 3, 302-308.

[95] Hoel, D.G./Sobel, M. (1972)
Proceedings of the sixth Berkeley Symposium on mathematical statistics and probability, Vol. IV, 53-69.

[96] Hoel, D.G./Sobel, M./Weiss, G.H. (1972)
A two-stage procedure for choosing the better of two binomial populations. Biometrika, Vol. 59, 2, 317-322.

[97] Hoel, D.G./Sobel, M./Weiss, G.H. (1975)
Comparison of sampling methods for choosing the best binomial population with delayed observations. J.Stat.Comput.Simul., Vol. 3, 299-313.

[98] Hoel, D.G./Sobel, M./Weiss, G.H. (1975)
A survey of adaptive sampling for clinical trials. Perspectives in biometrics, Vol. 1, 29-61, Academic Press, New York.

[99] Hoel, D.G./Weiss, G.H. (1974)
Comparison of methods for choosing the better of two negative exponential lifetime distributions. Biometry and Reliability, ed. F.Proschan and R.J.Serfling, pp.563-568, Philadelphia Society for Industrial and Applied Mathematics.

[100] Hogg, R.V. (1967)
Some observations on robust estimation. JASA 62, 1179-1186.

[101] Hogg, R.V. (1972)
More light on the kurtosis and related statistics. JASA 67, 422-424.

[102] Hogg, R.V. (1976)
A new dimension to nonparametric tests. Commun. Statist.-Theor.Math., A 5, 1313-1325.

[103] Hogg, R.V./Fisher, D.M./Randles, R.H. (1975)
A two-sample adaptive distribution-free test. JASA 70, 656-661.

[104] Hogg, R.V./Uthoff, V.A./Randles, R.H. (1972)
On the selection of the underlying distribution and adaptive estimation. JASA 67, 597-600.

[105] Hsi, B.P./Louis, T.A. (1975)
A modified play-the-winner rule for sequential trials. JASA, Vol. 70, No. 351, 644-647.

[106] Huyett, M./Sobel, M. (1957)
Selecting the best one of several binomial populations. The Bell systems technical journal, Vol. XXXVI, 537-576.

[107] Isbell, J.R. (1959)
On a problem of Robbins. AMS, Vol. 30, 606-610.

[108] Jojima, K. (1976)
A truncated sequential decision procedure for selecting the better of two treatments with a finite population. J. Kumamoto Women's University, 28, 35-44.

[109] Jojima, K. (1977)
A truncated sequential selection plan based on the play-the-winner sampling and inverse stopping rule in a finite population. J. Kumamoto Women's University, 39-47.

[110] Jöhnk, M.D. (1964)
Erzeugung von betaverteilten und gammaverteilten Zufallszahlen. Metrika, Vol. 8, 5-15.

[111] Kesten, H./Morse, N. (1959)
A property of the multinomial distribution. AMS 30, 120-127.

[112] Kiefer, J.E./Weiss, G.H. (1971)
A truncated test for choosing the better of two binomial populations. JASA, Vol. 66, 867-871.

[113] Kiefer, J.E./Weiss, G.H. (1974)
Truncated version of a play-the-winner rule for choosing the better of two binomial populations. JASA, Vol. 69, No. 347, 807-809.

[114] Ladd, D.W. (1975)
An algorithm for the binomial distribution with dependent trials. JASA, Vol. 70, No. 350, 333-340.

[115] Lehmann, E.L. (1951)
Consistency and unbiasedness of certain nonparametric tests. AMS 22, 165-179.

[116] Lehmann, E.L. (1963)
A class of selection procedures based on ranks. Mathematische Annalen 150, 268-275.

[117] Lehmann, E.L./D'Abrera, H.J.M. (1975)
Nonparametrics, Statistical methods based on ranks. Holden-Day, San Francisco, McGraw-Hill, New York.

[118] Loève, M. (1963)
Probability Theory. 3rd ed., Van Nostrand Reinhold Company, New York.

[119] McDonald, G.C. (1969)
On some distribution-free ranking and selection procedures. Department of Statistics, Purdue University, Mimeograph, Series No. 174.

[120] McDonald, G.C. (1972)
Some multiple comparison selection procedures based on ranks. Sankhyã, Series A 34, 53-64.

[121] McDonald, G.C. (1972)
The distribution of some rank statistics with applications in block design selection problems. Sankhyã, Series A 34, 188-204.

[122] McDonald, G.C. (1974)
Characteristics of three selection rules based on ranks in a small sample exponential case. Sankhyã, Series A 36, 261-266.

[123] Mahamunulu, D.M. (1967)
Some fixed-sample ranking and selection problems. AMS 38, 1079-1091.

[124] Maurer, H. (1974)
Datenstrukturen und Programmierverfahren. B.G.Teubner, Stuttgart.

[125] Maurice, R.J. (1959)
A different loss function for the choice between two populations. JRSS, Series B, 21, 203-213.

[126] Meeter, D.A./Srivastava, J.N., ed., (1975)
A two-armed bandit with terminal decision, A survey of statistical design and linear models. North-Holland publishing company.

[127] Mood, A.M. (1950)
Introduction to the theory of statistics. McGraw-Hill, New York.

[128] Mood, A.M. (1954)
On the asymptotic efficiency of certain nonparametric two-sample tests. AMS 25, 514-522.

[129] Nagel, K. (1970)
On subset selection rules with certain optimality properties. Department of Statistics, Purdue University, Mimeograph, Series No. 222.

[130] Nebenzahl, E./Sobel, M. (1972)
Play-the-winner sampling for a fixed sample size binomial selection problem. Biometrika, Vol. 59, 1-8.

[131] Nordbrock, E. (1976)
An improved Play-the-winner sampling procedure for selecting the better of two binomial populations. JASA, Vol. 71, No. 353, 137-139.

[132] Odeh, R.E. (1967)
The distribution of the maximum sum of ranks. Technometrics 9, 271-278.

[133] Olkin, J./Sobel, M. (1965)
Integral expressions for tail probabilities of the multinomial and negative multinomial distributions. Biometrika, Vol. 52, 167-179.

[134] Paulson, E. (1964)
A sequential procedure for selecting the population with the largest mean from k normal populations. AMS 35, 174-180.

[135] Paulson, E. (1964)
Sequential estimation and closed sequential decision procedures. AMS 35, 1048-1058.

[136] Paulson, E. (1967)
Sequential procedures for selecting the best one of several binomial populations. AMS, Vol. 38, 117-123.

[137] Paulson, E. (1969)
A new sequential procedure for selecting the best of k binomial populations. AMS, Vol. 40, 1865-1866.

[138] Potthoff, R.F. (1963)
Use of the Wilcoxon statistic for a generalized Behrens-Fisher problem. AMS 34, 1596-1599.

[139] Pradhan, M./Sathe,Y.S. (1973)
Play-the-winner sampling for a fixed sample size with curtailment. Biometrika, Vol. 60, 424-427.

[140] Pradhan, M./Sathe, Y.S. (1976)
Analytical remarks on Canner's minimax method for finding the better of two binomial populations. JASA, Vol. 71, No. 353, 239-241.

[141] Puri, M.L. (1964)
Asymptotic efficiency of a class of c-sample tests. AMS 35, 102-121.

[142] Puri, P.S./Puri, M.L. (1969)
Multiple decision procedures based on ranks for certain problems in analysis of variances. AMS 40, 619-632.

[143] Puri, M.L./Sen, P.K. (1967)
On some optimum procedures in two-way layouts. JASA 62, 1214-1229.

[144] Puri, M.L./Sen, P.K. (1969)
On the asymptotic normality of one-sample rank order test statistics. The Prob.Appl. 14, 163-166.

[145] Puri, M.L./Sen, P.K. (1971)
Nonparametric Methods in Multivariate Analysis. John Wiley, New York.

[146] Pyke, R./Smith, C.V. (1965)
The Robbins-Isbell two armed bandit problem with finite memory. AMS, Vol. 36, 1375-1386.

[147] Ramberg, J.S./Schmeiser, B.W. (1972)
An approximate method for generating symmetric random variables. ACM 15, 987-990.

[148] Ramberg, J.S./Schmeiser, B.W. (1974)
An approximate method for generating asymmetric random variables. ACM 17, 78-82.

[149] Randles, R.H. (1970)
Some robust selection procedures. AMS 41, 1640-1645.

[150] Randles, R.H./Hogg, R.V. (1973)
Adaptive distribution-free tests. Communications in Statistics 2(4), 337-356.

[151] Randles, R.H./Ramberg, J.S./Hogg, R.V. (1973)
An adaptive procedure for selecting the population with largest location parameter. Technometrics 15, 769-778.

[152] Rao, C.R. (1973)
Linear Statistical Inference and Its Applications. 2nd ed., J. Wiley & Sons, New York.

[153] Rényi, A. (1966)
Wahrscheinlichkeitsrechnung. Deutscher Verlag der Wissenschaften, Berlin.

[154] Rizvi, M.H./Woodworth, G.G. (1970)
On selection procedures based on ranks: Counterexamples concerning least favorable configurations. AMS 41, 1942-1951.

[155] Robbins, H. (1956)
A sequential decision procedure with a finite memory. Proc.Nat.Acad.of Sci. USA, 42, 920-923.

[156] Roy, J. (1972)
Evaluation of Gamma score. Sankhyã, Series B, Vol. 34, Pt. 1, 27-32.

[157] Rutherford, D.E. (1952)
Some Continuant determinants arising in physics and chemistry-II. The Royal Society of Edinburgh, Section A, Vol. 63, 232-241.

[158] Santner, T.J. (1975)
A restricted subset selection approach to ranking and selection problems. AS, Vol. 3, No. 2, 334-349.

[159] Schriever, K.-H. (1977)
A play-the-winner procedure for selecting the best of $k \geq 3$ binomial populations. Preprint Series Nr. 23/77, Institut für Statistik und Mathematische Wirtschaftstheorie, Universität Karlsruhe, Federal Republic of Germany.

[160] Schriever, K.-H. (1977)
A truncated vector-at-a-time procedure for selecting the best of two binomial populations. Preprint Series Nr. 24/77, Institut für Statistik und Mathematische Wirtschaftstheorie, Universität Karlsruhe, Federal Republic of Germany.

[161] Schriever, K.-H. (1977)
A vector-at-a-time procedure with a symmetrical stopping rule for selecting the best of two binomial populations. Preprint Series Nr. 27/77, Institut für Statistik und Mathematische Wirtschaftstheorie, Universität Karlsruhe, Federal Republic of Germany.

[162] Schriever, K.-H. (1978)
A truncated vector-at-a-time procedure for selecting the best of $k \geq 3$ binomial populations. Preprint Series Nr. 29/78, Institut für Statistik und Mathematische Wirtschaftstheorie, Universität Karlsruhe, Federal Republic of Germany.

[163] Schriever, K.-H. (1978)
A truncated play-the-winner procedure for selecting the best of $k \geq 3$ binomial populations. Bull.Math.Stat., Vol.18, No.1/2, Japan.

[164] Schriever, K.-H. (1978)
Nichtparametrische Selektionsverfahren. Karlsruhe.

[165] Sen, P.K. (1966)
On a distribution-free method of estimating asymptotic efficiency of a class of nonparametric tests. AMS 37, 1759-1770.

[166] Sen, P.K. (1970)
On some convergence properties of one-sample rank-order-statistics. AMS 41, 2140-2143.

[167] Sen, P.K./Ghosh, M. (1974)
Sequential rank tests for location. AMS 2, 540-552.

[168] Sen, P.K./Puri, M.L. (1967)
On the theory of rank order tests for location in the multivariate one sample problem. AMS 38, 1216-1228.

[169] Shirahata, S. (1976)
On the asymptotic normality of rank statistics for the two-sample problem. AS, Vol. 4, No. 2, 400-405.

[170] Slutsky, E. (1925)
Über stochastische Asymptoten und Grenzwerte. Metron 5, Nr. 3, S. 3.

[171] Sobel, M./Starr, N. (1975)
Selecting the coin with the greatest bias. Sankhyã, Vol. 37, Series A, Pt. 2, 197-210.

[172] Sobel, M./Weiss, G.H. (1969)
Play-the-winner sampling for selecting the better of two binomial populations. University of Minnesota, Technical report No. 123.

[173] Sobel, M./Weiss, G.H. (1969)
Play-the-winner rule and inverse sampling for selecting the better of two binomial populations. University of Minnesota, Technical report No. 124.

[174] Sobel, M./Weiss, G.H. (1969)
Play-the-winner rule and inverse sampling for selecting the best of $k \geq 3$ binomial populations. University of Minnesota, Technical report No. 126.

[175] Sobel, M./Weiss, G.H. (1970)
Inverse sampling and other selection procedures for tournaments with two or three players. In: Nonparametric Techniques in Statistical Inference, Cambridge Univ.Press, London and New York, pp. 515-543.

[176] Sobel, M./Weiss, G.H. (1970)
Play-the-winner sampling for selecting the better of two binomial populations. Biometrika, Vol. 57, 357-365.

[177] Sobel, M./Weiss, G.H. (1971)
Play-the-winner rule and inverse sampling in selecting the better of two binomial populations. JASA, Vol. 66, 545-551.

[178] Sobel, M./Weiss, G.H. (1971)
A comparison of play-the-winner and vector-at-a-time sampling for selecting the better of two binomial populations with restricted parameter values. Trabajos de Estadistica y de Investigacion Operativa XXII, 195-206.

[179] Sobel, M./Weiss, G.H. (1972)
Play-the-winner rule and inverse sampling for selecting the best of $k \geq 3$ binomial populations. AMS, Vol. 43, No. 6, 1808-1826.

[180] Sobel, M./Weiss, G.H. (1972)
Proceedings of the sixth Berkeley Symposium on mathematical statistics and probability, Vol. I, 717-736.

[181] Starr, N./Woodroofe, M. (1974)
Approximations in large sample ranking problems. Sankhyã, Vol. 36, Series B, Pt. 4, 400-405.

[182] Stein, Ch. (1946)
A note on cumulative sums. AMS, Vol. 17, 498-499.

[183] Sugimura, M. (1967)
Optimum designs for selecting one of two medical treatments, fixed sample size plan 2. Kumamoto J.Sci., Series A, 7, No. 4, 87-94.

[184] Sugimura, M. (1967)
Optimum designs for selecting one of two treatments, sequential plan 2. Kumamoto J.Sci., Series A, 8, No. 1, 11-19.

[185] Sugimura, M. (1967)
Optimum designs for selecting one of two treatments, general considerations. Kumamoto J.Sci., Series A, 8, No. 2, 63-83.

[186] Sugimura, M. (1967)
Optimum designs for selecting one of two medical treatments, fixed sample size plan 1. Mem.Fac.Gen.Educ. Kumamoto Univ., Ser.Nat.Sci., 2, 1-7.

[187] Sugimura, M. (1967)
Optimum designs for selecting one of two medical treatments, fixed sample size plan 4.1. J.Kumamoto Women's Univ., 19, 67-71.

[188] Sugimura, M. (1968)
Optimum designs for selecting one of two medical treatments, fixed sample size plan 5. J.Kumamoto Women's Univ., 20, 58-63.

[189] Sugimura, M. (1968)
Optimum designs for selecting one of two treatments, fixed sample size plan and sequential plan in a population with one parameter exponential distribution. Bull.Math.Stat. 13, No. 1~2, 1-23.

[190] Swanepoel, J.W.H. (1977)
Nonparametric elimination selection procedures based on robust estimators. South African.Stat.J. 27 41.

[191] Swanepoel, J.W.H./Geertsema, J.C. (1973)
Nonparametric sequential procedures for selecting the best of k populations. South African.Stat.J. 7, 85-94.

[192] Swanepoel, J.W.H./Venter, J.H. (1975)
On the construction of sequential selection procedures. South African. Stat.J. 103-118.

[193] Swanepoel, J.W.H./Venter, J.H. (1975)
A class of elimination procedures based on ranks. South African.Stat.J. 119-128.

[194] Taheri, H./Young, D.H. (1974)
A comparison of sequential sampling procedures for selecting the better of two binomial populations. Biometrika, 61, 3, 585-592.

[195] Tong, Y.L./Wetzell, D.E. (1979)
On the behaviour of the probability function for selecting the best normal population. Biometrika, 66, 1, 174-176.

[196] Uthoff, V.A. (1973)
The most powerful scale and location invariant test of the normal versus the double-exponential. AS 1, 170-174.

[197] Vorlickowa, D. (1970)
Asymptotic properties of rank tests under discrete distributions. Z. Wahrscheinlichkeitsth. verw. Gebiete 14, 275-289.

[198] Wald, A. (1947)
Sequential Analysis. Wiley, New York.

[199] Wald, A. (1950)
Statistical decision functions. Chelsea publishing Comp., New York.

[200] Wei, L.J. (1978)
The adaptive biased coin design for sequential experiments. AS, Vol. 6, No. 1, 92-100.

[201] Zaremba, S.K. (1962)
A generalization of Wilcoxon's Test. Monatshefte Math. 66, 359-370.

[202] Zelen, M. (1969)
Play-the-winner rule and the controlled clinical trial. JASA, 64, 131-146.